T0145424

Studies in Big Data

Volume 30

About this Series

The series "Studies in Big Data" (SBD) publishes new developments and advances in the various areas of Big Data- quickly and with a high quality. The intent is to cover the theory, research, development, and applications of Big Data, as embedded in the fields of engineering, computer science, physics, economics and life sciences. The books of the series refer to the analysis and understanding of large, complex, and/or distributed data sets generated from recent digital sources coming from sensors or other physical instruments as well as simulations, crowd sourcing, social networks or other internet transactions, such as emails or video click streams and other. The series contains monographs, lecture notes and edited volumes in Big Data spanning the areas of computational intelligence incl. neural networks, evolutionary computation, soft computing, fuzzy systems, as well as artificial intelligence, data mining, modern statistics and Operations research, as well as self-organizing systems. Of particular value to both the contributors and the readership are the short publication timeframe and the world-wide distribution, which enable both wide and rapid dissemination of research output.

More information about this series at http://www.springer.com/series/11970

Nilanjan Dey · Aboul Ella Hassanien
Chintan Bhatt · Amira S. Ashour
Suresh Chandra Satapathy
Editors

Internet of Things and Big Data Analytics Toward Next-Generation Intelligence

Editors
Nilanjan Dey
Techno India College of Technology
Kolkata, West Bengal
India

Amira S. Ashour
Tanta University
Tanta
Egypt

Aboul Ella Hassanien
Cairo University
Cairo
Egypt

Suresh Chandra Satapathy
Department of Computer Science and Engineering
PVP Siddhartha Institute of Technology
Vijayawada, Andhra Pradesh
India

Chintan Bhatt
Charotar University of Science and Technology
Changa, Gujarat
India

ISSN 2197-6503 ISSN 2197-6511 (electronic)
Studies in Big Data
ISBN 978-3-319-86864-6 ISBN 978-3-319-60435-0 (eBook)
DOI 10.1007/978-3-319-60435-0

Softcover reprint of the hardcover 1st edition 2017

Printed on acid-free paper

This Springer imprint is published by Springer Nature
The registered company is Springer International Publishing AG
The registered company address is: Gewerbestrasse 11, 6330 Cham, Switzerland

Preface

Internet of Things and big data are two sides of the same coin. The advancement of Information Technology (IT) has increased daily leading to connecting the physical objects/devices to the Internet with the ability to identify themselves to other devices. This refers to the Internet of Things (IoT), which also may include other wireless technologies, sensor technologies, or QR codes resulting in massive datasets. This generated big data requires software computational intelligence techniques for data analysis and for keeping, retrieving, storing, and sending the information using a certain type of technology, such as computer, mobile phones, computer networks, and more. Thus, big data holds massive information generated by the IoT technology with the use of IT, which serves a wide range of applications in several domains. The use of big data analytics has grown tremendously in the past few years directing to next generation of intelligence for big data analytics and smart systems. At the same time, the Internet of Things (IoT) has entered the public consciousness, sparking people's imaginations on what a fully connected world can offer. Separately the IoT and big data trends give plenty of reasons for excitement, and combining the two only multiplies the anticipation. The world is running on data now, and pretty soon, the world will become fully immersed in the IoT.

This book involves 21 chapters, including an exhaustive introduction about the Internet-of-Things-based wireless body area network in health care with a brief overview of the IoT functionality and its connotation with the wireless and sensing techniques to implement the required healthcare applications. This is followed by another chapter that discussed the association between wireless sensor networks and the distributed robotics based on mobile sensor networks with reported applications of robotic sensor networks. Afterward, big data analytics was discussed in detail through four chapters. These chapters addressed an in-depth overview of the several commercial and open source tools being used for analyzing big data as well as the key roles of big data in a manufacturing industry, predominantly in the IoT environment. Furthermore, the big data Learning Management System (LMS) has been analyzed for student managing system, knowledge and information, documents, report, and administration purpose. Since business intelligence is considered one of the significant aspects, a chapter that examined open source applications, such as

Pentaho and Jaspersoft, processing big data over six databases of diverse sizes is introduced.

Internet-of-Things-based smart life is an innovative research direction that attracts several authors; thus, 10 chapters are included to develop Industrial Internet of Things (IIoT) model using the devices which are already defined in open standard UPoS (Unified Point of Sale) devices in which they included all physical devices, such as sensors printer and scanner leading to advanced IIoT system. In addition, smart manufacturing in the IoT era is introduced to visualize the impact of IoT methodologies, big data, and predictive analytics toward the ceramics production. Another chapter is presented to introduce the home automation system using BASCOM including the components, flow of communication, implementation, and limitations, followed by another chapter that provided a prototype of IoT-based real-time smart street parking system for smart cities. Afterward, three chapters are introduced related to smart irrigation and green cities, where data from the cloud is collected and irrigation-related graph report for future use for farmer can be made to take decision about which crop is to be sown. Smart irrigation analysis as an IoT application is carried out for irrigation remote analysis, while the other chapter presented an analysis of the greening technologies' processes in maintainable development, discovering the principles and roles of G-IoT in the progress of the society to improve the life quality, environment, and economic growth. Then, cloud-based green IoT architecture is designed for smart cities. This is followed by a survey chapter on the IoT toward smart cities and two chapters on big data analytics for smart cities and in Industrial IoT, respectively. Moreover, this book contains another set of 5 chapters that interested with IoT and other selected topics. A proposed system for very high capacity and for secure medical image information embedding scheme to hide Electronic Patient Record imperceptibly of colored medical images as an IoT-driven healthcare setup is introduced including detailed experimentation that proved the efficiency of the proposed system, which is tested by attacks. Thereafter, another practical technique for securing the IoT against side channel attacks is reported. Three selected topics are then introduced to discuss the framework of temporal data stream mining by using incrementally optimized very fast decision forest, to address the problem classifying sentiments and develop the opinion system by combining theories of supervised learning and to introduce a comparative survey of Long-Term Evolution (LTE) technology with Wi-Max and TD-LTE with Wi-Max in 4G using Network Simulator (NS-2) in order to simulate the proposed structure.

This editing book is intended to present the state of the art in research on big data and IoT in several related areas and applications toward smart life based on intelligence techniques. It introduces big data analysis approaches supported by the research efforts with highlighting the challenges as new opening for further research areas. The main objective of this book is to prove the significant valuable role of the big data along with the IoT based on intelligence for smart life in several domains. It embraces inclusive publications in the IoT and big data with security issues, challenges, and related selected topics. Furthermore, this book discovers the technologies impact on home, street, and cities automation toward smart life.

In essence, this outstanding volume cannot be without the innovative contributions of the promising authors to whom we estimate and appreciate their efforts. Furthermore, it is unbelievable to realize this quality without the impact of the respected referees who supported us during the revision and acceptance process of the submitted chapters. Our gratitude is extended to them for their diligence in chapters reviewing. Special estimation is directed to our publisher, Springer, for the infinite prompt support and guidance.

We hope this book introduces capable concepts and outstanding research results to support further development of IoT and big data for smart life toward next-generation intelligence.

Kolkata, India Nilanjan Dey
Cairo, Egypt Aboul Ella Hassanien
Changa, India Chintan Bhatt
Tanta, Egypt Amira S. Ashour
Vijayawada, India Suresh Chandra Satapathy

Contents

Part I Internet of Things Based Sensor Networks

Internet of Things Based Wireless Body Area Network in Healthcare 3
G. Elhayatmy, Nilanjan Dey and Amira S. Ashour

Mobile Sensor Networks and Robotics 21
K.P. Udagepola

Part II Big Data Analytics

Big Data Analytics with Machine Learning Tools 49
T.P. Fowdur, Y. Beeharry, V. Hurbungs, V. Bassoo and V. Ramnarain-Seetohul

Real Time Big Data Analytics to Derive Actionable Intelligence in Enterprise Applications 99
Subramanian Sabitha Malli, Soundararajan Vijayalakshmi and Venkataraman Balaji

Revealing Big Data Emerging Technology as Enabler of LMS Technologies Transferability 123
Heru Susanto, Ching Kang Chen and Mohammed Nabil Almunawar

Performance Evaluation of Big Data and Business Intelligence Open Source Tools: Pentaho and Jaspersoft 147
Victor M. Parra and Malka N. Halgamuge

Part III Internet of Things Based Smart Life

IoT Gateway for Smart Devices 179
Nirali Shah, Chintan Bhatt and Divyesh Patel

Smart Manufacturing in the Internet of Things Era 199
Th. Ochs and U. Riemann

Home Automation Using IoT 219
Nidhi Barodawala, Barkha Makwana, Yash Punjabi and Chintan Bhatt

A Prototype of IoT-Based Real Time Smart Street Parking System for Smart Cities 243
Pradeep Tomar, Gurjit Kaur and Prabhjot Singh

Smart Irrigation: Towards Next Generation Agriculture 265
A. Rabadiya Kinjal, B. Shivangi Patel and C. Chintan Bhatt

Greening the Future: Green Internet of Things (G-IoT) as a Key Technological Enabler of Sustainable Development 283
M. Maksimovic

Design of Cloud-Based Green IoT Architecture for Smart Cities 315
Gurjit Kaur, Pradeep Tomar and Prabhjot Singh

Internet of Things Shaping Smart Cities: A Survey 335
Arsalan Shahid, Bilal Khalid, Shahtaj Shaukat, Hashim Ali and Muhammad Yasir Qadri

Big Data Analytics for Smart Cities 359
V. Bassoo, V. Ramnarain-Seetohul, V. Hurbungs, T.P. Fowdur and Y. Beeharry

Bigdata Analytics in Industrial IoT 381
Bhumi Chauhan and Chintan Bhatt

Part IV Internet of Things Security and Selected Topics

High Capacity and Secure Electronic Patient Record (EPR) Embedding in Color Images for IoT Driven Healthcare Systems 409
Shabir A. Parah, Javaid A. Sheikh, Farhana Ahad and G.M. Bhat

Practical Techniques for Securing the Internet of Things (IoT) Against Side Channel Attacks 439
Hippolyte Djonon Tsague and Bheki Twala

Framework of Temporal Data Stream Mining by Using Incrementally Optimized Very Fast Decision Forest 483
Simon Fong, Wei Song, Raymond Wong, Chintan Bhatt and Dmitry Korzun

Sentiment Analysis and Mining of Opinions 503
Surbhi Bhatia, Manisha Sharma and Komal Kumar Bhatia

A Modified Hybrid Structure for Next Generation Super High Speed Communication Using TDLTE and Wi-Max 525
Pranay Yadav, Shachi Sharma, Prayag Tiwari, Nilanjan Dey, Amira S. Ashour and Gia Nhu Nguyen

Part I
Internet of Things Based Sensor Networks

Internet of Things Based Wireless Body Area Network in Healthcare

G. Elhayatmy, Nilanjan Dey and Amira S. Ashour

Abstract Internet of things (IoT) based wireless body area network in healthcare moved out from traditional ways including visiting hospitals and consistent supervision. IoT allow some facilities including sensing, processing and communicating with physical and biomedical parameters. It connects the doctors, patients and nurses through smart devices and each entity can roam without any restrictions. Now research is going on to transform the healthcare industry by lowering the costs and increasing the efficiency for better patient care. With powerful algorithms and intelligent systems, it will be available to obtain an unprecedented real-time level, life-critical data that is captured and is analyzed to drive people in advance research, management and critical care. This chapter included in brief overview related to the IoT functionality and its association with the sensing and wireless techniques to implement the required healthcare applications.

Keywords Internet of things · Wireless body area network · Healthcare architecture · Sensing · Remote monitoring

G. Elhayatmy
Police Communication Department, Ministry of Interior, Cairo, Egypt
e-mail: gamal_elhayatmy@hotmail.com

N. Dey (✉)
Information Technology Department, Techno India College of Technology, Kolkata, West Bengal, India
e-mail: neelanjan.dey@gmail.com

A.S. Ashour
Department of Electronics and Electrical Communications Engineering, Faculty of Engineering, Tanta University, Tanta, Egypt
e-mail: amirasashour@yahoo.com

N. Dey et al. (eds.), *Internet of Things and Big Data Analytics Toward Next-Generation Intelligence*, Studies in Big Data 30,
DOI 10.1007/978-3-319-60435-0_1

1 Introduction

Internet of things (IoT) represents the connection between any devices with Internet including cell phone, home automation system and wearable devices [1, 2]. This new technology can be considered the phase changer of the healthcare applications concerning the patient's health using low cost. Interrelated devices through the Internet connect the patients with the specialists all over the world. In healthcare, the IoT allows the monitoring of glucose level and the heart beats in addition to the body routine water level measurements. Generally, the IoT in healthcare is concerned with several issues including (i) the critical treatments situations, (ii) the patient's check-up and routine medicine, (iii) the critical treatments by connecting machines, sensors and medical devices to the patients and (iv) transfer the patient's data through the cloud.

The foremost clue of relating IoT to healthcare is to join the physicians and patients through smart devices while each individual is roaming deprived of any limitations. In order to upload the patient's data, cloud services can be employed using the big data technology and then, the transferred data can be analyzed. Generally, smart devices have a significant role in the individuals' life. One of the significant aspects for designing any device is the communication protocol, which is realized via ZigBee network that utilizes Reactive and Proactive routing protocols. Consequently, the IoT based healthcare is primarily depends on the connected devices network which can connect with each other to procedure the data via the secure service layer.

The forth coming IoT will depend on low-power microprocessor and effective wireless protocols. The wearable devices along with the physician and the associated systems facilitate the information, which requires high secured transmission systems [3]. Tele-monitoring systems are remotely monitoring the patients while they are at their home. Flexible patient monitoring can be allowed using the IoT, where the patients can select their comfort zone while performing treatment remotely without changing their place. Healthcare industry can accomplish some severe changes based on numerous inventions to transfer the Electronic health records (EHRs) [4]. Connected medical devices with the Internet become the main part of the healthcare system. Recently, the IoT in healthcare offers IoT healthcare market depth assessment including vendor analysis, growth drivers, value chain of the industry and quantitative assessment. In addition, the medical body area networks (MBANs) which are worn devices networks on the patient's body to interconnect with an unattached controller through wireless communication link. This MBAN is used to record and to measure the physiological parameters along with other information of the patient for diagnosis.

The 5G (fifth generation) of communication technologies supports the IoT technologies in several applications especially in healthcare. It allows 100 times higher wireless bandwidth with energy saving and maximum storage utilization by applying big data analytics. Generally, wireless communication dense deployments are connected over trillions wireless devices with advanced user controlled privacy.

Wired monitoring systems obstacle the patients' movement and increase the errors chances as well as the hospital-acquired infections. The MBAN's facilitates the monitoring systems to be wirelessly attached to the patients using wearable sensors of low-cost. The Federal Communications Commission (FCC) has permitted a wireless networks precise spectrum that can be employed for monitoring the patient's data using the healthcare capability of the MBAN devices in the 2360–2400 MHz band [5].

2 IoT Based WBAN for Healthcare Architecture

The IoT based wireless body area network (WBAN) system design includes three tiers as illustrated Fig. 1 [6].

Figure 1 demonstrates that multiple sensor nodes as very small patches positioned on the human body. Such sensors are wearable sensors, or as in-body sensors that implanted under the skin that operate within the wireless network. Continuously, such sensors capture and transmit vital signs including blood pressure, temperature, sugar level, humidity and heart activity. Nevertheless, data may entail preceding on-tag/low-level handling to communication based on the computation capabilities and functionalities of the nodes. Afterward, the collected data either primarily communicated to a central controller attached the body or directly communicated through Bluetooth or ZigBee to nearby personal server (PS), to be remotely streamed to the physician's site for real time diagnosis through a WLAN (wireless local area network) connection to the consistent equipment for emergency alert or to a medical database. The detailed WBAN system block diagram is revealed in Fig. 2. It consists of sink node sensor nodes and remote observing station.

The detailed description for the WBAN system is as follows.

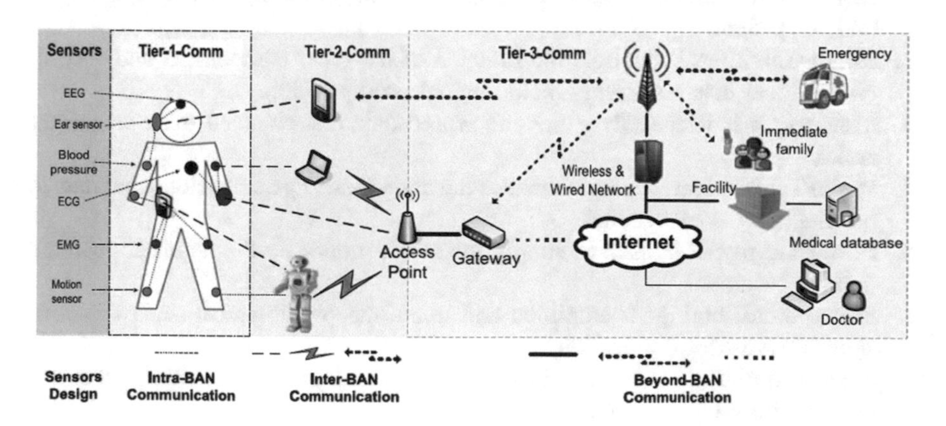

Fig. 1 IOT-based WBAN for healthcare architecture [6]

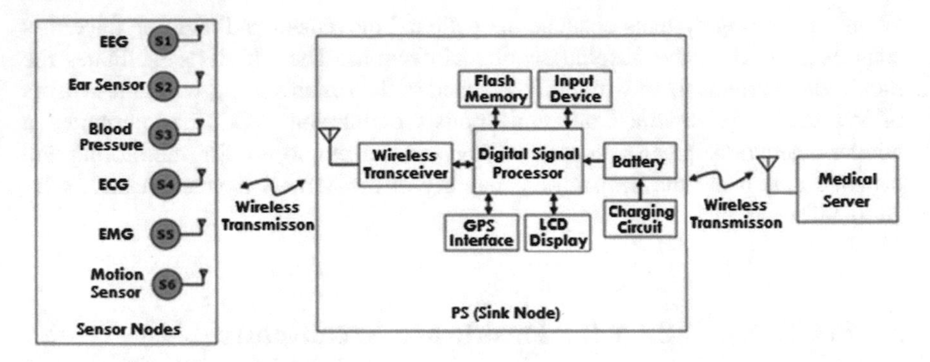

Fig. 2 WBAN system block diagram [7]

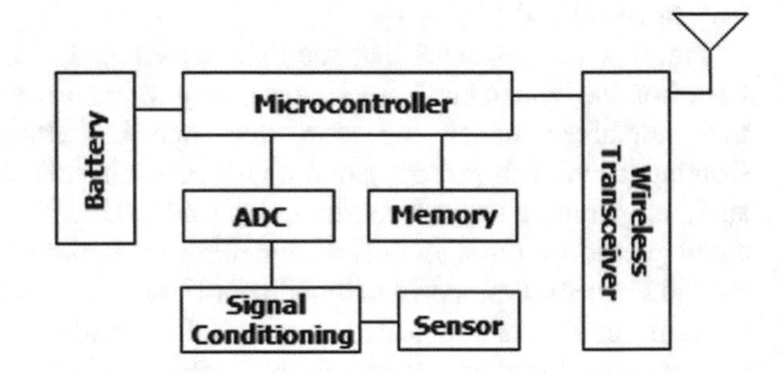

Fig. 3 Wireless sensor node block diagram [7]

2.1 Sensor Nodes

The sensor nodes have small size and a minute battery with limited power, communication and computing capabilities. The elementary smart sensor components are illustrated in Fig. 3.

The central components of the sensor nodes are:

1. **Sensor**: It encloses an embedded chip for sensing vital medical signs from the body of patient.
2. **Microcontroller**: It controls the function of the other components and accomplishes local data processing including data compression.
3. **Memory**: It is temporally stores the sensed data that obtained from the sensor nodes.
4. **Radio Transceiver**: It communicates the nodes and allows physiological data to be wirelessly send/received.
5. **Power supply**: It is used to supply the sensor nodes by the required powered through batteries.
6. **Signal conditioning**: It amplifies and filters the physiological sensed data to suitable levels for digitization.
7. **Analog to digital converter** (**ADC**): It produces digital signals from the analog ones to allow further required processes.

Furthermore, a sophisticated sensor that can be combined into the WBAN is the Medical Super Sensor (MSS), which has superior memory size, communication and processing abilities compared to the sensor nodes. The MSS utilized a RF to connect with other body sensors. In addition, Bluetooth or ZigBee can be utilized as a communication protocol to connect the obtained sensed data with the personal server. It gathers the multiple sensed vital signs by the body sensors and filters out all unnecessary data to reduce the data transmitted large volume (big data). Afterward, it stores the transmitted data temporarily, processes and transfers the significant data of patient to the PS over wireless personal realized by ZigBee/IEEE 802.15.4. This increases the inclusive bandwidth use and reduces the BSs power, where each node has to transmit the sensed data to collector which is MSS instead of the PS, where the MSS is closer to the BSs than the PS.

2.2 Personal Server (Sink Node)

The PS (body gateway) is running on a smart phone to connect the wireless nodes via a communication protocol by either ZigBee or Bluetooth. It is arranged to a medical server using the IP address server to interface the medical services. The personal servers is used also to process the generated dynamic signs from the sensor nodes and provides the transmission priority to the critical signs to be send through the medical server. It performs the analysis task of the vital signs and compares the patient's health status based on the received data by the medical server to provide a feedback through user-friendly graphical interface.

The PS hardware entails several modules including the input unit, antenna, digital signal processor, transceiver, GPS interface, flash memory, display, battery and charging circuit. The data received are supplementary processed for noise removal and factors measurements [7].

2.3 Medical Server

The medical server contains a database for the stored data, analyzing and processing software to deliver the system required service. It is also responsible about the user authentication. The measured data by the sensors are directed via the internet/intranet to medical personnel to examine it. The medical unit is notified for necessary actions, when there is deviation from the expected health records of the patient.

2.4 WBAN Communication Architecture

Typically, the WBAN communications design is divided into three components, namely the intra-BAN communications, inter-BAN and Beyond-BAN communications.

2.4.1 Intra-BAN Communications

The intra-BAN communications refers to about 2 m radio communications nearby the human body, which is sub-classified into connections among the body sensors or between portable PS and the body sensors. The intra-BAN communications design is grave due to the direct association with the BANs and the body sensors [6]. The essential operated battery and the low bit rate features of the prevailing body sensor devices lead to an interesting aspect to enterprise an energy effective MAC protocol with sufficient QoS. For wireless connection of sensors and PS challenges solving, several systems can be employed including the MITHril [8] and SMART [9]. These structures exploit cables to link multiple sensors with the PS. Instead, a codeblue [10] stipulates can be used to communicate the sensors directly with the access points (Aps) without a PS. In addition, a star topology can be used, where multiple sensors can forward the body signals to a PS to process the physiological data to an AP (e.g., WiMoCa [11]).

2.4.2 Inter-BAN Communication

The BAN is seldom works alone, dissimilar to the WSNs, which generally work as independent systems. The APs can be considered one of the main parts of the dynamic environment's infrastructure while managing emergency cases. The communication between the APs and PS is utilized in the inter-BAN communications. Correspondingly, the tier-2-network functionality is employed to communicate the BANs with various easy accessible networks, such as the cellular and Internet networks. The inter-BAN communication paradigms have the following categories: (i) infrastructure-based construction, which delivers large bandwidth with central control and suppleness and (ii) ad hoc-based construction that enables fast distribution in the dynamic environments including disaster site (e.g., AID-N [12]) and medical emergency response situations. Figure 4a, b illustrate the two structures respectively [6].

Infrastructure Based Architecture

Infrastructure-based and inter-BAN communications have a significant role in several BAN limited space applications, such as in office, in home and in hospital

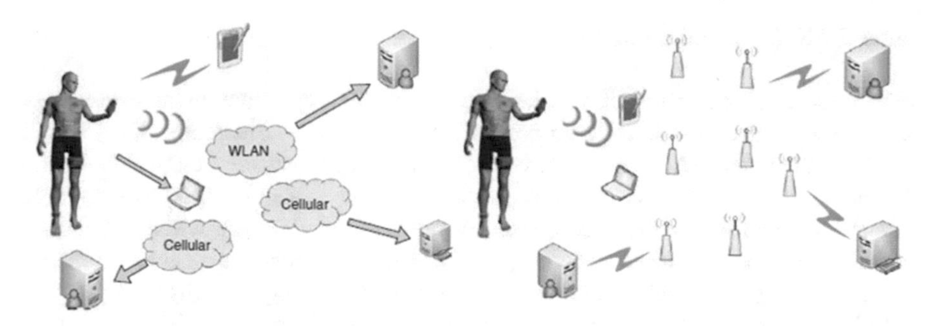

Fig. 4 Inter-BAN communication structure: **a** infrastructure-based structure; **b** ad hoc-based structure [6]

environments. The infrastructure-based networks allow security control and centralized management. Furthermore, the AP can act as database server in particular uses including SMART [9], CareNet [13].

Ad Hoc Based Architecture

Generally, multiple APs are organized to support the information transmission of the body sensors in the ad hoc-based construction. Consequently, the service coverage is in excess of the corresponding one in the infrastructure-based construction. These enable the users' movement around anywhere, emergency saving place and building, where the BAN coverage is limited to about 2 m. Thus, the ad hoc-based architecture of interconnection outspreads the system to about 100 m that allows a short-term/long-term setup. In this architecture setup, two classes of nodes can be used, namely router nodes around the BAN and sensor nodes in/around the human body.

Every node in the WSNs acts as a sensor/router node. The ad hoc-based architecture setup employs a gateway for outside world interface resembling the traditional WSN. Typically, all infrastructures share the same bandwidth, where there is only one radio. Consequently, the collisions possibility is definitely arise, where in some situations; the number of sensor/actuator nodes and the routers nodes is large in certain area. In order to handle such collision situations, an asynchronous MAC mechanism can be employed. A mesh structure is considered one form of the various APs of this system having the following characteristics:

A. Large radio coverage because of the multi-hop data distribution. Thus, superior support to the patient's mobility can be acquired, where during multi-hop data forwarding, the bandwidth is reduced.
B. Flexible and fast wireless arrangement is realized to speedily mount the emergency reply systems [10, 12].
C. Adaptation of the network can be simply extended without any effect on the whole network by adding new APs or any other requirements.

Inter-BAN Communication Technology

For inter-BAN communication, the wireless technologies are more established compared to the intra-BAN communications. It includes Bluetooth, WLAN and ZigBee. Since the BANs require provision low energy consumption protocols, the Bluetooth becomes a superior communications tool over a short range that is viable for BANs. The Bluetooth is considered a prevalent short range wireless communications protocol. ZigBee become popular due to its effective features, namely: (i) its low duty cycle, which allows offering extended battery life, (ii) its support to 128-bit security, (iii) it enables low-latency communications and (iv) for interconnection between nodes, it requires low energy consumption. Thus, several BAN applications deploy the ZigBee protocol due to its capability to support mesh networks.

2.4.3 Beyond-BAN Communication

A gateway device like the PDA is used to generate a wireless connection between the inter-BAN and beyond-BAN communications. The beyond-BAN tier communications can develop several applications and can be employed in different healthcare systems to enable authorized healthcare personnel for remote accessing to the patients' medical information through the Internet or any cellular network. One of the significant "beyond-BAN" tier components is the database, which retains the user's medical history. Thus, the doctor can admission the user's information as well as automatic notifications can be delivered to the patient's relatives based on through various telecommunications means.

The beyond-BAN communication design is adapted to the user-specific services' requirements as it is application-specific. Consequently, for example, an alarm can be alerted to the doctor via short message service (SMS) or email, if any irregularities are initiated based on the transmitted up-to-date body. Doctors can directly communicate their patients through video conference using the Internet. Afterward, remote diagnosis can occur through the video connection with the patient based on the transmitted patient's medical data information obtained from the BAN worn or stored in the database.

3 WBAN Topology

For frames exchanging through a relay node, the IEEE 802 Task Group 6 approved a network topology with one hub. This hub can be associated to all nodes via one-hop star topology or via two-hop extended star topology. Generally, the two-hop extended star topology is constrained in the medical implant communication service (MICS) band.

The beacon mode and non-beacon mode are the star topology communication methods that can be used. In the beacon mode, the network hub representing the star topology's center node switches the connection to describe the start and the end of a super-frame to empower the synchronization of the device and the network connotation control. The system's duty cycle known as the beacon period length can be identified by the user and founded on the WBAN's standard [14, 15]. The nodes required to be power up and elect the hub to obtain data. In the WBANs, cautious deliberations should be considered upon the one-hop or the two-hop topology choice.

4 Layers of WBAN

Generally, both the PHY (Physical) and MAC (Medium Access Control) layers are proposed by all permitted standards of 802.15.x. The IEEE 802.15.6 (WBAN) active collection has definite new MAC and PHY layers for the WBANs, which offer ultra-low power, high reliability, low cost, and low complexity. Typically, there may be a HME (hub management entity) or logical NME (node management entity) that connects the network management information with the PHY.

4.1 Physical Layer

The IEEE 802.15.6 PHY layer is responsible about the several tasks, namely the radio transceiver's deactivation/activation, transmission/reception of the data and Clear channel assessment (CCA) in the present channel. The physical layer selection is based on the application under concern including non-medical/medical and on-, in-and off-body. The PHY layer delivers a technique to transform the physical layer service data unit (PSDU) into a physical layer protocol data unit (PPDU). The NB PHY is accountable for the radio transceiver deactivation/activation, data transmission/reception, and CCA in the present channel. The PSDU should be pre-attached with a physical layer preamble (PLCP) and a physical layer header (PSDU) according to the NB specifications to create PPDU. After PCLP preamble, the PCLP header is directed through the data rates specified in its effective frequency band. The PSDU is considered the last PPDU module that comprises a MAC- header/frame body as well as a Frame Check Sequence (FCS) [16].

The HBC PHY offers the Electrostatic Field Communication (EFC) necessities, which cover preamble/Start Frame Delimiter (SFD), packet structure and modulation. For ensuring packet synchronization, the preamble sequence is sent four times, whereas the SFD sequence is only transmitted once [16]. The PHY header entails pilot information, data rate, synchronization, payload length, WBAN ID and a CRC designed over the PHY header.

The UWB physical layer is utilized to communicate the on-body and the off-body devices, in addition to communicating the on-body devices. Comparable signal power levels are generated in the transceivers in a UWB PHY. The UWB PPDU entails a PHY Header (PHR), a Synchronization Header (SHR) and PSDU. The SHR is made up of repetitions of 63 length Kasami intervals. It contains two subfields, namely (i) the preamble that is used for timing synchronization; frequency offset recovery and packet detection; and (ii) the SFD. The Ultra wideband frequencies provide higher throughput and higher data rates, whereas lower frequencies have less attenuation and shadowing from the body [17].

4.2 MAC Layer

On the PHY layer upper part, the MAC layer is defined based on the IEEE 802.15.6 working assembly to control the channel access. The hub splits the time axis or the entire channel into a super-frames chain for time reference resource allocation. It selects the equal length beacon periods to bound the super-frames [16]. For channel access coordination, the hub employed through one of the subsequent channel access modes:

(1) Beacon Mode with Beacon Period Super-frame Boundaries: In each beacon period, the hub directs beacons unless barred by inactive super-frames or limitations in the MICS band. The super-frame structure communication is managed by the hubs using beacon frames or Timed frames (T-poll).
(2) Non-beacon mode with superframe boundaries: It is incapable of beacons transmition. It is forced to employ the Timed frames of the superframe structure.
(3) Non-beacon mode without superframe boundaries: Only unscheduled Type II polled allocation in this mode is give by the hub. Thus, each node has to determine independently its own time schedule.

In each super-frame period, the following access mechanisms exist:

(a) Connection-oriented/contention-free access (scheduled access and variants): It schedules the slot allocation in one/multiple upcoming super-frames.
(b) Connectionless/contention-free access (unscheduled and improvised access): It uses the posting/polling for resource allocation.
(c) Random access mechanism: It uses either the CSMA/CA or the slotted Aloha approach for resource allocation.

5 WBAN Routing

For Ad Hoc networks [18] and WSNs [19], numerous routing protocols are designed. The WBANs is instead of node-based movement are analogous to MANETs with respect to the moving topology with group-based movement [20]. Furthermore, the WBAN has more recurrent changes in the topology and a higher moving speed, whereas a WSN has low mobility or static scenarios [20]. The routing protocols planned for WSNs and MANETs are unrelated to the WBANs due to the specific WBANs challenges [21].

5.1 Challenges of Routing in WBANs

There are several challenges of routing in WBANs including the following.

- Postural body movements, where the environmental obstacles, node mobility, energy management and the WBANs increased dynamism comprising frequent changes in the network components and topology that amplify the Quality of Service (QoS) complexity. Furthermore, due to numerous body movements, the link superiority between nodes in the WBANs varies with time [22]. Consequently, the routing procedure would be adapted to diverse topology changes.
- Temperature rise and interference at which the node's energy level should be considered in the routing protocol. Moreover, the nodes' transmission power required to be enormously low to minimize the interference and to avoid tissue heating [21].
- Local energy awareness is required, where the routing protocol has to distribute its communication data between nodes in the network to achieve balanced use of power and to minimize the battery supply failure.
- Global network lifetime at which the network lifetime in the WBANs definite as the time interval from the network starting till it is damaged. The network lifetime is significant in the WBANs associated to the WSNs and WPANs [23].
- Efficient transmission range is one of the significant challenges where in WBANs; the low RF transmission range indicates separating between the sensors in the WBANs [24].
- Limitation of packet hop count, where one-hop/two-hop communication is available in the WBANs in consistent with the IEEE802.15.6 standard draft for the WBANs. Multi-hop transmission offers stronger links that lead to increasing the overall system reliability. Large energy consumption can be achieved with the larger hops number [25].
- Resources limitations including energy, data capacity, and WBANs device lifetime, which is severely limited as they necessitate a small form factor.

6 Security in WBANs

Security is considered a critical aspect in all networks especially for the WBANs. The stringent resource restrictions with respect to the computational capabilities, communication rate, memory, and power along with the inherent security vulnerabilities lead to inapplicable certain security specifications to the WBANs. Consequently, the convenient security integration mechanisms entail security requirements' knowledge of WBANs that are delivered as follows [26]:

- Secure management at which the decryption/encryption processes involves secure management at the hub to deliver key distribution to the WBS networks.
- Accessibility of the patient's information to the physician should be guaranteed at all times.
- Data authentication at which the medical/non-medical applications necessitate data authentication. Verification becomes essential to both the WBAN nodes and the hub node.
- Data integrity at which the received data requirements should be guaranteed of not being changed by a challenger via appropriate data integrity using data authentication protocols.
- Data confidentiality at which data protection from revelation is realizable via data privacy.
- Data freshness which is essential to support both the data integrity and confidentiality.

A security paradigm for WBANs has been proposed by the IEEE 802.15.6 standard as illustrated in Fig. 5 comprising three security levels [27].

Figure 5 revealed that the main security levels are as follows:

(a) Level 0 refers to unsecured communication, which is considered the lowest security level, where the data is transmitted in unsecured frames and offers no measure for integrity, authenticity, validation and defense, replay privacy, protection and confidentiality.
(b) Level 1 is concerned with the authentication without encryption, where data is transmitted in unencrypted authenticated frames. It includes validation, authenticity, integrity and replay defense measurements. Nevertheless, it did not provide confidentiality or privacy protection.

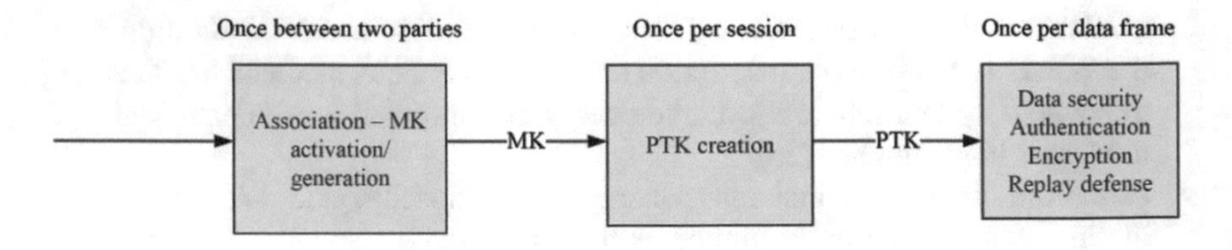

Fig. 5 IEEE.802.15.6 security framework

(c) Level 2 includes both encryption and authentication. Thus, it is considered as the highest security level at which messages are conveyed in encrypted and authenticated frames. The essential security level is chosen through the association process. In a WBAN, at the MAC layer prior to data exchange, all nodes and hubs have to go through definite stages.

Typically, the hub and node can exchange several frames, namely the connection task secure frames, connection request frame, security disassociation frame and the control unsecured frame.

7 WBAN Requirements in IEEE 802.15.6

The chief IEEE 802.15.6 standard requirements are as follows [21, 28–31]:

- The WBAN links have to support 10 Kb/s to 10 Mb/s bit rate ranges.
- The Packet Error Rate (PER) must be less than 10% for a 256 octet payload.
- In less than 3 s, the nodes must have the ability to being removed and to be added to the network.
- Each WBAN should has the ability to support 256 nodes.
- Reliable communication is required by the nodes even when the person is moving. In a WBAN, nodes may move individually relative each other.
- Latency, jitter and reliability have to be supported for WBAN applications. Latency should be <125 ms in the medical applications and <250 ms in thc non-medical applications, while jitter should be <50 ms.
- In-body and on-body WBANs have to be able to coexist within range.
- Up to 10 co-located WBANs which are randomly distributed.
- In a heterogeneous environment, the WBANs must be able to operate as different standards networks collaboratc among each other to receive the information.
- The WBANs have to incorporate the QoS management features.
- Power saving techniques must be incorporated to allow the WBANs from working under power constrained conditions.

8 Challenges and Open Issues of WBANs

The main challenges and open issues to realize the WBANs are:

- Environmental challenges:
  The WBANs suffer from high path loss due to body absorption. This should be reduced via multi-hop links and heterogeneous with different sensors at several positions. Due to the multi-path and mobility, the channel models become more

complex. In addition, antenna design leads to more challenging issues due to certain WBANs constraints related to the antenna shape antenna, size, material and malicious RF environment [32–36]. In fact, implant dictates the location of its antenna.

- Physical layer challenges:
  The PHY layer protocols have to be implemented for minimizing the power consumption without reliability. These protocols must be convenient for interference-agile places [37].
  Low power RF technology advancements can decease the peak power consumption leading to small production, low cost and disposable patches. The WBANs should be scalable and have about 0.001–0.1 mW peak power consumption instand-by mode and up to 30 mW in fully active mode [37].
  Interference is considered one of the significant WBAN systems drawbacks [38]. It occurs when peoples who wear WBAN devices and step into each other's range. The co-existence issues become more prominent with higher WBAN density. In addition, unpredictable postural body movements may facilitate the networks movement in and out of each other's range [17].
- MAC layer challenges:
  The IEEE 802.15.6 mechanisms do not build up the whole MAC protocol. Only the fundamental requirements to ensure the interoperability among the IEEE 802.15.6 devices [17].
  Furthermore, the MAC protocols must support the prolong sensor lifetime, WBAN applications energy efficiency requirement, save energy and allow flexible duty cycling. Generally, for WBANs, the proposed MAC protocols do not offer effective network throughput. Such protocols lead to delayed performance at varying traffic and challenging power requirements. The WBANs also have precise QoS necessities that required to be performed by the MAC proposal.
- Security challenges:
  Due to the resources limitation in terms of processing power, memory, and energy, the existing security techniques are incapable to WBANs.
- Transport (QoS) challenges:
  In WBANs applications, the QoS requirements must be achieved without performance degradation and complexity improvement. In WBANs, the limited memory requires efficient retransmission, error detection strategies and secure correction.

Currently, Smartphone devices are more user-accepted, inescapable and powerful. Correspondingly, mobile health (mHealth) technologies have qualified a minor change from implanting and/or wearing body sensors to carry a prevailing wireless device with multifunctional abilities [39]. In addition, Smart phones can be unconventionally used for sleep monitoring [40–43]. From the preceding discussion, the role of both IoT and big data is clarified. Both techniques has a role in different applications especially in healthcare [44–65].

9 Conclusion

In this chapter, an on-going research review of WBANs regarding the system architecture, PHY layer, routing, MAC layer, security are provided. Open issues in WBANs are also presented. In medical applications, the WBANs will permit incessant patients' monitoring that allows early abnormal conditions detection resulting in main developments in the life quality. Elementary vital signs monitoring can allow patients to perform normal activities. In instant, the technical research on this appreciated technology has noteworthy prominence in superior uses of accessible resources, which affect our future well-being.

References

1. Bhayani, M., Patel, M., & Bhatt, C. (2015). Internet of things (IoT): In a way of smart world. In *Proceedings of the International Congress on Information and Communication Technology, ICICT* (Vol. 1, pp. 343–350).
2. Bhatt, C., & Shah, T. (2014). The internet of things: Technologies, communications and computing. *CSI Communications*.
3. Fernandez, F., & Pallis, G. C. (2014). Opportunities and challenges of the Internet of Things for healthcare systems engineering perspective. In *EAI 4th International Conference on Wireless Mobile Communication and Healthcare* (*Mobihealth*), Athens, November 3–5.
4. http://www.cio.com/article/2981481/healthcare/how-the-internet-of-things-is-changing-healthcare-and-transportation.html
5. Amendment of the Commission's Rules to Provide Spectrum for the Operation of Medical Body Area Networks, ET Docket No. 89–59, First Report and Order and Further Notice of Proposed Rulemaking, 27 FCC Rcd 6422, 6423, para. 2 (2012).
6. Chen, M., Gonzalez, S., Vasilakos, A., Cao, H., & Leung, V. C. M. (2011). Body area networks: A survey. *Mobile Networks and Applications, 16,* 171–193.
7. Safeer, K. P., Gupta, P., Shakunthala, D. T., Sundersheshu, B. S., & Padaki, V. C. (2008). Wireless sensor network for wearable physiological monitoring. *Journal of Networks, 3*(5), 21–29.
8. Pentland, A. (2004). Healthwear: Medical technology becomes wearable. *Computer, 37*(5), 42–49.
9. Curtis, D., Shih, E., Waterman, J., Guttag, J., Bailey, J. et al. (2008). Physiological signal monitoring in the waiting areas of an emergency room. In *Proceedings of Body Nets 2008*. Tempe, Arizona.
10. Shnayder, V., Chen, B., Lorincz, K., Fulford-Jones, T. R. F., & Welsh, M. (2005). *Sensor networks for medical care*. Harvard University Technical Report TR-08-05.
11. Farella, E., Pieracci, A., Benini, L., Rocchi, L., & Acquaviva, A. (2008). Interfacing human and computer with wireless body area sensor networks: The WiMoCA solution. *Multimedia Tools and Applications, 38*(3), 337–363.
12. Gao, T., Massey, T., Selavo, L., Crawford, D., Chen, B., Lorincz, K., et al. (2007). The advanced health and disaster aid network: A light-weight wireless medical system for triage. *IEEE Transactions on Biomedical Circuits and Systems, 1*(3), 203–216.
13. Jiang, S., Cao, Y., Lyengar, S., Kuryloski, P., Jafari, R., Xue, Y., et al. (2008). CareNet: An integrated wireless sensor networking environment for remote healthcare. In *Proceedings of the 7th International Conference On Body Area Networks*. Tempe, Arizona.

14. Sukor, M., Ariffin, S., Fisal, N., Yusuf, S. S., & Abdullah, A. (2008). Performance study of wireless body area network in medical environment. In *Asia International Conference on Mathematical Modelling and Computer Simulation* (pp. 202–206).
15. IEEE standard for information technology- telecommunications and information exchange between systems- local and metropolitan area networks- specific requirements part 15.4: Wireless medium access control (MAC) and physical layer (PHY) specifications for low-rate wireless personal area networks (WPANs). (2006). IEEE Std 802.15.4-2006 (Revision of IEEE Std 802.15.4-2003) (pp. 1–305).
16. Kwak, K., Ullah, S., & Ullah, N. (2010). An overview of IEEE 802.15.6 standard. In *3rd International Symposium on Applied Sciences in Biomedical and Communication Technologies* (*ISABEL*) (pp. 1–6).
17. Boulis, A., Smith, D., Miniutti, D., Libman, L., & Tselishchev, Y. (2012). Challenges in body area networks for healthcare: The MAC. *IEEE Communication Magazine*, Vol. 50, p. 5.
18. Abolhasan, M., Wysocki, T., & Dutkiewicz, E. (2004). A review of routing protocols for mobile ad hoc networks. *Ad Hoc Networks, 2,* 1–22.
19. Akkaya, K., & Younis, M. (2005). A survey on routing protocols for wireless sensor networks. *Ad Hoc Networks, 3,* 325–349.
20. Cheng, S., & Huang, C. (2013). Coloring-based inter-WBAN scheduling for mobile wireless body area network. *IEEE Transactions on Parallel and Distributed Systems, 24*(2), 250–259.
21. Ullah, S., Higgins, H., Braem, B., Latre, B., Blondia, C., Moerman, I., et al. (2010). A comprehensive survey of wireless body area networks. *Journal of Medical Systems*, 1–30.
22. Maskooki, A., Soh, C. B., Gunawan, E., & Low, K. S. (2011). Opportunistic routing for body area network. In *IEEE Consumer Communications and Networking Conference* (*CCNC*) (pp. 237–241).
23. Tachtatzis, C., Franco, F., Tracey, D., Timmons, N., & Morrison, J. (2010). An energy analysis of IEEE 802.15.6 scheduled access modes. In *IEEE GLOBECOM Workshops* (*GC Wkshps*) (pp. 1270–1275).
24. Takahashi, D., Xiao, Y., Hu, F., Chen, J., & Sun, Y. (2008). Temperature-aware routing for telemedicine applications in embedded biomedical sensor networks. *EURASIP Journal on Wireless Communications and Networking* 26.
25. Braem, B., Latré, B., Moerman, I., Blondia, C., & Demeester, P. (2006). The wireless autonomous spanning tree protocol for multi hop wireless body area networks. In *Proceedings* of *3rd Annual International Conference on Mobile and Ubiquitous Systems: Networking Services* (pp. 1–8).
26. Saleem, S., Ullah, S., & Yoo, H. S. (2009). On the security issues in wireless body area networks. *JDCTA, 3*(3), 178–184.
27. IEEE p802.15 working group for wireless personal area networks (WPANs): Medwin MAC and security proposal documentation. (2009). IEEE 802.15.6 Technical Contribution.
28. Lewis, D. (2008). 802.15.6 call for applications in body area networks-response summary. 15-08-0407-05-0006.
29. Smith, D., Miniutti, D., Lamahewa, T. A., & Hanlen, L. (2013). Propagation models for body area networks: A survey and new outlook. *IEEE Antennas and Propagation Magazine*.
30. Khan, J. Y., Yuce, M. R., Bulger, G., & Harding, B. (2012). Wireless body area network (WBAN) design techniques and performance evaluation. *Journal of Medical Systems, 36*(3), 1441–1457.
31. Zhen, B., Patel, M., Lee, S., Won, E., & Astrin, A. (2008). Tg6 technical requirements document (TRD). IEEE p802.15-08-0644-09-0006.
32. Patel, M., & Wang, J. (2010). Applications, challenges, and prospective in emerging body area networking technologies. *Wireless Communication, 17,* 80–88.
33. Scanlon, W., Conway, G., & Cotton, S. (2007). Antennas and propagation considerations for robust wireless communications in medical body area networks. In *IET Seminar on Antennas and Propagation for Body- Centric Wireless Communications* (p. 37).

34. Serrano, R., Blanch, S., & Jofre, L. (2006). Small antenna fundamentals and technologies: Future trends. In *IEEE 1st European Conference on Antennas and Propagation* (*EuCAP*) (pp. 1–7).
35. Kiourti, A., & Nikita, K. S. (2012). A review of implantable patch antennas for biomedical telemetry: Challenges and solutions [wireless corner]. *IEEE Antennas and Propagation Magazine, 54*(3), 210–228.
36. Higgins, H. (2006). Wireless communication. In G. Z. Yang (Ed.), *Body sensor networks* (pp. 117–143). London: Springer.
37. Hanson, M., Powell, H., Barth, A., Ringgenberg, K., Calhoun, B., Aylor, J., et al. (2009). Body area sensor networks: Challenges and opportunities. *Computer, 42,* 58–65.
38. Yuce, M., & Khan, J. (2011). *Wireless body area networks: Technology, implementation, and applications*. Singapore: Pan Stanford Publishing.
39. Kumar, M., Veeraraghavan, A., & Sabharwal, A. (2015). Distance PPG: Robust non-contact vital signs monitoring using a camera. *Journal of Biomedical Optics Express, 6,* 1565–1588.
40. Hao, T., Xing, G., & Zhou, G. iSleep: Unobtrusive sleep quality monitoring using smartphones. In *Proceedings of the 11th ACM Conference on Embedded Networked Sensor Systems, Roma, Italy*, November 11–15, 2013.
41. Kwon, S., Kim, H., & Park, K. S. (2012). Validation of heart rate extraction using video imaging on a built-in camera system of a smartphone. In *Proceedings of the IEEE International Conference of Engineering in Medicine and Biology Society*, San Diego, CA, 28 August–1 September.
42. Hao, T., Xing, G., & Zhou, G. (2013). Unobtrusive sleep monitoring using smartphones. In *Proceedings of the 7th International Conference on Pervasive Computing Technologies for Healthcare*, Venice, Italy, 5–8 May.
43. Frost, M., Marcu, G., Hansen, R., Szaanto, K., & Bardram, J. E. (2011). The MONARCA self-assessment system: Persuasive personal monitoring for bipolar patients. In *Proceedings of the 5th International Conference on Pervasive Computing Technologies for Healthcare*, Dublin, Ireland, 23–26 May.
44. Baumgarten, M., Mulvenna, M., Rooney, N., & Reid, J. (2013). Keyword-based sentiment mining using twitter. *International Journal of Ambient Computing and Intelligence, 5*(2), 56–69.
45. Bureš, V., Tučník, P., Mikulecký, P., Mls, K., & Blecha, P. (2016). Application of ambient intelligence in educational institutions: Visions and architectures. *International Journal of Ambient Computing and Intelligence (IJACI), 7*(1), 94–120.
46. Fouad, K. M., Hassan, B. M., & Hassan, M. F. (2016). User authentication based on dynamic keystroke recognition. *International Journal of Ambient Computing and Intelligence (IJACI), 7*(2), 1–32.
47. Jain, A., & Bhatnagar, V. (2016). Movie analytics for effective recommendation system using Pig with Hadoop. *International Journal of Rough Sets and Data Analysis (IJRSDA), 3*(2), 82–100.
48. Kamal, M. S., Nimmy, S. F., Hossain, M. I., Dey, N., Ashour, A. S., & Santhi, V. (2016). ExSep: An exon separation process using neural skyline filter. In *IEEE International Conference on Electrical, Electronics, and Optimization Techniques* (*ICEEOT*) (pp. 48–53).
49. Klus, H., & Niebuhr, D. (2009). Integrating sensor nodes into a middleware for ambient intelligence. *International Journal of Ambient Computing and Intelligence (IJACI), 1*(4), 1–11.
50. Poland, M. P., Nugent, C. D., Wang, H., & Chen, L. (2011). Smart home research: Projects and issues. In *Ubiquitous developments in ambient computing and intelligence: Human-centered applications* (pp. 259–272). Hershey: IGI Global.
51. Curran, K., & Norrby, S. (2011). RFID-enabled location determination within indoor environments. In *Ubiquitous developments in ambient computing and intelligence: Human-centered applications* (pp. 301–324). Hershey: IGI Global.

52. Zappi, P., Lombriser, C., Benini, L., & Tröster, G. (2012). Collecting datasets from ambient intelligence environments. In *Innovative applications of ambient intelligence: Advances in smart systems* (pp. 113–127). Hershey: IGI Global.
53. Al Mushcab, H., Curran, K., & Doherty, J. (2012). An activity monitoring application for windows mobile devices. In *Innovative applications of ambient intelligence: Advances in smart systems*, (pp. 139–156). Hershey: IGI Global.
54. Graham, B., Tachtatzis, C., Di Franco, F., Bykowski, M., Tracey, D. C., Timmons, N. F., et al. (2013). Analysis of the effect of human presence on a wireless sensor network. In *Pervasive and ubiquitous technology innovations for ambient intelligence environments* (pp. 1–11). Hershey: IGI Global.
55. Chakraborty, C., Gupta, B., & Ghosh, S. K. (2015). Identification of chronic wound status under tele-wound network through smartphone. *International Journal of Rough Sets and Data Analysis (IJRSDA), 2*(2), 58–77.
56. Borawake-Satao, R., & Prasad, R. S. (2017). Mobile sink with mobile agents: Effective mobility scheme for wireless sensor network. *International Journal of Rough Sets and Data Analysis (IJRSDA), 4*(2), 24–35.
57. Panda, S. K., Mishra, S., & Das, S. (2017). An efficient intra-server and inter-server load balancing algorithm for internet distributed systems. *International Journal of Rough Sets and Data Analysis (IJRSDA), 4*(1), 1–18.
58. Dey, N., Dey, G., Chakraborty, S., & Chaudhuri, S. S. (2014). Feature analysis of blind watermarked electromyogram signal in wireless telemonitoring. In *Concepts and trends in healthcare information systems* (pp. 205–229). Berlin: Springer.
59. Dey, N., Dey, M., Mahata, S. K., Das, A., & Chaudhuri, S. S. (2015). Tamper detection of electrocardiographic signal using watermarked bio–hash code in wireless cardiology. *International Journal of Signal and Imaging Systems Engineering, 8*(1–2), 46–58.
60. Dey, N., Mukhopadhyay, S., Das, A., & Chaudhuri, S. S. (2012). Using DWT analysis of P, QRS and T components and cardiac output modified by blind watermarking technique within the electrocardiogram signal for authentication in the wireless telecardiology. *IJ Image, Graphics and Signal Processing (IJIGSP), 4*(7), 33–46.
61. Dey, N., Roy, A. B., Das, A., & Chaudhuri, S. S. (2012). Stationary wavelet transformation based self-recovery of blind-watermark from electrocardiogram signal in wireless telecardiology. In *International Conference on Security in Computer Networks and Distributed Systems* (pp. 347–357). Berlin: Springer.
62. Dey, N., Biswas, S., Roy, A. B., Das, A., & Chaudhuri, S. S. (2012). Analysis of photoplethysmographic signals modified by reversible watermarking technique using prediction-error in wireless telecardiology. In *International Conference of Intelligent Infrastructure, 47th Annual National Convention of CSI.*
63. Bhatt, C., Dey, N., & Ashour, A. S. (2017). Internet of things and big data technologies for next generation healthcare.
64. Kamal, S., Ripon, S. H., Dey, N., Ashour, A. S., & Santhi, V. (2016). A MapReduce approach to diminish imbalance parameters for big deoxyribonucleic acid dataset. *Computer Methods and Programs in Biomedicine, 131,* 191–206.
65. Solanki, V. K., Katiyar, S., BhashkarSemwal, V., Dewan, P., Venkatasen, M., & Dey, N. (2016). Advanced automated module for smart and secure city. *Procedia Computer Science, 78,* 367–374.

Mobile Sensor Networks and Robotics

K.P. Udagepola

Abstract The collaboration between wireless sensor networks and the distributed robotics has prompted the making of mobile sensor networks. However, there has been a growing enthusiasm in developing mobile sensor networks, which are the favoured family of wireless sensor networks in which autonomous movement assumes a key part in implementing its application. By introducing mobility to nodes in wireless sensor networks, the capability and flexibility of mobile sensor networks can be enhanced to support multiple mansions, and to address the previously stated issues. The reduction in costs of mobile sensor networks and their expanding capacities makes mobile sensor networks conceivable and useful. Today, many types of research are focused on the making of mobile wireless sensor networks due to their favourable advantage and applications. Allowing the sensors to be mobile will boost the utilization of mobile wireless sensor networks beyond that of static wireless sensor networks. Sensors can be mounted on, or implanted in animals to monitor their movements for examinations, but they can also be deployed in unmanned airborne vehicles for surveillance or environmental mapping. Mobile wireless sensor networks and robotics play a crucial role if it integrated with static nodes to become a Mobile Robot, which can enhance the capabilities, and enables their new applications. Mobile robots provide a means of exploring and interacting with the environment in more dynamic and decentralised ways. In addition, this new system of networked sensors and robots allowed the development of fresh solutions to classical problems such as localization and navigation beyond that. This article presents an overview of mobile sensor network issues, sensor networks in robotics and the application of robotic sensor networks.

Keywords Mobile sensor networks · Robots · Coverage · Localization · Robotic sensor networks · Network topology · Healthcare

---

K.P. Udagepola (✉)
Department of Information and Computing Sciences,
Scientific Research Development Institute of Technology Australia,
Loganlea, QLD 4131, Australia
e-mail: kalumu@srdita.com.au

N. Dey et al. (eds.), *Internet of Things and Big Data Analytics Toward Next-Generation Intelligence*, Studies in Big Data 30,
DOI 10.1007/978-3-319-60435-0_2

1 Introduction

The wireless connectivity within the network is facilitated by main components of Wireless Sensor Network (WSN) connect with an application platform at one end of the network. Any part of network can use one or more sensor/actuator devices. Figure 1 depicts the required components link with the real world and the application platform. It uses g and sensor (S). Figure 1 shows an advanced WSN of the basic model because it has a relay node (R) to connect both a gateway and a sensor to make a mesh network. Another hand it is facilitating reducing obstacles to make model efficiency (Fig. 2).

Innovative advances such as 4G networks and that of ubiquitous computing have triggered new interests in Multi Hop Networks (MHNs). Specifically, the automated organization of wireless MHNs that are composed of large motes, which can be mobile and static, and can likewise be utilized for computational and power, is of great interest. On the other hand, WSNs are some of the ordinary examples of these networks. Their topology dynamically changes when connectivity among the nodes varies with their mobility due on the time factor. E.g. Fig. 3 shows Multi-hop WSN architecture.

A large part of the research in WSNs is focused on networks whose nodes cannot be replaced and are stationary. Mobility in sensor nodes has been taken advantage of in order to improve or enable the overall communication coverage and sensing of these networks [1]. The credit for the creation of mobile sensor networks (MSNs) goes to WSNs and also to the interaction of distributed robotics [2]. MSNs are a class of networks where little sensing devices communicate in a collaborative way by moving in a space to observe and monitor environmental and physical conditions [3, 4]. MSN is composed of nodes and all nodes have computation, sensing, locomotion modules and communication. Each sensor node is also capable of navigating human interaction or autonomously [5]. MSNs have emerged as an important area for research and development [6].

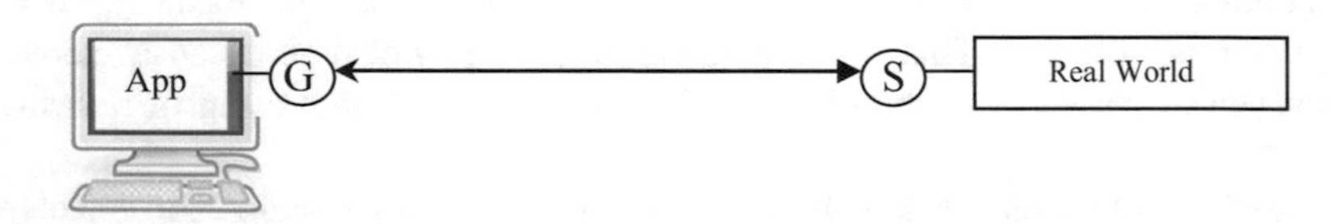

Fig. 1 Basic components of WSN

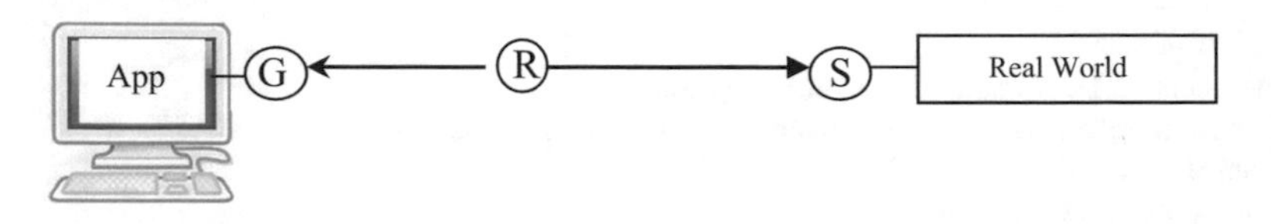

Fig. 2 Using relay node with basic model

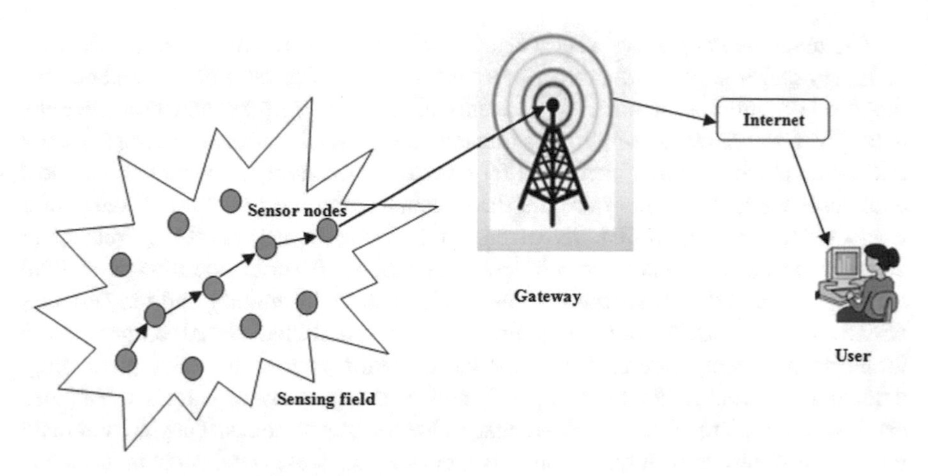

Fig. 3 Multi-hop WSN architecture

Even though MSNs are still developing, they can be used for monitoring the environment, disaster-prone areas and hazardous zones. It can also be used in monitoring healthcare, agriculture and defense. Mobile wireless sensor networks (MWSNs) can be used for both monitoring and control as many practical applications of MSNs that continue to emerge [7]. These include robotics, which is the science of technology having applications in various fields such as design, fabrication, and theory [8]. It can be considered as the area of technology that deals with the construction, operation and control of both robotic applications and computer systems. Furthermore, sensory feedback, as well as information processing, can be managed by robots. The main advantage of this technology is that it can replace humans in manufacturing processes, dangerous environments, or it can be made to resemble humans in terms of behavior, cognition or experience [8].

The word "robot" has its roots from "robota," which is a Czechoslo-vakian word meaning work robot. The word was first used in Karel Chapek's 1920s play Rossum Universal Robots. A leap forward in the autonomous robot technology happened in the mid-1980s with the work on behavior based robotics. This work was laid the basis for several robotic applications today [9]. Most of the problems encountered in traditional sensor networks may be addressed by integrating Mobile Robots (MRs), which are intelligent directly into sensor networks. MRs offer ways to interact and survey the environment in a more decentralized and dynamic way. The new system of robots and networked sensors has led to the emergence of new solutions for existing issues such as navigation and localization [10–12]. Mobile nodes can be utilized as intuitive self-controlled mobile robots or as intuitive robots whose sensor systems are capable of solving both environmental and navigational functions. In this way, sensor systems on robots are the dispersed systems. MRs carry sensors around an area to produce a detailed natural appraisal and sense phenomena [13–16].

The fundamental parts of a sensor node are a transceiver, a micro controller outer memory, multiple sensors and a power source [17]. The controller regulates the range of capabilities of other components in the sensor nodes and processes the data. The feasible option of wireless transmission media is infrared, radio frequency (RF) and optical communication. As far as external memory is concerned, the most applicable types of memory are the flash memory and the on-board memory of a micro controller. An availability of energy is the most important requirement to consider design and making a wireless sensor node. It should be always without interrupt the activation. Figure 4 shows DHT11 digital humidity and temperature, which is a blended sensor containing a calibrated digital signal output of the humidity and temperature. Sensor nodes consume energy for data processing, detection and communication; power is stored in capacitors or batteries. Batteries can be both rechargeable and non-rechargeable for sensor nodes. They are the main resource of a power supply. A sensor is a device that senses or detects motion, etc. It responds in a particular way [18–25]. The Analog to-digital converter (ADC) is making calibration match with the required data to a processor once sensor picked the data. Figure 5 presents a sensor node architecture that we discussed above.

A solution is given by the use of multi-robot systems for carrying sensors around the environment. It has received a considerable attention and can also provide some exceptional advantages as well. A number of applications have been addressed so far by Robotic Sensor Networks (RSNs) such as rescue, search and environmental monitoring. In WSNs, robotics can also be utilized to address several issues to advance performances such as responding to a particular sensor failure, node distribution, data aggregation and more. Similarly, to address the issues present in the field of robotics, WSNs play a crucial role in problems such as localization, path planning, coordination (multiple robots) and sensing [14, 27].

Today, the industry has many applications of sensor networks which are on the ground, in the air, underwater and underground. In mobile underwater sensor network (UWSN), mobility offers two main advantages. Firstly, floating sensors can increase system reusability. Secondly, it can help to enable dynamic monitoring as well as coverage. These features can be used to track changes in water aggregates

Fig. 4 Temperature humidity sensor module [26]

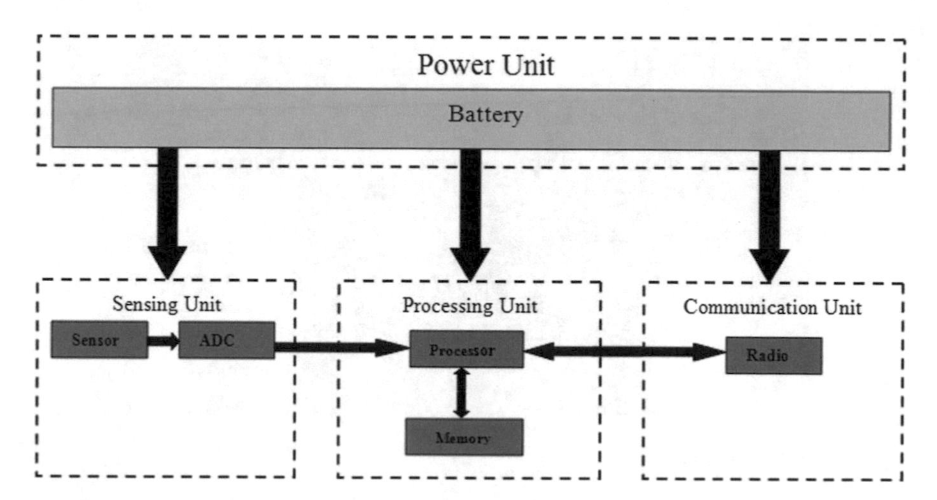

Fig. 5 A sensor node architecture

in this way providing 4D (space and time) environmental monitoring. As compared to ground-based sensor networks, mobile UWSN has to employ acoustic communications because radios do not work in hard water environments. Similarly, the underground sensor network makes a huge impact for monitoring number of characteristics at underground s as the properties of the soil, toxic substances and more. These sensor networks are buried completely underground and do not require any wire for connection. On the ground, they can be used for target tracking, environmental monitoring, forest fire detection, industrial monitoring and machine health monitoring. Wireless sensor nodes have been in service for a long time and are still used for different applications such as warfare, earthquake measurements and more.

National Aeronautics and Space Administration (NASA) embarked on the sensor webs project and smart dust project after the recent growth of small sensor nodes in 1998. The main aim of the smart dust project was to make self-controlling, sensing and corresponding possible within a cubic millimeter of space. The task drove numerous research activities incorporating real research focus in the center for embedded networked sensing (CENS) and Berkeley NEST. The term mote was coined by researchers working in these projects to indicate a sensor node, and the pod was the name used to refer to a physical sensor node in the NASA sensor webs project. In a sensor web, the sensor node can be another sensor web itself [17].

The crossbow radio/processor boards usually recognized as motes. It permits to wirelessly transmit many sensors scattered over a large area. This helps to receive t data to the base station. TinyOS is an operating system for the mote. This uses for low power wireless devices. E.g. ubiquitous computing, PAN, smart building, smart meter, sensor network and more. It controls radio transmission, power and networking transparent to the user. Subsequently, an ad hoc network initiates [18, 28].

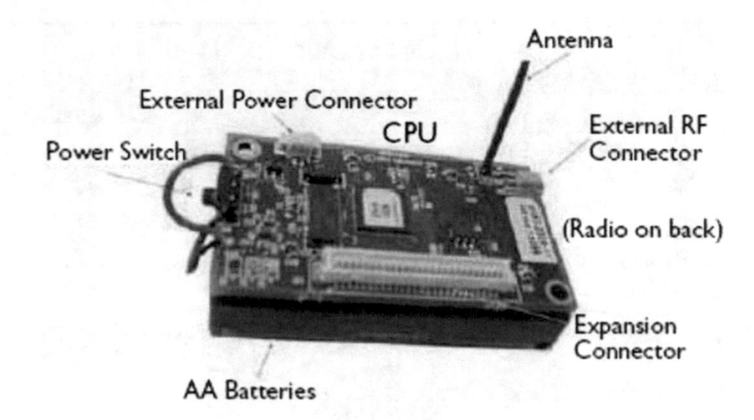

Fig. 6 Mica 2 processor

Fig. 7 Stargate processor [26]

The MICA2 (see Fig. 6) Mote is a third generation mote module with 512 KB of measurement (serial) flash memory, 128 KB of program flash memory and 4 KB of Programmable read-only memory.

Stargate (see Fig. 7) is a 400-MHz Intel PXA255 Xscale processor with 32 MB of flash memory and 64 MB of synchronous dynamic random-access memory. Different classes of sensors are available in the current market. E.g. barometric pressure, acceleration, seismic, acoustic, radar, light, temperature, relative humidity, magnetic camera, global positioning system (GPS) and more. Usually, sensors are categorized into three different kinds: passive, omnidirectional and narrow-beam (Wikipedia). Passive sensor has a self-activation characteristic is giving more powerful to fetch the data without actually manipulating the environment. Narrow beam sensor has a distinct direction of the measurement. Omni-directional sensor has no direction of the measurement [29].

Internet of Things (IoT) employs WSN systems to give lots benefits for numerous applications in real life. E.g. healthcare systems manufacturing (Sensors with Connectivity), Home systems, smart city,... etc. WSN systems are using data acquisition from various long-term industrial environments for IoT. Sensor interface device is acquiring sensor data from real time and makes a precious picture through WSN in IoT environment. This is a reason major manufactures pay attention to ongoing research on equipments in multi sensor acquisition interface [30].

2 Mobile Sensor Networks

MSN is a class of networks in which small devices capable of sensing their surroundings moved in a space over time to collaboratively monitor physical and environmental conditions [3, 29]. Worldwide researches conducted many investigations on MSNs because there could be a lot of current applications with adopted sensors. Potentially, the sensors have many capabilities such as environmental information sensing, locomotion, dead-reckoning and many more. The architecture of MSN can be broken down into node, server and client layer [5, 31]. The job of the node layer is to acquire most kinds of data as it is straight embedded into the current world. This layer also includes all the mobile and static sensor nodes. Server layer comprises a single board computer running a personal computer or server software. Any smart terminal can use at the client layer devices. Remote and local clients are also linked with the client layer. The detail is shown in Fig. 8. Mobility is an unrealistic or undesirable characteristic of a sensor node as it can address the objective challenges [3, 5, 32, 33]. References [3, 29, 34] analyzed the research issues on MSNs based on data management and communication. Our work is focused on communication issues which include coverage and localization issues.

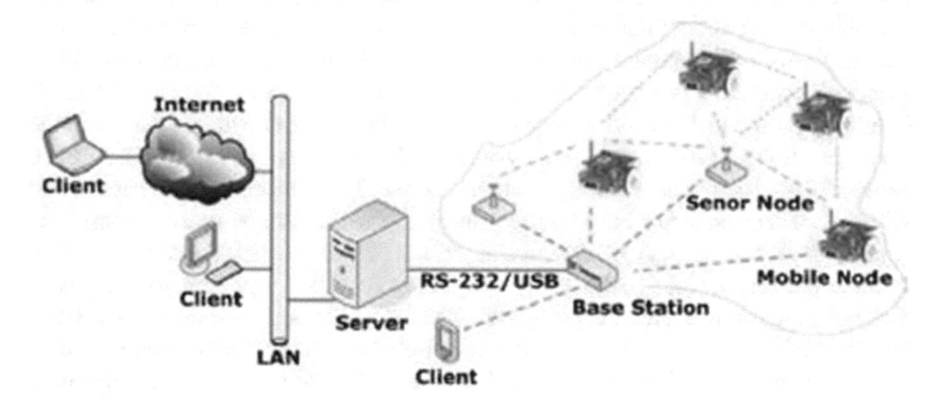

Fig. 8 The system architecture of a MSN [26]

2.1 Coverage

The degree of the quality of service is one of the methods to analyze the coverage. The quality of service can be also depended on upon the reach of a sensor network [35–37]. It can be seen that for all the applications of MSNs, network reach coverage is one of the most fundamental issues [38]. It decreases as a result of sensor failure and undesirable sensor deployment. Reference [39] defines coverage as the maintenance of spatial relationship, which adjusts to the exact local conditions to optimize the performances of some functions. Gage describes three coverage behavior types, which are blanket, barrier, and sweep. The aim of the blanket coverage is to bring about a fixed layout of nodes that minimizes the overall detection area. Likewise, the main goal of the barrier coverage is to reduce the chances of undiscovered penetration via the barrier. The concept of sweep coverage comes from robotics, which is less or more equivalent to the moving barrier. The lifetime of sensors is strongly affected by hardware defects, battery depletions and harsh external environments such as fire and the wind [3, 29]. In MSNs, already revealed territories get to be covered when sensors travel through and far from the zone. As a result, the already covered areas become uncovered. The zones covered by sensors change after some time, and more regions get to be secured at any rate once time goes [36, 40]. For robotic applications, Ref. [41] was a person to describe potential field techniques for tasks such as obstacle avoidance and local navigation. He introduced a similar concept 'Motor Schemas'. This uses the superposition of spatial vector fields to make behavior. Reference [42] used potential fields, but for the issue of deployment. He considered the issue of arranging mobile sensors in an anonymous environment where fields are constructed, i.e. each node is repelled by the other node. Also, throughout the environment as obstacles that force the network to spread [43]. In addition, the proposed potential field technique is distributable, scalable and requires not a prior map of the environment. In reference [44], for the uncovered areas by the sensor network, new nodes are always placed on the boundary of uncovered areas. The potential field technique is also able to find a suboptimal deployment solution and also makes sure that each node is in the line of sight with the other node. Thus, in order to increase the coverage [45], proposed algorithms needed to calculate the desired target positions where sensors should move and identify the coverage holes existing in the network. To find out the coverage holes [46] used the Voronoi diagram. It has designed three movement-assisted sensor deployment protocols. The concept called based) and Minimax. These concepts base on the principle of sensors moving from densely to sparsely deployed areas. A virtual force algorithm (VFA) was proposed by [46, 47] to increase the sensor field coverage by combining repulsive and attractive forces to determine randomly deployed virtual motion paths and sensor movements. The static sensors guide the mobile sensors to the position where the task is to occur and thus become aware of the arrival of tasks. References [45, 46] deal with the dynamic aspects of coverage in MSNs with characterized area coverage during a time interval, at specific time instants and the detection time of the randomly located target [48–53].

2.2 Localization

Much attention has been given to building mobile sensors lately. This has also led to the evolution of small-profile sensing devices capable of controlling their locomotion. Mobility has turned into an imperative territory of examination in mobile sensor systems. Mobility empowers sensor nodes to aim and locate dynamic situations such as vehicle movement, chemical clouds and packages [54, 55]. Localization is one of the main difficulties to achieve in mobile sensor nodes. Localization is the capacity of sensor nodes to calculate their current coordinates; and on mobile sensors, it is performed for navigational and tracking purposes. Thus, localization is needed in several areas such as health, military and others. The broad examination has been done so far on localization, and numerous positioning systems have been proposed to remove the need for GPS on each sensor node [56].

GPS is usually thought to be a decent answer for open air localization. Nonetheless, it is still costly and thus not utilized for a substantial number of gadgets in WSN. Some of the problems associated with GPS are as follows:

GPS does not work reliably in some situations: Because a GPS receiver needs line of sight to multiple satellites, its performance is not admirable in indoors. The receivers are accessible only for mote scale devices. GPS receivers are still expensive and undesirable for many applications [57]. The problem of using GPS is requiring a real environment to get measurements. Normal GPS shows 10–20 m of error at standard outdoor environments. This error can be minimized but should use a costly mechanism. Deploying large numbers of GPS in MSNs have possibilities and limits.

There [56] are two sorts of localization algorithms to be specific: centralized and distributed algorithms. Centralized location methods rely upon sensor nodes to send information to a base station. It is there that calculation is implemented in order to find out the position of every node. On the other hand, distributed algorithms need to not have a central base station and for determining their location. They relay with each node restricted data and information with nearby nodes [56]. References [3, 29] Localization algorithms in MWSNs are categorized into range-free, range and mobility-based methods. These methods vary in the information utilized for the idea of localization. Range-based methods employ range computations while range-free techniques operate only the content of messages [58–60]. Range-based methods also require costly hardware to measure the angle of signal arrival and the arrival period of the signal. As compared to range over free methods, these two methods are expensive because of their pricey hardware [3, 29]. While range free methods use local and hop count techniques for range-based approaches, these methods are very cost effective. Many localization algorithms have been proposed so far such as the elastic localization algorithm (ELA) and the mobile geographic distributed localization algorithm. Both of these algorithms assume non-limited storage in sensor nodes [3]. Two types of range-free algorithms introduced for the sensor networks. These are local and hop count techniques. The local technique based on high speed on a high-density seed. This method gives a chance to node to pick several seeds, which ever close. The hop count technique depends on a flooding network. In here, each node considers a location to calculate distance from the seeds' location. This method is correctly calculated if the seeds are static but in ad hoc situation, this is impossible. If the triangular regions need to separate environment use beaconing nodes, the approximate point-in-triangulation test (APIT) method is suitable. In this case, the grid algorithm calculates the maximum area and gives chance to a node to settle at the environment [61]. Hop count techniques propagate the location estimation throughout the network where the seed density is low. Figure 9 shows coordination, measurement and location estimation phase. Mobility-based method was to improve accuracy and precision of the localization method. Sequential Monte-Carlo Localization (SML) was proposed by [49] without additional hardware except for GPS [3]; and without decreasing the non-limited computational ability, many techniques using SML is also being proposed. In order to achieve accuracy in localization, researchers proposed many algorithms using the principles of Doppler shift and radio interferometry to achieve accuracy [3, 29]. References [54, 55] described the three phases typically used in localization, which are: coordination, measurement and position estimation.

To initiate the localization a group of nodes, coordinate first, a signal is then emitted by some nodes and then some property of the signal is observed by some other nodes. By transforming the signal measurements into position estimates, node position can then be determined. Reference [62] proposed three approaches which are Mobility Aware Dead Reckoning Driven (MADRD), Dynamic Velocity Monotonic (DVM) and Static Fixed Rate (SFR).

1. Mobility Aware Dead Reckoning Driven: This approach predicts future mobility with computes the mobility pattern of the sensors. The result of difference between the predicted mobility and expected mobility reaches the error threshold at the time the localization should be triggered [62].
2. Static Fixed Rate: This approach uses the performance of the protocol changes linked the mobility of sensors. In the fix time, every sensors appeal to their localization as the periodical way. If sensors are moving quickly, the glitch or laps will be high and Vs versa [63, 64].
3. Dynamic Velocity Monotonic: This is an adaptive protocol with the mobility of sensors localization called adaptively in DVM [65].

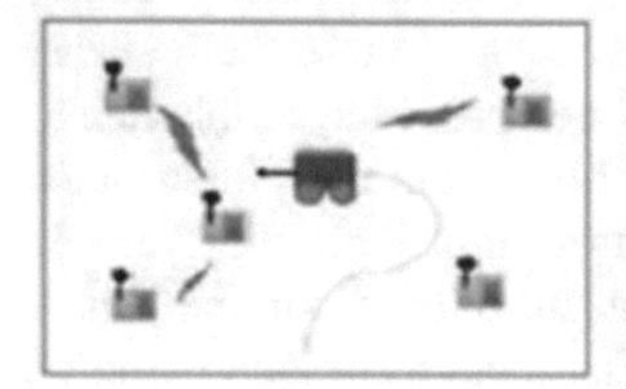
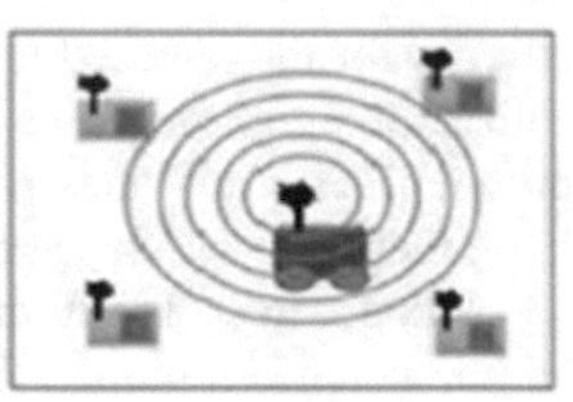
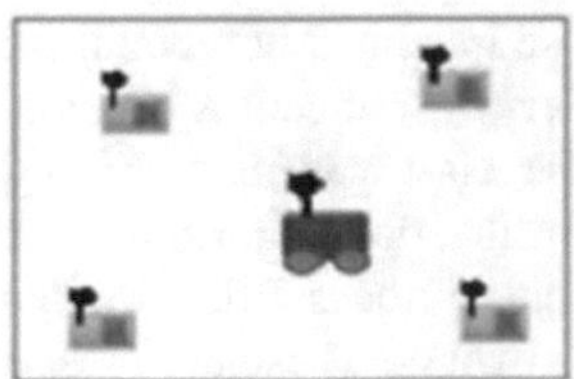

Fig. 9 Coordination, measurement and location estimation phase [26]

A new method called the Mobility Aware Interpolation (MAINT) was proposed by [65]. It estimates the current position of a mobile node with a better tradeoff with advantage of accuracy and energy consumption. This method uses interpolation to get best estimation in most cases.

In this section discusses a brief overview of the localization methods used by some researchers. Under mobile node localization, Ref. [66] proposed a system appropriate for the real environment in moving both anchors and unknown nodes. In this method used the history of anchor information to find out the current position. User's movement was module by the archived information; and for discovering new positions, movement models were also used. Reference [62] used in complex situations where anchors and nodes are mobile. The above three methods are used to resolve the localization problem. Once the sensor has at least two anchors in the neighborhood, the methods determine the exact position. Otherwise it gives a fairly accurate position and in that situation and it can compute the generated maximal error. The method also defines the intervals a node will invoke its localization. Reference [67] proposed a GPS free localization algorithm in MWSN's. To build the coordinate system, the proposed algorithm uses the distance between the nodes; also, nodes positions are computed in two dimensions. Based on Dead Reckoning, Refs. [63] and [64] put forward a series of methods for locating mobile sensors. Among all the forwarded methods, the Mobility Aware Dead Reckoning Driven finds the approximate the location of the sensor, instead of localizing the sensor each moment it manoeuvres. Inaccuracy in the approximated area is calculated every moment the localization is invoked and with time, the error in the estimation grows. Also, the next localization time is fixed depending on the value of this error. Rapid moving mobile sensors cause localization with a higher frequency for a particular degree of precision in position estimation. Instead of localizing the sensor [65] proposed a technique for guessing the location of a mobile sensor. His method gave higher precision results for a given cost of energy. Thus, the location of the sensor is only required when the sensor contains data, which is to be sent while an inactive sensor transmits neither data nor information. The most of it apply to the base because in order to decrease the complexity of computation in sensors. For solving the sets of equations [56] proposed the novel algorithm for MSN localization where they chose three nodes, which are neighbors to each other. Thus, if the answers are the equations are exclusive, it means that they are the points of the node; and for searching the position of the final node, a scan algorithm was also introduced called a Metric Average Localization Error (MALE) (which is the root-mean-square error of node locations divided by the total number of nodes) to evaluate the localization error.

3 Robotic Sensor Network Applications

Most of the issues encountered with traditional sensor networks may be addressed by integrating intelligent mobile robots directly into it. MRs present the way to investigate and interface with nature in dynamic and decentralized ways. Notwithstanding empowering mission capacities well past those provided by sensor networks allows these new systems of networked sensors and robots develop new answers to traditional problems such as localization and path finding [1]. Many problems in sensor networks can be solved by putting robotics into use. Problems such as node positioning and localization; identifying and responding to sensor failure; acting as a data mule and for nodes as a mobile battery charger is also possible. In addition, WSNs can take care of numerous issues in robotics such as navigation, localization mapping and sense [14]. There are many applications of WSNs in robotics such as advanced robotic sensing, multiple robot coordination, robot planning and navigation, and robot localization. Using WSNs helps emergency response robots to be conscious of conditions such as electromagnetic field monitoring, forest fire detection and others. These networks improve the sensing capability and can also help robots in finding the way to the area of interest. WSN's can be useful for organizing various robots and swarm robotics because the network can assist the swarm to share sensor data and track its members. To perform the coordinated tasks, WSNs send robots to different locations, and a swarm decides to depend on the localization of events; thus, enabling path planning and coordination for many robots to happen efficiently and optimally as well as directs the swarm members to the area of interest. In the localization part, there are many methods for localizing robots within a sensor network. Cameras have been put into use to identify the sensors mounted with infrared light to triangulate them in view of the space gotten from the pixel size. A modified SLAM algorithm has been utilized by some methods, which uses robots to localize themselves inside the surroundings the environment and later the compensation for SLAM sensor error is achieved by fusing the approximated area with the approximated area in the WSN based on Received Signal Strength Indicator (RSSI) triangulation [14]. References [68, 69] presented an intruder detection system, which uses both WSNs and MRs. In order to learn and detect intruders in a previously unknown environment, a sensor network uses an unsupervised fuzzy adaptive resonance theory (ART) neural network [70, 71].

In WSNs, robotics can also play a crucial role. They can be used for replacing broken nodes, repositioning nodes, recharging batteries and more. To increase the feasibility of WSNs, Ref. [72] used robots because they have actuation, but limited coverage in sensing while sensor networks lack actuation, but they can acquire data. In servicing WSN's, Ref. [73] examined the robot allotment for a task and its achievement. Problems were examined in its task allotment like multitasking and single-tasking robots in a network, and how to sort out their behavior to ideally benefit the system. The course in which a robot takes to administration hubs is analyzed in the robot errand satisfaction. The route in which a robot takes to service nodes is examined in the robot task fulfillment. To improve the robot localization [74] adapts sensor network models with information maps and then checks the capability of such maps to improve the localization. The node replacement application was created by [75] in which a robot would explore a sensor system based on RSSI from nearby nodes and sends a help signal if a mote begins to experience power shortage. It is through the network that the help signal would pass to guide the robot to change the node. On the other hand, robots can be used to recharge batteries as well. The problem of localization can also be solved using robots. They can be utilized to limit the nodes in the network and also in data aggregation. In a network, they can serve as data mules; data mules are robots that move around the sensor network to collect data from the nodes and then transport the collected data back to the sink node. It can also be used in performing aggregation operations on data [14, 27]. The utilization of multi-robot systems for moving sensors around the environment served as an answer, which has obtained appreciable attention and can provide some remarkable advantages [13]. When put into use, RSNs can be used for effective search and rescue, monitoring of electromagnetic fields and others. The search and rescue systems rapidly and precisely find casualties, model search and space; and with human respondents, it should maintain communication. Thus, in order to satisfy the main goal of a search and rescue system, the system ought to be capable of quickly and accurately trace casualties within the search space and should also be capable of handling a changing and likely unfriendly environment. For utilizing ad hoc networks [10, 11] presented an algorithmic framework consisting of a large number of robots and small, cheap, simple wireless sensors to perform proficient and robust target tracking. Without dependence on the magnetic compass or GPS localization service, they described a RSN for target tracking, focusing on algorithms, which are simple for information propagation and distributed decision making. They presented a robotic sensor network system that freely handles object tracking without component possessing localization capabilities. Their approach provides a way out with a little hardware hypothesis in relation to detection, localization, broadcast, memory or processing capacity while subject to a progressively evolving environment. Moreover, their framework adjusts actively to object to movement and inclusion/exclusion of network units. The network gradient algorithm grants a beneficial trade-off between power consumption and performance and requires a reasonable bandwidth [10]. The monitoring of electromagnetic fields is highly necessary for practice, particularly to ensure the safety of the general population living and working where these fields are present [76]. Reference [13] presented a specific RSN oriented to monitor EMFs. In this network, the activities of the system are being supervised by a coordinator computer while a number of explorers (MRs equipped with EMF sensors) navigate in the environment and perform EMF measurement tasks. The system architecture is hierarchical. The activities of the system are being supervised by a computer and to perform the EMF tasks a number of explorers (MRs equipped with EMF sensors navigate through the environment). The grid map of the environment is maintained by the system in which each cell can be either free or occupied by an obstacle, or by a robot. In addition, the map should also be known to the coordinator and the explorers. The environment is assumed to be static, and the map is used by the explorers to navigate in the environment and is also used by the coordinator to localize the EMF source [13, 77, 78].

4 IoT Concept and Applications

MIT added a new phrase "Internet of Things" to our dictionary at the start of 21st century. This phrase considers all the kind of good and services link with the Internet at the current world. E.g. Sensing, identification, communication, networking, and informatics devices and systems, and seamlessly connects. Figure 10 presents impressive details of the idea, and it was published in the Economist magazine in 2007.

Currently, IoT connects with human life through sensors and actuators, WSN and much more [80]. It is making huge differences in human life using direct applications. E.g. human body (checking on the baby, remember take medicine, track activity levels, monitor an aging family members, stay out doctor's office, smart walking sticks or smart canes [81] … etc.), home (heat home efficiency, control all house hold appliances, track dawn lost key, lighting home, avoid disasters, keep plants alive, discovery of public things [82] … etc.), City (keep city clean, Light Street more effectively, share findings, Intelligent Traffic Monitoring

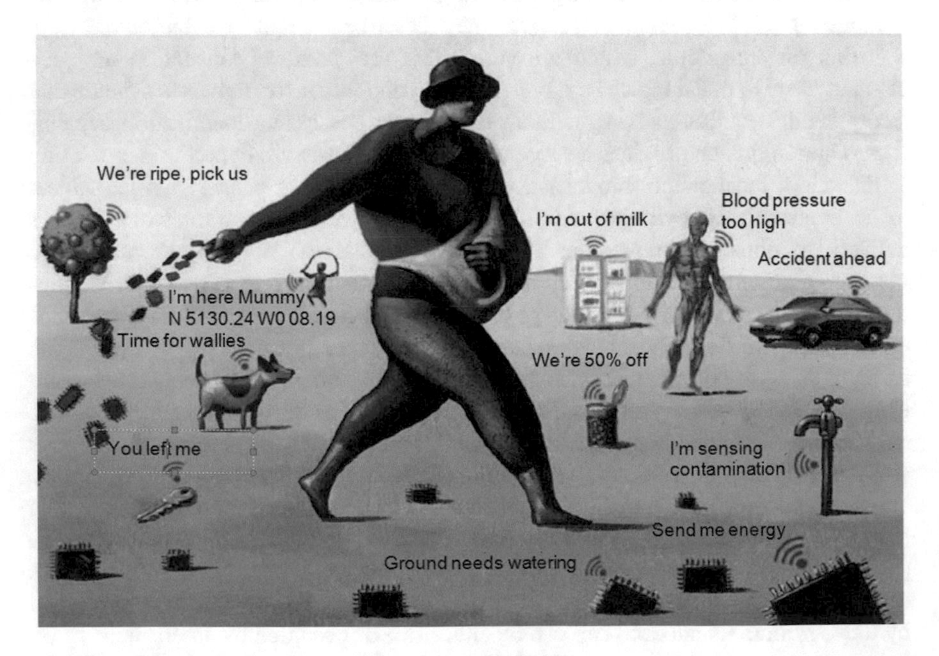

Fig. 10 A Impressive description of the vision of IoT [79]

System [83] … etc.) Industry (maintaining and repairing, stop guessing, monitoring, keep track assets [84] … etc.) environment (monitor pollution levels, track water, help protect wildlife, get advanced warning… etc.) … etc.

5 Coverage for Multi-Robots

The use of multi-robots holds numerous advantages over a single robot system. Their potential of doing work is the way far better than that of a single robot system [85, 86]. Coverage for multi-robot systems is an important field and is vital for many tasks like search and rescue, intrusion detection, sensor deployment, harvesting and mine clearing and more [87]. To get the coverage, the robots must be capable of spotting obstacles in the environment, and they should also swap their insight about their surroundings and have a tool to assign to dole out the scope of errands among themselves [55, 88]. The problem of deploying a MSN into an environment was addressed in [89, 90] with the task of maximizing sensor coverage; and also [89, 90] proposed two behaviors based techniques for solving the 2D coverage problems using multiple robots. Informative and molecular are the techniques proposed for solving coverage problems and both of these techniques has the same architecture. When robots are with-in the sensor range of each other, the informative approach is to assign local identities to them. This approach allows robots to spread out in a coordinated manner. It has ephemeral identification with temporary local identities and the mutual local information. The molecular approach does not have local identities, and robots do not perform any direct communication. Instead, each robot moves in a direction without communicating with its neighbors. Robot can select its own direction without support immediate sensed neighbors. Reference [51] then compares these algorithms with another approach known as the basic approach, which only seeks to maximize each individual robot's sensor coverage [87]. Both these approaches perform significantly better than basic approach and with the addition of a few robots the coverage area quickly maximizes. References [91–93] proposed (StiCo) coverage algorithm for multi-robot systems. This algorithm is based on the principle of stigmergic (pheromone-type) coordination known from the ant societies. These were a group of robots that coordinated indirectly via ant-like stigmergic communication. This algorithm does not require any prior information about the environment and also no direct robot to robot communication is required. Similar kind of approach was used by [93] for coverage in multi-robots in which a robot deposits a pheromone, which can then be detected by other robots, these pheromones come up with a decay rate that allows the continuous coverage of an area via implicit coordination [55, 88]. For multi-robot coverage, Refs. [55, 88, 94] proposed Boustrophedon decomposition algorithm in which the robots are at first dispersed through space and each robot is distributed essentially with a limited zone to cover and is then disintegrated into cells with a static cell width. By using the adjacency graph, the broken down area is described, which is incrementally developed and shared among all robots with no limitation; and robot correspondence is accessible. By sharing information regularly, task selection protocol performance is improved. By planting laser reference points in the earth, the problem of localization in the hardware experiment is overthrown utilizing the laser range finder to localize the robots as this is the major problem for guaranteeing accurate and consistent coverage. Reference [95] addressed strength and productivity in a group of multi-robot coverage algorithms in view of the spanning-tree coverage of estimated cell disintegration.

6 Localization for Robot

In mobile robotics, localization is a key component [96]. The process of determining a robot's position within the environment is called localization or it is a process that takes a map as an input estimates the current position of the robot, a set of sensor readings and then outputs the robot's current posed as a new estimate [97]. There are numerous technologies accessible for robot localization including GPS, active/passive beacons, odometer (dead reckoning), sonar and others. For robot localization and map count, Ref. [98] presented a method for using data from a range based sonar sensor. The robots position is determined by an algorithm, which correlates a local map with a global map. As a result, there is no need for pre-insight of the surrounding which is assumed, thus it utilizes sensor data to build the complete map progressively. The algorithm approximates the robot's location by computing the location known as feasible poses where the normal view of the robot matches approximately to the observed range sensor data. The algorithm chooses the most matching one among the feasible poses [92, 98]. It requires the robot's orientation information to make sure that the algorithm identifies the feasible poses. For location information, Vassilis 2000 used dead reckoning as another source, when connected with range sensor-based localization algorithm it can produce an almost real-time estimated location. References [99, 100] introduced a Monte Carlo Localization (MCL) for mobile robot position estimation. They used the Monte Carlo type methods and then combined the pros of their previous work in which they used grid based Markov localization with the performance and precision of Kalman filter based method. The MCL technique can manage ambiguities and subsequently can comprehensively localize the robot when contrasted with their previous grid based technique. The MCL technique has altogether decreased memory necessities while fusing sensor estimations at a significantly higher rate. Based on the condensation algorithm the MCL method was proposed in [101]. It localizes the robot all around by utilizing scalar brightness estimation when given a visual guide of the roof. Sensor information of a low feature is used by these probabilistic methods, specifically in the 2D plane and needs the robot to move around for probabilities to step by step focalize toward a pack. The pose of the robots was also computed by some researchers based on the appearance. All encompassing picture based model for robot localization was used by [102]. With the depth and 3D planarity data, the panoramic model was developed while the image matching is achieved by taking into account the planar patches. References [55, 103] utilized all encompassing pictures for probabilistic appearance based robot localization. Markov localization is applied to hundreds of training images for extracting the 15-dimensional feature vectors. In urban environments, the problem of a mobile robot localization was forwarded by [104] by utilizing feature correspondence between pictures taken by a camera on the robot and a computer aided design program, or a comparable model of its surroundings. For localization of cars in urban environments, Refs. [17, 105, 106] used an inertial measurement unit and a sensor suite consists of four GPS antennas. Humanoid robots are getting popular as a research tool as they offer a new viewpoint compared to a wheeled vehicle. A lot of work has been done so far on the localization for humanoid robots. In Order to estimate the location of the robot [107] applied a vision based approach and then compared the current image to be previously recorded reference images. In the local environment of the humanoid [108, 109] detects objects with given colors and shapes and then determines its pose relative to these objects. With respect to a close object [107, 109] localizes the robot to track the 6D pose of a manually initialized object relative to the camera by applying a model-based approach [26].

7 Wireless Medical Sensor Network

WSN has opened many doors and potential to be utilized in medical applications, which are known as Wireless Medical Sensor Networks (WMSNs). In 21st century, the medical instruments were enhanced and embedded with this technology. WMSNs are rapidly used conduct medical diagnoses physiological condition of patients. E.g. temperature, heart rate, blood pressure, oxygen saturation … etc. This type of technology is faster transmitting data to remote location without the need for human interaction; a doctor could read and analyze data to make quick decisions of patience conditions. WMSNs are giving more and more benefits for modern healthcare as continuous and ubiquitous monitoring (see Fig. 11). The doctors and their patients' direct interactions are going shorter and reducing health care coast. WMSNs are direct human involvement and facilitate mobility and demand high data rates, with reliable communication and multiple recipients.

In current healthcare sector is using application with WMSNs for the following purposes.

1. Monitoring of patients
   E.g. Oslo University Hospital uses a biomedical wireless sensor network (BWSN) for their patients.
2. Home and elderly care
   E.g. this purpose is using ZigBee Wireless Sensor Networks (ZWSN) and robots integrated [110].
3. Collection clinical data [111].

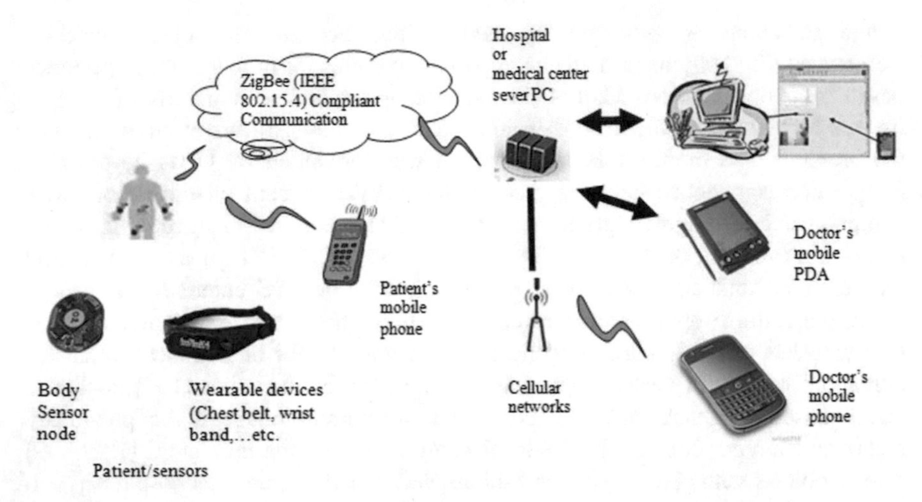

Fig. 11 Monitoring of patients in clinical data

8 The Challenges in the MSN and Their Limitations

Mainly MSN has two sets of challenges, which are hardware and environment. Without power is nothing running and same situation for the hardware as well. Another issue the power should be sufficient to run the system. If the system uses complex algorithms, then more power needs to complete the process. Based on these circumstances, the system should use low complexity algorithms, simple microcontrollers and radio. MSN should have used low cost materials to embed within the system. The major environmental components are topology and medium. The topology is going to vary in the circumstances. The medium is sharing too. The shared medium mandates that channel entry must be governed in some way. This issue is overcome by a medium access control (MAC) scheme. Designers are always using code division multiple access (CDMA), frequency divisions multiple access (FDMA) or carrier sense multiple access (CSMA) for better solutions. The changing topology of the network initiates from the nodes' mobilization. The result of this is giving unstable multi hop paths from the sensors.

MSNs are a unique network type, and they will require specific solutions to the research problems they have. The major issues that affect the design and performance of a MSN include MAC and routing protocols, localization techniques, security, physical layer transmission, resource management, quality of service and many more.

Current research and development is used much more mobile sensors to make new applications in MSNs. E.g. patrol defense, map productions, disaster managements and more. The major difference of WSNs and MSNs is static vs mobile sensors. Sensors are moving all the time. Sensor mobility makes a great impact on most existing protocols of WSNs. It could find three main challenges in MSNs include challenges for localization coverage services, data collection. MSN could be divided two types as follows controllable sensors and uncontrollable sensors. Sensor mobility is giving a lot of challenges in MSNs but it is giving good opportunities to support in improving many protocols such as localization, coverage and data collection [112].

9 Conclusion

The current trend is widely going with sensor network technology to solve the real-world problems. This paper gives a comprehensive review of sensor network working with a robotic paradigm. The sensor networking system is a form of capturing data from the source and transferring it into robotic devices to activate the required functions.

Focusing on WSNs, the paper describes the architecture of how sensor nodes are manipulated to retrieve and process data through wireless communication. Furthermore, it emphasizes the advantages of underwater WSN such as the system reusability and coverage whilst neglecting the use of wires for connections as underground wireless networks are installed via acoustic communications compare to ground based networks.

Prior studies have shown that MSN consists of capabilities such as locomotion, environmental information sensing and dead-reckoning to name a few. The design of MSN describes the use nodes, serves and client layers describing the complexity as opposed to simple WSN. MSN faces objective challenges due to unreliable characteristics such as communication and data management. The studies conducted on this paper particularly focus on the communication issues split into problems base on coverage and localization. MSN coverage issues are simply described as a degree of the quality of service; fundamental issue being the network reach coverage issues as it decreases due to sensor failure or undesirable sensor deployment. Speaking of localization issues, this paper focuses on the difficulties caused to sensor nodes due to errors introduce when calculating coordinates. GPS is a typical method of utilizing open air localization. However, studies show that GPS have major vulnerabilities as it needs the constant sight of multiple satellites thus indoor functionality of GPS is hindered greatly. As an alternative for GPS, this paper discusses two methods of localization algorithms as well as improvements that can be done to increase the reliability of localization.

Further studies were conducted in RSN application as a method of utilizing WSN. RSN is intelligent MRs, which is an improvement into a traditional sensor network. RSN studies have shown that it can provide answers to issues face by traditional sensor network such as localization and path finding by integrating to utilization of robots. A great example of RSN's ability compares to WSN is where robotics technology has the ability to replace broken nodes, repositioning nodes, recharging batteries, etc. Publication [69–71] shows the amazing capabilities of RSN such as battery recharge and mapping information to improve localization as opposed to WSN. Furthermore, multi-robot systems prove to be an improvement to the RSN as it provides further advantages. Some of the applications of RSN are they can be used for effective search and rescue, monitoring of electromagnetic fields and others. Further studies were conducted on RSN utilizing MRs equipped with EMF sensors to ensure all aspects RSN application is thoroughly described. As mentioned earlier, multi- robot is advantages in many ways compare to a single robot ensuring higher potential to maximize an area of coverage at a time. The studies described further into issues faced by RNS such as sensor coverage. Techniques and algorithms such as informative and molecular were investigated as solutions for coverage problems. Another issue covered in this study is the localization for robots as it is a key component in determining the robot's position within the localized environment; it is a process of determining the position of the robot by taking a map as an input. Further studies described techniques and algorithms relevant to improve the localization of the robot.

In summary, this paper broadly discusses MSN and WSN functions to think about future directions with robotic and WSN. It is a green light to combine new proactive based solution for the current issues which are discussed in this paper; it also provides examples of how sensor networking can lead to a further modernized future in the sensor technology field. These technologies could be adaptive to the medical instruments, so that they can be successfully integrated as part of our healthcare monitoring system to upgrade healthcare management.

Acknowledgements The author sincerely thanks Ryu JH, Reyaz A, Baasandorj B from Chonbuk National University, Jeonju, South Korea for their cooperation and contribution. The proof reading was given by Edward Rajah from American University of Nigeria.

References

1. Basagni, S., Carosi, A., & Petrioli, C. (2008). *Mobility in wireless sensor networks. Algorithms and protocols for wireless sensor networks*. New York: Wiley.
2. Chittedi, A. (2009). *Development of software system for constraint control of mobile sensors in wireless sensor networks*. ETD Collection for Tennessee State University. Paper AAI1473382.
3. Chunsheng, Z., Lei, S., Takahiro, H., Lei, W., & Shojiro, N. (2010). *Research issues on mobile sensor networks*. Paper presented at the 5th International ICST Conference on Communications and Networking in China (CHINACOM), Beijing.
4. Fei, X. (2011). *Coverage-awareness scheduling protocols for wireless sensor networks*. Ph. D. thesis, University of Ottawa.
5. Song, G., Yaoxin, Z., Fei, D., & Aiguo, S. (2008). A mobile sensor network system for monitoring of unfriendly environments. *Sensors, 8*(11), 7259–7274. doi:10.3390/s8117259.
6. Dimitriou, T., Alrashed, E. A., Karaata, M. H., & Hamdan, A. (2015). Imposter detection for replication attacks in mobile sensor networks. In *7th International Conference on New Technologies, Mobility and Security (NTMS)*, Paris, July 2015. Ieeexplore, pp. 1–5.
7. www.wikipedia.com. (2016). Accessed January 01, 2016.
8. Flanagin, C. (2016). *A survey on robotics system and performance analysis*. http://www1.cse.wustl.edu/~jain/cse567-11/ftp/robots/index.html. Accessed January 01, 2016.

9. Goodrich, M. A., & Schultz, A. C. (2007). Human-robot Interaction: A survey. *Foundations and Trends in Human-Computer Interaction, 1*(3), 203–275. doi:10.1561/1100000005
10. Joshua, R., & Elizabeth, S. (2006). Robot-sensor networks for search and rescue. *Paper presented at the In Proceedings IEEE International Workshop on Safety, Security and Rescue Robotics*, National Institute *of* Standards *and* Technology, Gaithersburg, August 22–25, 2006.
11. Huiyong, W., Minglu, Z., & Jingyang, W. (2009). An emergency search and rescue system based on WSN and mobile robot. *International Conference on Information Engineering and Computer Science (ICIECS)*, Wuhan, Dec 2009. Ieeexplore, pp. 1–4.
12. Atanasov, A., et al. (2010). Testbed environment for wireless sensor and actuator networks. *2010 fifth international conference on systems and networks communications (ICSNC)*, Nice, Aug 2010. Ieeexplore, pp. 1–6.
13. Amigoni, F., Fontana, G., & Mazzuca, S. (2007). Robotic sensor networks: an application to monitoring electro-magnetic fields. In I. Maglogiannis et al. (Ed.), *Proceedings of the 2007 conference on emerging artificial intelligence applications in computer engineering: Real word ai systems with applications in eHealth, HCI, information retrieval and pervasive technologies* 2007, pp. 384–393
14. Shue, S., & James, M. C. (2013). A survey of robotic applications in wireless sensor networks. *2013 Proceedings of IEEE Southeastcon*, Jacksonville.
15. Amigoni, F., Caglioti, V., & Fontana, G. (2004). A perceptive multirobot system for monitoring electro-magnetic fields. *2004 IEEE Symposium on Virtual Environments, Human-Computer Interfaces and Measurement Systems (VECIMS)*, July 12–14, 2004, pp. 95–100.
16. Amigoni, F., et al. (2005). Agencies for perception in environmental monitoring. In *Proceedings of the IEEE Instrumentation and Measurement Technology Conference (IMTC 2005)*, Ottawa, 2005, pp. 1266–1271.
17. Georgiev, A., & Allen, P. K. (2004). Localization methods for a mobile robot in urban environments. *IEEE Transactions on Robotics, 20*(5), 851–864.
18. Kasture, A., Sangli, R. A., & Thool, S. (2014). *Visualization of wireless sensor network by a java framework for security in defense surveillance*. Paper presented at the 2014 International Conference on Electronic Systems, Signal Processing and Computing Technologies (ICESC), pp. 256–261, Nagpur, January 11, 2014.
19. Abdelgawad, A., & Bayoumi, M. (2012). *Resource-aware data fusion algorithms for wireless sensor networks. Lecture Notes in Computer Science* (Vol 118, pp 1–15). US: Springer.
20. Holger, K., & Andreas, W. (2006). Single-node architecture, in protocols and architectures for wireless sensor networks. Chichester, UK.: Wiley. doi:10.1002/0470095121.ch2
21. Kashif, K., Madjid, M., Qi, S., & David, J. (2010). Security in wireless sensor networks. In P. Stavroulakis & M. Stamp (Eds.), *Handbook of information and communication security* (pp. 513–552). Heidelberg: Springer.
22. Gungor, V. C., & Hancke, G. P. (2009). Industrial wireless sensor networks: Challenges, design principles, and technical approaches. *IEEE Transactions on Industrial Electronics, 56* (10), 4258–4265.
23. Vehbi, C. G., & Gerhard, P. H. (2011). *Industrial wireless sensor networks, industrial communication systems* (pp. 1–15). Taylor and Francis Group LLC. doi:10.1201/b10603
24. Nagarajan, M., & Karthikeyan, S. (2012). *A new approach to increase the life time and efficiency of wireless sensor network*. Paper presented at 2012 International Conference on Pattern Recognition, Informatics and Medical Engineering (PRIME), pp. 231–235. Salem, Tamil Nadu, March 21–23, 2012.
25. Katta, S. B. (2013). *A study on sensor deployment and topology control in wireless sensor networks*. M.S. Dissertations & Theses, ProQuest, UMI Dissertations Publishing.
26. Ryu, J. H., Irfan, M., & Reyaz, A. (2015). A review on sensor network issues and robotics. *Journal of Sensors* 2015, Article ID 140217. doi:10.1155/2015/140217

27. Ahmed, S., Nikola, S., Lilyana, M., & Andon, T. (2015). Neural net tracking control of a mobile platform in robotized wireless sensor networks. *2015 IEEE International Workshop of Electronics, Control, Measurement, Signals and their Application to Mechatronics (ECMSM)* (pp. 1–6). Liberec, June 22–24, 2015.
28. Andrea, G., & Giandomenico, S. (2014). Service-oriented middleware for the cooperation of smart objects and web services. In G. Fortino & P. Trunfio (Eds.), *Internet of things based on smart objects* (pp. 49–68). Springer International Publishing Switzerland.
29. Chunsheng, Z., Lei, S., Takahiro, H., Lei, W., Shojiro, N., & Laurence, T. Y. (2014). A survey on communication and data management issues in mobile sensor networks. *Wireless Communications and Mobile Computing, 14*(1), 19–36.
30. Reddy, M. A., Anjaneyulu, D., Varma, D. P., & Rao, G. R. (2016). WSN in IOT environment interfacing with smart sensors using Arm7 with Zigbee for industries. *International Journal of Engineering Research and Development, 12*(8), 54–60.
31. Park, J. Y., & Ha, Y. S. (2008). *Multilevel localization for mobile sensor network platforms.* Paper Presented at International Multi Conference on Computer Science and Information Technology (IMCSIT 2008), pp. 711–718. Wisia, October 20–22, 2008.
32. Cecílio, J., & Furtado, P. (2014). *Wireless Sensors in Heterogeneous Networked Systems. Computer communications and networks* (pp. 1–22). Springer International Publishing,
33. Syrotiuk, V. R., Li, B., & Mielke, A. M. (2008). Heterogeneous wireless sensor networks. In Boukerche A (Ed.), *Algorithms and protocols for wireless sensor networks*. Hoboken: Wiley. doi:10.1002/9780470396360.ch2
34. Stavrou, E., Pitsillides, A., Hadjichristofi, G., & Hadjicostis, C. (2010). *Security in future mobile sensor networks issues and challenges*. Paper presented at Proceedings of the 2010 International Conference on Security and Cryptography (SECRYPT), pp. 1–9. Athens, July 26–28, 2010.
35. Meguerdichian, S., Koushanfar, F., Potkonjak, M., & Srivastava, M. B. (2001). Coverage problems in wireless ad-hoc networks. Paper presented at IEEE INFOCOM 2001. *Twentieth Annual Joint Conference of the IEEE Computer and Communications Societies* (pp. 1380–1387). Anchorage, AK.
36. Liu, B., Dousse, O., Nain, P., & Towsley, D. (2013). Dynamic coverage of mobile sensor networks. *IEEE Transactions on Parallel and Distributed Systems, 24*(2), 301–311.
37. Li, F., Yang, Y., & Wu, J. (2009). Mobility management in MANETs: Exploit the positive impacts of mobility. In S. Misra, I. Woungang, & S. C. Misra (Eds.), Chapter *guide to wireless ad hoc networks, part of the series computer communications and networks* (pp. 211–235). London: Springer.
38. Luo, J., & Zhang, Q. (2008). *Probabilistic coverage map for mobile sensor networks*. Paper Presented at the IEEE GLOBECOM 2008—2008 IEEE Global Telecommunications Conference, New Orleans, LO, 30 November–4 December 2008.
39. Gage, D. W. (1992). *Command control for many-robot systems*. Paper presented at the 19th Annual AUVS Technical Symposium, Huntsville, Alabama, June 22–24, 1992.
40. Arkin, R. C. (1987). *Motor schema based navigation for a mobile robot: An approach to programming by behavior*. Paper presented at the 1987 IEEE International Conference Proceedings on Robotics and Automation, March 1985.
41. Khatib, O. (1985). *Real-time obstacle avoidance for manipulators and mobile robots*. Paper Presented at the 1985 IEEE International Conference Proceedings on Robotics and Automation, March 1985.
42. Howard, A., Mataric, M. J., & Sukhatme, G. S. (2002). Mobile sensor network deployment using potential fields: A Distributed, scalable solution to the Area coverage problem. In H. Asama et al. (Ed.), *Distributed autonomous robotic systems* (Vol. 5, pp. 299–308). Japan: Springer.
43. Poduri, S., Sukhatme, G. S. (2004). *Constrained coverage for mobile sensor networks*. Paper presented at the 2004 IEEE International Conference on Robotics and Automation, April 26, 2004–May 1, 2004.

44. Howard, A., Mataric, M. J., & Sukhatme, G. S. (2002). An incremental self deployment algorithm for mobile sensor networks. *Autonomous Robots, 13*(2), 113–126.
45. Wang, G., Guohong, C., & Tom, L. P. (2006). Movement assisted sensor deployment. *IEEE Transactions on Mobile Computing, 5*(6), 640–652.
46. Zou, Y., & Chakrabarty, K. (2003). *Sensor deployment and target localization based on virtual forces*. Paper presented at the INFOCOM 2003. INFOCOM 2003. Twenty-Second Annual Joint Conference of the IEEE Computer and Communications, San Francisco, March 30, 2003–April 3, 2003.
47. Roya, S., Karjeeb, J., Rawat, U. S., Dayama Pratik, N., & Deyd, N. (2016). Symmetric key encryption technique: A cellular automata based approach in wireless sensor networks. *Procedia Computer Science, 78*, 408–414. *International Conference on Information Security & Privacy (1C1SP2015)*, December 11–12, 2015, Nagpur, India.
48. Batalin, M. A., et al. (2004). Call and response: Experiments in sampling the environment. In *SenSys '04 Proceedings of the 2nd International Conference on Embedded Networked Sensor Systems* (pp. 25–38). ACM, New York.
49. Benyuan, L., Peter, B., & Olivier, D. (2005). Mobility improves coverage of sensor networks. In *Proceeding MobiHoc '05 Proceedings of the 6th ACM International Symposium on Mobile Ad hoc Networking and Computing* (pp. 300–308). ACM, New York.
50. Bisnik, N., Abouzeid, A. A., & Isler, V. (2007). Stochastic event capture using mobile sensors subject to a quality metric. *IEEE Transactions on Robotics, 23*(4), 676–692.
51. Maxim, A. B., & Gaurav, S. S. (2004). Coverage, exploration and deployment by a mobile robot and a communication network. *Telecommunication Systems, 26*(2), 181–196.
52. Wang, G., et al. (2005). Sensor relocation in mobile sensor networks. In *24th Annual Joint Conference of the IEEE Computer and Communications Societies*. Miami, March 13–17, 2005.
53. Wang, Y., & Zhengdong, L. (2011). Intrusion detection in a K-Gaussian distributed wireless sensor network. *Journal of Parallel and Distributed Computing, 71*(12), 1598–1607.
54. Isac, A., & Xenofon, D. K. (2009). A survey on localization for mobile sensor networks. In F. Richard & D. K. Xenofon (Eds.), *Mobile entity localization and tracking in gps-less environnments* (Vol. 5801, pp. 235–254)., Lecture Notes in Computer Science Heidelberg: Berlin.
55. Se, S. (2005). Vision-based global localization and mapping for mobile robots. *IEEE Transactions on Robotics, 21*(3), 364–375.
56. Ganggang, Y., & Fengqi, Y. (2007). A localization algorithm for mobile wireless sensor networks. In *IEEE International Conference on Integration Technology*, Shenzhen, 20–24 March 2007.
57. Isaac, A. (2010). *Spatio-temporal awareness in mobile wireless sensor networks*. Ph.D. thesis, Vanderbilt University Nashville.
58. Bram, D., Stefan, D., & Paul, H. (2006). Range-based localization in mobile sensor networks. In R. Kay & K. Holger (Eds.), *Wireless sensor networks* (Vol. 3868, pp. pp 164–179). *Lecture Notes in Computer Science*.
59. Shigeng, Z., et al. (2008). *Locating Nodes in Mobile Sensor Networks More Accurately and Faster*. In *SECON '08. 5th Annual IEEE Communications Society Conference on Sensor, Mesh and Ad Hoc Communications and Networks* (pp. 37–45). San Francisco, 2008.
60. Chaurasia, S., & Payal, A. B. (2011). *Analysis of range-based localization schemes in wireless sensor networks: A statistical approach*. Paper presented at the 13th International Conference on Advanced Communication Technology (ICACT), Seoul, February 13–16, 2011.
61. Lingxuan, H., David, E. (2004). *Localization for mobile sensor networks*. Paper presented at the MobiCom 04, Philadelphia, Pennsylvania, September 26, 2004–October 1, 2004.
62. Clement, S., Abder, R. B., & Jean-Claude, K. (2007). *A distributed method to localization for mobile sensor networks*. Paper presented at the IEEE Wireless Communications and Networking Conference, Kowloon, March 11–15, 2007.

63. Tilak, S., et al. (2005). *Dynamic localization protocols for mobile sensor networks*. Paper Presented at the 24th IEEE International Performance, Computing, and Communications Conference, April 7–9, 2005.
64. Yassine, S., & Najib, E. K. (2012). A distributed method to localization for mobile sensor networks based on the convex hull. *International Journal of Advanced Computer Science and Applications, 3*(10), 33–41.
65. Buddhadeb, S., Srabani, M., & Krishnendu, M. (2006). Localization control to locate mobile sensors. In K. M. Sanjay et al. (Eds.), *Distributed computing and internet technology* (Vol. 4317, pp. 81–88). *Lecture Notes in Computer Science*.
66. Jiyoung, Y., Jahyoung, K., & Hojung, C. (2008). *A localization technique for mobile sensor networks using archived anchor information*. Paper presented at the 5th Annual IEEE Communications Society Conference on Sensor, Mesh and Ad Hoc Communications and Networks, San Francisco, June 16–20, 2008.
67. Capkun, S., Hamdi, M., & Hubaux, J. P. (2001). *GPS free positioning in mobile ad hoc networks*. Paper presented at the Proceedings of the 34th Annual Hawaii International Conference on System Sciences, Maui, January 6, 2001.
68. Aboelela, E. H., & Khan, A. H. (2012). *Wireless sensors and neural networks for intruders detection and classification*. Paper presented at the 2012 International Conference on Information Networking (ICOIN), Bali, February 1–3, 2012.
69. Yuan, Y. L., & Parker, L. E. (2008). *Intruder detection using a wireless sensor network with an intelligent mobile robot response*. Paper presented at the IEEE Southeast Conference, Huntsville, April 3–6, 2008.
70. Martínez, J. F., et al. (2010). Pervasive surveillance-agent system based on wireless sensor networks: Design and deployment. *Measurement Science and Technology*, *21*(12), 124005.
71. Troubleyn, E., Moerman, I., & Demeester, P. (2013). QoS challenges in wireless sensor networked robotics. *Wireless Personal Communications, 70*(3), 1059–1075.
72. LaMarca, A., et al. (2002). Plantcare: An investigation in practical ubiquitous systems. In G. Borriello & L. E. Holmquist (Eds.), *UbiComp 2002: Ubiquitous computing* (Vol. 2498, pp. 316–332)., Lecture Notes in Computer Science Berlin, Heidelberg: Springer.
73. Xu, L., et al. (2012). Servicing wireless sensor networks by mobile robots. *IEEE Communications Magazine, 50*(7), 147–154.
74. Schaffert, S. M. (2006). *Closing the loop: Control and robot navigation in wireless sensor networks*. Ph.D. thesis, University of California.
75. Sheu, J., Hsieh, K., & Cheng, P. (2008). Design and implementation of mobile robot for node replacement in wireless sensor networks. *Journal of Information Science and Engineering, 24,* 393–410.
76. Huiyong, W., Minglu, Z., & Jingyang, W. (2009). *An emergency search and rescue system based on WSN and mobile robot*. Paper presented at the 2009 International Conference on Information Engineering and Computer Science, Wuhan, December 19–20, 2009.
77. Zhu, A., & Yang, S. X. (2010). *A survey on intelligent interaction and cooperative control of multi-robot systems*. Paper presented at the 2010 8th IEEE International Conference on Control and Automation (ICCA), Xiamen, June 9–11, 2010.
78. Cooper-Morgan, A., & Yen-Ting, L. (2011). *Cost efficient deployment networked camera sensors*. Proquest, Umi Dissertation Publishing.
79. Pang, Z. (2013). *Technologies and architectures of the internet-of-things (IoT) for health and well-being*. Doctoral Thesis in Electronic and Computer Systems KTH Royal Institute of Technology Stockholm, Sweden.
80. Alcaraz, C., Najera, P., Lopez, J., & Roman, R. (2010). Wireless sensor networks and the internet of things: Do we need a complete integration? *1st International Workshop on the Security of the Internet of Things (SecIoT'10).*
81. Ang, L., Seng, K. P., & Heng, T. Z. (2016). Information communication assistive technologies for visually impaired people. *International Journal of Ambient Computing and Intelligence, 7*(1), 45–68. doi:10.4018/IJACI.2016010103

82. Kimbahune, V. V., Deshpande, A. V., & Mahalle, P. N. (2017). Lightweight key management for adaptive addressing in next generation internet. *International Journal of Ambient Computing and Intelligence (IJACI), 8*(1), 50–69. doi:10.4018/IJACI.2017010103
83. Roy, P., Patra, N., Mukherjee, A., Ashour, A. S., Dey, N., & Biswas, S. P. (2017). Intelligent traffic monitoring system through auto and manual controlling using PC and android application. In N. Dey, A. Ashour, & S. Acharjee (Eds.), *Applied video processing in surveillance and monitoring systems* (pp. 244–262) Hershey, PA: IGI Global. doi:10.4018/978-1-5225-1022-2.ch011
84. Graham, B., Tachtatzis, C., Di Franco, F., Bykowski, M., Tracey, D. C., Timmons, N. F., et al. (2011). Analysis of the effect of human presence on a wireless sensor network. *International Journal of Ambient Computing and Intelligence, 3*(1), 1–13. doi:10.4018/jaci.2011010101
85. Locchi, L., Nardi, D., & Salrno, M. (2013). Reactivity and deliberation: A survey on multi-robot systems. *Balancing reactivity and social deliberation in multi-agent systems* (Vol. 2103, pp. 9–32). *Lecture Notes in Computer Science*.
86. Baasandorj, B., et al. (2013). Formation of multiple-robots using vision based approach. In J. Wang (Ed.), *Applied mechanics and materials* (Vol. 419, pp. 768–773). Switzerland: Trans Tech Publications.
87. Walenz, B. (2016). *Multi robot coverage and exploration: A survey of existing techniques*. https://bwalenz.files.wordpress.com/2010/06/csci8486-walenz-paper.pdf. Accessed January 01, 2016.
88. Kong, C. S., Peng, N. A., & Rekleitis, I. (2006). *Distributed coverage with multi-robot system*. Paper presented at the Proceedings of the 2006 IEEE International Conference on Robotics and Automation, Florida, May 15–19, 2006.
89. Maxim, A. B., & Gaurav, S. S. (2002) Spreading out: A local approach to multi-robot coverage. *Distributed autonomous robotic systems* (Vol. 5, pp. 373–382). Tokyo: Springer.
90. Kataoka, S., & Honiden, S. (2006). *Multi-robot positioning model: Multi-agent approach*. Paper presented at the International Conference on Computational Intelligence for Modelling, Control and Automation, 2006 and International Conference on Intelligent Agents, Web Technologies and Internet Commerce, Sydney, November 28–December 1, 2006.
91. Ranjbar-Sahraei, B., Weiss, G., & Nakisaee, A. (2012). Stigmergic coverage algorithm for multi-robot systems (demonstration). In *Proceedings of the 11th International Conference on Autonomous Agents and Multiagent Systems*, Richland, June 4–8, 2012.
92. Parzhuber, O., & Dolinsky, D. (2004). Hardware platform for multiple mobile robots. In D. W. Gage (Ed.), *SPIE Proceedings Mobile Robots XVII* (Vol. 5609), December 29, 2004.
93. Wagner, I. A., Lindenbaum, M., & Bruckstein, A. M. (1999). Distributed covering by ant-robots using evaporating traces. *IEEE Transactions on Robotics and Automation, 15*(5), 918–933.
94. Binh, H. T. T., Hanh, N. T., & Van Quan, L., et al. (2016). Improved cuckoo search and chaotic flower pollination optimization algorithm for maximizing area coverage in wireless sensor networks. *Neural Comput & Application* (pp. 1–13). doi:10.1007/s00521-016-2823-5
95. Hazon, N., & Kaminka, G. A. (2005). *Redundancy, efficiency and robustness in multi-robot area coverage*. Paper presented at the Proceedings of the 2005 IEEE International Conference on Robotics and Automation, April 18–22, 2005.
96. Royer, E., et al. (2007). Monocular vision for mobile robot localization and autonomous navigation. *International Journal of Computer Vision, 74*(3), 237–260.
97. Brown, R. J., & Donald, B. R. (2000). Mobile robot self-localization without explicit landmarks. *Algorithmica, 26*(3), 515–559.
98. Varveropoulos, V. (2016). Robot localization and map construction using sonar data. http://rossum.sourceforge.net. Accessed January 25, 2016.
99. Dellaert, F., et al. (1999). *Monte Carlo localization for mobile robots*. Paper presented at the Proceedings of the 1999 IEEE International Conference on Robotics and Automation, Detroit, May 10–15, 1999.

100. Ueda, R., et al. (2002). *Uniform Monte Carlo localization-fast and robust self-localization method for mobile robots*. Paper presented at the Proceedings of the IEEE International Conference on Robotics and Automation, Washington DC, May 11–15, 2002.
101. Dellaert, F., et al. (1999). *Using the condensation algorithm for robust, vision based mobile robot*. Paper presented at the IEEE Computer Society Conference on Computer Vision and Pattern Recognition, Fort Collins, Jun 23–25, 1999.
102. Cobzas, D., & Hong, Z. (2001) *Cylindrical panoramic image based model for robot localization*. Paper presented at the Proceedings 2001 IEEE/RSJ International Conference on Intelligent Robots and Systems, Maui, October 29–November 03, 2001.
103. Kröse, B. J. A., Vlassis, N., & Bunschoten, R. (2002). Omnidirectional vision for appearance-based robot localization. In D. H. Gregory et al. (Eds.), *Sensor based intelligent robots* (Vol. 2238, pp. 39–50). *Lecture Notes in Computer Science*.
104. Talluri, R., & Aggarwal, J. K. (1996). Mobile robot self location using model image feature correspondence. *IEEE Transactions on Robotics and Automation, 12*(1), 63–77.
105. Nayak, R. A. (2000). *Reliable and continuous urban navigation using multiple GPS antenna and a low cost IMU*. MS thesis, University of Calgary.
106. Cindy, C., et al. (2012). Virtual 3D city model for navigation in urban areas. *Journal of Intelligent and Robotic Systems, 66*(3), 377–399.
107. Ido, J., Shimizu, Y., & Matsumoto, Y. (2009). Indoor navigation for a humanoid robot using a view sequence. *Journal International Journal of Robotics Research archive, 28*(2), 315–325.
108. Cupec, R., Schmidt, G., & Lorch, O. (2005). *Experiments in vision guided robot walking in a structured scenario*. Paper presented at the Proceedings of the IEEE International Symposium on Industrial Electronics, Dubrovnik, June 20–23, 2005.
109. Hornung, A., Wurm, K. M., & Bennewitz, M. (2010). *Humanoid robot localization in complex indoor environments*. Paper presented at the 2010 IEEE/RSJ International Conference on Intelligent Robots and Systems (IROS), Taipei, October 18–22, 2010.
110. Song, B., et al. (2011). Zigbee wireless sensor networks based detection and help system for elderly abnormal behaviors in service robot intelligent space. *Applied Mechanics and Materials, 48–49,* 1378–1382.
111. Minaie, A., et al. (2013). *Application of wireless sensor networks in health care system*. Paper presented at the 120th ASEE Annual Conference & Exposition, Atlanta, June 23–26, 2013.
112. Natalizio, E., & Loscr, V. (2013). Controlled mobility in mobile sensor networks: advantages, issues and challenges. *Article in Telecommunication Systems, 52*(4), 1–8. doi:10.1007/s11235-011-9561-

Part II
Big Data Analytics

Big Data Analytics with Machine Learning Tools

T.P. Fowdur, Y. Beeharry, V. Hurbungs, V. Bassoo and V. Ramnarain-Seetohul

Abstract Big data analytics is the current, trending hot-topic in the research community. Several tools and techniques for handling and analysing structured and unstructured data are emerging very rapidly. However, most of the tools require high expert knowledge for understanding their concepts and utilizing them. This chapter presents an in-depth overview of the various corporate and open-source tools currently being used for analysing and learning from big data. An overview of the most common platforms such as IBM, HPE, SAPRANA, Microsoft Azure and Oracle is first given. Additionally, emphasis has been laid on two open-source tools: H2O and Spark MLlib. H2O was developed by H2O.ai, a company launched in 2011 and MLIB is an open source API that is part of the Apache Software Foundation. Different classification algorithms have been applied to Mobile-Health related data using both H2O and Spark MLlib. Random Forest, Naïve Bayes, and Deep Learning algorithms have been used on the data in H2O. Random Forest, Decision Trees, and Multinomial Logistic Regression Classification algorithms have been used with the data in Spark MLlib. The illustrations demonstrate the

T.P. Fowdur (✉)
Department of Electrical and Electronic Engineering, Faculty of Engineering, University of Mauritius, Réduit, Moka, Mauritius
e-mail: p.fowdur@uom.ac.mu

Y. Beeharry
Faculty of Information, Communication and Digital Technologies, University of Mauritius, Réduit, Moka, Mauritius
e-mail: y.beeharry@uom.ac.mu

V. Hurbungs
Department Software and Information Systems, Faculty of Information, Communication and Digital Technologies, University of Mauritius, Réduit, Mauritius
e-mail: v.hurbungs@uom.ac.mu

V. Bassoo · V. Ramnarain-Seetohul
Department of Information Communication Technology, Faculty of Information, Communication and Digital Technologies, University of Mauritius, Réduit, Mauritius
e-mail: v.bassoo@uom.ac.mu

V. Ramnarain-Seetohul
e-mail: v.seetohul@uom.ac.mu

N. Dey et al. (eds.), *Internet of Things and Big Data Analytics Toward Next-Generation Intelligence*, Studies in Big Data 30,
DOI 10.1007/978-3-319-60435-0_3

flows for developing, training, and testing mathematical models in order to obtain insights from M-Health data using open source tools developed for handling big data.

Keywords Open-source tools · H2O · Spark MLlib

1 Introduction

There are several conventional corporate tools which can be used to process data efficiently. This brings about useful insights for every department ranging from profit forecasting in sales to inventory management. The conventional architecture consists of a data warehouse where the data typically resides and Standard Query Language (SQL)-based Business Intelligence (BI) tools which execute the tasks of processing and analysing this same data [1]. In order to perform data analysis in a more advanced manner, companies have used dedicated servers to which the data was moved to perform data and text mining, statistical analysis and predictive analytics. The criticality of business analysis and profit making had brought about the wide usage [2] of tools like R [3] and Weka [4] in the deployment of machine learning algorithms and models. However, the evolution and growth of traditional data into what is commonly termed as Big Data nowadays has rendered the conventional tools unusable due to the major challenges resulting from the features of Big Data.

Volume, variety and velocity, which are the three main features of big data, have caused it to surpass the capabilities of its ingestion, storage, analysis and processing by conventional systems. Big Data means very large data sets, with both structured and unstructured data, that come at high speed from different sources. In addition, not all incoming data are valuable; some of it may not be of good quality carrying a certain level of uncertainty. The analysis conducted by the International Data Corporation's annual Digital Universe [5] has shown that the expected growth in data volume will be around 44 zettabytes (4.4×10^{22} bytes) by the year 2020 which is approximately ten times larger than it was in 2013. This growth can be explained by the increasing adoption of social media (e.g. Facebook, Twitter), Internet of Things (IoT) sensors, click streams from websites, video camera recordings and smartphones communications.

Handling huge amounts of data requires equipment and skills. Many organizations are already in possession of these equipment and expertise for managing and analysing structured data only. However, the fast flows and increasing volumes of data leave them with the inability to "mine" and extract intelligent actions in a timely manner. In addition to the fast growth of the volume of data which cannot be handled for traditional analytics, the variety of data types and speed at which the data arrives requires novel solutions for data processing and analytics [2]. Compared to conventional structured data of organisations, Big Data does not always fit into neat tables of rows and columns. New data types pertaining to structured and unstructured data have come into existence and which are processed to yield business insights. The very common toolkits for machine learning such as Weka [4] or R [3] have not been conceived for handling such workloads. Despite having distributed applications of some available algorithms, Weka does not match the level of the initial tools developed for handling data at the terabyte scale. Therefore, traditional analytics tools are not well adapted to capture the full value of large data sets like Big Data.

Machine learning allows computers to learn without being explicitly programmed; the programmes have the ability to change when exposed to different sets of data. Machine learning techniques are therefore more appropriate to capture hidden insights from Big Data. The whole process is data-driven and executes at the machine layer by applying complex mathematical calculations to Big Data. Machine learning requires large data sets for learning and discovering patterns in data. Commonly used machine learning algorithms are: Linear Regression, Logistic Regression, Decision Tree, Naive Bayes and Random Forest. There are several tools that have been developed for working with Big data for example Jaspersoft, Pentaho, Tableau and Karmasphere. These tools alleviate the problem of scalability by providing solutions for parallel storage and processing [6].

Businesses are increasingly adopting Big Data technologies to remain competitive and therefore, the demand for data management is on the rise given the advantages associated with analytics. Emerging big data applications areas are Healthcare, Manufacturing, Transportation, Traffic management, Media and Sports among others. Extensive research and development is also being carried out in the field of Big Data, analytics and machine learning. In [7], the authors highlight an expandable mining technique using the k-nearest neighbor (*K*-NN) classification approach to demonstrate training data with reduced data sets. This reduced data set ensures faster data classification and storage management. In [8], a methodology to clean raw data using the Fishers Discrimination Criterion (FDC) is proposed. The objective was to ease the data flow across a social network such as Facebook to enhance user satisfaction using machine learning techniques. Excessive data sets may result in the field of biological science (e.g. DNA information) and this adds to the difficulty of extracting meaningful information. A system to reduce similar information from large data sets was therefore proposed; the study analysed processing techniques such as NeuralSF and Bloom Filter applied to biological data [9]. Another biological study attempts to train a neural network using nine parameters to categorise breast cancer as cancerous or non-cancerous [10]. This study applies machine learning techniques on breast cancer dataset to accurately classify this type of cancer with an accuracy of 99.27%. Other example of neural network training is the performance comparison of NN-GA with NN, SVM and Random Forest classifiers to predict the categorisation of a pixel image [11]. In the field of computing and intelligence, related works include cryptographic key management [12] and intelligent activities of daily living (ADLs) assistance systems [13].

Big Data undoubtedly brings many benefits to businesses. However, there are many challenges associated with Big Data and related concepts. First of all, Big

Data changes the whole concept of data storage and manipulation. Businesses must now handle a constant inflow of data and this involves modifying the data management strategies of organisations. Secondly, privacy issues are associated with Big Data since the information is being captured from different sources. In the process, sensitive data may be processed without the knowledge of users. Thirdly, businesses have to invest in new technologies since the whole software environment is changing with Big Data. The latter implies super computing power for intensive data storage, processing capabilities and new software or hardware to handle analytics and machine learning. Overall, Big Data brings forward a new approach to handle organisational data and therefore, businesses must properly balance the challenging aspects of Big Data.

1.1 Big Data Analytic Tools

An overview of some existing corporate-level Big Data platforms is given in the following sub-sections.

1.1.1 Big Data Analytic Tools

IBM solution is currently helping businesses to achieve tangible benefits by analyzing all of their data available. IBM enables rapid return on investment across several areas such as Financial Services, Transportation, Healthcare/Life Sciences, Telecommunications, Energy and Utilities, Digital Media, Retail and Law Enforcement [14]. A few challenges and benefits of IBM in three areas are summarised in Table 1.

IBM gives enterprises the ability to manage all their data with a single integrated platform, multiple data engines and a consistent set of tools & utilities to operate on Big Data. IBM delivers its Big Data platform through a "Lego block" approach [15] and the platform has four core capabilities [16]:

Table 1 Big data challenges and benefits [14]

Area	Big data challenge	Benefit
Healthcare	Analyzing streaming patient data	Reduction in patients' mortality by 20%
Telecommunication	Analyzing networking and call data	Reduction in processing time by 92%
Utilities	Analyzing petabytes of untapped data	99% improved accuracy in placing power generation resources

1. **Analytics using Hadoop:** It involves the analysis and processing of data in any format using clusters of commodity hardware.
2. **Stream Computing:** It involves real-time analysis of huge volumes of data on the fly with response times of the order of millisecond.
3. **Data Warehousing:** It involves advanced analytics on data stored in databases and the delivery of proper actionable insights.
4. **Information Integration and Governance:** It involves the processes of understanding, transforming, cleansing, delivering and governing reliable information to critical business initiatives [16].

Supporting Platform Services include Visualization & Discovery, Application Development, Systems Management and Accelerators [16]:

- **Visualization & Discovery:** It involves the exploration of complex and huge sets of data.
- **Application Development:** It involves the development of applications pertaining to big data.
- **Systems Management:** It involves the securing and optimizing of the performance of systems for big data.
- **Accelerators:** It involves the analytic applications that accelerate the development and implementation of big data solutions (Fig. 1).

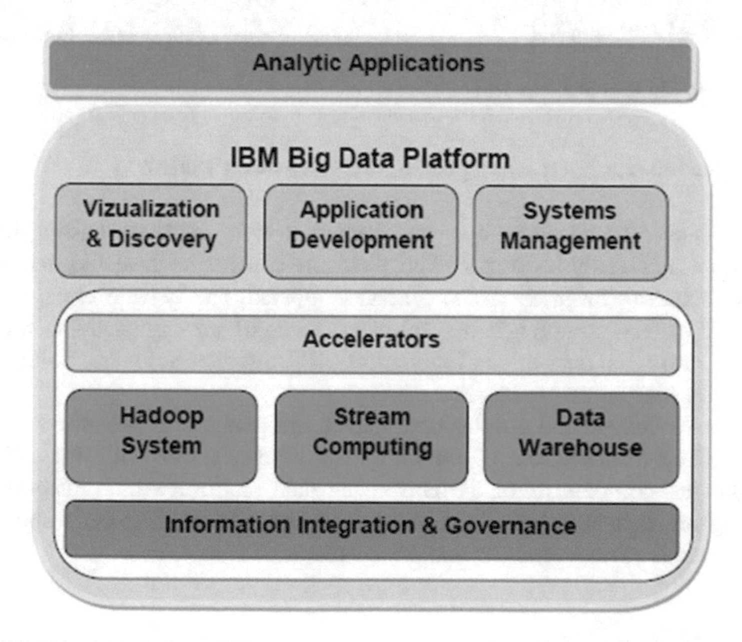

Fig. 1 IBM big data platform [17]

Table 2 Big data software, solutions and services [19]

Big data software	Solutions	Services
SQL analytics	Vertical advanced analytics	Platform for advanced analytics to obtain valuable insights from big data
Data analytics	IDOL	Unstructured data analytics across multiple languages
Big data verticals	Big Data industry solutions	Entrerprise related issues
Big data cloud	Haven on demand	Big data platform to analyse data and create applications/services on the cloud

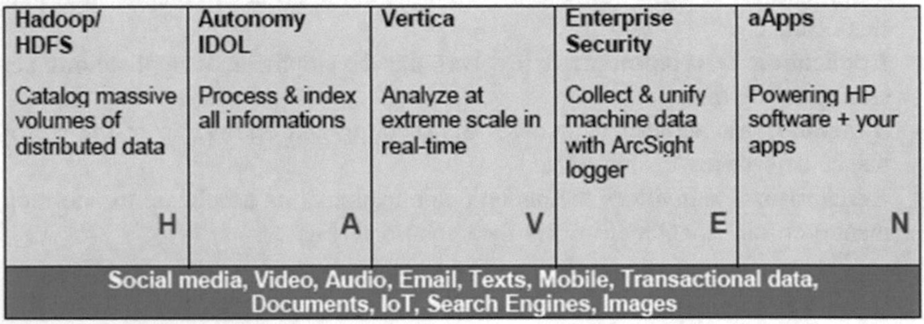

Hadoop/ HDFS	Autonomy IDOL	Vertica	Enterprise Security	aApps
Catalog massive volumes of distributed data	Process & index all informations	Analyze at extreme scale in real-time	Collect & unify machine data with ArcSight logger	Powering HP software + your apps
H	A	V	E	N
Social media, Video, Audio, Email, Texts, Mobile, Transactional data, Documents, IoT, Search Engines, Images				

Fig. 2 Haven big data platform [20]

1.1.2 Hewlett-Packard Enterprise (HPE) Big Data Platform

HPE big data is based on an open architecture which improves existing business intelligence and analytics assets. The platform consists of scalable and flexible models, including on-premises, as-a-service, hosted, and private cloud solutions [18]. HPE offers comprehensive solutions to help an organization bridge to tomorrow through a data-driven foundation. HPE big data Software, Solutions and Services are highlighted in Table 2.

HP Haven OnDemand enables exploration and analysis of massive volumes of data including unstructured information such as text, audio, and video. Figure 2 highlights the components of the Haven big data platform which integrates analytics engines, data connectors and community applications. The main objective of this platform is further analysis and visualisation of data.

1.1.3 SAP HANA Platform

SAP HANA Vora [21] is a distributed computing and an in-memory solution which can aid organisations process big data to obtain insights profitable to the business growth. Organisations use it for the quick and easy running of interactive analytics on both Hadoop and enterprise data. SAP HANA Vora single solution [21] for computation of all data includes Science, Predictive Analytics, Business Intelligence and Vizualisation Apps. Capabilities of the SAP HANA Platform can be categorised under the following groups [22]:

- **Application Services:** Making use of open standards for development such as: Java Script, HTML5, SQL, JSON, and others, to build applications on the web.
- **Database Services:** Performing advanced analytics and transactions on different types of data stored in-memory.
- **Integration Services:** Managing data collected from various sources and developing strategies to integrate all of them together.

According to [23], the SAP HANA Platform is most suitable to perform analytics in real-time, and develop and deploy applications which can process real-time information. The SAP HANA in-memory database is different from other database engines currently available on the market. This type of database provides high speed interactivity and fast response time for real-time applications such as Sales & Operational Planning, Cash Forecasting, Manufacturing scheduling optimization, point-of-sale data and social media data [23] (Fig. 3).

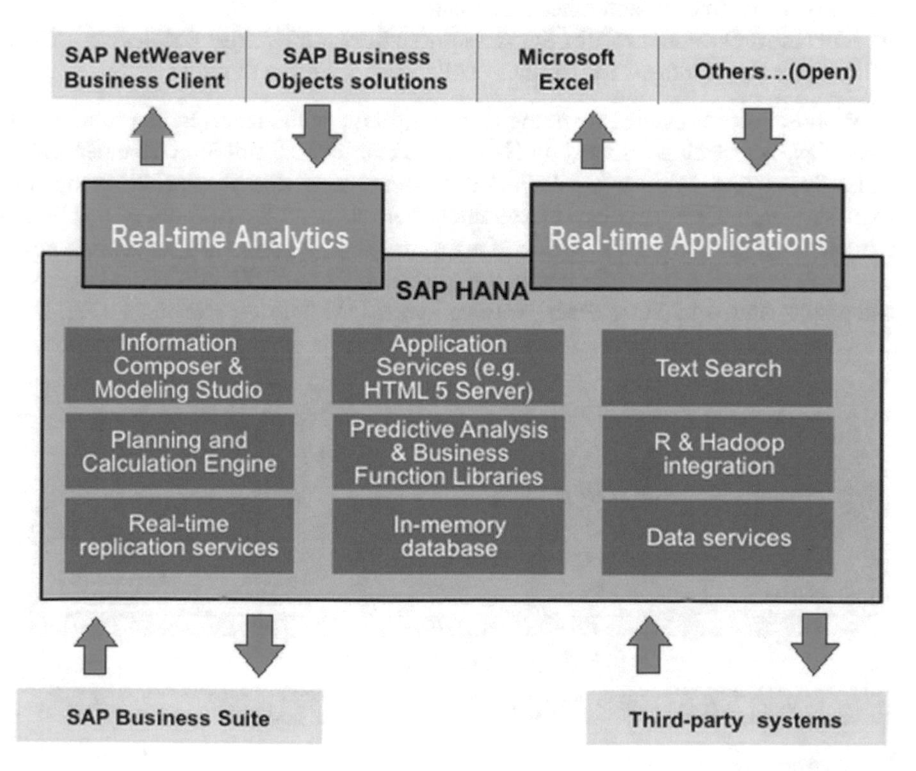

Fig. 3 SAP HANA platform [23]

1.1.4 Microsoft Azure

Microsoft Azure is an enterprise-level cloud platform which is flexible and open and allows for the fast development, deployment, and management of application across Microsoft data centers [24]. Azure complements existing IT environments by providing services in the cloud. Azure Big data and analytics services and products consist of [25]:

- **HDInsight:** R Server Hadoop, HBase, Storm, and Spark clusters.
- **Data Lake Analytics:** A service for performing analytics in a distributed environment which eases the processing of big data.
- **Machine Learning:** Very powerful tools on the cloud platform which allow predictive maintenance.

The important constituents of the Azure Services Platform are as follows [26]:

- **Windows Azure:** Networking and computation, storage with scalability, management and hosting services.
- **Microsoft SQL Services:** Services for storage and reporting using databases.
- **Microsoft.NET Services:** Services for implementing .NET Frameworks.
- **Live Services:** Services for distributing, maintaining and synchronizing files, photos, documents, and other information.
- **Microsoft Dynamics CRM Services and Microsoft SharePoint Services:** Fast solution development for Business collaboration in the cloud (Fig. 4).

World-class companies are doing amazing things with Azure. In its promotional campaign, Azure cloud is used by Heineken to target 10.5 million consumers [27]. Rolls-Royce and Microsoft collaborate to create new digital capabilities by not focussing on the infrastructure management but rather on the delivery of real value to customers [27]. Over 450 million fans of the Real Madrid C.F. worldwide have been able to access the stadium with the Microsoft Cloud [27]. NBC News chooses Microsoft Azure to Host Push Notifications to Mobile Applications [27]. 3M

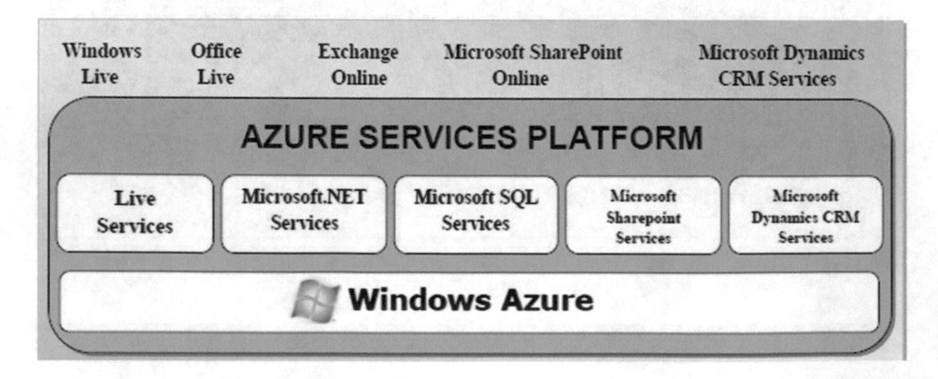

Fig. 4 Azure services platform [26]

accelerates mobile application development and benefits from real-time insight with Azure cloud [27].

1.1.5 Oracle Big Data

Oracle provides the power of both Spark and Hadoop ready to be incorporated with the existing data of enterprises which were previously using Oracle applications and Database. The service provided is highly performant, secure and fully automated. Oracle Big Data offers an integrated set of products to organise and analyse various data sources to find new insights and take advantage of hidden relationships. Oracle Big Data features include [28, 29]:

- **Provisioning and Elasticity:** On demand Hadoop and Spark clusters
- **Comprehensive Software:** Big Data Connectors, Big Data Spatial and Graph, Advanced Analytics
- **Management:** Security, Automated life-cycle
- **Integration:** Object Store Integration, Big Data SQL (Fig. 5).

According to [31], Oracle helps Dubai Smart Government Establishment (DSG) in providing smart government services in the region by proposing new strategies and overseeing processes at the level of government entities. Using Oracle Big Data, the public health services of England has developed a fraud and error detection solution by analysing hidden patterns in physician prescription records [31].

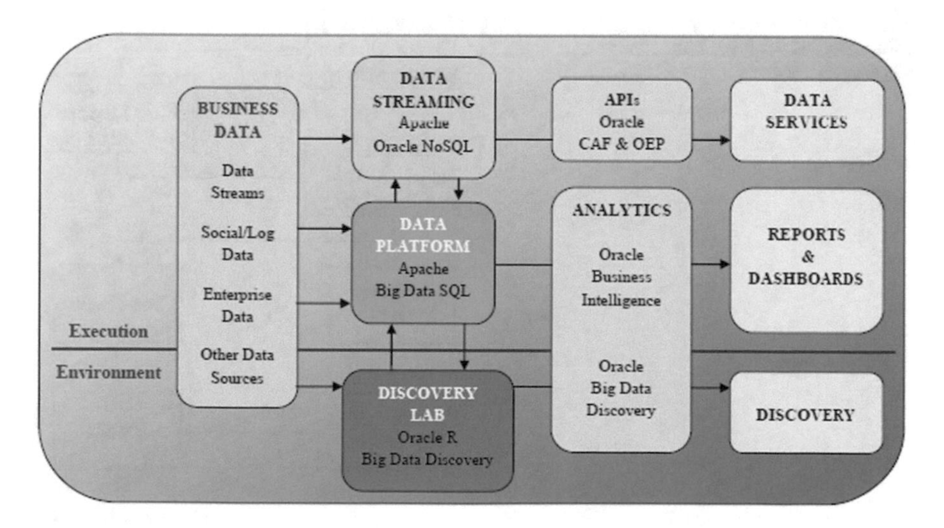

Fig. 5 Oracle solution for big data [30]

1.1.6 Other Big Data Platforms

Other Big Data platforms and Big Data Analytics Software include Jaspersoft BI Suite, Talend open studio, Dell Big Data analytics, Pentaho Business analytics, Redhat Big Data, Splunk, Tableau, Karmasphere Studio and Analyst, Skytree Server.

1.2 Machine Learning Tools for Big Data

Several open-source tools have emerged in recent years for handling massive amounts of data. In addition to managing the data, obtaining insights from the latter is of utmost importance. As such, machine learning techniques add value in all the fields having huge amounts of data ready to be processed. In this chapter two of the open source tools namely H20 [32] and MLIB [33], which facilitate the collaboration of data and computer scientists, will be described in details.

The framework for big data analytics with machine learning is shown in Fig. 6. The structured data from the big data source is imported into H2O and Spark on Databricks. The Data frame obtained is then divided into a training set and a test set. The training set is used to train the machine learning model. Based on the algorithm used, the model is tested using the test set by predicted the labels from the features. The corresponding accuracy is finally computed.

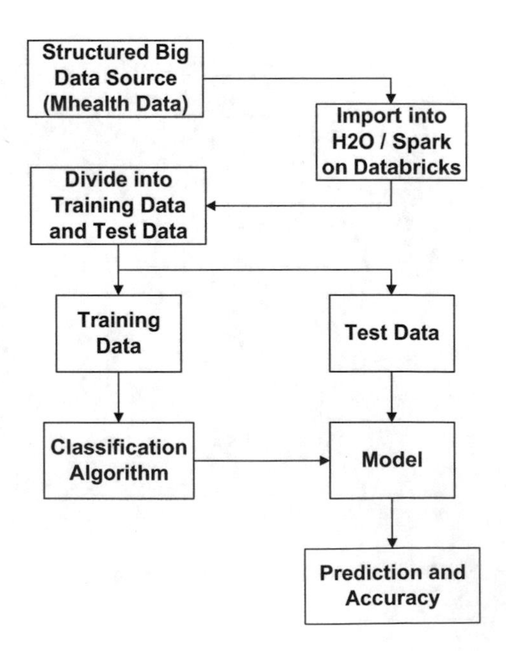

Fig. 6 Framework for big data analytics with machine learning

1.2.1 H2O

H2O is an open-source software used with the trending big-data analytics [32]. The start-up company H2O.ai launched in 2011 in Silicon Valley [34] has brought H2O into existence. As part of pattern mining in data, several models can be fitted by H2O. It provides a fast open source software that supports parallel in-memory machine learning, predictive analytics and deep learning. It is based on pure Java and Apache v2 Open Source. It provides simple deployment with a single jar and automatic cloud discovery. H2O allows data to be used without sampling and provides reliable predictions quicker. The main aim of the H2O project is to bring forth a cloud computing interface, analytics, giving the necessary resources for data analytics to its users. H2O is open-source and has a free license. The profits of the company originate mainly from the service provided to customers and tailored extensions. According to VentureBeat in November 2014, H2O's clients included Nielsen, Cisco, eBay, and PayPal [35–37].

The chief executive of H2O is Sri Satish Ambati had assisted the setting up of a firm for big-data called Platfora. The software for the Apache Hadoop distributed file system is implemented by Platfora [38]. However, the R programming language did not provide satisfactory performance on large data-sets. Ambati therefore began developing the H2O software. The pioneer of the S language at Bell Labs, John Chambers who is also part of the R's leading development team, supported the work [37, 39, 40].

0xdata was co-founded by Ambati and Cliff Click. Cliff held the post of the Chief Technical Officer (CTO) of H2O and had a significant input in H2O's products. Click gave a helping hand in writing the HotSpot Server Compiler and even collaborated with Azul Systems for the implementation of a big-data Java virtual machine (JVM) [41]. In February 2016, Click moved away from H2O [42]. The author of Grammar of Graphics, Leland Wilkinson, serves as Chief Scientist and provides visualization leadership [32].

Three professors at the Stanford University [43] have been listed as Mathematical Scientists by H2O's Scientific Advisory Council. One of them is Professor Robert Tibshirani, who partners with Bradley Efron on bootstrapping [44]. He also specialises on generalized additive models as well as statistical learning theory [45, 46]. The second one is Professor Stephen P. Boyd who specialises in convex minimization and applications in statistics and electrical engineering [47]. The third is Professor Trevor Hastie who is a collaborator of John Chambers on S [40], and also an expert on generalized additive models and statistical learning theory [45, 46, 48].

H2O Architecture and Features

The main features of the H20 software are given in Table 3 [34].

Table 3 Features of the H20 software [34]

H2O components	
Component	**Description**
H2O cluster	It supports clusters with multiple nodes. The system is modeled by taking the concept of shared memory into consideration. This is done since all the computations are performed in memory. Each node sees only a few rows from the dataset and the cluster size can grow without any limit
Distributed key value store	A key is employed to reference the objects such as models, data frames, and results in the H2O cluster
H2O frame	Distributed data frames are supported. Distributed arrays are used across all the nodes for the columns of the dataset. The whole dataset is visible to each node by using the HDFS
Distributed key-value store	
Feature	**Description**
Peer-to-peer	The H2O Key-Value Store is a typical peer-to-peer scattered hash table. There is neither a central key dictionary nor a name-node
Pseudo-random hash	Each key is associated to a home-node. The homes are selected using a pseudo-random algorithms for each key. This technique enables load-balancing
Key's home node	A key's "home" controls the starting and stopping of writes performed. Caching of Keys can be performed anywhere. A write is complete only upon reaching "home"
Data in H2O	
Characteristic	**Description**
Highly compressed	Data is compressed up to 4 times more as compared to gzip and is read from NFS and HDFS
Speed	The speed is dependent on the memory rather than the CPU. If linear data is performed, the speed is similar to that of Fortran or C. The speed is a ratio of the data volume to the memory bandwidth. It is approximately 50 GB/s and depends on the hardware
Data shape	Its operation is fast for a Table width <1 k works for a width <10 k but is slow for <100 k. The Table length is limited only by memory
Processing data in H2O	
Operation	**Description**
Map reduce	It provides a convenient way for developing parallel codes and it supports a relatively fast and efficient method. It also supports distributed fork/join and parallel map and inside every node, the conventional fork/join is supported
Group by	It supports groups of the order of one million on billions of rows, and executes Map/Reduce operations on the elements of the group
Others	All the simple data frame operations of H2O have been integrated in R and Python. Within the algorithms, transactions such as imputation and one-hot encoding of categories are executed

(continued)

Table 3 (continued)

Communication in H2O	
Feature	Description
Network communication	The disparate processes or machine memory spaces need connectivity to JVMs. The network speed of the connection can be fast or slow and there can even be dropped packets & sockets. Even TCP can fail in some cases and retries might be required
Reliable RPC	H2O builds a dependable RPC which restarts dropped connections at the RPC level. Recovery of the cluster can be achieved by pulling cables from a running cluster and re-plugging them
Optimizations	Compression of data are achieved in several ways: to save network bandwidth, UDP packets are used for transmitting short messages and TCP provides congestion control for larger messages

The functionalities for Deep Learning using H2O are as follows [35]:

- Undertaking of tasks pertaining to classification and regression by making use of supervised training models.
- Operations in Java using refined Map/Reduce model and which increase the efficiency in terms of speed and memory.
- Making usage of cluster with multiple nodes or a single node running distributed and Multi-threaded computations simultaneously.
- Availability of the options to set annealing, momentum, and learning rate. In addition, in order to achieve convergence as fast as possible, a completely adaptive and automatic learning rate per-neuron is also available.
- Availability of model averaging, and L1 and L2 Loss function and regularization to avoid over fitting of the model.
- Possibility of selecting the model from a user-friendly web-based interface as well as using scripts in R equipped with the H2O package.
- Availability of check-points to help achieve lowered run times as well as tune the models.
- Availability of the automatic process of discarding missing values in the numerical and categorical datasets during both the pre-processing and post-processing stages.
- Availability of the automatic feature which handles the balance between computation and communication over the cluster to achieve the best performance. Additionally, there is the possibility of exporting the designed model in java code and use in production environments.
- Availability of additional features and parameters for adjustments in the model automatic encoders for handling unsupervised learning models and ability to detect anomalies.

A typical architecture of the H2O software stack is shown in Fig. 7 [49].

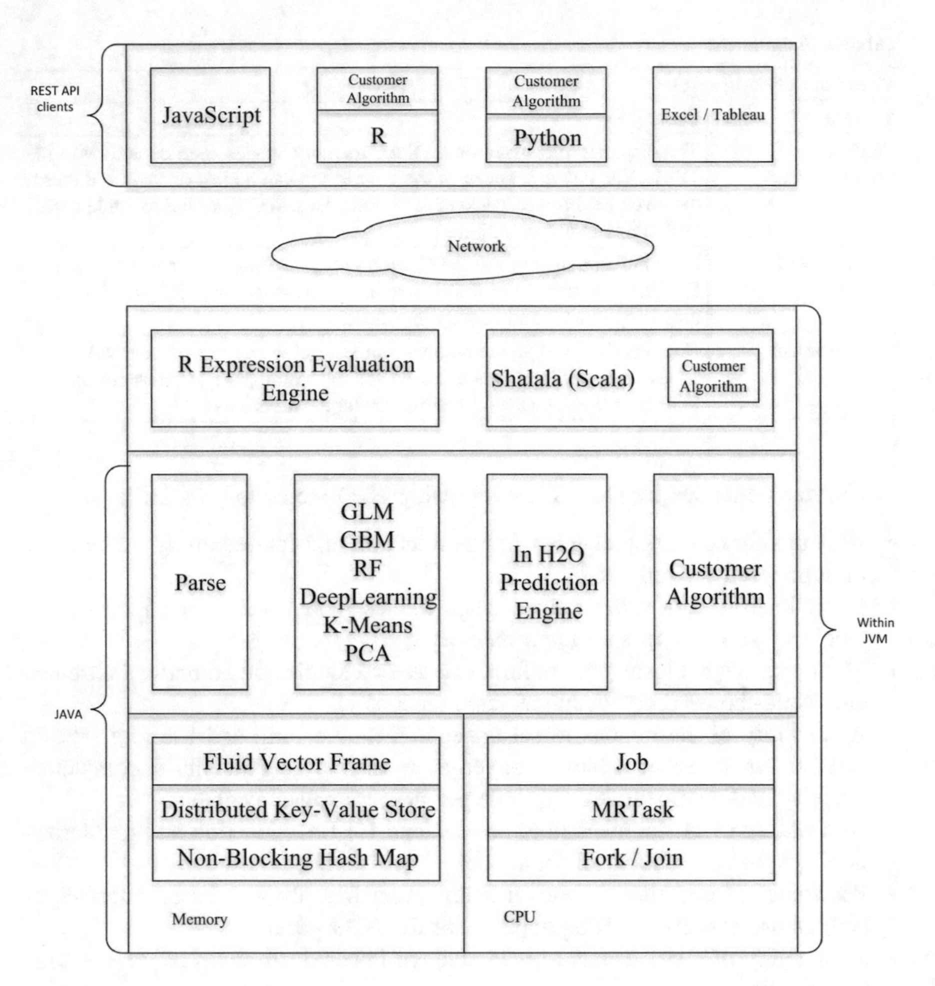

Fig. 7 H2O software architecture [49]

A socket connection is used by all the REST API clients for communication with H2O. JavaScript is the language used to code the embedded H2O Web UI which uses the standard REST API. The H2O R package "library(h2o)" can be used in R scripts. Instead of executing H2O using 'apply' or 'ddply', R functions can be written by the users. The way the system is at present, the REST API must be used directly in Python scripts. The commonly used Microsoft Excel is also equipped with an H2O worksheet. This enables importing large volumes of data into H2O and the running of algorithms such as GLM directly from Excel. Visualization in Tableau is also possible by pulling results directly from H2O.

One or more nodes combined make up an H20 cloud. The JVM process, which consists of a single node, is divided into three layers: core infrastructure, algorithms, and language.

The Shalala Scala and R layer comprise of a language layer consisting of engine for evaluation of expressions. The R REST client front-end is master to the evaluation layer of R. The Scala layer is different from the R layer in the sense that algorithms and native programs using H2O can be written directly. The algorithms layer is made up of all the default algorithms in H2O [49]. For example, datasets are imported using the parse algorithm. It also includes GLM and algorithms for predictive analytics and model evaluation. The bottom (core) layer manages Memory and CPU resources.

The elementary unit of data storage which the users are exposed to is an H2O Data Frame. The engineering term referring to the capability of adding, removing, and updating columns fluidly in a frame is "Fluid Vector". It is the column-compressed store implementation.

The distributed Key-Value store are essentially atomic and distributed in-memory storage and spread across the cluster. The Non-blocking Hash Map is used in the Key-Value store implementation. CPU Management is based on three main parts. First, there are jobs which are large pieces of work having progress bars and which can be monitored using the web user interface. One example of a job is Model creation. Second there is the MapReduce Task (MRTask). This is an H2O in-memory MapReduce Task and not a Hadoop MapReduce task. Finally there is Fork/Join which is a modified version of the JSR166y lightweight task execution framework.

H2O can support billions of data rows in-memory even if the cluster size is relatively small. This is possible by the use of sophisticated in-memory compression techniques. The H2O platform has its own built-in Flow web interface so as to make analytic workflows become user-friendly to users who do not have engineering background. It also includes interfaces for Java, Python, R, Scala, JSON and Coffeescript/JavaScript. The H2O platform was built alongside and on top of both Hadoop and Spark Clusters. The deployment is typically done within minutes [35, 36, 43]. The software layers that are associated with the running of an R program to begin a GLM on H2O are shown in Fig. 8.

Figure 8 shows two different sections. The section on the left depicts the steps which run the R process and the section on the right depicts the steps which run in the H2O cloud. The TCP/IP network code lies above the two processes allowing them to communicate with each other. The arrows with solid lines demonstrate a request originating from R to H2O and the arrows with dashed lines demonstrate the response from H2O to R for the request. The different components of the R program are: R script, R package for R script, (RCurl, Rjson) packages, and R core runtime. Figure 9 shows the R program retrieving the resulting GLM model [49].

The H2OParsedData class consists of an S4 object which is used to represent an H2O data frame in R. The big data object in H2O is referenced by an @key slot in the S4 object. The H2OParsedData class is used by the package in R to perform summarization and merging operations. These operations are transmitted over to the H2O cloud by making use of an HTTP connection. The operation is performed on the H2O cloud platform and the result is returned as a reference which is in turn stored in another S4 object in R.

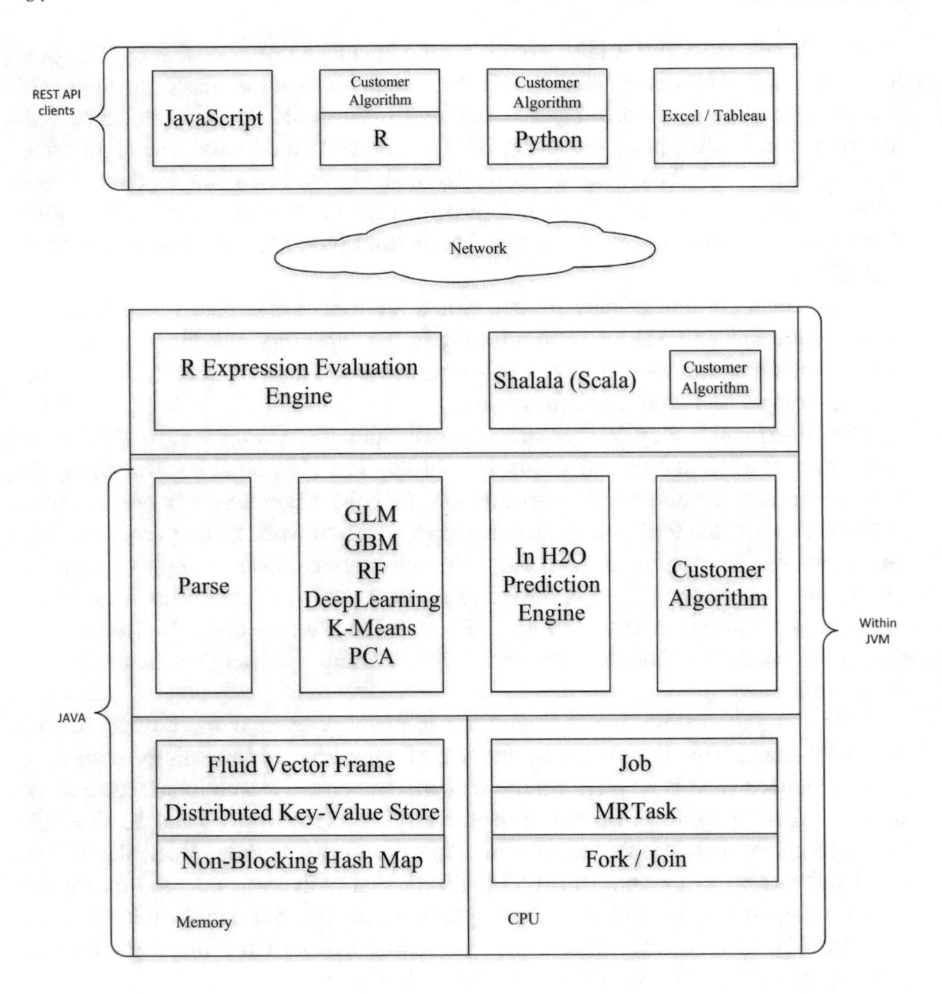

Fig. 8 R script starting H20 GLM

H2O Algorithms

Several common data analysis and machine learning algorithms are supported by H2O. These algorithms can be classified as per Table 4 [36, 49].

H2O Deployment and Application Example

H2O can be launched by opening a command prompt terminal, moving to the location of the jar file, and executing the command: java—jar h2o.jar as shown in Fig. 10.

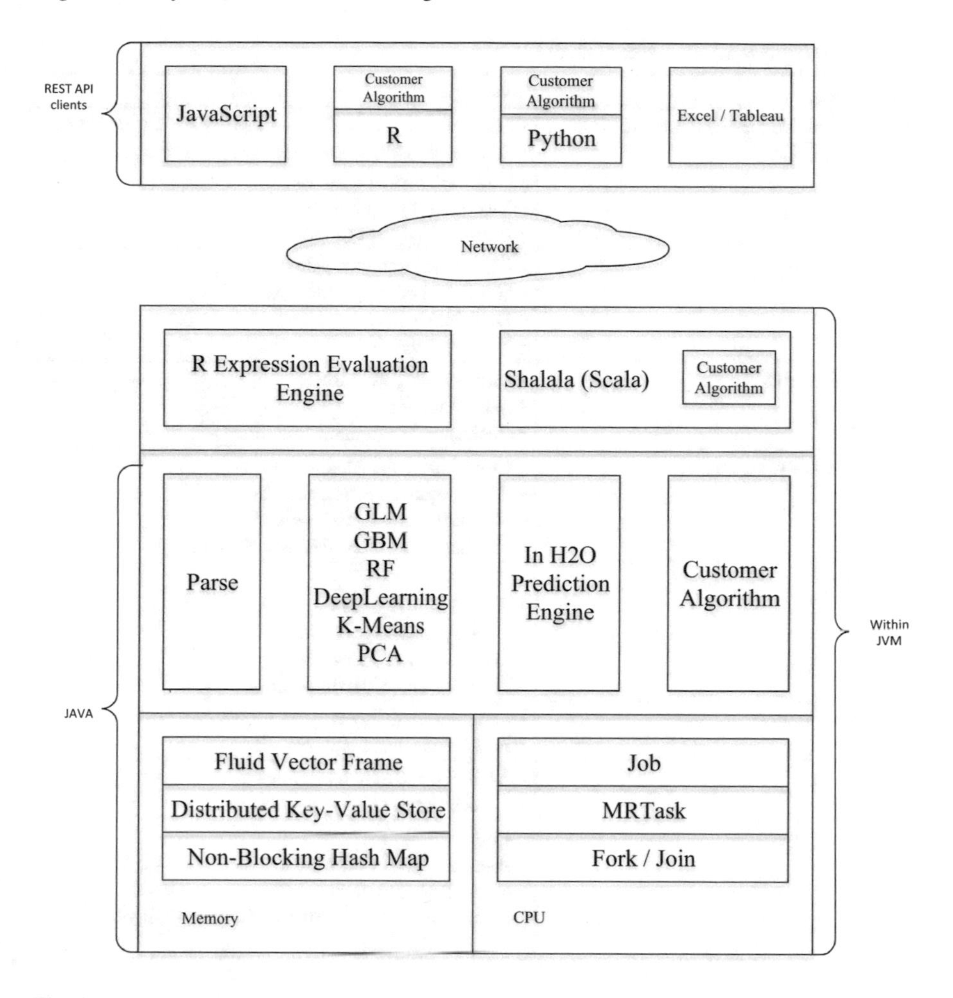

Fig. 9 R Script retrieving the resulting GLM model

To access the web interface, type-in the url: https://localhost:54321/flow/index.html in the web browser. The start-up page shows a list of functionalities supported in H2O. This can also be viewed by using the *assist* option as depicted in Fig. 11.

In this section, the Mobile Health (MHEALTH) dataset from [50–52] has been used to demonstrate classification machine learning algorithms with H20.

The dataset consists of recordings of body motions for volunteers while performing different physical activities. The twelve activities recorded are:

L1 Standing still
L2 Sitting and Relaxing
L3 Lying down
L4 Walking
L5 Climbing stairs

Table 4 Overview of H2O algorithms [49]

Category	Techniques
Regressions/Generalized Linear Modeling (GLM)	• GLMNet • Distributions: Gaussian, Binomial, Poisson, Gamma, Tweedie • Bayesian regression • Multinomial regression
Classifications	• Distributed random forest • Gradient boosting machine • Distributed trees • Naïve Bayes
Neural networks	• Multi-layer perceptron • Auto-encoder • Restricted Boltzmann machines
Solvers and optimization	• Generalized ADMM solver • BFGS (quasi newton method) • Ordinary least square solver • Stochastic gradient descent MCMC
Clustering	• K-means, • K-nearest neighbors • Locality sensitive hashing • Dimensionality reduction • Singular value decomposition
Recommendation	• Collaborative filtering • Alternating least squares
Time series	• ARIMA, ARMA modeling • Forecasting
Data munging	• Plyr • Integrated R-environment • Slice, Log transform • Anonymizing/Obfuscating (for personalized or confidential data)

L6 Waist bends forward
L7 Frontal elevation of arms
L8 Knees bending (crouching)
L9 Cycling
L10 Jogging
L11 Running
L12 Jump front and back.

In addition to these known classes during which experimentation is performed, there is also a Null class with a value of '0' for data recorded when none of the above recordings are being performed. The feature values are mainly obtained from sensors: acceleration from chest sensor, electrocardiogram sensor, accelerometer sensor, gyro sensor, and magnetometer sensor. The accelerometer, gyro, and magnetometer sensors are placed on the arms and ankles of the subjects.

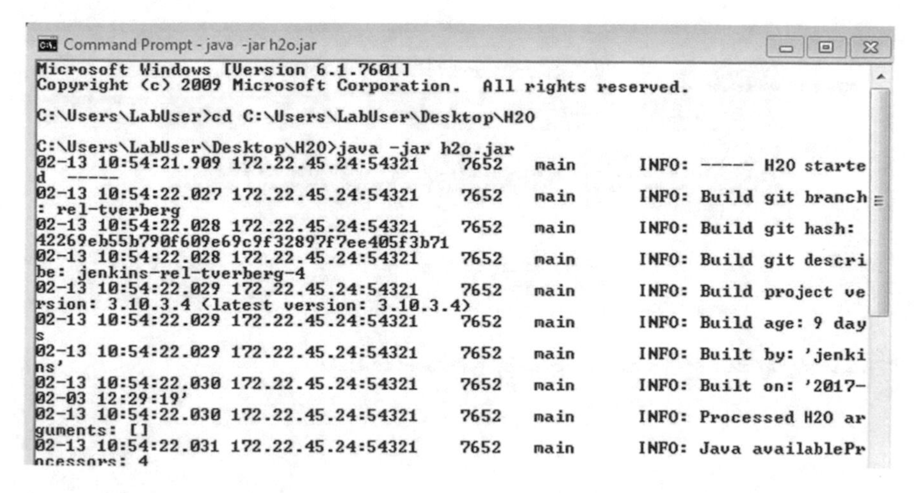

Fig. 10 Launching of H2O through the command prompt

```
assist
```

Assistance

Routine	Description
importFiles	Import file(s) into H_2O
getFrames	Get a list of frames in H_2O
splitFrame	Split a frame into two or more frames
mergeFrames	Merge two frames into one
getModels	Get a list of models in H_2O
getGrids	Get a list of grid search results in H_2O
getPredictions	Get a list of predictions in H_2O
getJobs	Get a list of jobs running in H_2O
buildModel	Build a model
importModel	Import a saved model
predict	Make a prediction

Fig. 11 Assistance list for operations available in H2O

The common dataset to be used with the classification algorithms can be imported using the *importFiles* option in the list on the homepage of H2O as depicted in Fig. 12.

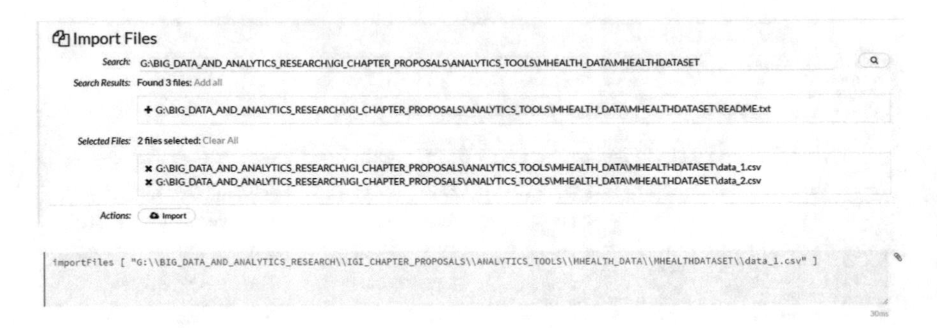

Fig. 12 Import data set on the cluster to be processed in H2O

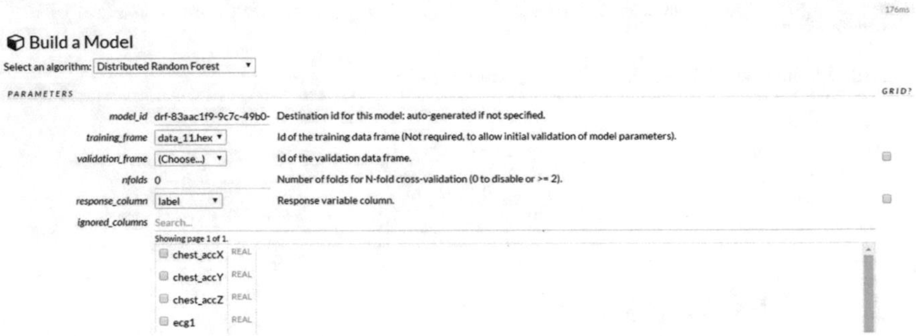

Fig. 13 Building a distributed random forest model

As can be observed from Fig. 12, two csv-files are loaded in H2O. The file ‘data1.csv’ is used for training of the different models and ‘data2.csv’ is used for testing.

Illustration 1: Random Forest Classification in H2O

The model is then built using the training data set as input. The Distributed Random Forest (DRF) is used with the “label” column selected as the response column. These are shown in Fig. 13. The equation being modelled is: label $\sim$ chestAccX + chestAccY + ….

The output metrics on the training model is shown in Fig. 14.

With the training model obtained, prediction can be performed on the test set data. A comparison can then be performed on the predicted and already known “label”. Figure 15 shows the step where the model is used to predict the “label” for the test set.

The predicted “label” can be merged with the test data set and compared with that already known. Figure 16 shows the confusion matrix for the test data set.

An analysis of the percentage difference between the predicted and already known “label” for the test data set can be performed as shown in Table 5.

▾ OUTPUT - TRAINING_METRICS

model	drf-dfb6e02f-cd40-4383-88e2-09c055796131
model_checksum	-3755657836809008128
frame	data_1_NB.hex
frame_checksum	30919538483876304
description	Metrics reported on Out-Of-Bag training samples
model_category	Multinomial
scoring_time	1487413838996
predictions	·
MSE	0.061304
RMSE	0.247597
nobs	982273
r2	0.994981
logloss	0.206972
mean_per_class_error	0.113143

Fig. 14 Output metrics for training model

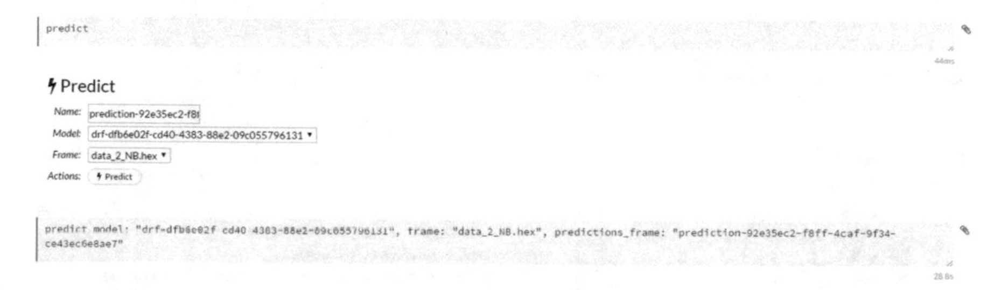

Fig. 15 Prediction using training model on test data set

▾ PREDICTION - CM - CONFUSION MATRIX

Null	climbing_stairs	cycling	frontal_elevation_arms	jogging	jump_front_and_back	knees_bending	lying_down	running	sitting_and_relaxing	standing_still	waist_bends_forward	walking	frontal_elevation_arms	Error	Rate
160041	226	196	0	764	144	38	1848	2431	0	0	57	243	0	0.0326	5,387 / 185,428
4139	2005	0	0	0	0	0	0	0	0	0	0	0	0	0.6737	4,139 / 6,144
4559	0	1585	0	0	0	0	0	0	0	0	0	0	0	0.7420	4,559 / 6,144
0	0	0	0	0	0	0	0	0	0	0	0	0	0	·	0 / 0
2637	0	0	0	3506	1	0	0	0	0	0	0	0	0	0.4294	2,638 / 6,144
1762	0	0	0	5	332	0	0	0	0	0	0	0	0	0.8418	1,767 / 2,099
5608	0	0	0	0	0	228	0	0	0	0	0	0	0	0.9609	5,608 / 5,836
3196	0	0	0	0	0	0	2948	0	0	0	0	0	0	0.5202	3,196 / 6,144
453	0	0	0	32	3	0	0	5656	0	0	0	0	0	0.0794	488 / 6,144
6144	0	0	0	0	0	0	0	0	0	0	0	0	0	1.0	6,144 / 6,144
6144	0	0	0	0	0	0	0	0	0	0	0	0	0	1.0	6,144 / 6,144
5178	0	0	0	0	0	0	0	0	0	0	147	0	0	0.9724	5,178 / 5,325
5176	0	0	0	0	0	0	0	0	0	0	0	968	0	0.8424	5,176 / 6,144
5632	0	0	0	0	0	0	0	0	0	0	0	0	0	1.0	5,632 / 5,632
210668	2231	1721	0	4307	480	266	4296	8087	0	0	204	1211	0	0.2401	56,056 / 233,472

Fig. 16 Confusion matrix for test data set

Illustration 2: Naïve Bayes Classification in H2O

The same principle as with the Random Forest model is applied to the Naïve Bayes model. The "label" column selected as the response column as shown in Fig. 17. The output metrics on the training model is shown in Fig. 18.

Table 5 Analysis of percentage difference between predicted and already known “label”

Number of training data	982,273
Number of subjects	10
Number of test data	233,472
Correct predictions	177,416
Incorrect predictions	56,056
Percentage accuracy	75.99%

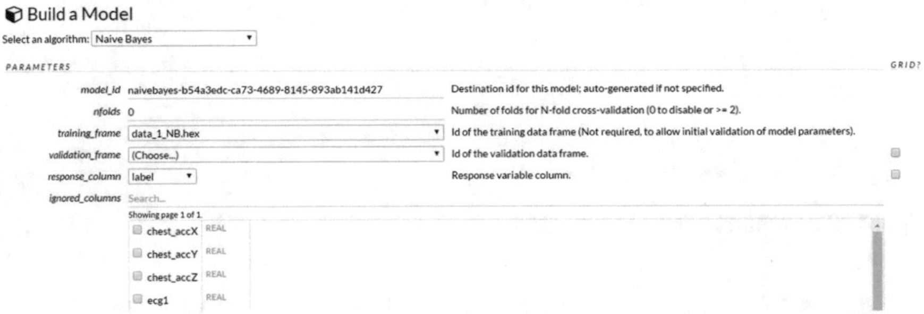

Fig. 17 Building a Naïve Bayes model

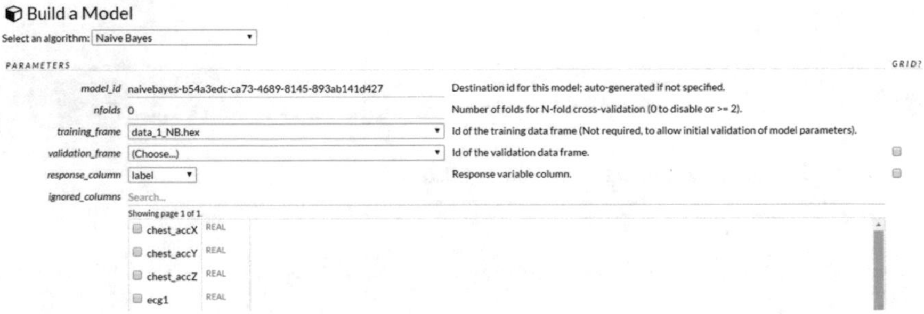

Fig. 18 Output metrics for training model

Figure 19 shows the step where the model is used to predict the “label” for the test set.

The predicted “label” can be merged with the test data set and compared with that already known. Figure 20 shows the confusion matrix for the test data set.

An analysis of the percentage difference between the predicted and already known “label” for the test data set can be performed as shown in Table 6.

Illustration 3: Deep Learning Classification in H2O

The same principle as with the Random Forest model and Naïve Bayes is applied to the Deep Learning model. The “label” column selected as the response column as shown in Fig. 21. The output metrics on the training model is shown in Fig. 22.

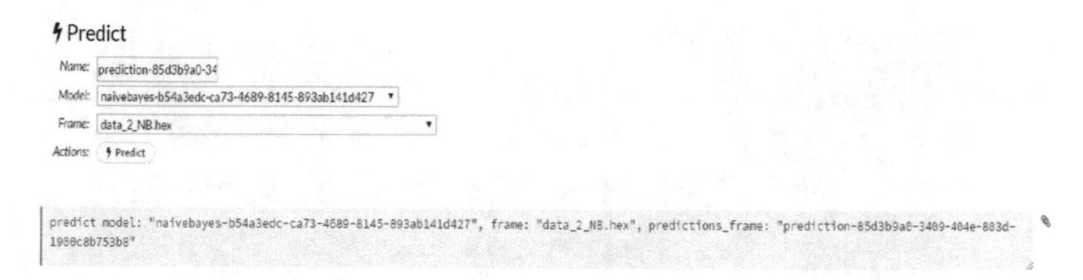

Fig. 19 Prediction using training model on test data set

▾ PREDICTION · CM · CONFUSION MATRIX

Null	climbing_stairs	cycling	frontal_elevation_arms	jogging	jump_front_and_back	knees_bending	lying_down	running	sitting_and_relaxing	standing_still	waist_bends_forward	walking	frontal_elevation_arms	Error	Rate
59643	4682	3216	23888	9800	3293	7933	1973	4084	5360	17875	11718	11963	0	0.6395	105,785 / 165,428
2712	3390	0	3	12	6	793	0	2	0	0	59	167	0	0.4482	2,754 / 6,144
66	1	6071	0	0	0	6	0	0	0	0	0	0	0	0.0119	73 / 6,144
0	0	0	0	0	0	0	0	0	0	0	0	0	0	-	0 / 0
889	0	0	0	4664	485	0	0	185	0	0	0	1	0	0.2409	1,480 / 6,144
379	0	0	0	1074	416	0	0	29	0	0	0	1	0	0.8018	1,683 / 2,099
3	36	44	43	0	0	4594	0	0	0	82	1034	0	0	0.2128	1,242 / 5,836
0	0	0	0	29	0	0	6114	1	0	0	0	0	0	0.0049	30 / 6,144
27	0	0	0	1206	429	0	0	4583	0	0	0	0	0	0.2541	1,561 / 6,144
0	0	0	109	0	0	0	0	0	6035	0	0	0	0	0.0177	109 / 6,144
0	2	0	1	0	0	0	0	0	3	5958	180	0	0	0.0303	186 / 6,144
189	27	0	145	0	0	383	0	0	0	59	4597	5	0	0.1367	728 / 5,325
346	842	0	0	6	0	70	0	3	1	0	204	4772	0	0.2233	1,372 / 6,144
28	0	0	5306	0	0	3	0	0	21	259	15	0	0	1.0	5,632 / 5,632
63222	8980	9331	29495	16690	4629	13782	6087	8887	11420	24133	17807	16909	0	0.5253	122,655 / 233,472

Fig. 20 Confusion matrix for test data set

Table 6 Analysis of percentage difference between predicted and already known "label"

Number of training data	982,273
Number of subjects	10
Number of test data	233,472
Correct predictions	176,202
Incorrect predictions	57,270
Percentage accuracy	75.47%

Build a Model

Select an algorithm: Deep Learning

PARAMETERS GRID?

model_id	deeplearning-aa9a026f-b38e-4c39-a088-ce105d058ef3	Destination id for this model; auto-generated if not specified.
training_frame	data_1_NB.hex	Id of the training data frame (Not required, to allow initial validation of model parameters).
validation_frame	(Choose...)	Id of the validation data frame.
nfolds	0	Number of folds for N-fold cross-validation (0 to disable or >= 2).
response_column	label	Response variable column.
ignored_columns	Search...	

Showing page 1 of 1.

chest_accX REAL
chest_accY REAL
chest_accZ REAL
ecg1 REAL
ecg2 REAL

Fig. 21 Building a deep learning model

▾ OUTPUT - TRAINING_METRICS

model	deeplearning-aa9a026f-b38e-4c39-a088-ce105d058ef3
model_checksum	-8629438079940385792
frame	·
frame_checksum	0
description	Metrics reported on temporary training frame with 10032 samples
model_category	Multinomial
scoring_time	1487415393998
predictions	·
MSE	0.116052
RMSE	0.340664
nobs	10032
r2	0.990465
logloss	0.381634
mean_per_class_error	0.335862

Fig. 22 Output metrics for training model

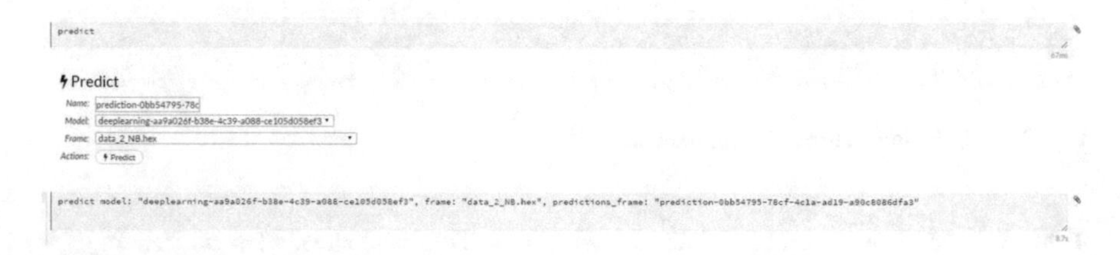

Fig. 23 Prediction using training model on test data set

▾ PREDICTION - CM - CONFUSION MATRIX

Null	climbing_stairs	cycling	frontal_elevation_arms	jogging	jump_front_and_back	knees_bending	lying_down	running	sitting_and_relaxing	standing_still	waist_bends_forward	walking	frontal_elevation_arms	Error	Rate
152366	3234	665	9	1262	371	879	1679	2671	19	432	42	1799	0	0.0790	13,062 / 165,428
1847	4277	7	0	0	0	3	1	0	0	0	0	9	0	0.3039	1,867 / 6,144
2386	0	3758	0	0	0	0	0	0	0	0	0	0	0	0.3883	2,386 / 6,144
0	0	0	0	0	0	0	0	0	0	0	0	0	0	·	0 / 0
1290	0	0	0	4831	21	0	0	2	0	0	0	0	0	0.2137	1,313 / 6,144
1480	3	0	0	142	470	2	0	2	0	0	0	0	0	0.7761	1,629 / 2,099
5174	134	0	0	0	0	528	0	0	0	0	0	0	0	0.9095	5,308 / 5,836
3072	0	0	0	0	0	0	3072	0	0	0	0	0	0	0.5000	3,072 / 6,144
383	0	0	0	5	125	0	0	5631	0	0	0	0	0	0.0835	513 / 6,144
6144	0	0	0	0	0	0	0	0	0	0	0	0	0	1.0	6,144 / 6,144
6144	0	0	0	0	0	0	0	0	0	0	0	0	0	1.0	6,144 / 6,144
5325	0	0	0	0	0	0	0	0	0	0	0	0	0	1.0	5,325 / 5,325
2495	1586	0	0	0	0	0	0	0	0	0	0	2063	0	0.6642	4,081 / 6,144
5621	0	0	0	0	0	0	0	0	0	11	0	0	0	1.0	5,632 / 5,632
193727	9234	4430	9	6240	987	1412	4752	8306	19	443	42	3871	0	0.2419	56,476 / 233,472

Fig. 24 Confusion matrix for test data set

Figure 23 shows the step where the model is used to predict the “label” for the test set.

The predicted “label” can be merged with the test data set and compared with that already known. Figure 24 shows the confusion matrix for the test data set.

An analysis of the percentage difference between the predicted and already known “label” for the test data set can be performed as shown in Table 7.

Table 7 Analysis of percentage difference between predicted and already known "label"

Number of training data	982,273
Number of subjects	10
Number of test data	233,472
Correct predictions	176,996
Incorrect predictions	56,476
Percentage accuracy	75.81%

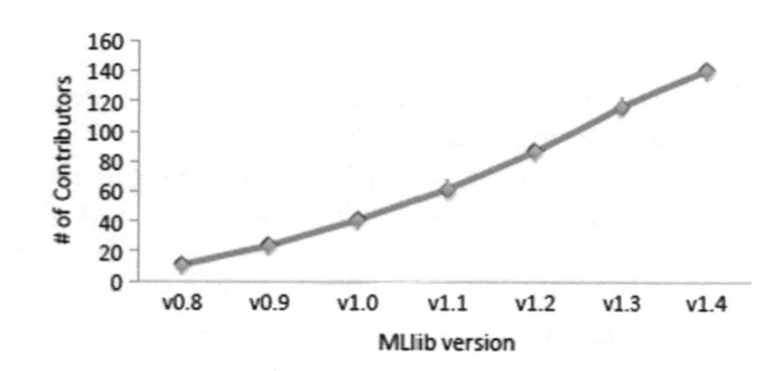

Fig. 25 Growth in MLlib contributors [55]

1.2.2 MLlib

Spark MLlib is an open source API that is part of the Apache Software Foundation. Spark MLlib and DataFrames provide several tools. These tools ease the integration of existing workflows developed on tools such as R and Python, with Spark. For example, users are provided with possibility to use the R syntax they are familiar with and call the appropriate MLlib algorithms [53].

The UC Berkeley AMPLab pioneered Spark and the latter was made open-source in 2010. The major aim for the design of Spark is efficient iterative computing. Machine learning algorithms have also been packaged with the early releases of Spark. With the creation of MLlib, the lack of scalable and robust learning algorithms was overcome. As part of the MLbase project [54], MLlib was developed in 2012 and was made open-source in September 2013. When MLlib was first launched, it was incorporated in Spark's version 0.8. The Apache 2.0 license packages Spark and MLlib as an open-source Apache project. Moreover, as of Spark's version 1.0, Spark and MLib have been released on a three month cycle [5].

Eleven contributors helped developing the initial version of MLlib at UC Berkeley. The original version provided only a limited group of common machine learning algorithms. With the initiation of MLlib, the number of contributors towards the project has grown exponentially. The number of contributors as from the Spark 1.4 release has grown to more than 140 contributors from more than 50 organizations in a time span of less than two years. Figure 25 depicts the number of contributors of MLib against the release versions. The implementation of several other functionalities was motivated by the adoption rate of MLib [55].

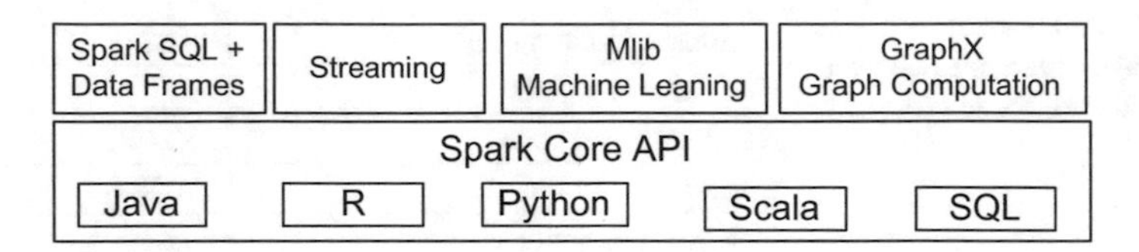

Fig. 26 The spark ecosystem

Architecture and Features

Several components within the Spark ecosystem are beneficial for MLlib. Figure 26 illustrates the Spark ecosystem.

As observed in Fig. 25, Spark supports programming platforms such as R, SQL, Python, Scala and Java. Additionally, it has several libraries which can provide functionalities such as graph computations, stream data processing, and real-time interactive query processing in addition to machine learning [53].

A general execution engine is provided for data transformation is provided by Spark core at the very low level. These transformations involve more than 80 operators for performing feature extraction and data cleaning. There are other high-level libraries included in Spark and leveraged by MLlib. The functionality of data integration is provided by Spark SQL which aid in the simplification of data preprocessing and cleaning. Support for the fundamental DataFrame abstraction in Spark.ml package is also provided.

Large-scale graph processing is supported by GraphX [56]. The implementation of learning algorithms that can be modeled as large, sparse graph problems, e.g., LDA [57, 58] is also supported. Additionally, Spark Streaming [59] provides the ability to handle and process live data streams to users, thereby enabling the development of online learning algorithms, as in Freeman [60]. The improvements in MLlib are brought up by performance enhancements of the high-level libraries in Spark core.

E-client distributed learning and prediction are supported by the several optimizations included in MLlib. A careful use of blocking is made by the ALS algorithm for recommendations in order to lower the collection of JVM garbage overhead and to leverage high-level operations of linear algebra. Ideas such as data-dependent feature discretization for the reduction of communication costs, and parallelized learning within and across trees using ensembles of trees have been used by Decision trees from the PLANET project [61]. Optimization algorithms are used to learn the Generalized Linear Models. These algorithms perform the gradient computation in parallel by making use of c++ based linear algebra libraries for worker computations. E-client communication primitives are very beneficial for many algorithms. The driver is prevented from being a bottleneck by tree-structured aggregation, and large models are quickly distributed to workers by Spark broadcasts.

Performing machine learning in practice involves a set of stages in an appropriate sequence. The stages in the proper order are: pre-processing of data, extraction of features, fitting of models, and validation. Native support for the multiple set of functionalities needed for the construction of the pipeline are not provided by most of the machine learning libraries. The process of putting together an end-to-end pipeline has severe implications in terms of cost and labor for the network overhead when dealing with large-scale datasets. A package is added in MLlib with the aim to address these concerns inspired by the previous works of [62–64], and [65], and by leveraging on Spark's rich ecosystem. The implementation and setting of cascaded learning pipelines is supported by the spark.ml package via a standard set of high-level APIs [66].

In order to ease the combination of several algorithms in a single workflow, the APIs for machine learning algorithms have been standardized by MLlib. The Scikit-Learn project has brought forward the pipeline concept. The key concepts introduced by the pipelines API are: DataFrame, Transformer, Estimator, Pipeline and Parameter. These are discussed next.

DataFrame:

DataFrame from the Spark SQL is used by the ML API as an ML dataset. The dataset can contain several data types and a DataFrame can consist of several columns comprising of feature vectors, text, labels, and predicted labels.

Transformer:

A Transformer is an algorithm with the capability of acting upon a DataFrame and converting it to another DataFrame. A machine learning model, for example, is a Transformer which converts a DataFrame consisting of features into predictions.

Estimator:

The role of an Estimator algorithms is to learn from the training DataFrame and produce a mathematical model. A Transformer is produced when an Estimator is fit on a DataFrame.

Pipeline:

The role of a Pipeline is to concatenate several Estimators and Transformers which specify a workflow.

Parameter:

The parameters specified for all Transformers and Estimators share a common API [67]. In machine learning, it is common to run a sequence of algorithms to process and learn from data. For example, a simple text document processing workflow might include the following stages:

- Splitting of each document's text into words.
- Converting each word into a numerical feature vector.
- Learning of a prediction model by making use of the feature vectors and labels.

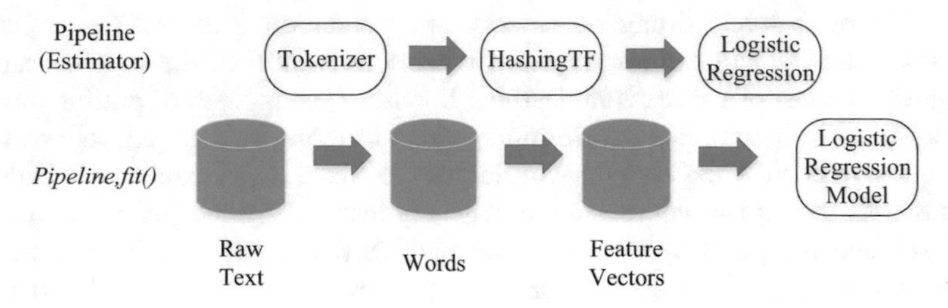

Fig. 27 Training time usage of a pipeline

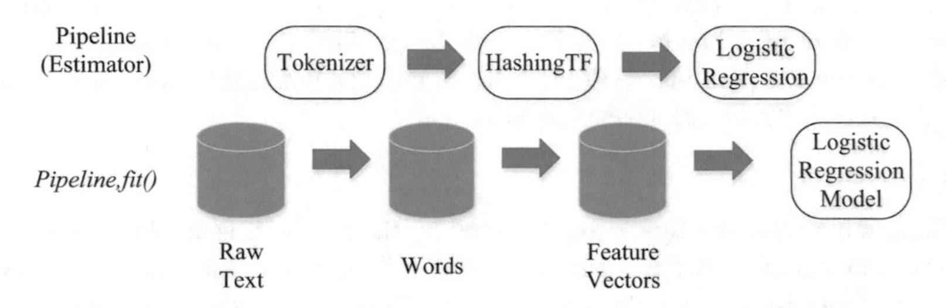

Fig. 28 Pipeline transformer

This type of workflow is represented as a Pipeline by MLlib. The Pipeline consists of a sequence of Estimators and Transformers to be executed in a predefined and specific order. A transformation operation is performed on the input DataFrame as it flows through each stage of the Pipeline. The *transform*() method is called on the DataFrame at the Transformer stages while the *fit*() method is called on the DataFrame for the Estimator stages. Consider for example the simple text document workflow. Figure 27 illustrates the *training time* usage of a Pipeline [67].

A Pipeline with three stages are depicted in Fig. 27. The first and the second stages (Tokenizer and HashingTF) are Transformers and the third stage (Logistic Regression) is an Estimator. The data flow through the Pipeline is represented in Fig. 27 and the cylinders indicate DataFrames. The Pipeline.fit() method is applied on the original DataFrame comprising of the raw text documents and labels. The Tokenizer.transform() method performs a splitting on the raw text documents to convert them into words and adding a new column to the DataFrame with the words. The HashingTF.transform() method performs a conversion operation on the words to form feature vectors, thereby adding a new column to the DataFrame with these vectors. The Pipeline calls LogisticRegression.fit() to produce the machine learning model since LogisticRegression is an Estimator. Further stages of the Pipeline would call the model's transform() method on the DataFrame before passing the latter to the other stage [67]. A Pipeline is an Estimator which produces a Transformer following the running of the Pipeline's fit() method. Figure 28 depicts the Pipeline model used for the test phase.

Figure 28 shows the Pipeline model which consists of the same number of stages as the original Pipeline with the difference that all the Estimators have become

Transformers in this case. The model's *transform*() method is executed on the test dataset and the dataset is updated at each stage before being passed to the next. The Pipeline and models aid in ensuring that the training and test data undergo the same feature processing steps [67].

MLlib Algorithms

The different categories of algorithms supported by MLIB are as follows:

A. Extracting, transforming and selecting features:

The algorithms for working with features are roughly divided into the following groups [68]:

- Extraction: Extracting features from "raw" data.
- Transformation: Scaling, converting, or modifying features.
- Selection: Selecting a subset from a larger set of features.
- Locality Sensitive Hashing (LSH): This class of algorithms combines aspects of feature transformation with other algorithms.

B. Classification and regression:

The classification and regression algorithms are given as follows [69]:

Classification:

- Naive Bayes
- Decision tree classifier
- Random forest classifier
- One-vs-rest classifier
- Multilayer perceptron classifier
- Gradient-boosted tree classifier.

Regression:

- Isotonic regression
- Logistic regression
  - Binomial logistic regression
  - Multinomial logistic regression
- Linear regression
- Generalized linear regression
- Survival regression
- Decision tree regression
- Random forest regression
- Gradient-boosted tree regression.

The Clustering algorithms are given as follows [70]:

- K-means
- Bisecting k-means
- Latent Dirichlet allocation (LDA)
- Gaussian Mixture Model (GMM).

MLIB also supports function for Collaborative Filtering [71], ML Tuning: model selection and hyperparameter tuning [72].

MLlib Deployment and Application Example

MLlib can be used locally in a Java programming IDE like Eclipse by importing the corresponding jar file in the workspace. In this work, the environment already available in the online Databricks community edition has been used [53]. Users simply need to sign up and use the platform.

In this section, the Mobile Health (MHEALTH) dataset from [50–52] has been used to demonstrate classification machine learning algorithms with MLIB.

The classification algorithms used in this section are: Random Forest, Decision Tree, and Multinomial Logistic Regression [36]. Naïve Bayes is not used in this case because the current version does not support negative values for input features. The sensors in the case of the MHEALTH dataset [51, 52, 73] output both positive and negative values. The common dataset to be used with the classification algorithms can be imported using the *Create Table* option on the Databricks homepage as depicted in Fig. 29.

Given that the community edition is being use, there is a limitation on the resources provided which would not allow the modelling using the large data set as with H2O on the local machine. Therefore, part of the MHEALTH dataset is uploaded to Databricks for the presentation of the modelling with the different classification algorithms.

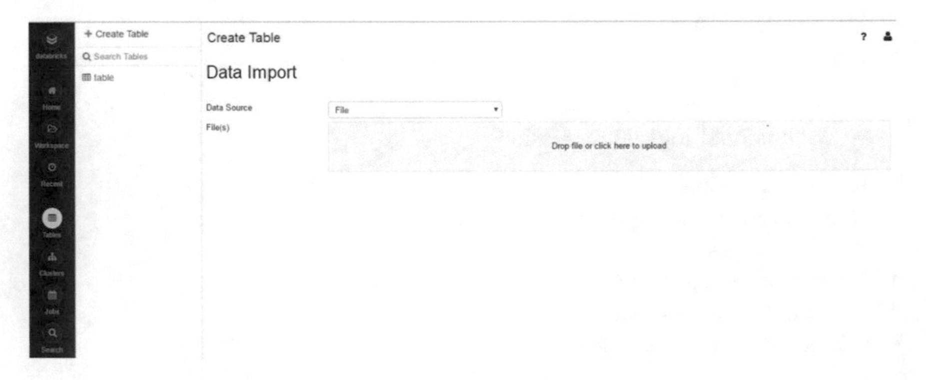

Fig. 29 Import data set on the cluster to be processed in databricks

```
import org.apache.spark.mllib.classification.NaiveBayes
import org.apache.spark.mllib.linalg.Vectors
import org.apache.spark.mllib.regression.LabeledPoint
import org.apache.spark.mllib.tree.RandomForest
import org.apache.spark.mllib.tree.model.RandomForestModel
import org.apache.spark.mllib.classification.LogisticRegressionWithLBFGS
import org.apache.spark.mllib.evaluation.MulticlassMetrics
import org.apache.spark.mllib.tree.DecisionTree
import org.apache.spark.mllib.tree.model.DecisionTreeModel

val data = sc.textFile("/FileStore/tables/7lxvhelo1487641043634/data_2_mhealth.csv", 6)

import org.apache.spark.mllib.classification.NaiveBayes
import org.apache.spark.mllib.linalg.Vectors
import org.apache.spark.mllib.regression.LabeledPoint
import org.apache.spark.mllib.tree.RandomForest
import org.apache.spark.mllib.tree.model.RandomForestModel
import org.apache.spark.mllib.classification.LogisticRegressionWithLBFGS
import org.apache.spark.mllib.evaluation.MulticlassMetrics
import org.apache.spark.mllib.tree.DecisionTree
import org.apache.spark.mllib.tree.model.DecisionTreeModel
data: org.apache.spark.rdd.RDD[String] = /FileStore/tables/7lxvhelo1487641043634/data_2_mhealth.csv MapPartitionsRDD[1142] at textFile at <console>:71
Command took 0.21 seconds -- by yogesh536@hotmail.com at 2/23/2017, 7:05:15 PM on My Cluster
```

Fig. 30 Import the libraries

```
//To find the headers
val header = data.first;

//To remove the header
val data1 = data.filter(_(0) != header(0));

(1) Spark Jobs

header: String = chest_accX,chest_accY,chest_accZ,ecg1,ecg2,la_accX,la_accY,la_accZ,la_gyroX,la_gyroY,la_gyroZ,la_magX,la_magY,la_magZ,rla_accX,rla_accY,rla_acc
Z,rla_gyroX,rla_gyroY,rla_gyroZ,rla_magX,rla_magY,rla_magZ,movement
data1: org.apache.spark.rdd.RDD[String] = MapPartitionsRDD[499] at filter at <console>:62
Command took 0.81 seconds -- by yogesh536@hotmail.com at 2/23/2017, 6:58:22 PM on My Cluster
```

Fig. 31 Filter header from RDD

The MLlib algorithms used in this section require the input data to be in the form of a Resilient Distributed Dataset (RDD) of LabeledPoints. The way the data is read needs to be very specific to be able to proceed with the modelling. Figure 30 shows all the necessary library imports to be used and the Scala code to read data from the csv file as an RDD. The headers are located and removed so as to obtain only the numerical data in the RDD and further processing and application of Machine Learning algorithms as shown in Fig. 31. The data has to be parsed to obtain the label and feature vectors (RDD of LabeledPoints) as shown in Fig. 32. The data is then split into the training and test sets to be used for each classification model. The training and test sets are given 60 and 40% of the dataset respectively. The allocation is done using a random shuffling as shown in Fig. 33.

This process is identical for all the algorithms shown in the following sub-sections.

Illustration 1: Random Forest Classification with MLlib and Spark on Databricks

The model is then built using the training data set as input as shown in Fig. 34.

With the training model obtained, prediction can be performed on the test set data. A comparison can then be performed on the predicted and already known

```
> val parsedData = data1.map( line => {
    val parts = line.split(',')
    LabeledPoint(parts(parts.length-1).toDouble, Vectors.dense(parts.slice(0, parts.length-2).map(_.toDouble)))
  })

parsedData: org.apache.spark.rdd.RDD[org.apache.spark.mllib.regression.LabeledPoint] = MapPartitionsRDD[501] at map at <console>:62
Command took 0.38 seconds -- by yogesh536@hotmail.com at 2/23/2017, 6:58:28 PM on My Cluster
```

Fig. 32 Parse RDD to obtain RDD of LabeledPoints with label and feature vector

```
> val splits = parsedData.randomSplit(Array(0.6, 0.4), seed = 11L)
  val training = splits(0)
  val test = splits(1)

splits: Array[org.apache.spark.rdd.RDD[org.apache.spark.mllib.regression.LabeledPoint]] = Array(MapPartitionsRDD[502] at randomSplit at <console>:70, MapPartitionsRDD[503] at randomSplit at <console>:70)
training: org.apache.spark.rdd.RDD[org.apache.spark.mllib.regression.LabeledPoint] = MapPartitionsRDD[502] at randomSplit at <console>:70
test: org.apache.spark.rdd.RDD[org.apache.spark.mllib.regression.LabeledPoint] = MapPartitionsRDD[503] at randomSplit at <console>:70
Command took 0.22 seconds -- by yogesh536@hotmail.com at 2/23/2017, 6:58:31 PM on My Cluster
```

Fig. 33 Split into training and test sets

```
> val numClasses = 13
  val categoricalFeaturesInfo = Map[Int, Int]()
  val numTrees = 10 // Use more in practice.
  val featureSubsetStrategy = "auto" // Let the algorithm choose.
  val impurity = "gini"
  val maxDepth = 20
  val maxBins = 32

  val modelRF = RandomForest.trainClassifier(parsedData, numClasses, categoricalFeaturesInfo, numTrees, featureSubsetStrategy, impurity, maxDepth, maxBins)

▸ (23) Spark Jobs
numClasses: Int = 13
categoricalFeaturesInfo: scala.collection.immutable.Map[Int,Int] = Map()
numTrees: Int = 10
featureSubsetStrategy: String = auto
impurity: String = gini
maxDepth: Int = 20
maxBins: Int = 32
modelRF: org.apache.spark.mllib.tree.model.RandomForestModel =
TreeEnsembleModel classifier with 10 trees

Command took 1.24 minutes -- by yogesh536@hotmail.com at 2/23/2017, 6:58:35 PM on My Cluster
```

Fig. 34 Training of random forest model

"label". Figure 35 shows the step where the model is used to predict the "label" for the test set and compute the test error. The computation of the mean squared error is shown in Fig. 36.

The confusion matrix can also be obtained as shown in Fig. 37.
The overall statistics can be computed as shown in Fig. 38.
The precision by label can be obtained as shown in Fig. 39.
The recall by label can be obtained as shown in Fig. 40.
The false positive rate by label can be obtained as shown in Fig. 41.

```
> val labelAndPreds = test.map { point =>
    val prediction = modelRF.predict(point.features)
    (point.label, prediction)
  }
  val testErr = labelAndPreds.filter(r => r._1 != r._2).count.toDouble / test.count()
  println("Test Error = " + testErr)
  println("Learned classification forest model:\n" + modelRF.toDebugString)

▸ (2) Spark Jobs

Test Error = 0.028756228543139722
Learned classification forest model:
TreeEnsembleModel classifier with 10 trees
```

Fig. 35 Prediction using training model on the test data set

```
> val labelAndPreds = test.map { point =>
    val prediction = modelRF.predict(point.features)
    (point.label, prediction)
  }
  val testErr = labelAndPreds.filter(r => r._1 != r._2).count.toDouble / test.count()
  println("Test Error = " + testErr)
  println("Learned classification forest model:\n" + modelRF.toDebugString)

▸ (2) Spark Jobs

Test Error = 0.028756228543139722
Learned classification forest model:
TreeEnsembleModel classifier with 10 trees
```

Fig. 36 Computation of the mean squared error

```
> // Instantiate metrics object
  val metrics = new MulticlassMetrics(labelAndPreds)

  // Confusion matrix
  println("Confusion matrix:")
  println(metrics.confusionMatrix)

▸ (2) Spark Jobs

Confusion matrix:
63353.0  17.0    26.0    10.0    7.0     3.0     12.0    12.0    ... (13 total)
455.0    2427.0  0.0     0.0     0.0     0.0     0.0     0.0     ...
283.0    0.0     2428.0  0.0     0.0     0.0     0.0     0.0     ...
362.0    0.0     0.0     2441.0  0.0     0.0     0.0     0.0     ...
330.0    0.0     0.0     0.0     2447.0  0.0     0.0     0.0     ...
4.0      0.0     0.0     0.0     0.0     2437.0  0.0     0.0     ...
141.0    0.0     0.0     0.0     0.0     0.0     2066.0  0.0     ...
173.0    0.0     0.0     0.0     0.0     0.0     0.0     2219.0  ...
287.0    0.0     0.0     0.0     0.0     0.0     0.0     0.0     ...
198.0    0.0     0.0     0.0     0.0     0.0     0.0     0.0     ...
197.0    0.0     0.0     0.0     0.0     0.0     0.0     0.0     ...
128.0    0.0     0.0     0.0     0.0     0.0     0.0     0.0     ...
15.0     0.0     0.0     0.0     0.0     0.0     0.0     0.0     ...
metrics: org.apache.spark.mllib.evaluation.MulticlassMetrics = org.apache.spark.mllib.evaluation.MulticlassMetrics@38d05f87
Command took 11.14 seconds -- by yogesh536@hotmail.com at 2/23/2017, 7:45:52 PM on My Cluster
```

Fig. 37 Confusion matrix

```
> // Overall Statistics
val accuracy = metrics.accuracy
println("Summary Statistics")
println(s"Accuracy = $accuracy")

▸ (1) Spark Jobs
Summary Statistics
Accuracy = 0.9712437714568602
accuracy: Double = 0.9712437714568602
Command took 5.83 seconds -- by yogesh536@hotmail.com at 2/23/2017, 7:46:23 PM on My Cluster
```

Fig. 38 Overall statistics

```
> // Precision by label
val labels = metrics.labels
labels.foreach { l =>
  println(s"Precision($l) = " + metrics.precision(l))
}

▸ (1) Spark Jobs
Precision(0.0) = 0.9609713921669751
Precision(1.0) = 0.9930441898527005
Precision(2.0) = 0.9894050529747351
Precision(3.0) = 0.9959200326397388
Precision(4.0) = 0.9971475142624286
Precision(5.0) = 0.9987704918032787
Precision(6.0) = 0.9942252165543792
Precision(7.0) = 0.994621246077992
Precision(8.0) = 0.9969578444154715
Precision(9.0) = 1.0
Precision(10.0) = 0.9995840266222962
Precision(11.0) = 0.9992076069730587
Precision(12.0) = 0.9975278121137207
labels: Array[Double] = Array(0.0, 1.0, 2.0, 3.0, 4.0, 5.0, 6.0, 7.0, 8.0, 9.0, 10.0, 11.0, 12.0)
Command took 10.43 seconds -- by yogesh536@hotmail.com at 2/23/2017, 7:46:47 PM on My Cluster
```

Fig. 39 Precision by label

Illustration 2: Decision Tree Classification with MLlib and Spark on Databricks

The model is built using the training data set as input as shown in Fig. 42.

With the training model obtained, prediction can be performed on the test set data. A comparison can then be performed on the predicted and already known "label". Figure 43 shows the step where the model is used to predict the "label" for the test set and compute the test error. The computation of the mean squared error is shown in Fig. 44.

The confusion matrix can also be obtained as shown in Fig. 45.

The overall statistics can be computed as shown in Fig. 46.

The precision by label can be obtained as shown in Fig. 47.

```
// Recall by label
labels.foreach { l =>
  println(s"Recall($l) = " + metrics.recall(l))
}
```

```
Recall(0.0) = 0.9984397654920255
Recall(1.0) = 0.8421235253296322
Recall(2.0) = 0.8956104758391737
Recall(3.0) = 0.8708526578665715
Recall(4.0) = 0.8811667266834714
Recall(5.0) = 0.9983613273248668
Recall(6.0) = 0.9361123697326688
Recall(7.0) = 0.927675585284281
Recall(8.0) = 0.8888027896164278
Recall(9.0) = 0.9238754325259516
Recall(10.0) = 0.9242307692307692
Recall(11.0) = 0.9516981132075472
Recall(12.0) = 0.9817518248175182
```

Fig. 40 Recall by label

```
// False positive rate by label
labels.foreach { l =>
  println(s"FPR($l) = " + metrics.falsePositiveRate(l))
}
```

```
FPR(0.0) = 0.08731801676451624
FPR(1.0) = 1.8881126647933628E-4
FPR(2.0) = 2.8822277403334516E-4
FPR(3.0) = 1.109680855785876E-4
FPR(4.0) = 7.765525504204478E-5
FPR(5.0) = 3.315723159221026E-5
FPR(6.0) = 1.322867977775818E-4
FPR(7.0) = 1.3255713764954101E-4
FPR(8.0) = 7.74867719010826E-5
FPR(9.0) = 0.0
FPR(10.0) = 1.1071867491889857E-5
FPR(11.0) = 2.2156000398808007E-5
FPR(12.0) = 2.17162339707048E-5
```

Fig. 41 False positive rate by label

```
> // Train a DecisionTree model.
  //  Empty categoricalFeaturesInfo indicates all features are continuous.
  val numClasses = 13
  val categoricalFeaturesInfo = Map[Int, Int]()
  val impurity = "gini"
  val maxDepth = 5
  val maxBins = 32

  val modelDT = DecisionTree.trainClassifier(parsedData, numClasses, categoricalFeaturesInfo,impurity, maxDepth, maxBins)

▸ (8) Spark Jobs
numClasses: Int = 13
categoricalFeaturesInfo: scala.collection.immutable.Map[Int,Int] = Map()
impurity: String = gini
maxDepth: Int = 5
maxBins: Int = 32
modelDT: org.apache.spark.mllib.tree.model.DecisionTreeModel = DecisionTreeModel classifier of depth 5 with 63 nodes
```

Fig. 42 Training of decision tree model

```
> // Evaluate model on test instances and compute test error
  val labelAndPreds = test.map { point =>
    val prediction = modelDT.predict(point.features)
    (point.label, prediction)
  }
  val testErr = labelAndPreds.filter(r => r._1 != r._2).count().toDouble / test.count()
  println("Test Error = " + testErr)
  println("Learned classification tree model:\n" + modelDT.toDebugString)

▸ (2) Spark Jobs
Test Error = 0.20314467439382689
Learned classification tree model:
DecisionTreeModel classifier of depth 5 with 63 nodes
```

Fig. 43 Prediction using training model on the test data set

```
> val testMSE = labelAndPreds.map{ case(v, p) => math.pow((v - p), 2)}.mean()
  println("Test Mean Squared Error = " + testMSE)

▸ (1) Spark Jobs
Test Mean Squared Error = 9.233482925989286
testMSE: Double = 9.233482925989286
```

Fig. 44 Computation of the mean squared error

```
> // Instantiate metrics object
  val metrics = new MulticlassMetrics(labelAndPreds)

  // Confusion matrix
  println("Confusion matrix:")
  println(metrics.confusionMatrix)
```

```
▸ (2) Spark Jobs
Confusion matrix:
63484.0  19.0    26.0    8.0     1.0     5.0     8.0     7.0     ... (13 total)
361.0    2425.0  0.0     0.0     0.0     0.0     0.0     0.0     ...
257.0    0.0     2428.0  0.0     0.0     0.0     0.0     0.0     ...
334.0    0.0     0.0     2443.0  0.0     0.0     0.0     0.0     ...
333.0    0.0     0.0     0.0     2453.0  0.0     0.0     0.0     ...
2.0      0.0     0.0     0.0     0.0     2435.0  0.0     0.0     ...
115.0    0.0     0.0     0.0     0.0     0.0     2070.0  0.0     ...
168.0    0.0     0.0     0.0     0.0     0.0     0.0     2224.0  ...
239.0    0.0     0.0     0.0     0.0     0.0     0.0     0.0     ...
230.0    0.0     0.0     0.0     0.0     0.0     0.0     0.0     ...
237.0    0.0     0.0     0.0     0.0     0.0     0.0     0.0     ...
159.0    0.0     0.0     0.0     0.0     0.0     0.0     0.0     ...
7.0      0.0     0.0     0.0     0.0     0.0     0.0     0.0     ...
```

Fig. 45 Confusion matrix

Fig. 46 Overall statistics

```
> // Overall Statistics
  val accuracy = metrics.accuracy
  println("Summary Statistics")
  println(s"Accuracy = $accuracy")
```

```
▸ (1) Spark Jobs
Summary Statistics
Accuracy = 0.972761222139713
accuracy: Double = 0.972761222139713
```

The recall by label can be obtained as shown in Fig. 48.

The false positive rate by label can be obtained as shown in Fig. 49.

Illustration 3: Multinomial Logistic Regression Classification with MLlib and Spark on Databricks

The model is built using the training data set as input as shown in Fig. 50. With the training model obtained, prediction can be performed on the test set data. A comparison can then be performed on the predicted and already known "label".

```
// Precision by label
val labels = metrics.labels
labels.foreach { l =>
  println(s"Precision($l) = " + metrics.precision(l))
}
```

▸ (1) Spark Jobs

```
Precision(0.0) = 0.9629584685859903
Precision(1.0) = 0.9922258592471358
Precision(2.0) = 0.9894050529747351
Precision(3.0) = 0.9967360261117911
Precision(4.0) = 0.9995925020374898
Precision(5.0) = 0.9979508196721312
Precision(6.0) = 0.9961501443695862
Precision(7.0) = 0.9968623935454953
Precision(8.0) = 0.9973924380704041
Precision(9.0) = 1.0
Precision(10.0) = 1.0
Precision(11.0) = 0.998811410459588
Precision(12.0) = 0.992583436341162
labels: Array[Double] = Array(0.0, 1.0, 2.0, 3.0, 4.0, 5.0, 6.0, 7.0, 8.0, 9.0, 10.0, 11.0, 12.0)
```

Fig. 47 Precision by label

```
// Recall by label
labels.foreach { l =>
  println(s"Recall($l) = " + metrics.recall(l))
}
```

```
Recall(0.0) = 0.8309440462299272
Recall(2.0) = 0.6726342710997443
Recall(3.0) = 0.7667610953729934
Recall(5.0) = 0.49122807017543857
Recall(6.0) = 0.7913513513513514
Recall(7.0) = 0.6051660516605166
Recall(8.0) = 0.640325865580448
Recall(9.0) = 0.8121845710165826
Recall(10.0) = 0.6209311907704985
Recall(11.0) = 0.5853827502597853
```

Fig. 48 Recall by label

The confusion matrix can also be obtained as shown in Fig. 51.
The overall statistics can be computed as shown in Fig. 52.
The precision by label can be obtained as shown in Fig. 53.
The recall by label can be obtained as shown in Fig. 54.
The false positive rate by label can be obtained as shown in Fig. 55.
The F-measure by label can be computed as shown in Fig. 56.
The weighted statistics can be computed as shown in Fig. 57.

```
// False positive rate by label
labels.foreach { l =>
  println(s"FPR($l) = " + metrics.falsePositiveRate(l))
}
```

```
FPR(0.0) = 0.2918159667478859
FPR(2.0) = 9.731543624161074E-4
FPR(3.0) = 1.6714581801163334E-4
FPR(5.0) = 0.025974025974025976
FPR(6.0) = 0.014631389003630672
FPR(7.0) = 0.006551452737531732
FPR(8.0) = 0.008058454191722674
FPR(9.0) = 0.0016639858006545O1
FPR(10.0) = 0.009912478451133801
FPR(11.0) = 0.009263373022925183
```

Fig. 49 False positive rate by label

```
// Run training algorithm to build the model
val modelLRLBFGS = new LogisticRegressionWithLBFGS()
  .setNumClasses(13)
  .run(training)

// Compute raw scores on the test set
val predictionAndLabels = test.map { case LabeledPoint(label, features) =>
  val prediction = modelLRLBFGS.predict(features)
  (prediction, label)
}
```

```
(50) Spark Jobs
modelLRLBFGS: org.apache.spark.mllib.classification.LogisticRegressionModel = org.apache.spark.mllib.classification.LogisticRegressionModel: intercept = 0.0, numFeatures = 264, numClasses = 13, threshold = 0.5
predictionAndLabels: org.apache.spark.rdd.RDD[(Double, Double)] = MapPartitionsRDD[441] at map at <console>:65
```

Fig. 50 Training of decision tree model

1.2.3 Comparative Analysis of Machine Learning Algorithms

The Machine Learning techniques used in this work together with the open source tools are: Naïve Bayes, Random Forest, Deep Learning, Decision Tree, and Multinomial Logistic Regression. The description, advantages, and limitations are given in Table 8.

1.3 Conclusions and Discussions

The aim of this Chapter was to present the various analytic tools which can be used for the processing of big data, and machine learning tools for analysing and

```
> // Instantiate metrics object
val metrics = new MulticlassMetrics(predictionAndLabels)

// Confusion matrix
println("Confusion matrix:")
println(metrics.confusionMatrix)
```

```
▸ (2) Spark Jobs
Confusion matrix:
58887.0  879.0  423.0  890.0   321.0  140.0  475.0   543.0  ... (13 total)
1488.0   956.0  0.0    0.0     0.0    0.0    0.0     0.0    ...
2454.0   0.0    0.0    0.0     0.0    0.0    0.0     0.0    ...
0.0      0.0    0.0    2451.0  0.0    0.0    0.0     0.0    ...
2356.0   4.0    0.0    0.0     94.0   0.0    0.0     0.0    ...
2209.0   13.0   0.0    0.0     35.0   30.0   5.0     15.0   ...
790.0    0.0    0.0    0.0     0.0    0.0    1287.0  0.0    ...
1470.0   0.0    1.0    0.0     0.0    0.0    0.0     760.0  ...
998.0    0.0    0.0    0.0     0.0    0.0    10.0    0.0    ...
59.0     0.0    1.0    0.0     0.0    0.0    0.0     1.0    ...
1177.0   0.0    10.0   0.0     0.0    0.0    0.0     15.0   ...
1095.0   6.0    79.0   0.0     52.0   26.0   4.0     2.0    ...
339.0    37.0   5.0    3.0     84.0   0.0    29.0    30.0   ...
```

Fig. 51 Confusion matrix

Fig. 52 Overall statistics

```
> // Overall Statistics
val accuracy = metrics.accuracy
println("Summary Statistics")
println(s"Accuracy = $accuracy")
```

```
▸ (1) Spark Jobs
Summary Statistics
Accuracy = 0.7597584993381332
accuracy: Double = 0.7597584993381332
```

obtaining insights from big data. Essentially, the use of open-source tools: H2O and Spark MLlib, by developing classification based machine learning algorithms for M-Health data, has been illustrated. Random Forest, Naïve Bayes, and Deep Learning algorithms have been used on the data in H2O. Random Forest, Decision Trees, and Multinomial Logistic Regression Classification algorithms have been used with the data in Spark MLlib. The main contribution in this Chapter is that the illustrations demonstrate the flows for developing, training, and testing

```
// Precision by label
val labels = metrics.labels
labels.foreach { l =>
  println(s"Precision($l) = " + metrics.precision(l))
}
```

▸ (1) Spark Jobs

```
Precision(0.0) = 0.8031286653391888
Precision(1.0) = 0.5044854881266491
Precision(2.0) = 0.0
Precision(3.0) = 0.7329545454545454
Precision(4.0) = 0.16040955631399317
Precision(5.0) = 0.15306122448979592
Precision(6.0) = 0.7110497237569061
Precision(7.0) = 0.5563689604685212
Precision(8.0) = 0.6561224489795918
Precision(9.0) = 0.7551679586563308
Precision(10.0) = 0.4979114452798663
Precision(11.0) = 0.5974963890226288
Precision(12.0) = 0.20903954802259886
labels: Array[Double] = Array(0.0, 1.0, 2.0, 3.0, 4.0, 5.0, 6.0, 7.0, 8.0, 9.0, 10.0, 11.0, 12.0)
```

Fig. 53 Precision by label

```
// Recall by label
labels.foreach { l =>
  println(s"Recall($l) = " + metrics.recall(l))
}
```

```
Recall(0.0) = 0.8932287716530656
Recall(1.0) = 0.3911620294599018
Recall(2.0) = 0.0
Recall(3.0) = 1.0
Recall(4.0) = 0.03830480847595762
Recall(5.0) = 0.012295081967213115
Recall(6.0) = 0.6193455245428297
Recall(7.0) = 0.340654415060511
Recall(8.0) = 0.5588874402433724
Recall(9.0) = 0.9729504785684561
Recall(10.0) = 0.49584026622296173
Recall(11.0) = 0.4916798732171157
Recall(12.0) = 0.09147095179233622
```

Fig. 54 Recall by label

```
> // False positive rate by label
  labels.foreach { l =>
    println(s"FPR($l) = " + metrics.falsePositiveRate(l))
  }
```

```
FPR(0.0) = 0.5347682732560294
FPR(1.0) = 0.010378557612600165
FPR(2.0) = 0.00573702536892721
FPR(3.0) = 0.009870893575628951
FPR(4.0) = 0.005438567401757586
FPR(5.0) = 0.0018346798704671802
FPR(6.0) = 0.005757312226857917
FPR(7.0) = 0.006682251235003529
FPR(8.0) = 0.007437815886468472
FPR(9.0) = 0.008374210084404967
FPR(10.0) = 0.013279566922609513
FPR(11.0) = 0.009248299131589137
FPR(12.0) = 0.0030398436651829334
```

Fig. 55 False positive rate by label

```
> // F-measure by label
  labels.foreach { l =>
    println(s"F1-Score($l) = " + metrics.fMeasure(l))
  }
```

```
F1-Score(0.0) = 0.8457859358841779
F1-Score(1.0) = 0.4406545286932473
F1-Score(2.0) = 0.0
F1-Score(3.0) = 0.8459016393442622
F1-Score(4.0) = 0.0618421052631579
F1-Score(5.0) = 0.022761760242792105
F1-Score(6.0) = 0.6620370370370371
F1-Score(7.0) = 0.422574367528496
F1-Score(8.0) = 0.6036141750762731
F1-Score(9.0) = 0.85033642480451
F1-Score(10.0) = 0.4968736973739058
F1-Score(11.0) = 0.5394479460986742
F1-Score(12.0) = 0.12725709372312985
```

Fig. 56 F-measure by label

```
> // Weighted stats
println(s"Weighted precision: ${metrics.weightedPrecision}")
println(s"Weighted recall: ${metrics.weightedRecall}")
println(s"Weighted F1 score: ${metrics.weightedFMeasure}")
println(s"Weighted false positive rate: ${metrics.weightedFalsePositiveRate}")

Weighted precision: 0.7066478199852897
Weighted recall: 0.7597584993381331
Weighted F1 score: 0.726725431961217
Weighted false positive rate: 0.38160579557566016
```

Fig. 57 Weighted statistics

Table 8 Description, advantages, and limitations of machine learning algorithms used

Algorithm	Description	Advantages	Disadvantages/limitations
Naïve Bayes	It is a multiclass classification algorithm which is simple and assumes that every pair of features is independent. It calculates the conditional probability distribution for every feature given label, inside each pass to the training data. After that it employs the Bayes' theorem to obtain the conditional probability distribution of the label given an observation and uses it for prediction	– The training of this algorithm can performed in a very efficient manner – Classifiers based on Naive Bayes have performed satisfactorily in several non-trivial real-world applications although they rely heavily on assumptions	– The assumption that any feature pair is independent for a specific output class, is very strong for a given data distribution – An estimation for the probability of any value of a feature has to be made using a statistical approach. Consequently, when data available are scarce, probabilistic estimates tend towards either the lower limit i.e. 0 or the upper limit i.e. 1. Fluctuations in numerical results then occur and poor results are obtained
Random forest	Random forests form part of a group of learning techniques for classification, regression and similar related operations. The rationale behind these methods is the construction of multiple decision trees at the training instant. For classification problems they output the mode of	– Provides very good accuracy in the classification of several data sets – It executes with good efficiency efficiently on big databases – It can handle Thousands of input variables can be supported deleting any variable	When applied to specific databases with noisy classification/regression operations, it can lead to an over fitting problem

(continued)

Table 8 (continued)

Algorithm	Description	Advantages	Disadvantages/limitations
	the classes that were being classified. For regression problems, they output the mean prediction or regression of the individual trees		
Deep learning	In Deep learning a concatenation of multiple layers that are capable of non-linear processing are employed for feature extraction and transformation. The output of layer $n - 1$ is fed to layer n as input. As such a hierarchical representation is obtained whereby the features of higher levels are derived from lower levels. Both supervised and unsupervised algorithms can be used for applications such as pattern analysis and classification	– Provides significantly better performance for problems in speech recognition, language, vision, and gaming – Decreases the dependence on time consuming processes such as feature engineering – It is equipped with techniques such as convolutional neural networks, recurrent neural networks and long short-term memory. These allow it to easily adapt to problems such as Vision and language	– Huge amounts of data are required so that it really outperform other techniques – The training is very time-consuming and requires major computational costs
Decision tree	The algorithm uses a tree-like model for the representation of classification examples. The tree is learned by performing recursive splits at each node based on a threshold value. The recursive process is halted when a maximum depth is reached or when the condition that splitting a node no longer adds value to predictions is reached	– Very little data preparation is required – Statistical tests can be used for the validation of the models – Has very good performance with large volumes of data – Has the capability of handling categorical as well as numerical data types – Interpretation can be done in an easy way	– The trees deviate severely with little changes in training data, which can cause predictions to experience big changes – The accuracy of trees tend to be lower than other approaches – There is always the risk of overfitting data

(continued)

Table 8 (continued)

Algorithm	Description	Advantages	Disadvantages/limitations
Multinomial logistic regression	It is a classification technique which extends the logistic regression mechanism to cases where more than 2 classes need to be dealt with. The model is specifically used to compute and predict the probabilities of the different outcomes/categories based on the dependent features in the training set	– There is the possibility of using Ordinary Least Squares (OLS) regression with multiple categorical labels – Categories can be deleted from the set	– The OLS model can make predictions without sense due to the non-continuous nature of the dependent variable – Information, power, and data is lost with deletions as well as causing the samples to be biased

mathematical models in order to obtain insights from M-Health data using open source tools developed for handling big data.

H2O being configured locally, provides the possibility to use a larger dataset as compared to MLlib on Databricks since an online community version of the latter has been used. The M-Health dataset used with H2O has 982,273 data rows which have been used for model training purposes and 233,472 data rows used for model testing purposes. Based on these data-sets, the accuracies of the Random Forest, Naïve Bayes, and Deep Learning models are 75.99, 75.47, and 75.81% respectively.

The MHealth data-set used with MLlib on Databricks consists of 233,472 data rows out of which 70% is used for model training purposes and the remaining used for model testing purposes. Based on these data-sets, the accuracies of the Random Forest, Decision Tree, and Multinomial Logistic Regression models are 97.12, 97.28, and 75.98% respectively.

References

1. Russom, P. (2011). Executive summary: Big data analytics. Renton: The Data Warehouse Institute (TWDI).
2. Landset, S., Khoshgoftaar, T. M., Ritcher, A. M., & Hasanin, T. (2015). A survey of open source tools for machine learning with big data in the Hadoop ecosystem. *Journal of Big Data, 2*(24), 1–36.
3. The R Foundation. (2017). *The R project for statistical computing*. Retrieved January 22, 2017 from https://www.r-project.org/
4. Machine Learning Group at the University of Waikato Weka 3. (n.d.). *Data mining software in Java*. Retrieved January 22, 2017 from http://www.cs.waikato.ac.nz/ml/weka/

5. International Data Corporation. (2017). *Discover the digital universe of opportunities, rich data and the increasing value of the Internet of things*. Retrieved January 22, 2017 from https://www.emc.com/leadership/digital-universe/index.htm
6. Vidhya, S., Sarumathi, S., & Shanthi, N. (2014). Comparative analysis of diverse collection of big data analytics tools. *World Academy of Science, Engineering and Technology: International Journal of Computer, Electrical, Automation, Control and Information Engineering, 8*(9), 1646–1652.
7. Kamal, S., Ripon, S. H., Dey, N., Ashour, A. S., & Santhi, V. (2016). A MapReduce approach to diminish imbalance parameters for big deoxyribonucleic acid dataset. *Computer Methods and Programs in Biomedicine, 131,* 161–206.
8. Kamal, S., Dey, N., Ashour, A. S., Ripon, S. H., & Balas, V. E. (2016). FbMapping: An automated system for monitoring Facebook data. *Neural Network World*. doi:10.14311/NNW.2017.27.002
9. Kamal, S., Nimmy, S. F., Hossain, M. I., Dey, N., Ashour, A. S., & Santhi, V. (2016). ExSep: An exon separation process using neural skyline filter. In *International conference on electrical, electronics, and optimization techniques* (*ICEEOT*).
10. Bhattacherjee, A., Roy, S., Paul, S., Roy, P., Kaussar, N., & Dey, N. (2016). Classification approach for breast cancer detection using back propagation neural network: A study. In *Biomedical image analysis and mining techniques for improved health outcomes* (p. 12). IGI-Global.
11. Chatterjee, S., Ghosh, S., Dawn, S., Hore, S., & Dey, N. (in press). Optimized forest type classification: A machine learning approach. In *3rd international conference on information system design and intelligent applications*. Vishakhapatnam: Springer AISC.
12. Kimbahune, V. V., Deshpandey, A. V., & Mahalle, P. N. (2017). Lightweight key management for adaptive addressing in next generation internet. *International Journal of Ambient Computing and Intelligence (IJACI), 8*(1), 20.
13. Najjar, M., Courtemanche, F., Haman, H., Dion, A., & Bauchet, J. (2009). Intelligent recognition of activities of daily living for assisting memory and/or cognitively impaired elders in smart homes. *International Journal of Ambient Computing and Intelligence (IJACI), 1*(4), 17.
14. IBM. (n.d.). *Big data at the speed of business* [Online]. Retrieved January 22, 2017 from https://www-01.ibm.com/software/data/bigdata/
15. Zikopoulos, P., Deroos, D., Parasuraman, K., Deutsch, T., Corrigan, D., & Giles, J. (2013). *Harness the power of big data: The IBM big data platform*. New York: Mc-Graw Hill Companies.
16. IBM. (n.d.). *IBM big data platform* [Online]. Retrieved January 22, 2017 from https://www-01.ibm.com/software/in/data/bigdata/enterprise.html
17. IBM. (n.d.). *Bringing big data to the enterprise* [Online]. Retrieved March 12, 2017 from https://www-01.ibm.com/software/sg/data/bigdata/enterprise.html
18. Hewlett-Packard Enterprise. (2017). *Big data services: Build an insight engine* [Online]. Retrieved January 22, 2017 from https://www.hpe.com/us/en/services/consulting/big-data.html
19. Hewlett-Packard Enterprise. (2017). *Big data software* [Online]. Retrieved March 13, 2017 from https://saas.hpe.com/en-us/software/big-data-analytics-software.
20. Hewlett-Packard. (2017). *HAVEN big data platform for developers* [Online]. Retrieved March 15, 2017 from http://www8.hp.com/us/en/developer/HAVEn.html?jumpid=reg_r1002_usen_c-001_title_r0002
21. SAP. (2017). *Unlock business potential from big data more quickly and easily* [Online]. Retrieved January 22, 2017 from http://www.sap.com/romania/documents/2015/08/f63628c9-3c7c-0010-82c7-eda71af511fa.html
22. SAP. (2017). *Unlock business potential from your Big Data faster and easier with SAP HANA Vora* [Online]. Retrieved January 22, 2017 from http://www.sap.com/hk/product/data-mgmt/hana-vora-hadoop.html

23. SAP Community. (n.d.). *What is SAP HANA?* [Online]. Retrieved March 15, 2017 from https://archive.sap.com/documents/docs/DOC-60338
24. Predictive Analytics Today. (n.d.). *50 big data platforms and big data analytics software* [Online]. Retrieved January 22, 2017 from http://www.predictiveanalyticstoday.com/bigdata-platforms-bigdata-analytics-software/
25. Microsoft. (2017). *Big data and analytics* [Online]. Retrieved January 22, 2017 from https://azure.microsoft.com/en-us/solutions/big-data/
26. Microsoft Developer. (2017). *Microsoft azure services platform* [Online]. Retrieved March 15, 2017 from https://blogs.msdn.microsoft.com/mikewalker/2008/10/27/microsoft-azure-services-platform/
27. Microsoft Azure. (2017). *Take a look at these innovative stories by world class companies* [Online]. Retrieved March 24, 2017 from https://azure.microsoft.com/en-gb/case-studies/
28. Oracle. (2017). *Big data features* [Online]. Retrieved January 22, 2017 from https://cloud.oracle.com/en_US/big-data/features
29. Oracle. (2017). *Big data in the cloud* [Online]. Retrieved January 22, 2017 from https://cloud.oracle.com/bigdata
30. Pollock, J. (2015). *Take it to the limit: An information architecture for beyond Hadoop* [Online]. Retrieved March 24, 2017 from https://conferences.oreilly.com/strata/big-data-conference-ca-2015/public/schedule/detail/40599
31. Evosys. (2017). *Oracle big data customers success stories* [Online]. Retrieved March 24, 2017 from http://www.evosysglobal.com/big-data-2-level
32. H2O.ai. (n.d.). *Fast scalable machine learning API* [Online]. Retrieved February 17, 2017 from http://h2o-release.s3.amazonaws.com/h2o/rel-tverberg/4/index.html
33. The Apache Software Foundation. (2017). *Apache spark MLlib* [Online]. Retrieved March 21, 2017 from http://spark.apache.org/mllib/
34. LeDell, E. (2015). *High performance machine learning in R with H2O*. Tokyo: ISM HPC on R Workshop.
35. Candel, A., Parmar, V., LeDell, E., & Arora, A. (2016) [Online]. Retrieved February 24, 2017 from http://docs.h2o.ai/h2o/latest-stable/h2o-docs/booklets/DeepLearningBooklet.pdf
36. Nykodym, T., & Maj, P. (2017). *Fast analytics on big data with H2O* [Online]. Retrieved February 20, 2017 from http://gotocon.com/dl/gotoberlin2014/slides/PetrMaj_and_Tomas Nykodym_FastAnalyticsOnBigData.pdf
37. Novet, J. (2014). *0xdata takes $8.9M and becomes H2O to match its open-source machine-learning project* [Online]. Retrieved March 21, 2017 from http://venturebeat.com/2014/11/07/h2o-funding/
38. Cage, D. (2013). *Platfora founder goes in search of big-data answers* [Online]. Retrieved March 21, 2017 from http://blogs.wsj.com/venturecapital/2013/04/15/platfora-founder-goes-in-search-of-big-data-answers/
39. Wilson, A. (1999). *ACM Honors Dr. John M. Chambers of bell labs with the 1998 ACM software system award for creating S system* [Online]. Retrieved March 21, 2017 from http://oldwww.acm.org/announcements/ss99.html
40. Chambers, J., & Hastie, T. (1991). *Statistical models in S*. Brooks Cole: Wadsworth.
41. Schuster, W. (2014). *Cliff click on in-memory processing, 0xdata H20, efficient low latency java and GCs* [Online]. Retrieved March 21, 2017 from https://www.infoq.com/interviews/click-0xdata
42. Click, C. (2016). *Winds of change* [Online]. Retrieved March 21, 2017 from http://www.cliffc.org/blog/2016/02/19/winds-of-change/
43. H2O. (2016). *H2O* [Online]. Retrieved March 21, 2017 from http://0xdata.com/about/
44. Efron, B., & Tibshirani, R. (1994). *An introduction to the bootstrap*. New York: Chapman & Hall/CRC.
45. Hastie, T. J., & Tibshirani, R. J. (1990). *Generalized additive models*. Boca Raton: Chapman & Hall/CRC.
46. Hastie, T., Tibshirani, R., & Friedman, J. H. (2011). *The elements of statistical learning*. New York: Springer.

47. Boyd, S., & Vandenberghe, L. (2004). *Convex optimization* [Online]. Retrieved March 21, 2017 from http://web.stanford.edu/~boyd/cvxbook/bv_cvxbook.pdf
48. Wikepedia. (2017). *H2O (Software)* [Online]. Retrieved March 21, 2017 from https://en.wikipedia.org/wiki/H2O_(software)
49. 0xdata. (2013). *H2O software architecture*. Retrieved March 21, 2017 from http://h2o-release.s3.amazonaws.com/h2o/rel-noether/4/docs-website/developuser/h2o_sw_arch.html
50. Banos, O., Garcia, R., & Saez, A. (2014). *MHEALTH dataset data set. UCI machine learning repository: Center for machine learning and intelligent systems* [Online]. Retrieved February 24, 2017 from https://archive.ics.uci.edu/ml/datasets/MHEALTH+Dataset
51. Banos, O., Garcia, R., Holgado, J. A., Damas, M., Pomares, H., Rojas, I., Saez, A., & Villalonga, C. (2014). mHealthDroid: a novel framework for agile development of mobile health applications. In *Proceedings of the 6th international work-conference on ambient assisted living and active ageing (IWAAL)*, Belfast, Northern Ireland.
52. Banos, O., Villalonga, C., Garcia, R., Saez, R., Damas, M., Holgado, J. A., et al. (2015). Design, implementation, and validation of a novel open framework for agile development of mobile health applications. *Biomedical Engineering OnLine, 14*(S2:S6), 1–20.
53. Databricks. (n.d.). *Making machine learning simple: Building machine learning solutions with databricks* [Online]. Retrieved March 21, 2017 from http://cdn2.hubspot.net/hubfs/438089/Landing_pages/ML/Machine-Learning-Solutions-Brief-160129.pdf
54. Kraska, T., Talwalkar, A., Duchi, J., Grith, R., Franklin, M., & Jordan, M. (2013). Distributed machine-learning system. In *Conference on innovative data systems research*.
55. Meng, X., Bradley, J., Yavuz, B., Sparks, E., Venkataraman, S., Liu, D., et al. (2016). Machine learning in apache spark. *Journal of Machine Learning Research, 17,* 1–7.
56. Gonzalez, J. E., Xin, R. S., Dave, A., Crankshaw, D., Franklin, M. J., & Stoica, I. (2014). raphX: Graph processing in a distributed data framework. In *Conference on operating systems design and implementation*.
57. Blei, D. M., Ng, A. Y., & Jordan, M. I. (2003). Latent dirichlet allocation. *Journal of machine Learning Research, 3,* 993–1022.
58. Bradley, J. (2015). *Topic modeling with LDA: MLlib meets GraphX* [Online]. Retrieved March 21, 2017 from https://databricks.com/blog/2015/03/25/topic-modeling-with-lda-mllib-meets-graphx.html
59. Zaharia, M., Das, T., Li, H., Hunter, T., Shenker, S., Stoica, & I. (2013). Discretized streams: Fault-tolerant streaming computing at scale. In *Symposium on operating systems principles*.
60. Freeman, J. (2015). *Introducing streaming k-means in Apache Spark 1.2* [Online]. Retrieved March 21, 2017 from https://databricks.com/blog/2015/01/28/introducing-streaming-k-means-in-spark-1-2.html
61. Panda, B., Herbach, J. S., Basu, S., Bayardo, R. J. (2009). Planet: Massively parallel learning of tree ensembles with mapreduce. In *International conference on very large databases*.
62. Pedregosa, F., Varoquaux, G., Gramfort, A., Michel, V., Thirion, B., Grisel, O., et al. (2011). Scikit-learn: Machine learning in python. *Journal of machine Learning Research, 12,* 2825–2830.
63. Buitink, L., Louppe, G., Blondel, M., Pedregosa, F., Mueller, A., Grisel, O., et al. (2013). API design for machine learning software: Experiences from the scikit-learn project. In *European conference on machine learning and principles and practices of knowledge discovery in databases*.
64. Sparks, E. R., Talwalkar, A., Smith, V., Kottalam, J., Pan, X., Gonzalez, J. E., et al. (2013). MLI: An API for distributed machine learning. In *International conference on data mining*.
65. Sparks, E. R., Talwalkar, A., Haas, D., Franklin, M. J., Jordan, M. I., & Kraska, T. (2015). Automating model for large scale machine learning. In *Symposium on cloud computing*.
66. Meng, X., Bradley, J., Sparks, E., & Venkataraman, S. (2015). *ML pipelines: A new high-level API for MLlib*. Retrieved March 21, 2017 from https://databricks.com/blog/2015/01/07/ml-pipelines-a-new-high-level-api-for-mllib.html
67. Apache Spark. (n.d.). *ML pipelines* [Online]. Retrieved March 21, 2017 from http://spark.apache.org/docs/latest/ml-pipeline.html

68. Apache Spark. (n.d.). *Extracting, transforming and selecting features* [Online]. Retrieved March 21, 2017 from http://spark.apache.org/docs/latest/ml-features.html
69. Apache Spark. (n.d.). *Classification and regression* [Online]. Retrieved March 21, 2017 from http://spark.apache.org/docs/latest/ml-classification-regression.html
70. Apache Spark. (n.d.). *Clustering* [Online]. Retrieved March 21, 2017 from http://spark.apache.org/docs/latest/ml-clustering.html
71. Apache Spark. (n.d.). *Collaborative filtering* [Online]. Retrieved March 21, 2017 from http://spark.apache.org/docs/latest/ml-collaborative-filtering.html
72. Apache Spark. (n.d.). *ML tuning: Model selection and hyperparameter tuning* [Online]. Retrieved March 21, 2017 from http://spark.apache.org/docs/latest/ml-tuning.html
73. Najafabadi, M. M., Villanustre, F., Khoshgoftaar, T. M., Seliya, N., Wald, R., & Muharemagic, E. (2015). Deep learning applications and challenges in big data analytics. *Journal of Big Data, 2*(1), 1–21.

Real Time Big Data Analytics to Derive Actionable Intelligence in Enterprise Applications

Subramanian Sabitha Malli, Soundararajan Vijayalakshmi and Venkataraman Balaji

Abstract Big data is a large-scale data management and analysis technology. The implementation of this technological advantage helps the enterprises and scientific community to concentrate on the innovative transformation. The extraordinary efficacy of this technology surpasses the competency of traditional relational database management system (RDBMS) and bestows diversified computational techniques to steer the storage bottleneck, noise detection, heterogeneous dataset, etc., it entails various analytic and computational methods to extract the actionable insights from the mountains of data accelerated from different causes. Enterprise resource planning (ERP) or systems applications and products (SAP in data processing) framework coordinates key procedures into a solitary framework and improves the complete inventory network by transmitting customer relationship management (CRM) and supply chain management (SCM) arrangements. Despite the fact that numerous business processes accessible in the enterprise, two continuous business use cases of an enterprise are discussed in this paper. The first model is a machine created data processing model. The general design and the outcomes of various analytics cases discussed in this model. The second model is a business arrangement of heterogeneous human-created information model. The data analytics of this model determines this kind of useful data required for decision making in that specific area. It also gives different new perspectives on the big data analytics and computation methodologies. The last segment communicates the challenges of big data.

Keywords Big data · Big data analytics · IoT · Smart manufacturing

S. Sabitha Malli (✉)
Information Systems at ZF Electronics TVS (India) Pvt Ltd.
and Research Scholar at Bharathiar University, Coimbatore, India
e-mail: drsabimurali@gmail.com

S. Vijayalakshmi
Thiagarajar College of Engineering, Madurai, India
e-mail: svlcse@tce.edu

V. Balaji
ZF Electronics TVS (India) Pvt Ltd., Madurai, India
e-mail: balaji.v@zftvs.com

N. Dey et al. (eds.), *Internet of Things and Big Data Analytics Toward Next-Generation Intelligence*, Studies in Big Data 30,
DOI 10.1007/978-3-319-60435-0_4

1 Introduction

In recent years, the internet of things (IoT) and big data are the most talked technology on the universe. These two advances regularly cooperate to collect information. IoT will tremendously raise the volume of information accessible for exploration by all types of the associations. Be that as it may, there are some key difficulties to overcome before the upcoming benefits are totally figured it out. The IoT and big data are clearly expanding quickly, and are altered to change over a few areas of industry and in everyday life. The IoT is a colossal inflow of huge data because of the connectivity of sensors. The eventual fate of the big data analysis depends on the IoT. The enterprises can convey noteworthy business experiences by deriving actionable business insights from the huge available data.

The key idea is to implement the IoT on the Enterprise Systems in smart manufacturing. The manufacturing system standard must be contemplated to classify the requirements of the decision support system in the distributed environment. Enterprise model and current IT setup are explored to distinguish the technological gaps in executing IoT as an IT framework of a smart factory. IoT paves a path to advance the manufacturing enterprises by improving their systems in the globalized and distributed environment. In any case, the IoT activities are in the fundamental phase in various enterprises and further research is a prerequisite in the implementation zone. The inside and remotely accumulated information through the connected devices conjecture the capability of the big data.

The IoT is denoted as Internet links to the physical world and everyday objects. This development opens up tremendous open doors. Smart objects play the noteworthy part in the vision of the IoT since embedded frameworks and information technology is straightforwardly included in the transformation. IoT can be portrayed as the connectivity between the physical universes, sensors inside or associated with the things and the web through remote or wired system connections.

The term big data alludes to the huge volume of data management and providing knowledge through investigation that surpasses the competency of conventional relational database management system (RDBMS). Big data are not just moving to information stockpiling and administration strategies, additionally, gives capable ongoing analysis and visual displays which consequently infer the quality information required for the business. Big data gets to be important for many organizations. The actions require another scope of uses which can deal with expanding the volume of data and persistently created from different information sources. Big data handles the data that can't be utilized and handle in a traditional way. A traditional database management system possesses less memory and it is difficult to resolve the errors in the data set and computation is exceptionally straightforward. Be that as it may, in big data case, it requires extraordinary focus for data cleaning and the calculation procedure. Constant streaming of data requires decision making about which part of the data streams to be caught for data analysis purpose. Big data analytics ought to be felicitated in liberating competitive advantage in the business environment for the betterment of the business.

In the traditional framework, the insights of various results like sales analysis and inventory levels can be caught through accessible BI (Business Intelligence) tools. Blend of traditional and big data determines action oriented, insightful analytical results required for the business. Planning and forecasting applications similarly extract the knowledge from big data. Enterprises require data analytics to derive insights from this multifarious big data. The term Analytics are for the most part used to get information based choice making. The analysis is utilized for corporate and scholastic related investigates. Despite the fact that both are distinctive sorts of analysis, the common data included in the corporate investigation require learning in data mining, business factual strategies, and visualization to address the queries of corporate pioneers have. The analytics assume an import role in inferring significant insights for finance and operations. It ought to investigate the request identified by customers, products, sales, and so on. Fusion of big business information and open information helped with anticipating the customer behavior in the choice of selection of materials. In any case, if there should arise an occurrence of scholarly research, they require examination to explore the theory and to formulate new hypotheses.

Industry 4.0 is a modern revolution and is the path of IoT and smart manufacturing. Coordinating IoT and big data is a multi-disciplinary movement and it requires a specialized skill set to get the most extreme advantages from the framework. An intelligent network can be setup in connecting in the manufacturing framework, control and relate each other autonomously with altogether reduced impedance by administrators. It has the concrete ability to impact the key requirements of the business and has the pending to renovate industry divisions and economies.

Big data analytics is a method of exploding huge data sets that holds different types of data i.e. big data to reveal all hidden patterns, unknown relationships, market trends, consumer choices and other helpful business related data. The outcomes of the analytics can prompt to more effective marketing, new business opportunities, and improved service to the customer, enhanced operational effectiveness, advantages over rival enterprises and other business benefits.

The main aim of big data analytics is to help organizations to evolve more useful business decisions by empowering data scientists, predictive modelers, and other analytics experts to analyze huge volumes of data from various transactions, and additionally other types of data that might be undiscovered by more traditional Business Intelligence (BI) programs. That data could include web server logs, social media, online trading, social network, internet click data, emails from customers, survey replies, mobile phone call details and machine data generated by sensors and connected to the IoT.

This chapter discusses the usage of IoT, big data and different investigative instruments and strategies for utilizing the gigantic volume of organized and unstructured data produced in the manufacturing environment. Big data analytics play a significant role in deriving the astonishing benefits by providing intelligent insights from the available data to leverage the business and operational knowledge. It gives authentic historical trend and also web monitoring for better decision making in various operational levels of the enterprise. Two use cases of an enterprise is taken as a sample and discussed in this paper. Both the cases accelerate the huge volume of data. First one deals about the data liberated from various machines in the IoT environment. It produces continuous and large volume of data in very short span of time. Another one is the human produced information through an enterprise business framework. The relationship between IoT and big data is discussed in Sect. 2. Architecture, framework and tools related to big data discussed in Sect. 3. Sect. 4 discusses the motivation and importance of data analytics. The industry use-cases, implementation challenges, architecture and significance of analytics are discussed in Sect. 5. Various challenges are discussed in Sect. 6. This chapter concluded in Sect. 7.

2 Relationship Between the IoT and Big Data

The IoT is becoming the next industrial revolution. As indicated by Gartner, revenue generated through IoT products and related services will surpass $300 billion in the year 2020. IoT will generate a huge amount of revenue and data, its effect will be felt over the entire big data world, compelling the enterprises to update current processes and tools, suitable technology to develop to accommodate this extra data volume and get the advantage of the insights from the newly captured data. The huge amount of data that the IoT generates would be useless without the analytic power of the big data. The IoT and big data are linked and bound together by technology and economics. There's no rule saying that IoT and big data must be joined at the hip, however, it makes sense to see them as natural partners because there is no use to operate without predictive analysis for complex machines or devices. That requires big data for the analytics.

The "exponential growth of data" created by IoT makes big data essential. Without the best data gathering, it is not possible for the organizations to analyze the data liberated by sensors. The data generated by machines and devices are often in a raw and simple format, in order to make it useful for analytical decisions the data further needs to be organised, transformed and enriched.

3 Big Data Analytics Architecture, Framework and Tools

Like big data, the big data analytics is also depicted by three primary characteristics: volume, velocity, and variety. There is undoubtedly data will keep on being generated and gathered, consistently leading to the incredible amount of data. Secondly, this data is being collected in real time at a rapid speed. This is the indication of velocity. Third are the different types of data being gathered in a standard format and stored in spreadsheets or relational databases. Considering the 3Vs (volume, velocity, variety), the analytic techniques have also developed to suit these attributes to scale up to the complex and refined analytics needed. Some researchers and professionals have considered a fourth characteristic: veracity. The effect of this is data assurance. Thus the collected data and the analytics are highly reliable and error free.

The big data analytics is entirely different from the traditional business intelligent (BI) tools. The architecture, tools, algorithms and techniques decide the value of the big data analytics. For example, the National Oceanic and Atmospheric Administration (NOAA) used big data analytics to help weather condition, atmosphere and environment and analysis of patterns, and traditional applications. National aeronautics and space (NASA) use analytics for aeronautical and different sorts of research. Pharmaceutical industries are utilizing analytics for drug discovery, analysis of clinical experimental data, side effects and reactions etc., Banking sector uses for investments, loans, customer experiences, etc. Insurance, healthcare, media companies are also using big data for analytics purpose. The architecture and platform, tools, methodologies, and interfaces related issues to be addressed to harness and maximize the potential of big data analytics.

Figure 1 depicts the common architecture of the big data analytics. The first vertical represents different types of sources for big data. The data can be from internal and external sources, often in different formats, located in a different location in numerous traditional and other applications. All this data need to be collected for analytics purpose. The collected raw data required to be transformed. In general, different types of sources liberate big data. The second vertical in the above architecture indicates different services used to call, retrieve and process the data. In a data warehouse, all the data from various sources are collected and made available for further processing. The third one specified different tools and

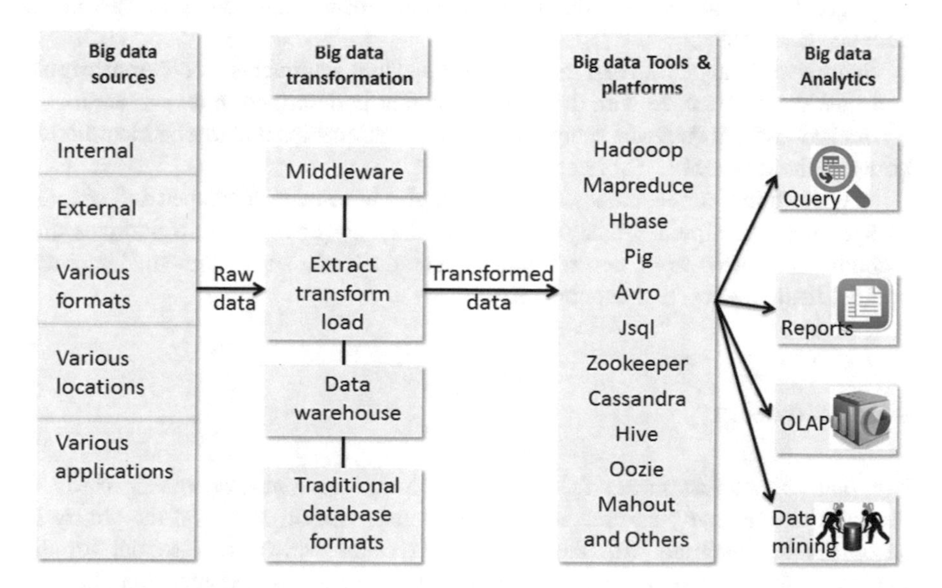

Fig. 1 Common big data analytics architecture

platforms used for big data. Most of the big data tools are open-sourced. The last one represents the technologies used in big data analytics. It includes queries, reports, online analytic processing (OLAP) and data mining. Visualization is the main outcome from the entire data analytics model. Different techniques and technologies have been used and adapted to aggregate, process, analyze and visualize big data. These techniques and technologies are derived from various fields.

Different data analytics techniques are currently in use. Following are the general data analytics techniques

- Association rule learning
- Classification tree analysis
- Genetic algorithms
- Machine learning
- Regression analysis
- Sentimental Analysis
- Social network analysis.

Association Rule Learning is a classification learning technique. It includes the analysis of the data available and discovering the relations between the data. The outcome is the grouping of data with similar characters together.

Classification tree analysis classify the large historical text data chronologically.

Genetic Algorithm is used to identify the most frequently watched videos, TV shows, and other media.

Machine learning can be used to determine the probable outcome of the data. Therefore, it is used in the predictive analysis.

Regression Analysis takes the use of independent variables and how they affect dependent variables.

Sentiment Analysis is used to looks at the actual sentiments of different people and then cross reference with the experience that is described in text or audio.

Social Network Analysis analyzes different relationships that can be identified in between the different social interactions.

A wide variety of techniques and technologies have been developed and adapted to aggregate, manipulate, analyze, and visualize big data. These techniques and technologies draw from several fields, including statistics, computer science, applied mathematics, and economics.

4 Motivation

Big data is not just meant the shift from traditional database management; it involves an effective real-time analytics and visualization tools and the ability to automatically integrate with the traditional systems which are essential for the business support systems/enterprise resource business/customer relationship management (CRM)/management information system. Insights from various data analytics bring the connectivity between the traditional system and the big data to bring the imperative results. The intelligent insights about customers involve their behavioral changes, customer care, and campaign management. Ultimately this will improve the customer's experience in their products. In the previous years, individuals who have the exposure can figure well to catch the business ground. In today's scenario, right business people diagnose more insights to lessen business esteem from the mountains of information. Data analytics will help them in determining better esteem suitable for producing best business analysis outcomes.

The IoT is turning out to be exceptionally noteworthy in encouraging access to different devices and machines in the enterprise environment. This transformation drives towards the digitization. The introduction of IoT converts the traditional production model into most intelligent and effective production environment. Facilitating connection with today's outside partners keeping in mind the end goal to relate with an IoT-implemented manufacturing framework is the main change venture against a smart manufacturing plant. An IoT-based model, claim the hierarchical and closed factory automation pyramid by permitting the above-specified stakeholders to track their services in multiple tier flat production systems [1]. This implies the framework can keep running in a common physical world as opposed to running in an entwined and seriously connected way. The shared physical world provides a space for innovative application development. Enterprises are involved in getting the best esteem from the data using business analytics, cloud administrations and numerous others. Significant difficulties connected with the cyber-physical framework incorporate the reasonableness, network connectivity and the interoperability of engineering frameworks. The absence of a typical modern strategy for production management results to tailor made programming or utilization of a manual methodology. Additionally, there is a necessity of a bringing together hypothesis of nonhomogeneous component and communication systems [1].

The ambiance intelligence concept discussed in [2]. Smart lecture halls and smart university campuses and related architectures are depicted in the paper. In [3], a MapReduce based K-NN classifier method introduced. DNA dataset consists of 90 million pairs. This dataset is used for the reduction. The imbalance data of the DNA dataset reduced to accurate results without losing its accuracy. [4] Discusses a keyword-based sentiment mining using twitter messages. This methodology used to offer a better understanding of the user preferences. This will help the people for marketing campaigns and development activities. Facebook data generates a huge volume of data. A new FbMapping system [5] developed to monitor the Facebook data. Emotions are superfluous and dangerous for superior rational thinking [6, 7]. By researching history and theories connected to emotions, information processing tool [7] is required [7]. The article [8] talks about the analysis of social inferences of scientific and technical developments in the monitoring technologies and social research. The usage of IoT and big data technologies in the healthcare organizations and research were discussed in [9–11].

In 2012, a report titled "The future of manufacturing—Opportunities to drive economic growth" was submitted by the world economic forum. The significance of export manufacturing in the nation is all around examined in the report [12]. The business analyst guaranteed that they are going into the third industrial revolution which concentrates on the digitization of manufacturing additionally called as smart manufacturing. The key component of the smart manufacturing is the IoT. Despite the fact that the M2M correspondence, computerization, PLC, PC based controller and sensor use is as of now actualized in numerous enterprises, for the most part, it is disengaged from IT and operational frameworks. Consequently, opportune choice making and activities are absent in numerous ventures. Taking after a part is vital for each association to drive towards the information investigation.

4.1 Data to Insight

Data is completely useful just when it is deciphered into noteworthy bits of knowledge. The vast majority of the enterprises use information for settling on capable choices.

Right people, right time and right data are the three fundamental key elements required for compelling choice making in the business environment. Figure 2 represents key decision-making elements required in the industrial environment. The inner triangle in the figure indicates various decisions to be taken in the organization. The quicker decisions can be taken by providing right data to the right people at the right time. Three basic required factors are represented as people, data and time. Availability to be known to the people at the correct time, the demand can be derived from the available data and to be known to the people. Further, the collected data required to be analyzed in real time.

Insights brought forth from analytics lead in coercive decision making. The most effective decision making involves a medley of data sources and gives a synoptic

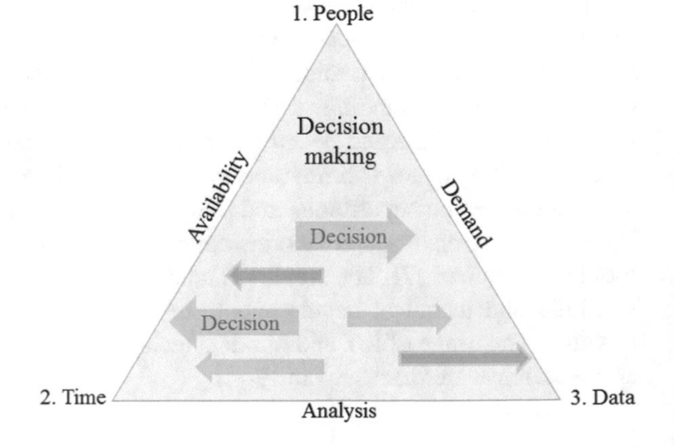

Fig. 2 Key decision making factors

view of the business. In big data, here and there irrelevant information can get to be a key component. Enterprises must comprehend the essential data relationship between various data sources and data types.

4.2 Shift from Data Management to Value Added Services

Today, from an innovation perspective the huge information is a colossal volume, continual information accessibility and structured and unstructured information. A decent data analytics framework ought to change an expansive volume of information into significant and knowledge experiences. That will prompt enhanced decision making in business environment.

To exploit the full potential of data analytics the system has to be designed with real-time and predictive analytics to support automatic decisions for continuous outcomes from machines. The real-time analytics provides valuable insights into the operations. It enhances the operational productivity. This is all that much valuable on account of execution observing and device administration. Big data analytics have gone into different ventures and it gets esteem from the tremendous volume of information and responds continuously.

Today, use of big data is getting to be ordinary in numerous commercial ventures and business regions. The utilization of big data analysis is attractive in numerous activities

Smart cities—provides an inventive mindset about how urban areas operate. Smart cities are intended to meet the compelling administration in vitality requests, preventive social insurance approach, transportation systems, e-voting and so on, require successful effective big data management.

Medical science—medical facility liberates and process a large volume of medical information and data generated from medical devices has been accelerating big data utilization. Expansive data set involved in the field of DNA sequence, medical imaging, molecular analysis, patient information, and investigation and so on, Deriving insights from this sort of gigantic information set will help the doctors to take a timely decision.

Mobile networks—tremendous changes are occurring in the world of mobile technology. The usages of smartphones are increasing day by day. Big data analytics are utilized to infer the insights to accomplish the highest quality of network by analyzing traffic planning, hardware maintenance, prediction of faulty devices and so on.

Manufacturing industries generally interface different sorts of sensors in their machines to screen the efficiency of the devices and this helps to prevent the problems in maintenance. The ultimate aim in digitization is to improve the quality in every phase of the manufacturing. The choice of sensor relies on upon the application nature and the product type. As a rule, right data to the right individuals at the ideal time is a key point of smart manufacturing.

The next section illustrates the use cases studied in the manufacturing industry.

5 Uses of Big Data in Industrial Environment

The initial phase in the vision of M2M (Machine-to-Machine) communication or the smart manufacturing is to comprehend the present manufacturing setup. It is viewed as that the IoT framework can change the ordinary manufacturing setup into smart manufacturing. The information system is the key transformation factor in steering towards the next generation manufacturing enterprises.

This section portrays the two use cases of usage of big data in a manufacturing industry. One use case depicts the machine integrated data analytics model and another one is the human coordinated enterprise business model.

5.1 Usecase #1: IoT (Internet of Things) in the Manufacturing Industry

IoT is a stage for digitization in the process and product enhancement. The principal prerequisite of Industry 4.0 is including the flavors of IoT and smart manufacturing. The intelligence network in the manufacturing setup essentially diminishes the intervention of the operators and supports to work autonomously. IoT will help the decision makers to infer actions and enhance the production efficiency and visibility of the production line information. It provides a real-time feedback from operations of the manufacturing plant. This gives a chance to make a move promptly when the arrangement is veering off from reality.

This section clarifies how IoT is realized in the manufacturing industry. The general architecture of the sensor connectivity with machines portrayed in the below Fig. 3.

The general architecture incorporates five stages. The main stage speaks to the machines associated with different sensors and actuators for getting signals. The received signal transmitted through the gateway. The network speaks to a remote or

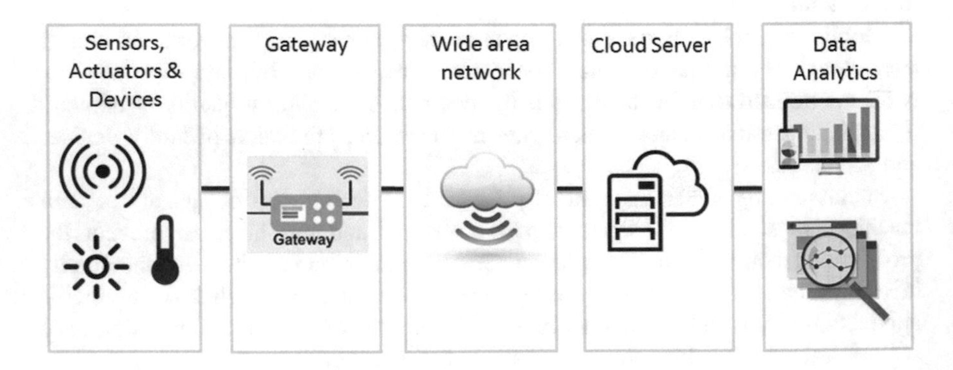

Fig. 3 Machine associated design

wired method. The data further transmitted for further decision making. The innovative big data analytics system is the ultimate outcome of the entire IoT framework.

This section clarifies different techniques utilized for data gathering as a part of IoT, the transformation of gathered information into required data format and data analysis methods. After the implementation of IoT in enterprises, tremendous growth in the volume and the detail of the data generated by machines has expanded enormously. Analyzing huge data sets finds a new method of producing improvement and innovation. Big data analytics provide an opportunity to derive the insights from a machine-generated big data. This gives a chance to change over the organizations more light-footed and to answer the inquiries which were considered beyond our reach some time recently.

The general structure of the IoT connected in a manufacturing industry displayed in the above Fig. 4. The starting stage is building up a sensor network with machines. At the point when machine triggers the sensor, it passes the signals to the data acquisition device. Then the data collected in the acquisition kit transferred to the data analytics software.

An exceptional data analytics tool developed and deployed to offer the people in the enterprise at various levels to infer quality decision, based on the real-time data

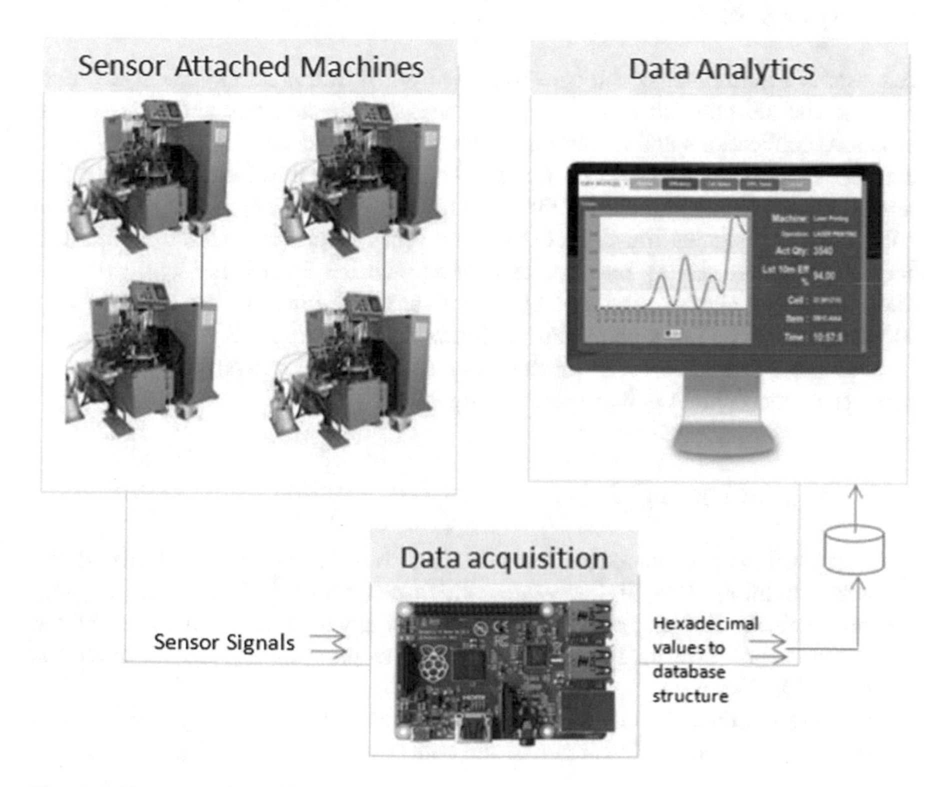

Fig. 4 IoT connected machines

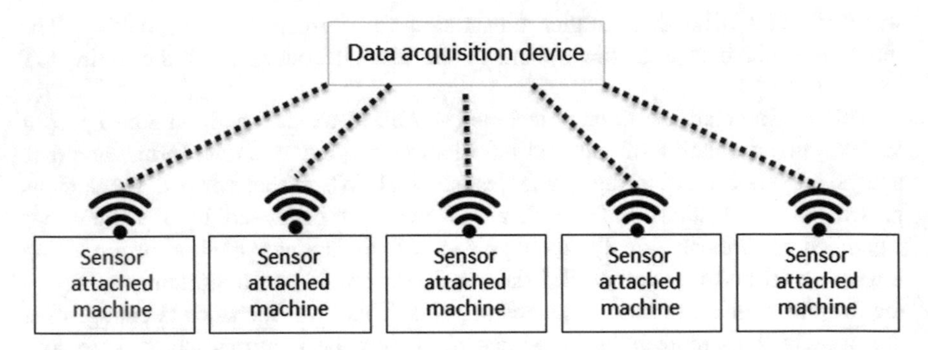

Fig. 5 Transmitting signals from various machines to data acquisition device

captured from various machines. The accompanying steps included in data processing.

The general structure contains three major stages. All the significant stages are discussed below.

5.1.1 Connected devices

A sensor is a transducer which converts the physical factors into identical electrical signals. The selection of sensor is based on the characteristics and types of the products, applications and machines. Various types of sensors are available in the market like IR (Infra Red) sensor, Reed sensor, metal detector sensor, etc., selective sensors further to be connected to the machines based on the requirement of the data collection. The signals transmitted from machines further transferred to the data acquisition device. Every machine associated with the sensors is identified separated node. The signals produced by sensors are transmitted to the common data acquisition device. The signal transmission is depicted in Fig. 5.

A sensor attached in every machine is responsible for converting the physical parameters and behaviors into electrical signals.

5.1.2 Data Acquisition

Data acquisition is a procedure for changing over the electrical signals of the physical conditions into digital values which can be controlled by a computer system. Typically it changes over the simple analog waveforms transmitted by the sensors in digital form for further processing. The anlog to digital transmission is represented in Fig. 6.

Handling the data produced by the machines is a major challenge in the manufacturing environment. The input of the data acquisition is the signals received

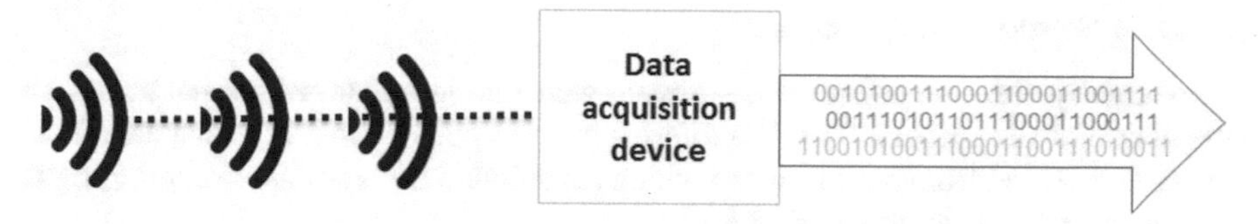

Fig. 6 Analog to digital transmission

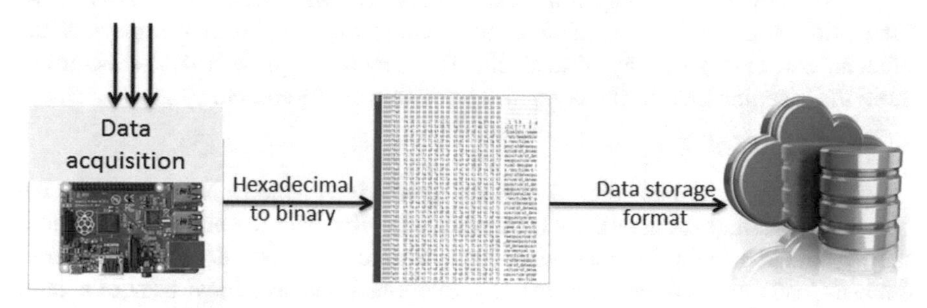

Fig. 7 Data acquisition method

from machines and the output is the hexadecimal values transferred from the data acquisition device. The general structure of the data acquisition related activities explained in Fig. 7.

Data acquisition is a significant phase in the smart manufacturing. It liberates enormous volume of data at a very high speed. It transmits data at each millisecond during its operation. This data is exceptionally enormous, complex and holds significant data. The data stream is high, thus it requires an effective and efficient strategy for acquisition and transformation. The complete acquisition process executes in the following phases.

- **Data Acquisition and Recording**

This equipment goes about as an interface between the sensor signals and a personal computer (PC) framework. The key component of the data acquisition device is the continuous signals transmitted from different machines. The data transfer is carried out by an interface. It gathers the data in every 20 ms. The data acquisition applications are carried out by software programs. The data interpretation commands are device specific and it varies device to device. The hexadecimal format further changed over to binary form to distinguish dynamic and latent machines and the relating status. Every pin is associated with one machine. However, the greatest challenge is the receipt of the right information. A buffer is introduced in the software to eliminate the issues during data transfer. In some cases, the data generated by the sensors are partial and mundane. The output generated from one sensor is not quite the same as others. Real-time analysis of data requires online and speedier data process.

- **Data Selection and Cleansing**

Normally, the data collected is gathered won't be in a required format ready for the data analysis purpose. The data filtering process pulls out the essential data from the data received from the machines which are suitable for analysis. Gathering right information is a technical challenge.

- **Data Representation**

The strategy for data analysis is a challenging task, which requires a high level of integration and aggregation of data in an automated way which empowers the efficient and extensive way of analysis. This process requires a data structure to store the machine data in the computer justifiable configuration.

- **Data analytics**

The IoT liberated data helps the organization to infer actionable insights with the help of an intelligent analytic tool. Big data analytics help the enterprises to outfit their data and benefit the enterprises to determine new open doors. This smarter analysis leads to better decision making, intelligent business moves, more gainful operations and customer loyalty.

The query process is altogether not the same as traditional RDBMS. In this case, the data are gathered from the machines. Sometimes, due to external disturbances, some noisy data might enter in the data set. Identification and removal of such data are highly recommended for big data. Querying options have to be applied in an intelligent way to get the experiences from the available data set. It ought to provide action-oriented real-time solutions. At that point, the gathered information will be put away in a table for further information investigation.

Analytics is an imperative stage in the M2M correspondence. This is an interface in the middle of human and machine. The output of the interface ought to be in a client justifiable format. The decision makers ought to comprehend the visual form of analysis and should derive actionable intelligent insights. Figure 7 is the screenshot of the overall machine running status in a plant. Each square demonstrates the status of the machine. White shade demonstrates that the machine running status. The machines alluded in white shading boxes don't have any issue. Gray ones show that the machine is running in lower productivity. This mirrors the efficiency of the administrator and in addition machine. The black shading squares demonstrate the idle state of the machine. The idle condition might be purposeful or accidental. The underneath screenshot is shown in a big size screen in the manufacturing plant. So that everybody in the plant can see the unmoving condition and can make a move quickly. In Fig. 8, the idle condition is accounted for the machines M12, M24, M42 and M44. The machines M21, M32, M31, and M46 are in gray shade, it shows that the machines are running at lower efficiency. No issues reported in all the staying white shaded machines.

The general machine status determined Fig. 8 is shown at the manufacturing plant on a greater size screen. The workers in the manufacturing plant can view the working status of each machine. Action can be made quickly because of this visibility. This makes the production people take a swift move. A Good visualization framework will convey the best result of the inquiries in a better understanding way. Figure 9 is a specimen screenshot of the machine running status. This graph demonstrates the idleness and the running state of the machine. This output is helpful for the supervisors take quick actions. It indicates not just the present status of the machine, also the running status of the machine.

In Fig. 9, the initial values are idle in the graph and the status changed after the machine started. It is well indicated by the trend in the graph. This screen is useful to view the trend of all the machines.

Fig. 8 Overall machine status

Cell # [CellID] | Volume | Efficiency | Cell Status | Eff% Trend | Layout

M11	M12	M13	M14	M15	M16
M21	M22	M23	M24	M25	M26
M27	M28	M29	M30	M31	M32
M31	M32	M33	M34	M35	M36
M41	M42	M43	M44	M45	M46

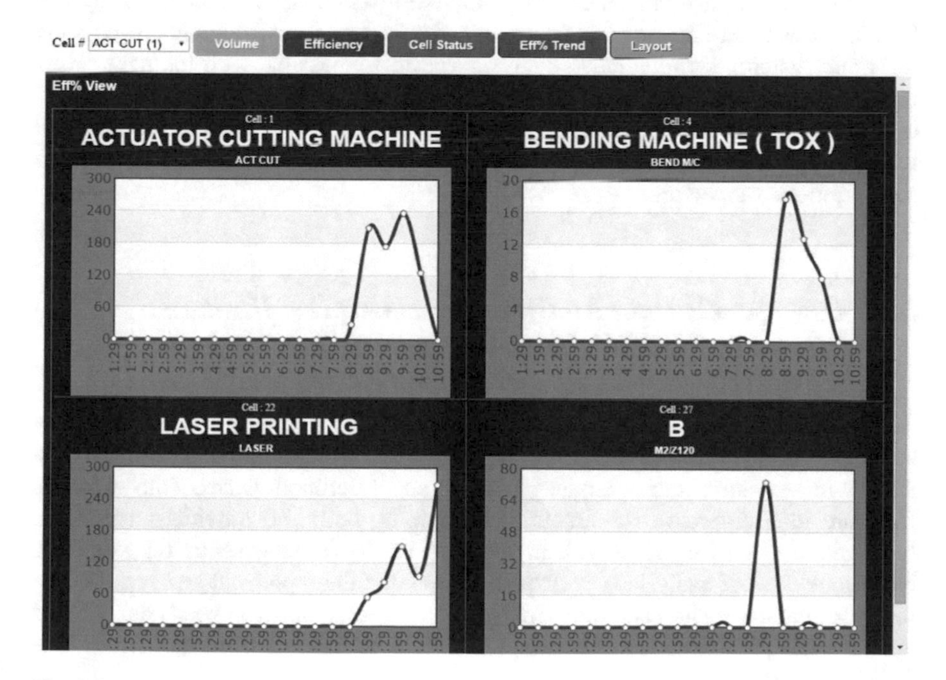

Fig. 9 Performance of all machines

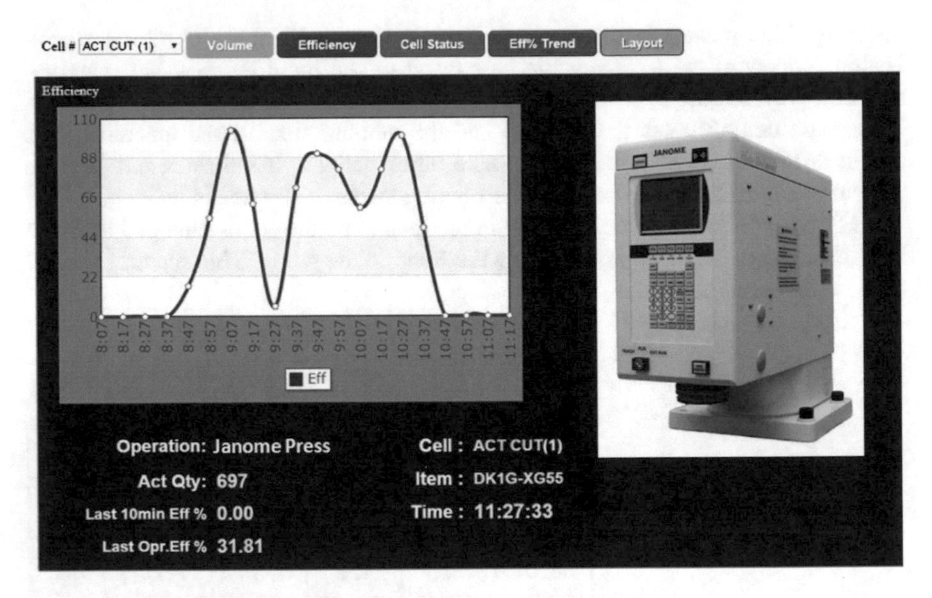

Fig. 10 Individual machine running status

The running status of a single machine is represented in Fig. 10.

The single machine status provides additional information like product name, quantity produced and efficiency of the running machine. The screen shots utilized here are specimen shots of the framework. Various types of outputs are possible with this system. The predictive analysis should be possible with the help of the real-time generated data and maintenance activities can be initiated immediately after receiving the wrong pulses from the system. This sort of machine coordinated information can diminish the human data entry by exchanging the right information to the current framework.

5.2 *Usecase #2: Data Analytics for Enterprise Business System in the Manufacturing Industry*

The enterprise resource data provide insights about sales, inventory and efficiency analysis. The data dwells in various types of databases. In this use case, the data resides in structured query language (SQL) and Microsoft Access (MS Access). Both are in a different data structure. Combining both and providing actionable intelligence is an important task in the case of the heterogeneous type of data. Numerous what-if analysis is profoundly valuable to comprehend and break down the information inside and out. In financial areas, statistical analysis, what-if analysis and predictive analysis brings a variety of insights from the tremendous volume of data.

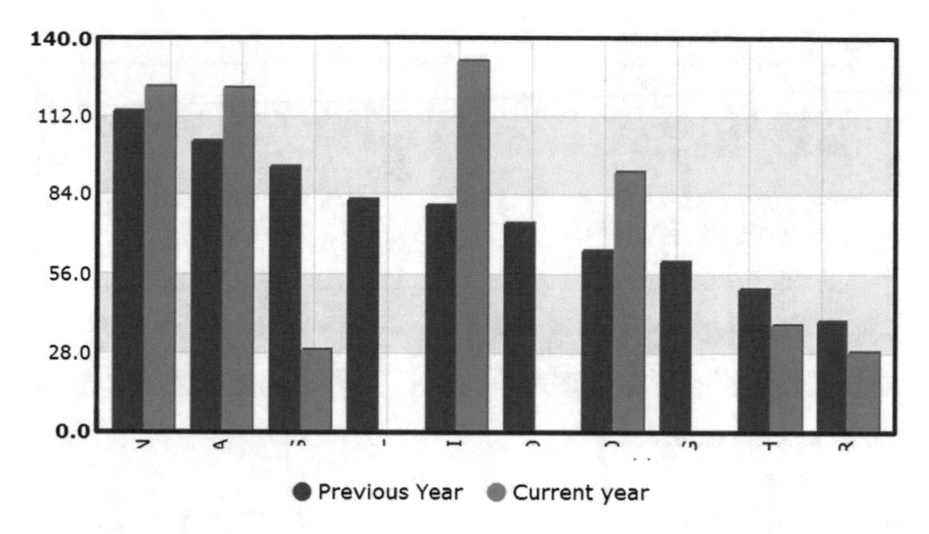

Fig. 11 Top customer sales comparison

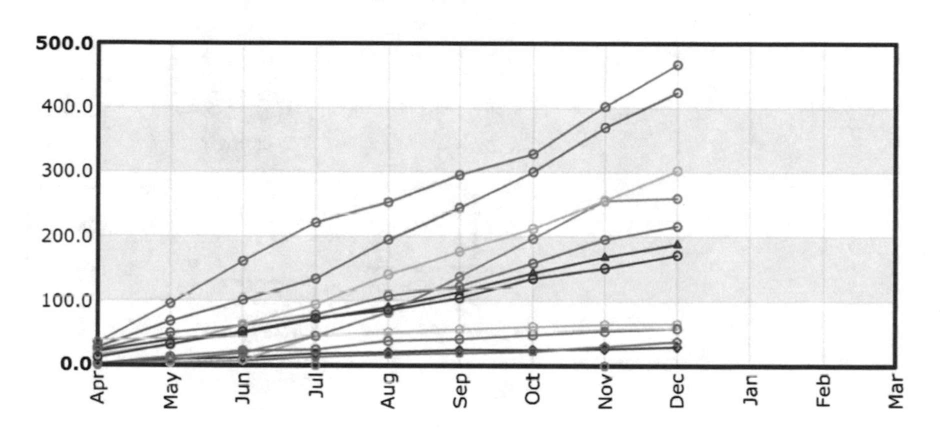

Fig. 12 Product wise sales

The top level management requires decision oriented analysis to step further. Predictions in time stay away from numerous issues in the decision making. Figure 11 lists the comparison of sales of a top customer in a year ago versus this year. For security reason, the customer names are hidden in the graph. Rather than breaking down the information with a few tables, this analytics tool gets the bits of knowledge a successful way.

Figure 12 represents the product wise sales. It plots a graph from the data gathered in the enterprise business framework. This helps to understand the sales of various product categories. The enterprises can make better decisions in the product category where they require more clarity of mind.

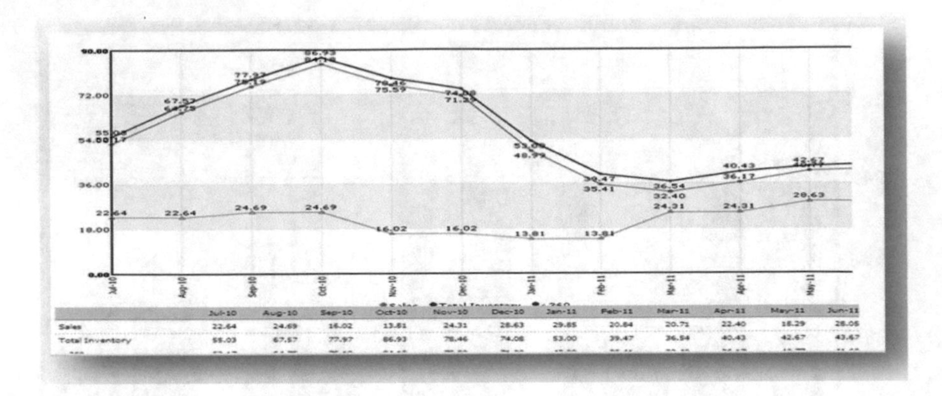

Fig. 13 Inventory analysis

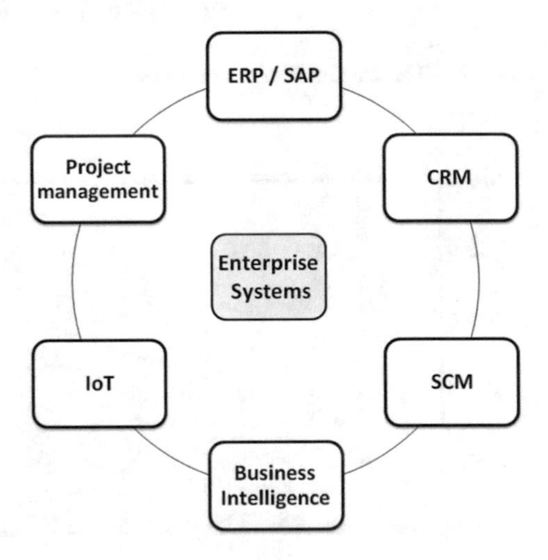

Fig. 14 Enterprise system structure

Figure 13 shows the month to month inventory analysis. This data analytics derived from the available value of the enterprise business system. The enterprise system contains a vast dataset of all the information about the enterprise. Data analytics are utilized to channel and plot data which are valuable for business decision making.

Sales and customer related data are plotted in Figs. 11 and 12 and materials related information is plotted in Fig. 13. This data is accessible in the cloud. Thus the enterprises can view and take choices from any place.

In both the use cases, the analytics helps the top level management to take the decision on time and real time.

Figure 14 describes various system involved in a typical manufacturing enterprise.

An enterprise resource planning (ERP) or systems applications and products (SAP) framework helps the enterprises to manage the vital parts of the business. It integrates all the enterprise business processes into a solitary arrangement. The broadened arrangement of ERP/SAP with CRM and supply chain management (SCM) systems enhances the complete supply chain in the manufacturing sector.

The introduction of IoT and improvements in the information technology provides more prominent chance to grow better relationships with customers. The ultimate aim of every organization is to improve the profitability; the CRM system paves a path for a better relationship with lowering communication cost.

A complete purpose of every affiliation is to upgrade the efficiency, the CRM system clears a path for a better relationship with cutting down correspondence cost.

SCM is the management of the relationship, keeping in mind the end goal to exchange more noteworthy customer value at a lesser expense in the supply.

IoT use cases are the shrewd method of getting inputs. Human interventions are not required by IoT interface. It gathers information automatically from the machine.

Business intelligence is a method of excerpting, altering, analyzing, and managing huge data using mathematical models to derive information and knowledge to derive valuable business decisions.

Even though the data analytics generates a lot of insights from the huge volume of data, it has also some challenges. These challenges are discussed in the next section.

6 Challenges of Big Data Analytics

Handling huge volume of data is a biggest challenge in big data analytics. Conversion of unstructured data to structured data is a major threat and then afterward purifying of data before applying analytics is additionally a major challenge. Content-based information accessible in the conventional framework gives information identified with product, customer, suppliers, the quality of the products and so forth. The innovative initiatives are going on in numerous enterprises to fulfill the Industry 4.0 necessity. IoT is required to meet this prerequisite. As examined in Use case #1, the framework ought to be in a position to anticipate precisely and support people for better choice setting in real time. MNC's (Multi National Company) have begun changing their procedures to meet the big data challenges.

6.1 Technology Shifts

Present technology stretches out adequately to some for data storage and querying. But it requires special focus in the territory of IoT and handling of machine produced information. Following things to be followed beyond the data mining techniques:

a. Selection of adequate method and outline scientifically
b. Improvised most recent algorithm to enhance its efficiency and scalability
c. Deriving business value from the big dataset.

Simply moving towards the data analytics is not helpful until we comprehend and derive the business value from the huge dataset. Embracing new strategies, simulation techniques and special tools for data analytics will help to derive useful insights in enterprises. Enterprises should be ready to embrace those progressions.

6.2 Data Visualization

Big data is not just to create the largest database. It is about creating high-performance computing and analytics. Picking more appropriate big data analytics tool is critical. It should, they bring visualize results of the joined quality of information generated by the machine and interpretation by machine. Its key advantages incorporate

- Problem-solving/inferring choices and significance of data analysis.
- An automated algorithm to accelerate the machine generated data analysis.
- Adding human values through visual interfaces.

6.3 Heterogeneous and Imperfect Data

Handling of structured and unstructured data is a major challenge on big data. Especially amid diagnostic stage phase, the machine should be in a position to understand how to handle the data. In human data consumption, the heterogeneity can easily be tolerated. Purification of defective data is a challenge in big data analytics. Indeed, even after data cleansing, some blemished and filthy data exist in the data set. The most serious issue is dealing with this amid the data analysis stage.

6.4 Scalability

Management of huge database and continuously growing data has been a challenging issue for a long time. Be that as it may, the most recent advancements in network communication, sensor devices and healthcare systems creating tremendous measures of information. At first, this issue was mitigated by introducing high-end processors, storage devices, and parallel data processing. The next shift is moving towards cloud computing, this idea is sharing of resources. Providing hardware solutions is not just adequate for data processing. It requires a new dimension of data management in the area of data preprocessing, query handling algorithm, database designing and error handling methods.

6.5 Real Time Execution

Speed is an imperative element in the real-time data execution. In various applications, the result might require quickly. In our use case #1, the machine is associated with data acquisition devices for data analytics purpose. Online decisions like machine stoppage alarms and efficiencies are configured on the machine. So quick action required on this. Immediate execution is required in case of online shopping, banking transactions, sensors, etc., To answer the queries in real time, scanning the complete set of data is not possible. The proper index structure will solve this problem.

KPMG survey [13] reported that 85% of the people reported that analytics speed is the primary benefit. 65% of the people expected to identify the insights and 59% expected to achieve insights in a micro level.

Today, the biggest challenge for most of the enterprises is converting the mountains of data into insights and converts the insights to actionable business advantage. The KPMG International [13] conducted a survey on big data and analytics to many executives in the industries. From survey, they have distinguished the following are the biggest challenged in data analytics

- Selection of correction solution for exact data analysis
- Discovering correct risk parameters and measures
- Real-time activity
- Selection of required data
- Providing data analytics in every zone of the business.

Deriving actionable insights turn out to be exceptionally troublesome when technology changes. But always a new, superior technology will emerge to predict the growth opportunities in the market. Even though data analytics have many challenges, every organization requires predictive analytics to grab the unknown patterns in a very large and multifaceted data.

7 Conclusion

This chapter has talked about the key roles of big data in a manufacturing industry, particularly in the IoT environment and as a strategic player in the agile business environment.

The success of many organizations demand new skills, as well as new perspectives on how to handle the big data, could advance the speed of business processes. One vital variable of this innovative procedure is the new analytics tools that have evolved together with new progressive business models. In this chapter, the innovative abilities of the growing big data phenomenon investigated and addressed various issues concerning its methodological for changes. The key findings are substantiated by depicting the real time cases.

Different issues concerning big data's innovation techniques and modeling strategies incorporated by numerous large business organizations. In reality, it is plainly seen that the enormous information is currently incorporated into the business procedure of numerous associations, not on account of the buzz it draws in transport for its creative abilities to totally change the business scene. Although there the innovative techniques of big data are ever evolving, we were able to cover few significant ones, which are clearing the path for the improvement of new products and services for many organizations.

We are in the era of big data. Data oriented organization is extremely efficient in anticipating the behaviors of customers, financial position, and SCM. Through better analysis, the enterprises can extract better insights which will enhance the profitability by offering the right products would require more profound experiences. The business choices would require deeper insights. The technical challenges discussed in this paper to be addressed to get the full advantage of the stored data.

Even though Data analytics is a powerful tool for decision making, the reliability of the data is very important. There are various future research directions in the Data analytics model.

The vision of many government and manufacturing industries in worldwide is moving towards the smart manufacturing to achieve Industry 4.0. The fundamental guideline to this vision is the thoughts of cyber-physical operation where the industrial facility is intensely instrumented to be application oriented, data directed and digitized. Overall equipment effectiveness (OEE) is a well-known factory metric used to cater a measure of the achievement of any work center. OEE additionally gives enterprises a structure for deliberating about IoT application—reconstructing efficiency, utilization, and quality.

References

1. Vermesan, O., & Friess, P. (Eds.). (2014). *Internet of things-from research and innovation to market deployment* (pp. 74–75)., River Publishers Series in Communication Aalborg: River Publishers.

2. Bureš, V. (2016). Application of ambient intelligence in educational institutions: Visions and architectures. *International Journal of Ambient Computing and Intelligence (IJACI), 7*(1), 94–120.
3. Kamal, S., Ripon, S. H., Dey, N., Ashour, A. S., & Santhi, V. (2016). A MapReduce approach to diminish imbalance parameters for big deoxyribonucleic acid dataset. *Computer Methods and Programs in Biomedicine, 131*(C), 191–206.
4. Baumgarten, M., Mulvenna, M., Rooney, N., & Reid, J. (2013). Keyword-based sentiment mining using Twitter. *International Journal of Ambient Computing and Intelligence, 5*(2), 56–69.
5. Kamal, S., Dey, N., Ashour, A. S., Ripon, S., Balas, V. E., & Kaysar, M. S. (2016). FbMapping: An automated system for monitoring Facebook data. *Neural Network World, 27* (1), 27
6. Brun, G., Doguoglu, U., & Kuenzle, D. (2013). Epistemology and emotions. *International Journal of Synthetic Emotions, 4*(1), 92–94
7. Alvandi, E. O. (2011). Emotions and information processing: A theoretical approach. *International Journal of Synthetic Emotions (IJSE), 2*(1), 1–14.
8. Odella, F. (2016). Technology studies and the sociological debate on monitoring of social interactions. *International Journal of Ambient Computing and Intelligence, 7*(1), 1–26.
9. Internet of Things and Big Data Technologies for Next Generation Healthcare. ISBN 978–3-319-49736-5.
10. Kamal, M. S., Nimmy, S. F., Hossain, M. I., Dey, N., Ashour, A. S., & Santhi, V. (2016). ExSep: An exon separation process using neural skyline filter. In *International conference on electrical, electronics, and optimization techniques (ICEEOT)*. doi:10.1109/ICEEOT.2016.7755515
11. Zappi, P., Lombriser, C., Benini, L., & Tröster, G. (2010). Collecting datasets from ambient intelligence environments. *International Journal of Ambient Computing and Intelligence, 2* (2), 42–56.
12. Building Smarter Manufacturing With The Internet of Things (IoT), Lopez Research LLC 2269 Chestnut Street #202 San Francisco, CA 94123 T(866) 849–5750W. www.lopezresearch.com
13. Going beyond the data: Achieving actionable insights with data and analytics, KPMG CAPITAL. https://www.kpmg.com/Global/en/IssuesAndInsights/ArticlesPublications/Documents/going-beyond-data-and-analytics-v4.pdf

Revealing Big Data Emerging Technology as Enabler of LMS Technologies Transferability

Heru Susanto, Ching Kang Chen and Mohammed Nabil Almunawar

Abstract We are living in the information age where almost every aspect of our life, directly or indirectly, relies on information and communication technology (ICT). The uses of ICT through big data has increased which therefore ended to be everything can directly go through online and people are now able to upload, retrieve, store their information and collect information to big data. Through big data learning management system (LMS), student managed and stored their intangible assets such as knowledge and information, documents, report, and administration purpose. LMS is basically application software that is capable and designed to provide electronic learning, and has been acknowledged to yield an integrated platform providing content, the delivery as well as management of learning, while supplying accessibility to a wide range of users that include students, content creators, lecturers as well as administrators. Universities aim to successfully implement a LMS in order to ease the learning process. This successful implementation lead to Universities make Business Process Re-engineering for their learning activity. Throughout the years, successful implementations of LMS have proven to be a very beneficial tool, providing ease and convenience. LMS is used not only in academic institutions such as schools and universities, but is also popularly used in a number of private corporations to provide online learning, training and is also capable of automating the process of record-keeping as well as employee registration. The objectives of this study are to reveal big data as enabler

H. Susanto (✉)
Department of Information Management, Tunghai University, Taichung, Taiwan
e-mail: susanto.net@gmail.com; heru.susanto@lipi.go.id

H. Susanto
Computational Science & IT Governance, The Indonesian Institute of Sciences, Jakarta, Indonesia

C.K. Chen · M.N. Almunawar
School of Business and Economics, University of Brunei, Bandar Seri Begawan, Brunei Darussalam
e-mail: chinkang.chen@ubd.edu.bn

M.N. Almunawar
e-mail: nail.almunawar@ubd.edu.bn

N. Dey et al. (eds.), *Internet of Things and Big Data Analytics Toward Next-Generation Intelligence*, Studies in Big Data 30,
DOI 10.1007/978-3-319-60435-0_5

of LMS as Business Process Re-engineering bring users specifically, various benefits of its multi-function ability.

Keywords Learning management system · Business core re-design · E-learning · Business Process Re-engineering · Big Data · Information security

1 Introduction

Learning management system (LMS) is popularly known in the communities of higher educational institutions as an online portal or platform that connects both lecturers and students together. It provides a local avenue where classroom materials or activities that are required to be shared, to be done easily and conveniently. Through big data technology it also represents a portal that allows both lecturers and students to interact outside of the classrooms without physically being in front of each other, making it possible for both parties to have online discussions that would otherwise be too time consuming if these discussions were instead conducted inside the classroom.

Today's era of big data observes a rapid increase in availability, ease in access as well as connectivity reach particularly in terms of internet access, especially in most urban locations, which is also where most universities are situated in. The internet can be defined as a vast computer network that links smaller computer networks all throughout the globe, scoping commercial, educational, governmental, as well as other networks, which all make use of the same set of communicative protocols.

This widespread ease of access in connecting to the internet has helped university students to increase their level of independency in their education, as lecturers normally hand out lecture notes, as well as any further information deemed necessary, to be left for the students to discover and digest information on their own. However, the university level of learning is still a two-way process, a culture where lecturers share their knowledge, and in return, students voice out their opinions and thoughts in discussions. Thus, it is important for university students to constantly broaden their scope of knowledge by frequently searching for information. Learning management systems represents the tool to ease this whole process.

The world has developed in various different ways and the most frightening development is in terms of technology which still continue to grow even until today. As the world become more modernized, the use of high technology in many ways is increasing every year which it give convenient to the society. For the past few years, institutions have begun using different methods of education to improve efficiency and distribution of learning methods such as the usage of technology to improvise the conventional teaching methods, in this case, the LMS or in general, E-learning. In organized enterprises, they concerned must working together, give encouragement to each other, support and feedback expected that the use of Internet in educational purposes will be increasing in the near future. Information and

Communication Technology (ICT) in education will create new possibilities because of new technologies that can also create changes in attitudes and leadership significantly. In order to make ICT functions smooth and efficiently, a team in charge of ICT will have to be able to cope and work together with new network of people. Nowadays, to make full usage of modern technology has become the problem to transform the organization faces by universities, government and industry [1, 2]. The program like LMS used by most of the companies and government agencies to manage information, staff development and internal communication. This means that distribution of information to staff members and internal communication is made easier and more efficient in government agencies and private sectors. In this modern era, the use of internet is obtainable and easy to get access [3]. This makes the use of LMS all the more important and should be featured in learning institutions and colleges because almost all nations have access to internet which is the main key in order to implement LMS.

LMS has evolved as the years went by since its first introduction. It has been undergoing a lot of changes and improvements in order to make the system to be more efficient and more user friendly which will also be talked about in the latter part of the report. There has been a myriad of research being done regarding LMS or E-learning in general. The report will make use of those research to uncover the benefits, uses, its implementation in developed and developing countries and differences between them and also the different types of LMS that are being used in the past and today.

1.1 LMS Aspects

LMS has a wide and broad definitions, Eke [4] defined that it is a way of learning either via the internet or any other electronic devices. Moreover, the definition of E-learning is where it symbolizes all or part of the aspect of academicals way of learning that are based on the tools such as the electronic media and the other devices that are used as a term to improve the ways to access self-training, communication and collaboration which in the end ease the assumption that act as a new ways of understanding and developing learning [5]. Whereas, according to Garrison [6], he defined that E-learning is an electronically ways that can be divided into two; (1) Asynchronous and (2) Synchronous communication as a reason to represent and to assure the accuracy of the knowledge. Therefore, this had proved the similarity of the definitions for E-learning in which E-learning is a way of learning which basically needed any electronic media and Internet to access the learning system in order for some students or teachers to learn and teach. Hence, E-learning is more than only a technology as a means to access course material [7].

1.1.1 Synchronous and Asynchronous Communication

Synchronous and asynchronous are two wide categories of LMS. In order to understand what is the relationship between both of these with E-learning, the explanation of synchronous and asynchronous will be continued under this sub-topic. First of all, synchronous is where it uses a learning model that initiates a classroom course, lecture or meeting using Internet technologies in which all of the participants need to be available at the same time so that it can be a live interaction between the participants and the lecturer [4]. Eke also described that asynchronous learning as a web-based version of computer based-training (CBT) which is typically offered on a CD-ROM or across an organization's local area network in which the participants can easily assess the course everywhere and every time they are available.

1.1.2 Benefits and Potentials of E-learning to the Students

LMS could be accessible 24/7. Some students may not be able to attend classes which cause miss some of important lessons. They can also learn it in their own way because the way that the teachers teach them may affect the understanding of another student which finally led to confusion. Hence, if the students do not understand, they can review the lecture notes back during their personal study. When it comes to assignment submissions, students can submit their assignments any time within the duration time given, they do not have to wait for the next class to submit everything. It will also help the students who take part time studies as they can study while working. Furthermore, E-learning can also help the students have more time to do their study and by E-learning, it can help the students to gain more knowledge from other sources.

2 Literature Review

The Learning Management System can be defined as a software that provide benefits or advantages to users to make work more efficient. LMS maintains all records, covering educational training courses that are being issued throughout the internet. As the software is accessible via online, not only LMS is being used by large organizations throughout the world, but also higher learning institutions. Today, there is an increasing number of free educational courses being published out the internet for society and the world. This education and training is known as E-learning or E-training.

As stated by Tao et al. [8], this new learning environment, which is centered around electronic networks, has given learners across universities globally, the ability to receive support on an individualised level, while also being able to have learning schedules that they would find more suitable to them. This allows a high degree of interaction and collaboration between instructors, or teachers, as well as students, as compared to the traditional environment for learning. This new way to learn is characterized through the utilisation of multimedia constructs, made the process of gaining knowledge more active, interesting, as well as enjoyable [9–11].

Martin and Madigan [12], said that almost all people around the world handles a computer everyday, every time and thus, interactions and learnings are, being highly recommended to the base of E-learning as well as E-training, which also basically helps promoting a green environment, especially with the reduced amount of paperwork and the usage of paper in general with the use of LMS.

In a study conducted by Falvo and Johnson in 2007, for the past years, LMS gained enough use and popularity such that other alternative means to deliver contents and information to the classroom, such as live satellite broadcasting or closed circuit television is rapidly diminishing in relevance. This correlates with the increasing span of the ease in which the internet can be accessed, as well as the increasing number of consumer-friendly computers that are being sold worldwide. The introduction of LMS, and this increase in access, inevitably led to the rapid increase in the number of students and teachers towards the online learning environment.

The growing trend of E-learning is widely spreading across its entire educational system ranging from both public and private educational institutions [13]. It was discovered that there were three main factors that encourage these academic institutions to get involved in the implementation or the development of learning management systems, namely the increasing number of academic students nationwide, the increasing need of life long educational learning, and the widespread need for students to be prepared with the adequate knowledge and skills that is required in order to succeed in the knowledge economy [14].

A significant number of learning management systems in Europe are either purchased from small commercial vendors, or are developed by the learning institutions themselves [15]. Although the implementation of these learning management systems vary from institution to institution, the usage of such a software has grown to become a standard at colleges and universities.

It was also found out that both distant and on-campus students found the collaborative tools found in learning management systems to be the most valued, especially when they have the need to share learning during group tasks [16]. However, different students may have different needs from coursework as well as experiences within the learning management system. It was found that on-campus residents value higher the interactive content that can be found within a learning management system. On the other hand, students who commute to campus value more being able interact and share experiences from peer to peer within the learning management system [17].

The implementation and the design of a learning management system also has to take international students or students who wish to learn another language with the use of the learning management system, into consideration. When offered these types of coursework and courses that involve language, it should be taken into consideration how well the learning management system is conventional to communicative language learning theories. Therefore, placing emphasis on developing a learning management system to be able to cater well and facilitate the needs of all types of learners, teachers, or any users, is undoubtedly of great significance [18].

3 Managing Change

LMS, has brought about a huge impact not only to certain groups of people or society, but throughout the whole academic world. Even though learning management systems, in major ways, could bring positive impacts towards society, however, there are certain groups of people which face difficulties in the attempt to adapt to these technological changes, especially those who originate from a less tech-savvy country. Most large organisations and as well as the government proceed with the use of learning management system, in order to manage their extensive size of important documents, organization development and staff. Generally, the effect of this framework is mostly on the learning group population for example students in higher institutions, where students can be expected to be independent in their research and gaining of knowledge. Thus, through making learning materials available online, it would only improve learning results only for specific forms of collective assessment.

Also, with respect to the knowledge to be gained and digested, the level of clarity, the offer of explanations as well as the confidence rating of making interpretations, may not be as effective compared to the traditional way of learning in the classroom. It can be argued that the learning process would be much easier with full use of face to face encounters with teachers or lecturers, especially in cases where the core structure of certain courses requires constant, independently-driven motives to conduct online research. This can go both ways, as instructors or lecturers also might not know the extent of how well the learner has absorbed the information tasked to be researched upon.

Moreover, since e-learning as well as having to access learning management systems require computers as well as internet connections at their preferred places of study, it might be somewhat burdening because having these may prove to be beyond the financial capabilities of some learners. However, a solution to this might possibly be having to lend these learners the necessary equipment, but this would undoubtedly lead to costs incurred by the institution.

Furthermore, Al-Busaidi and Al-Shihi [19], stated that the available tools from learning management systems are highly dependent on their will, and dedication to their chosen courses. There are a number of obstacles that have been identified as they were encountered by practitioners of the learning management systems. Al-Senaidi et al. [20] has stated that the common justifications for a lack of skills in the usage of learning management systems is the insufficient amount of time spent in the training and support. This lack of assistance can cause a problem on how to manage the usage of this level of technology that learning management systems are based on, especially since users may originate from a less tech-savvy country.

According to Brown and Johnson [21], LMS also has other attractive miscellaneous features that can provide instructors with assistance in creating records of students, duty rosters and others that aim to have a more organized class, creating an instant group conversation, not to mention that learning management systems can have the ability to provide forms of student assessments, such as a quiz, or a test, through these online platforms.

It is undeniable that by following Fig. 1, above by using E-learning, it is more easier and more accessible for students to retrieve educational informations for their learnings. Not only that, lecturers and students can communicate and discuss about relevant tasks and subjects more easily through email or simply by messengers through their phones. But in the traditional system, all information and notice given are more to a physical standard hand-operated way. All that are being taught or said by the lecturer are not being fully accounted by all the students, but at least not all. The effectiveness are being aid by the quickness and misdeed [22]. The effectiveness is referring to on how the system are managed in order to obtain and complete given objectives.

Another benefit that implementing a learning management system can provide is through the utilisation of online forums, which can be supported by the learning management system platform. It provides opportunities for improved relations between learners and instructors. Through having these online forum discussions, any unwanted barriers that have the potential to hinder a learner's participation, such as social anxiety for example, can be eliminated, as discussion is done online. It aims to promote an environment where students can interact with one another, exchanging as well as respecting their different point of views towards any relevant matters.

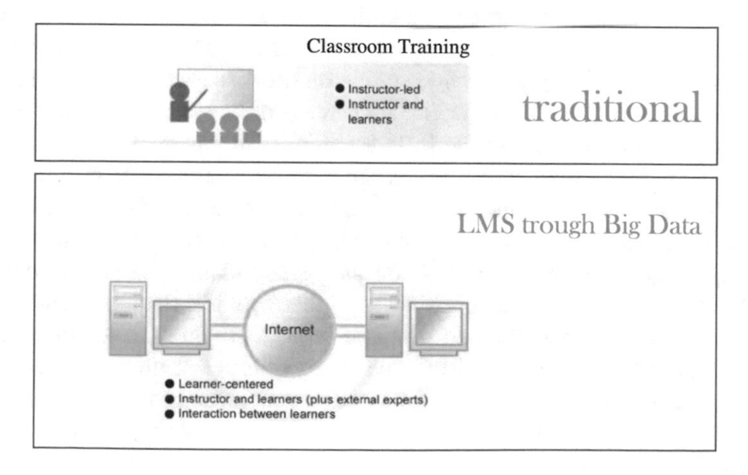

Fig. 1 LMS and traditional

On the perspective of the institution, fewer costs may also be incurred in such a way that less administration staff has to be employed, and therefore any costs associated with hiring more staff would be eliminated, such as office space, stationeries and wages. Also, with learning management systems, students better recognize the content associated with their chosen courses and therefore have the ability to know of any assignments due, or any content that needs to be reviewed by them prior to a given date. This allows self-pacing on behalf of students. Being able to study and complete assignments at their own pace and speed, no matter slow or fast, increases satisfaction and reduces stress [23]. On the other hand, certain notes given by the teachers might have limited information and could be difficult for some students to understand the content. In order to solve these problems, the students can just use YouTube website to watch and learn from a video education or use E-books to understand more on what they are learning. Search engines like Google also provide various websites that gives useful information and examples.

Furthermore, it makes students do their assignments easier because students do not have to continuously go to library every day to search for journals or books for their references to use in their assignment work. The student can just use Google books to find a variety of textbooks and use Google Scholar to find journals or any related reports from other researchers. This will be more convincing in term of time taken to do their research. All of the information will be available in a very short time in which we do not have to wait for few days. Other than that, new information is always being updated every single day in the internet. Therefore, the information needed would be up-to-date.

Less expensive and lighter to carry for the students, sometimes school textbooks are available on Google Book or E-book, students can just print out certain topics that are needed for study rather than buying the whole textbook at bookstores which it will cost more money. In terms of less weight to carry, usually students will carry along with them thick books or lots of files that contain notes to their lectures. Now, the students can just carry their devices such as laptop or smart phones for them to easily access to their notes for example, they can access the notes using Portable Document Format (PDF) files or even online textbooks such as E-books. E-learning can increase in students' participation during online discussions by which lecturers just need to open a group page for the students to discuss about certain topics that they did not understand during the lecture sessions. In this circumstance, this may increase the students confidence level as some of them might be to shy to voice out their opinion in class and sometimes the period of the lecture during the day is not enough for the students to discuss more and to give opinions at the same time. With E-learning, it can help the students to do their study alone without any disruption from others especially in the campus whereby there are lots of students at there that probably can cause a disruption for him/her to do their study. E-learning can also help the students to do work discussion easily through online with their small groups. It is also a convenience way for some students who have a part time job and are unavailable to do their assignments or study on time as they do not need to be rushing to do their work.

In terms of presenting their assignment, it leads to a better presentation as the students can show more visually to others so that it is easy to understand what they are presenting on that day. Before the existence of E-learning students usually will just talk or present their presentation using papers in which if they made a mistake they need to rewrite everything back and in the end lead to a waste of time for the students to rewrite back what they think is the best for the presentation. E-learning will also teach the student how to become a critical thinker, because they are so many information available on the internet from small information to big information that is usually hard to find but not all information from internet is true. This is when critical thinker takes action on applying whether the information that they gathered is useful enough to use or vice versa. This is because not all information that is gathered comes from a trusted source. It is important to check the author and the publisher before choosing the articles, journals or books.

3.1 The Benefits

Using E-learning can help the lecturers to check if the students commit plagiarism, whether the students are using reliable resources properly such as paraphrasing and in text citation. If the lecturers only required the students to submit their assignments in hard copy form, it would be difficult for the lecturers to check the numbers of plagiarism that are made by the current students. As for example Turnitin is an application that is used to check the plagiarism that the students might did. Also, in addition, E-learning can give an additional time for the lecturers to teach whereby in some universities the lecture session might be in a short time only in which the lecturer might face a difficulty to elaborate more on the topics that they were studying on that particular day.

Furthermore, E-learning can help both the lecturers and the students to be easily connected for them to ask questions and to discuss any topics that the students did not understand. The reason of this is because some students refused to voice out their opinions during the class session instead they prefer to email or chat with the lecturers personally after the lectures. In this way, it can help the teachers to easily recognised what are the students' problems so that the lecturers know what to do in the next lecture session.

Nevertheless, E-learning can also help both of the lecturers to save cost as some of the lecturers are usually use more than one textbook to prepare for class, some textbooks cost a lot of money and with the existence of E-learning, lecturers can just view the textbook for free on E-books which is if applicable. Printing and photocopying the notes for the students might lead to a high cost to the lecturers. Therefore, in order to save cost, the teachers can easily email the notes to the students and ask them to print it for themselves if they wanted to.

It is convenience for the lecturers as they do not have to worry about the class capacity for the students or if the teachers have urgent cases or meeting during the lesson hours. In terms of convenience for large number of student, Quizzes can easily be done with an online application whereby the students who came late can access to the quiz if they are not in class. Other than that, E-learning can help to reduce error make by human such as spelling and grammar error. Lecturers can substitute the way the class will be held into a video so that the students are well prepared on what will happen before the class session starts. This is an alternatives way to teach the students as some of them want to study visually about certain subject instead of just listening to what the lecturers are saying during the lesson hours. This will also help to increase the student concentration in class. With this substitution, it is a different ways of learning style that can cover what the lecturers are lacking out at because not all students have the same ways of learning styles as some of them prefer learning by visual, kinetic or aural and a lot more. E-learning can lead to the lecturers to give a consistent message to the students that he/she is teaching. This is because if the lecturers tend to have more than one class with other students, the would be a higher chance where the lecturers would have missed out some points that he/she told the other students on the class. With the uses of E-learning, all of the students will get the same information without anything that will be missed out in the future lecture.

3.1.1 Benefit of E-learning to Environmental

With the advance usage of E-learning this could help the environment to be green and the carbon footprint can be reduced as for examples E-learning can reduce the consuming of papers as the students and even the worker need to use a lot of paper to print for their notes or paper work. If we consume many papers, this means that more trees will be cut down for the process of producing more papers in the future for the uses of other people.

3.1.2 Benefit of E-learning to Business

E-learning can benefit the business also as it will help the business to save their cost to train a new staff and the travel cost that are needed to train some employees. To easily recruit the new comers, the company can just give a video or a web address that the workers can access for their training session. If the companies keep on transferring some of the employees to go training, there could be some possibilities that the cost can be high and lead to the companies to face a huge number of losses [24].

4 Big Data Emerging Technology

Data is numeric figures for a particular of statistics or facts. Nowadays, there is huge amount of data being generated by smart phone, social media and so on. Big Data is where the number of users keeps on increasing that forms a data. Big data is a collection of data from traditional and digital sources hence, it is too large and complex to manage it using current computing technologies [25–27]. There is no doubt that there are several difficulties when it comes to managing and analysing data. This is because the complexity of big data which require extra tools and big data analysis. Big Data involved unstructured and structured data. Unstructured data can be refer as an information that is not organized and easily interpreted such as documents, pictures, videos [28]. Social media posts are also part of unstructured data such as Twitter tweets, Facebook and more. In the other hand, structured data is more into many different data formats, types and it is stored in traditional relational database videos [28]. Traditional relational database is collection set of organized data stored in by showing the relation between the information. Furthermore, structured data is the data that can be obtained from human interaction between machines, for example, using web application.

According to Gartner (as cited in [29]), Big Data is often interpreted in many different ways therefore it can be describe using 3V's; volume, velocity and variety. SAS (System Automation Software) the software company introduced another two characteristics: complexity and variability (sas.com, 2013, as cited in [29]). Big data is such a big thing that volume, velocity, variety, complexity and variability of data is too massive to handle with. Therefore, it requires another tools so that big data specialists able to analyze the data to provide us insight of the Big Data.

In 2016, it is estimated that big data will drive $232 billion in spending according to Gartner (IT researcher). The scales of data, which often describe as big data volume, are created from various ranges of sources such as unstructured and structured data. For example, text messages, pictures, videos, TV, machine-to-data and more [29]. Velocity is the speed of streaming data. Data needs to be gathered from various sources therefore it needs to be processed and analysed. Sometimes, velocity is more important than volume because important information needs to be interpreted from what big data has provided. So that interested parties such as companies and health care will be able to make decision and act quickly without wasting time. Variety is information that comes in various types of formats for example unstructured data like Facebook contents, email and structured data such as cash transaction. Variability is about data flows that it can be greatly inconsistent which related with social media trends, daily, seasonal and other reasons. The disparity of data flow is depends with certain developmental indicators [29]. Big Data methods usually assume that interrelation can be considered practical indications of relations among the variables [30]. Therefore, any items or events that showing an abnormal behavior for example, unusually high rate flow of data of some categories may be linked with some kind of problem. Lastly, data complexity should not be confused with data variety. Data variety is more into different type of data formats such as videos, photos, whereas data complexity is about different type of data sources such as social media, blogs and so on. Therefore, these data need to find their connection, corresponding, organizing and transforming between other data sources across the systems. For example, by organizing and finding connection between the data from diverse data sources such as governments, corporations and so on helps to discover the issues.

In order for us to fully utilize big data, there are two types of big data technology that is operational and analytical systems [31]. Operational systems provide interactive workloads and offer operational capabilities in actual time where data is primarily put in and kept whereas analytical systems analyse most of the data for complex analysis on past data. These two technologies are often bringing into action by combine these two together to enhance the quality of the data.

Operational big data, NoSQL big data systems for example document databases have evolved to address a broad set of applications and are optimized for more specific applications such as graph database [31]. It is also to use effectively the new cloud computing architectures so that it can be run efficiently and inexpensively. Therefore, this makes operational big data easier to manage, faster to apply and cheaper. On the other hand, analytical big data which known as Map Reduce systems are used to address the limitations of traditional databases system such as their lack of ability to measure the resources beyond single server. With increasing volume of data, these analytical workloads provide valuable information for business. Hence, although operational and analytical big data system slightly different but both does provide precious information if it is use for good cause.

4.1 The Benefits

Big Data is important for businesses, education, health care, banking, government and manufacturing uses. Big Data is use to gather data, storage and retrieval of data in order to assist our findings. Therefore, it helps us to collect and analyse the data to get the right information in order to distinguish and solve the issues. It is also helps to make fair decision and act wisely without taking any longer time.

From businesses perspectives, big data helps to targeting and gain more understanding about their products, customers, competitors and so on. When the big data is efficiently and effectively analysed and processed, hence it can lead to lower costs, increase sales, improvement in customer services and products. For example, retailers would be able to know who their customers are and why they choose their product. By using social media for example, Instagram, Facebook and E-commerce sites such as eBay, help sellers to understand who did not buy their products and the reasons behind it. Furthermore, customers now can give their reviews about a certain products through online which mean companies can use that information to improve their product weaknesses. Netflix is one of the examples of using big data to become one of the successful innovative companies. Netflix have attracted millions of users with its original high-quality programming that it knows will be fond of by large audiences, [32].

In the case of farming activities and agriculture, big data can be used to improve agricultural productivity. Studies shows that farm operations become ineffective which can reduce farmer's productivity by up 40% because of late planting or weeding, the lack of proper land and harvesting techniques. Therefore, to efficiently use big data, data can be collected to observe and analyses by the providers of prescriptive-planting technology about soil conditions, seeding rates, crop yields and other variables from farmers' tractors and drones is combined with detailed records on weather patterns and crop performance. Next, the data is deciphered by algorithms and human experts, changed the data into useful information to advice farmers, and sent directly to their machines. This is to command the machines to set optimum amount of pesticides, herbicides, fertilizer, and other applications. Furthermore, a corn and soybean farmer in Iowa used a $30,000 drone to study how the yield in his 900-acrefarm that is affected by changes in terrain and other factors. This explained that farmers who have implemented big data technology which is known as data-driven prescriptive planting, based on analysis of nutrients in soil and other factors, it have reported a remarkable increase in productivity. In addition, this shows that even small modification in planting depth or the distance between rows of crops can lead to a significant increase agricultural productivity.

4.2 Privacy

Big Data collects patterns or behaviour of human or natural environment. It collects different types of data that includes health records. This can be beneficial for a private or public hospital or clinic to predict future health problems and to find ways to prevent future diseases, which benefits the public. However, this can make it as the disadvantage because it will expose the health problem that a patient faces which considered as personal information. Some people would prefer not to expose their health problems because they are fear of being judge by other people. This may lead to a serious problem as there would be a chance for the victims to instance attempting suicide.

Furthermore, big data can create inconveniences for the customer. For example, when buying vehicle insurance, the company need Social Security number where they can use this to get their customers' profile or other information such as their e-mail address. Once they know the contact details, the company will start spamming their customers' e-mail by offering and promoting the company's products. These can make the customers to feel dissatisfied which finally result to a huge lost in customer relationship, if the customers find it inconvenience. Fortunately, this can also be an advantage because some of the customers may find it useful because they will be aware about the company's products which cause improvement in customer relationship. Moreover, some people may be wary of what information has been collected. For instance, two big companies of social networking that use big data are Google and Facebook [33]. The users of Google and Facebook do not know on what information has been collected. This may create the feeling of fear from the users as the information can be used for identity theft. In addition, big data also collects most information of their customer's transactions that are made online. So, the customers' information such as their transactions and their behaviour of browsing on the company's websites can be seen from the big data technologies. Not only big data exist in online websites or large online companies, but it also used in mobile phones. This means that people will know the wealth status of a person by knowing what brand of mobile phones they use. Since mobile phones are wireless and easy to bring anywhere they go, big data also collects the location of people via the mobile phones. Practically, big data in this case are not secure.

4.3 Data Size

Most big data contains lack of structured data, contain variety of information, information that last for a short period which makes it difficult to organised and search [34]. Storing and using data could also be a problem for some companies. Using the data is hard because they need to learn several techniques including learning a machine, processing language, predicting the future, learning on mapping the neural networks and social networks [35]. Once the company received the data, they have to sort it out to make it organized. This is because the data is stored in multiple locations and it is organized in many different ways [36]. This means that the employees of an organization have to relocate, organized and merge the data. Therefore, the management of an organization have to train the employees on how to sort out the data. This may takes time and needs to invest on training their employees.

4.4 Shifting Paradigm

Some companies might upgrade to big data as part of their decision making and may no longer using the traditional ways to make decision such as making decision in democratic style. Upgrading new equipment or tools in an organization give some negative impacts on the employees. This is because some of the employees may not be interested to learn and reluctant to change their culture of making decisions, even though the upgrades give more convenience to them. This may lead to a reduction on motivation level, productivity and an increase in numbers of labour turnover. As big data technologies are difficult and take time to learn, this may make them difficult to accept the change on the organization. It is the work of the chief executive officer (CEO) or other managers to convince their subordinates to learn about big data technologies.

4.5 Security

In terms of security, big data may not be secure. This is because some people misuse the information by doing criminal activities such as identity theft. In addition, curiosity also leads to criminal activities where people accessing the data unauthorized [37, 38]. This may cause to a leakage of confidential information especially in the government sector that uses big data technologies. Big Data also can be use by the police to track down criminal activities such as drug distribution, fraud and bribes. However, the police can exploit this information by receiving bribes from the criminals to let them not facing the punishments that they deserve [38]. They receive bribes, by looking at the data of the pattern of the criminal to make sure it is secure to receive it.

4.6 Solutions of the Issues of Big Data

Transparency from the company is required in using customer's information. For example, the vehicle company that needs Social Security number. They can clarify to the customer why they need the Social Security number because this contains personal information. If the company need to contact the customers to notify about their products, they have to ask permission from the customer itself. In addition, the company also has to provide a freedom for their customers to stop receiving any information about their products anytime and anywhere they want. This is to avoid lost in customer relationship and also to make customers secure.

4.7 Convincing

These apply to the change in management from the traditional ways of making decision to using big data. The most important ways is to convince the employees who do not accept the change in management or culture. One of the ways to convince them is by explaining verbally and practically on the difference of outcome between the new technology used and the traditional ways. This can be done by the CEO or other employees such as manager to convince other employees or subordinates who are unwilling to accept the change of the culture. Moreover, the organization can discuss about the change of the management with their subordinates. This is to prevent from getting a high labour turnover, which could lead to the organization to be in a risk.

4.8 Understanding the Security

As security is one of the disadvantages of big data, it can be secure by using several ways such as understanding the security of the information that they collect [39]. It is important for the management of an organization to be aware of the security of the information that they collect because this involves with their customers [40].

Even though the information can improve their products and services quality, but if they do not understand well on the security, they are putting the organization in danger because it may have a chance of being sued by the customers for exposing their personal information. Other than that, the organization can set verification to access the information such as asking for username and password. This is to prevent from unauthorised access to the information [41].

5 The Challenge: LMS Through Big Data

Big Data is a whole load of information that is gathered through online sites when connecting technology that acts as a medium, to internet. It is formed from a collection of small and many structured of different data and information worldwide. One of the information or data that is not doubted to be found in the big data is E-Learning sites that have been posted or uploaded online. Whereas LMS is a new development concept of pedagogy in this modern era, which may vanish the traditional way of learning and teaching; face-to-face interactions, papers, pens and pencils. But there will be different people experiencing different collection of materials [42]. Moreover, different LMS programmes offer different courses. Consequently, the big data is storing the uploaded materials that are used for E-Learning purposes. Fortunately, E-Learning requires users to have the skills when using this modern and highly advanced technology, because LMS needs the technology itself and good internet connection. A complete set of the skills, the technology and the connection will allow the user to go for LMS. LMS has its own sites on the internet for E-learners to refer to. This is called as a Web page or websites. This Web page has its specific web address. All postings and uploads that are made, whether they are formed in words or numerical format, that are keyed in through online will automatically be saved in the big data. This is why when a user is entering the specified of particular web address; the process to get to the information will take roughly less than a few seconds to show up on the user's screen computer. It is because the information is already being set up for the specific web address. This is how E-Learning and big data are related.

LMS may sometimes require people to sign in or login before they can use or access to the materials. The reason is for a safety rights of every users and to keep track on the user's activities that are being sorted according to their identification or username. The procedure requires a space for storing every user's information. This is when big data comes in handy. Every information or materials used online will automatically transfer to the big data. At the end of the learning programme, marks or grades and the results of assignment will be delivered in a simple form of LMS environment [43]. This marked up the relationship between big data and E-Learning. Another relationship between big data and E-Learning is that information or sources in the E-Learning system are stored in the big data as for the users' reference. Despite of the importance of internet connection in the system, it also needs an advanced technology and IT skills for the use of E-Learning. This includes the web layouts and features that have already been formatted and displayed in the Hypertext Mark-up Language (HTML) page. Which will then is saved or stored in the big data once the page and the layouts are see by IT experts or companies. HTML uses tags or codes for the graphics, headings, list, or body text, which concluded that HTML are the three components of text, tags and references to files [44]. This shows how the data are coded and formatted accordingly or in sequence. In addition, the HTML tag is coded different documents in it. HTML tag will be referred to the specific website or web address that involves or contained in the big data.

Big Data represents the whole picture of many and variety of different contents and contexts of information and data of worldwide internet used. E-Learning sites will only be bits in the storage room. E-Learning has many functions; auto-correction, improvements, can gain and access wide knowledge and many few. Ideally, E-Learning will determine the impact and effects of students and the school systems. Furthermore, analysing through big data can strategically identify any problems or errors, which an immediate transformation can be made after the detection or trial and errors (Fig. 2). This advanced procedure can eventually be done and improved by opting for alternatives or using different method and system after discovering faults and mistakes. This is connected to the big data storage. Unfortunately, big data can also be challenging in retrieving, storing, processing the sorted data, causing from the continuously growing information for millions of bytes daily in the big data per seconds [45]. Errors may occur if data is corrupted and is inconvenient and no longer can be used.

Moreover, the purpose and the goal of LMS can give endless opportunities in the 21st century in referring to the big data technology and ideology [6]. As changes of learning are growing much faster now, there will be possibility for other new and innovative ideas to be taken into account. Hence, the involvement of big data gives

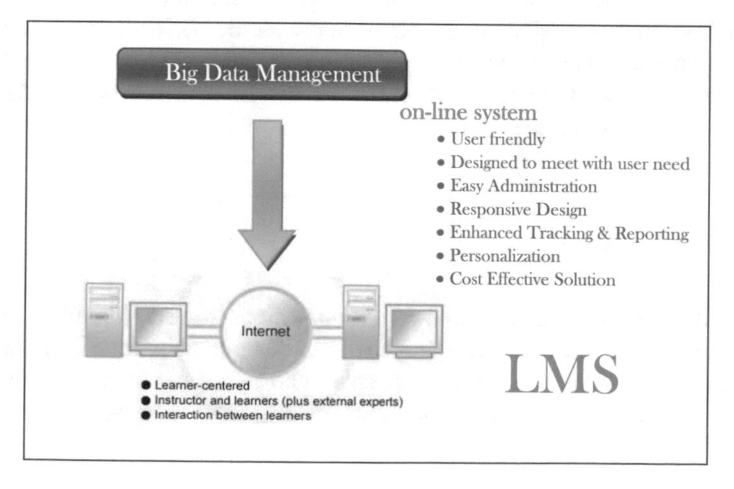

Fig. 2 LMS trough Big Data

people wide access to E-Learning Because it also diverse the idea of learning; distant learning or distant communication. It is arguable for the revolution of higher education [6]. Moreover, the duty of the big data not only just focus on the term education, it also covers other matters as well for example, health care and many other. E-technologies do not change the learning style, in fact they are actually removing constraints in terms of accessing, social behaviours and economy friendly [42]. This idea has the same meaning as to the idea of big data. Storage of information helps users to a wide access into relevant and reliable sources, to socialise by communicating through online. In addition, E-Learning is economy friendly, by means that users are now working in a paperless world [46]. Advanced technologies befriended humans and encouraging them to the most easy and available access of almost the cheapest education on the internet. Eventually, these good changes and transformations will prepare the world to be very IT oriented and minded. Technology varies rapidly from time to time; in terms of structure, multifunction, and other particular changes. It is very risky to stick or being loyal to one dusty and old-functioned technology. Therefore, education sectors should be responsible to keep on track on which technology is used by the corporate sector [47]. Other alternatives that can be taken into account, the authorities could keep pace on the newest and latest technology that is displayed and available, but should be cost lesser than the budget limit. Eventually this transformation will increase the capacity use for big data storage. As more and different new ways of how a technology can be capable of doing or has been differentiated the ability to function in advance. In addition, how the modern world is highly depends on technologies in many other activities; distant communication with other branches of a business, controlling traffic congestions for road safety, health pharmacy and many more. Lastly, big data encourages good decision making for the business to grow healthier in a market. In a way, that E-learners are learning through online, businesses can use both E-Learning and big data to introduce or promote and what are the benefits that can be retrieved in the area. Firstly, the use of E-Learning in the business is to contribute ideas and motivating people to invest in the business sector; uploading videos on how to succeed and participating in a business field. While, big data acts as storing the appropriate information for the use of the business and its people through internet and online [48].

6 Time Collision

Having all learners to be available at the same time in an online class might be impossible. This is because not everyone is available at the time slots that were given. As for examples, the in-service people (who are mostly having worked to do) may face difficulty to participate the online class during their working hour. This was stated by Codreanu and Vasilescu [49], in which the learners or the participants have to be managed with the work, family or even vacation schedule changes that need to be made in order for them to be online at the same time as others for discussing questions and answers as many as possible in a live online dialogue.

6.1 Proper Training to Use E-learning

Both lecturers and the students need to be trained properly on how to use Information Technology (IT) in order for them to easily access using E-learning for the students to learn. Without a proper training, there could be a high complication on using E-learning in the future. Even if we asked the most experience lecturer to use E-learning in a sudden without a proper training, the lecturer can become frustrated to start something that is new for the lecturer [50]. Maddix also states that through a constant practice, E-learning requires having the right facilities, knowing the right procedures, and developing the exact abilities. Therefore, a right ways of continuous training may consume less time for both of them to succeed using the E-learning system.

6.2 Academic Fraud

Antoine [7] stressed out that a question would be appeared as to whether the student who is getting the credit is actually the person completing the assignment which this can also be known as an academic fraud. There is a possibility that the students who are applying for the courses are not the one who completed every assignments and quizzes that were given by the lecturer. Hence, if this happen, the student will take an opportunity to search for people who can do all of his/her work in order to be considering as a graduate student. In other words, traditional face-to face ways of learning is much more safety as the lecturers can easily recognised their students.

6.3 Technical Expertise

Eke [4] reported that a lacked of technical expertise would be the issues of using E-learning. Imagine that in a middle time of E-learning session and suddenly you face a technical problem, it would take a lot of time to fix the issue if there is a lack of expertise because there could be a chance that you are not the only person who faced the problems itself. It may take several days or even months if there are a limited number of expertise in the office.

6.4 Confusion of Lecturers and Learners

Another issue is that, the lecturers and the learners may face a confusing on using the E-learning [4]. If confusion and misunderstanding occurs between lecturers and the learners, there would be a chance that the learners might do a mistake on the format of assignment or reports, as there is limited communication between them. If the learners have done the wrong format, the chance for him/her to fail the course is very high.

6.5 Limited to Access to the Online Resources

In addition of this issue, not all libraries have a complete books or resources that are needed for the learners to do the research. Even if they try to do the research through online, some journals or the resources need to be paid first before accessing through them. As we know, some students may not be able to afford the amount that are needed to access the journals. Hence, according to Eke [4] ones again, that the uses of a huge amount of money to access the resources could be the lack of information gathered for researchers especially when involving the students. Therefore, this can be concluded that the learners may not be able to finish their research for the assignments.

6.6 Solutions of the Issues of E-learning

E-learning is not only helping the students to do their work but also lecturers or even the cooperation organisations. E-learning can have a lot of advantages and some disadvantages. The disadvantages can be overcome if they use E-learning correctly. E-learning users should consider what they want the sources for their work or assignments and not just copying everything from the internet or Google as some of the sources cannot be trusted. Before copying everything users need to filter which information they can use and whether the resources are trustable enough. Most importantly E-learning can reduce cost and this will be a huge advantage to everyone who use E-learning [51, 52].

7 Conclusion

LMS is quite related to big data, since it online are going to use data or information that stored in big data, such as videos that include the lectures and more education materials from big data to support the learning processes. All aspects or perspectives of information have its own advantages which is same goes to the big data and LMS accordingly. The most significant outcome the implementation of LMS has level of convenience of contribution. In today's era of internet communication, where information can be disseminated in an instant over the internet, LMS prove to be an essential tool for university students and instructors alike. Not online can they keep in line with their assignments, the tools to complete their assignments, and various other assessments, university instructors also share the same level of ease, having an easier time in reaching out to their students whenever necessary outside of class hours. Although there are some difficulties to be faced in using a LMS, universities should provide the appropriate level of training and guidance for all users and then on, it would only be a matter of time where issues in access were only a memory, where the enhancement of their learning process remains to be the ultimate goal to be achieved.

References

1. Wiberg, M. (2012). Interaction per se: Understanding the ambience of. *Innovative Applications of Ambient Intelligence: Advances in Smart Systems*, p. 71.
2. Kamal, S., Ripon, S. H., Dey, N., Ashour, A. S., & Santhi, V. (2016). A MapReduce approach to diminish imbalance parameters for big deoxyribonucleic acid dataset. *Computer Methods and Programs in Biomedicine, 131,* 191–206.
3. Baumgarten, M., Mulvenna, M., Rooney, N., & Reid, J. (2013). Keyword-based sentiment mining using Twitter. *International Journal of Ambient Computing and Intelligence, 5*(2), 56–69.
4. Eke, H. N. (2010). The perspective of E-learning and libraries in Africa: Challenges and opportunities. *Library Review, 59*(4), 274–290. doi:10.1108/00242531011038587
5. Sangrà, A., Vlachopoulos, D., & Cabrera, N. (2012). Building an inclusive definition of E-learning: An approach to the conceptual framework. *International Review of Research in Open and Distance Learning, 13*(2). Retrieved from http://search.proquest.com/docview/1634473719?accountid=9765
6. Garrison, D. R. (2011). *E-learning in the 21st century*. New York, NY: Routledge.
7. Antoine, J. E. (2011). *E-learning: A student's perspective a phenomenological investigation* (Order No. 3494459). Available from ProQuest Central (921887410). Retrieved from http://search.proquest.com/docview/921887410?accountid=9765
8. Tao, Y. H., Yeh, C. R., & Sun, S. I. (2006). Improving training needs assessment processes via the internet: System design and qualitative study. *Internet Research, 16*(4), 427–449.
9. Liaw, S. S., & Huang, H. M. (2003). Exploring the world wide web for on-line learning: A perspective from Taiwan. *Educational Technology, 40*(3), 27–32.
10. Alvandi, E. O. (2011). Emotions and information processing: A theoretical approach. *International Journal of Synthetic Emotions (IJSE), 2*(1), 1–14.
11. Odella, F. (2017). Technology studies and the sociological debate on monitoring of social interactions. In *Biometrics: Concepts, methodologies, tools, and applications* (pp. 529–558). IGI Global.
12. Martin, A., & Madigan, D. (2006). *Digital literacies for learning*. London: Facet Publishing. ISBN 9781856045636.
13. Ndubisi, N. O. (2004). Factors influencing E-learning adoption intention: Examining the determinant structure of the decomposed theory of planned behaviour constructs (pp. 252–262).

14. Harasim, L. (1995). *Learning networks; A field guide to teaching and learning online*. Cambridge, MA: MIT.
15. Falvo, D. A., & Johnson, B. F. (2007). The use of learning management systems in the United States. *TechTrends, 51*(2), 40–45.
16. Heirdsfield, A., Walker, S., Tambyah, M., & Beutel, D. (2011). Blackboard as an online learning environment: What do teacher education students and staff think? *Australian Journal of Teacher Education, 36*(7), 1.
17. Lonn, S., Teasley, S. D., & Krumm, A. E. (2010). Who needs to do what where?: Using learning management systems on residential vs. commuter campuses. *Computers and Education, 56,* 642–649. doi:10.1016/j.compedu.2010.10.006
18. Wang, Y., & Chen, N. S. (2009). Criteria for evaluating synchronous learning management systems: Arguments from the distance language classroom. *Computer Assisted Language Learning, 22*(1), 1–18.
19. Al-Busaidi, A., & Al-Shihi, H. (2010). Instructors' acceptance of learning management systems: A theoretical framework. *Communications of the IBIMA, 2010,* 10.
20. Al-Senaidi, S., Lin, L., & Poirot, J. (2009). Barriers to adopting technology for teaching and learning in Oman. *Computers and Education, 53*(3), 575–590.
21. Brown, A., & Johnson, J. (2007). Five advantages of using learning management system.
22. Lindgaard, G. (1994). *Usability testing and system evaluation: A guide for designing useful computer system* (1st ed.). UK: Chapman & Hall.
23. Algahtani, A. F. (2011). *Evaluating the effectiveness of the e-learning experience in some universities in Saudi Arabia from male students' perceptions*. Durham theses, Durham University.
24. Almunawar, M., Susanto, H., & Anshari, M. (2013b). A cultural transferability on IT business application: iReservation system. *Journal of Hospitality and Tourism Technology, 4*(2), 155–176.
25. López, M. (2014). *Right-time experiences: Driving revenue with mobile and big data*. Somerset, NJ, USA: Wiley. Retrieved from http://www.ebrary.com
26. Bhatt, C., Dey, N., & Ashour, A. S. (2017). *Internet of things and big data technologies for next generation healthcare*. Switzerland: Springer.
27. Mason, C. (2015). Engineering kindness: Building a machine with compassionate intelligence. *International Journal of Synthetic Emotions (IJSE), 6*(1), 1–23.
28. Hurwitz, J., Halper, F., & Kaufman, M. (2013). *Big data for dummies*. Somerset, NJ, USA: Wiley. Retrieved from http://www.ebrary.com
29. Kshetri, Nir. (2014). The emerging role of big data in key development issues: Opportunities, challenges, and concerns. *Big Data and Society*. doi:10.1177/2053951714564227
30. Hannu, K. Ã., Jussila, J., & Jaani, V. Ã. (2013). Social media use and potential in business-to-business companies innovation. *International Journal of Ambient Computing and Intelligence (IJACI), 5*(1), 53–71.
31. Mongo D. B. (n.d.). *What is big data?* Selecting a big data technology: Operational vs. analytical. Retrieved from https://www.mongodb.com/big-data-explained
32. Fast Company Staff. (2015). *The World's top 10 most innovative companies of 2015 in big data. Most innovative companies.* Retrieved from http://www.fastcompany.com/3041638/most-innovative-companies-2015/the-worlds-top-10-most-innovative-companies-in-big-data
33. *Why big data has some big problems when it comes to public policy*. (2014). San Francisco: Newstex. Retrieved from http://search.proquest.com/docview/1618317749?accountid=9765
34. Toub, C., & Alliance B. (2012). Big data provides big risk and big opportunity. *Institutional Investor*. Retrieved from http://search.proquest.com/docview/1030124978?accountid=9765
35. Morton, J., Runciman, B., & Gordon, K. (2014). Big data: Opportunities and challenges. Swindon, GBR: BCS Learning & Development Limited. Retrieved from http://www.ebrary.com
36. Smith, K. (2015). Big data big concerns. *Best's Review, 7*, 58–61. Retrieved from http://search.proquest.com/docview/1729782816?accountid=9765

37. Susanto, H., Almunawar, M. N., & Tuan, Y. C. (2011a). Information security management system standards: A comparative study of the big five. *International Journal of Electrical Computer Sciences IJECSIJENS, 11*(5), 23–29
38. Susanto, H., Almunawar, M. N., Tuan, Y. C., Aksoy, M., & Syam, W. P. (2011b). Integrated solution modeling software: A new paradigm on information security review and assessment.
39. Susanto, H., & Almunawar, M. N. (2012). Information security awareness within business environment: An it review.
40. Susanto, H., & Bin Muhaya, F. (2010). Multimedia information security architecture framework. In *5th International Conference on Future Information Technology (FutureTech)* (pp. 1–6). IEEE.
41. Susanto, H., Almunawar, M. N., & Tuan, Y. C. (2012b). I-Sol framework: As a tool for measurement and refinement of information security management standard.
42. Horton, W. K. (2001). *Leading e-learning*. Unites States of America: ASTD.
43. Holmes, B., & Gardner, J. (2006). *Challenges of assessment for E-learning. E-learning: Concepts and practice*. London: Sage Publications Ltd.
44. Stair, R. M., & Reynolds, G. W. (2014). *Principles of information systems* (Vol. 11). USA: Course Technology.
45. Suneja, V. (2015). Issues with big data analytics–As it is.
46. Rosenberg, M. J. (2001). *E-learning: Strategies for delivering knowledge in the digital age*. USA: The McGraw-Hill Companies, Inc. (from paper to online).
47. Anderson, T. (2008). Characteristics of interactive online learning media. In P. J. Fahy (Ed.), *The theory and practice of online learning* (Vol. 2, p. 167). Canada: AU Press, Athabasca University.
48. Almunawar, M. N., Anshari, M., & Susanto, H. (2013a). Crafting strategies for sustainability: How travel agents should react in facing a disintermediation. *Operational Research, 13*(3), 317–342.
49. Vasilescu, C., & Codreanu, A. (2013). E-learning behaviors and their impact on andragogy. In Conference proceedings of» eLearning and Software for Education «*eLSE* (No. 01, pp. 126–137). Universitatea Nationala de Aparare Carol I.
50. Maddix, M. A. (2013). E-learning in the 21st century: A selected top ten book reviews since 2001. *Christian Education Journal, 10*(1), 163–179. Retrieved from http://search.proquest.com/docview/1344162401?accountid=9765
51. Arkorful, V. (2014). The role of E-learning, the advantages and disadvantages of its adoption in higher education. *International Journal of Education and Research, 2*, 397–403.
52. Susanto, H., Almunawar, M. N., & Tuan, Y. C. (2012a). Information security challenge and breaches: novelty approach on measuring ISO 27001 readiness level. *International Journal of Engineering and Technology, 2*(1) (UK: IJET Publications).

Performance Evaluation of Big Data and Business Intelligence Open Source Tools: Pentaho and Jaspersoft

Victor M. Parra and Malka N. Halgamuge

Abstract Despite the recent increase in the utilisation of the tools of "Big Data" and "Business Intelligence" (BI), the investigation that has been carried out regarding the inferences related to its implementation and performance is relatively scarce. Analytical tools have a significant impact on the development and sustainability of a company since the evaluation of the clients information are critical aspects and crucial in the progress towards a competitive market. All corporations at certain phase in their life cycle, require to implement different and improved data processing systems in order to optimize the decision making procedures. Enterprises utilise BI outcomes to pull together records that has been extracted from consolidated analyses from signals in the data grouping of BI information scheme. This, in turn, gives a marked advantage to companies in the development of activities based on predictions, and also to compete with competitors in the market. Business Intelligence applications are precise implements that resolve this matter for companies that require, data storing and administration. The chapter examines the most recommended Business Intelligence open source applications currently available: Pentaho and Jaspersoft, processing big data over and done with six databases of diverse sizes, with a special focus on their extract transform and load and Reporting procedures by calculating their routines through computer algebra systems. Moreover, the chapter correspondingly makes available a complete explanation of the structures, features, and comprehensive implementation scenario.

V.M. Parra (✉)
Charles Sturt University, Melbourne, Australia
e-mail: 11581338@studygroup.edu.au

M.N. Halgamuge
The University of Melbourne, Melbourne, Australia
e-mail: malka.nisha@unimelb.edu.au

N. Dey et al. (eds.), *Internet of Things and Big Data Analytics Toward Next-Generation Intelligence*, Studies in Big Data 30,
DOI 10.1007/978-3-319-60435-0_6

1 Introduction

Business Intelligence (BI) software adapts stored data of a company's clientele profile and turns it into information that forms the pool of knowledge to create a competitive value and advantage in the market it is in [1].

Additionally, BI is used to back up, improve the business with practical data, and utilise the study of the data, to continually increase the organisations competitiveness. Section of this examination is to make available appropriate and opportune outcomes, for managements and to make the decision based on factual information, so their decision-making is based on concrete evidence.

Howard Dresner, from Gartner Group [2], was the first to wedge the term BI, by means of a term to describe an assembly of concepts and procedures to support the decision making, by using information found upon facts.

BI system gives enough data to use and evaluate the needs and desires of customers. Additionally it allows to [3]: (i) Create information for divisions or total ranges in a corporation, (ii) Design a database for clients, (iii) Build settings for decision making, (iv) Share materials among parts or sections of a corporation, (v) Sandbox approaches of multidimensional strategies and proposals, (vi) Mine, transform and process data, (vii) Provide an original method to decision making and (viii) Increase the feature of client service.

The advantages of systemizing BI comprise the combination of data from numerous sources of information [4], generating users profiles for data managing, decreasing the reliance on the IT division, reducing the period of getting information, improves the examination, and even enhances the readiness to access real-time information in accordance with precise existing business standards.

The Gartner Magic Quadrant released for business intelligence platforms in 2015 [5] has emphasized the variations being made by the BI area in order to speedily expand systems that might be implemented by corporate operators and specialists to obtain information from the collected data. Usually, BI is assumed as a group of procedures, tools and technologies utilised to convert data into information and then into a personal profile of customers that is then produced into valuable structured data ready to be used by different areas of company enterprise [6].

The Internet of Things (IoT) is a simple concept: we are going to make our products smart. Through the implementation of microchips and sensors with internet connection, we will fill our lives with connected products, whether clothing, a factory production chain or the irrigation system of a garden.

Therefore, one of the consequences of the arrival of IoT is the generation of data [7]. Such data, from the aforementioned increase in the number of connected devices, should be managed not by users but by companies. To them was Howard Baldwin writing for Forbes: "We are going to have data thrown in from all directions: from applications, from machinery, from railroad tracks, from transport containers, from power stations. If that does not put you thinking about how to manage those data providers in real time, nothing will" [8].

The good news is that the Internet of Things is being implemented in a progressive way and has not yet begun to produce a scandalous amount of data, whose management is dedicated to the big data.

Big Data is the technological trend based on the massive analysis of data that cannot be processed or analysed using traditional tools (that is, they usually have to be managed by a specific platform: "Business Intelligence"). Unlike IoT, big data has been with us for some time. Thus there are analytical open source tools such as Pentaho and Jaspersoft, designed for the management and analysis of large volumes of data and information, which has become one of the most valued and requested fields of study of recent times.

Being able to gather and manage data thanks to big data is a basic requirement for the arrival of IoT. In fact, without a correct structure of big data in the company, the implementation of the IoT seems to be complicated, since it is worth nothing to generate data of high value for both companies and users if they cannot be managed and analysed in a correct way. Without big data, the internet of things will be nothing more than something in our imagination.

Consequently, big data will assist in the growth of better methods that permit (BI) tools to be used to collect information, for instance [9]: (i) Transform and examine large amount of information; (ii) Rise the world of data to take into account when decision-making: and intrinsic chronological data of the business, to include data from external sources; (iii) Make available an instant answer to the sustained delivery of actual time data of the systems and the opportunities of interconnections among equipment; (iv) Dealing by way of structures of composite and varied data; (v) and finally, to segregate from the material restrictions of data storing and process by using increasable solutions and great accessibility at a reasonable costs.

This chapter offers an investigational examination of the contrasts of two most suggested and accessible BI open source applications at present: Pentaho and Jaspersoft, processing big data over and done with six databases of diverse sizes, with an exceptional emphasis on ones ETL and Reporting procedures by calculating their routines with Computer Algebra Systems. The Pentaho and Jaspersoft tools, apart from being two of the open source BI applications currently with the best performance in the market, have as intrinsic characteristics, the ability to handle and process big data with respectable result metrics.

The Pentaho and Jaspersoft tools are world leaders in Open Source Business Intelligence systems. They offer a wide range of tools oriented to the integration of information and the intelligent analysis of the data [10, 11]. They have powerful capabilities for managing ETL processes (extracting, transforming and loading data), interactive reports (Reporting), multidimensional information analysis (OLAP) or data mining.

On the other hand, most of the big data and Business Intelligence efforts are aimed at extracting and analysing data of interest, which can provide valuable insights for strategic decision making [4]. Business Analytics applications offer integration solutions for big data projects. Thus, the tools selected considerably reduce the time for the design, development and implementation of big data solutions, better than other traditional tools, from extraction and transformation of data to analysis (design of scorecards, reports, etc.).

In conclusion, Pentaho and Jaspersoft provide the necessary tools to analyse large volumes of data and thus identify recurring patterns between them.

In addition, it is possible to apply all the power of analysis, reporting or dashboards for the data obtained. Finally, Data Integration incorporates the necessary connectors for leading analytic database distributors

The purpose of this work is to analyse, assess these tools, and outline how they increase the quality of data, which inadvertently assist us to recognize the market conditions to make upcoming calculations based on tendencies.

Section 2 defines the competencies and components of both Pentaho and Jaspersoft BI Open Sources. Section 3 presents the computer algebra systems SageMath and Matlab. This is followed by the materials and methods (Sect. 4) used in the analysis and experimentation, particularly the ETL and Reporting measurements and how they were implemented. In Sect. 5, the results of the study for CPUtime as a variable of the "size" from the initial data for the ETL and Reporting procedures from both Business Intelligence Open Sources, applying two different Computer Algebra Systems. Section 6 contains the discussion of the experimentation. Section 7, the conclusion of the study.

2 Pentaho and Jaspersoft Business Intelligence Open Sources

2.1 Pentaho

The Pentaho BI project is a continuing open source communal initiative that offers companies with a great application for their business intelligence requirements. Taking advantage of the richness of open source technologies that influence the open source development community, this application is capable to transform more rapidly than commercial sellers do [10]. Consequently, Pentaho offers an option that exceeds trademarked Business Intelligence solutions in numerous parts such as architecture, standards support, functionality and simplicity of deploy. This business intelligence tool developed under the philosophy of free software for business decision making and management has a platform composed of different programs that meet the requirements of BI. It offers solutions for information management and analysis, including multidimensional OLAP analysis, reporting, data mining and the creation of dashboards for the user [12]. This platform has been developed under the Java programming language and has an implementation environment based on Java, making Pentaho a very flexible solution to cover a high range of business needs. Pentaho denes itself as a solution-oriented and process-centric BI platform that includes all the key components required to implement process based solutions as it has been designed from the ground up. The solutions that Pentaho offers basically consist of an integrated analysis and reporting tools infrastructure with a business process workflow engine [13].

The platform is able to execute the necessary business rules, expressed in the form of processes, activities, and it is able to present and deliver the appropriate information at the required time. Pentaho compromises proper solutions through the range of resources to improve and preserve the processes of BI projects from the Extract Transform and Load usage data integration to the dashboards through Dashboard Designer [10]. This tool has constructed its application Business Intelligence incorporating diverse current and documented creditworthiness projects. Data Integration was recognized as Kettle; certainly, it kept this one old name as an informal term. Mondrian is an additional element of Pentaho that holds its individual object.

2.2 *Pentaho Components*

2.2.1 Pentaho Data Integration (Previously Kettle)

This is the module that control the ETL procedures. However, Extract, Transform and Load artefacts are most commonly implemented in data warehouses applications [14], as Pentaho Data Integration is able correspondingly to be implemented for additional commitments:

- Transferring data among programs or databases
- Distributing data from databases to flat files
- Filling data vastly into databases
- Data debugging
- Incorporating components

Incorporating components of Pentaho Data Integration is informal to implement. Each procedure is formed through a graphical instrument where users can determine what to do deprived of creating an algorithm to specify how to do it; due to, it might be said that Pentaho Data Integration is focused on metadata. Pentaho Data Integration is able to be implemented as a separate artefact, or it might be implemented together with the Pentaho Suite. As an Extract Transform and Load instrument, this is the most widespread open source application offered. It supports a massive collection of input and output presentations, comprising text les, data pages, and different sets of database engines. Furthermore, the modification competences of Pentaho Data Integration permits to operate data by means of almost zero restrictions. This component is one of the best commonly implemented ETL solutions and well appreciated in the market [12]. This one has a positive feedback on its, strength, and robustness that becomes a much-suggested tool. It permits renovations and function in a very easy and instinctive method, as it is illustrated in Fig. 1.

Similarly, the Data Integration projects are very simple to keep up.

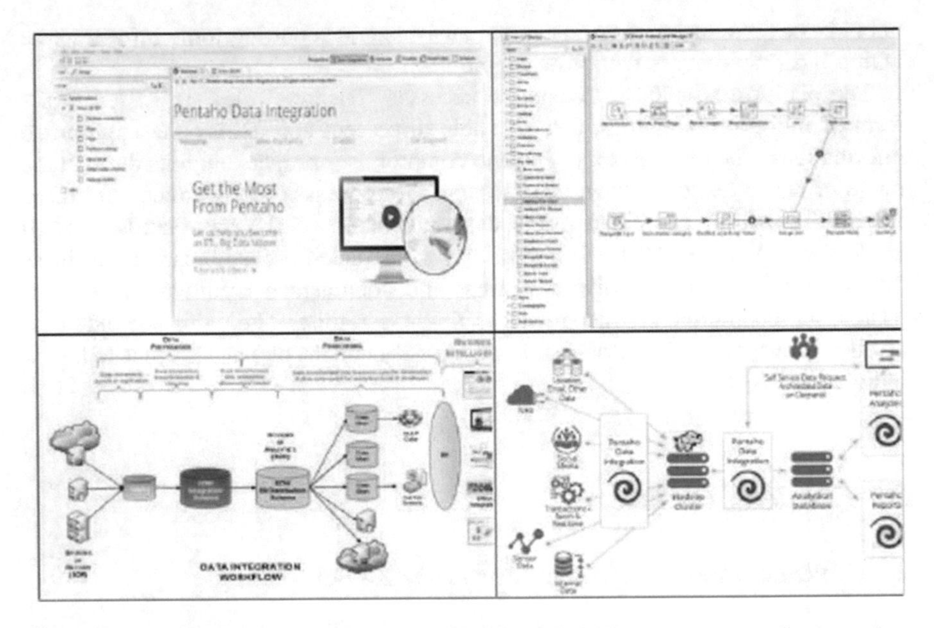

Fig. 1 Pentaho data integration interface and Workflow, the ETL component permits renovations and function in a very easy and instinctive method

2.2.2 Web Application-BI Server

Pentaho BI Server provides the end user web server and platforms. This can interact with the Business Intelligence solution previously created with Pentaho Reporting, Pentaho Analytics, Pentaho Services, DashBoard, Pentaho Data Integration and Pentaho Data Mining. This way, the user can navigate the data, adjusting the view of the data, the filters of visualization, adding or removing the fields of aggregation [13]. Data can be represented in an SVG or Flash form, dashboards widgets, or also integrated with data mining systems and web portals (portlets). In addition, with Microsoft Excel Analysis Services, dynamic data can be analysed in Microsoft Excel (using the OLAP server Mondrian). This hundred percent Java2EE component permits to administrate altogether BI resources [14]. The web application has a BI user interface accessible where reports are kept, OLAP views and dashboards as that one is shown in Fig. 2. Furthermore, this component gives admission to an administration support that permits the management and supervision for both application and usage.

A Report can be designed using the Report Designer or Report Design Wizard and it can deploy a report using the Pentaho BI Server.

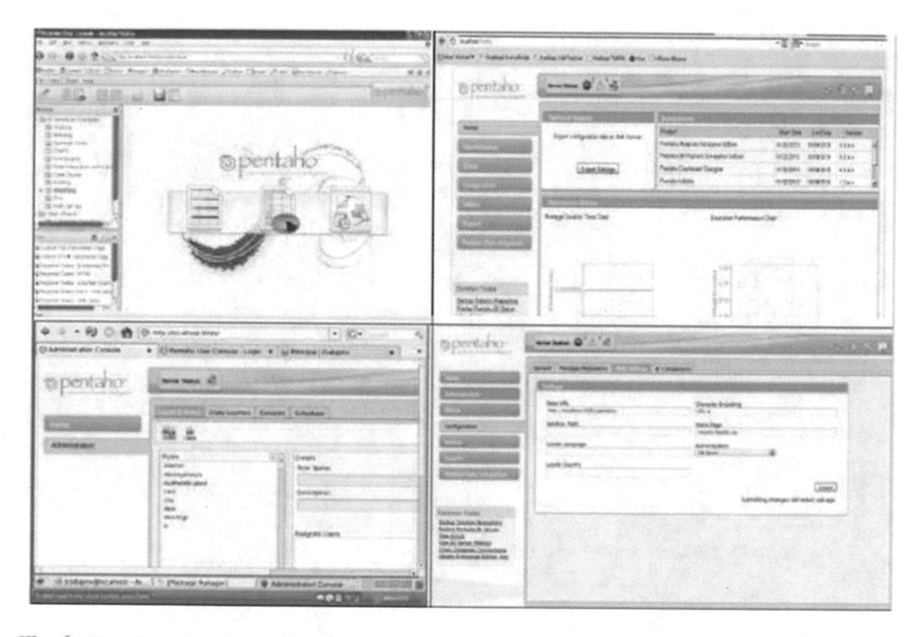

Fig. 2 Pentaho server user interface to administrate the stored BI assets, OLAP views, and dashboards. Moreover, the availability of a management permits the management and supervision for both application usage

2.2.3 Pentaho Reporting

Most organizations use reports to record, visualize and analyse results. Therefore, reports are considered a major need in Business Intelligence. Pentaho Reporting enables organizations to easily access, format, and distribute information to employees, customers, and associates. The reason for this is that Pentaho provides access to OLAP or XML-based relational data sources, as well as offering multiple output formats such as PDF, HTML, Excel or even plain text [10]. It also allows bringing this information to the end users via the web, e-mail, corporate portals or own applications. Pentaho Reporting allows increasing the reporting platform as needs grow. The Pentaho Report Designer is an independent tool that is a part of Pentaho Reporting, which simplifies the reporting process, allowing report designers to create documents quickly and visually rich, based on the project Reports from Pentaho JFreeReport. The report designer offers a familiar graphical environment with intuitive and easy-to-use tools and a very accurate and flexible reporting structure to give the designer the freedom to generate reports that are totally tailored to their needs.

This module arranges for a complete reporting solution. Enhancing the whole features required in any reporting situation, as illustrated in Fig. 3. In summary, this component is the old figure of JFreeReport [13] that offers an instrument for

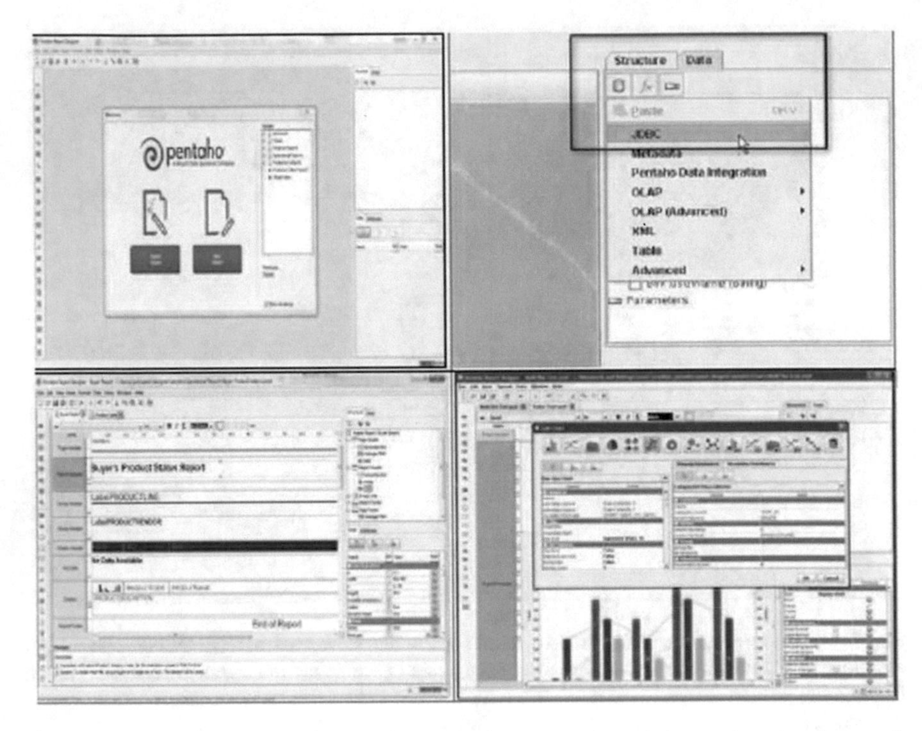

Fig. 3 Pentaho reporting interface with a complete reporting solution, enhancing the whole features required in any reporting situation

reporting, an implementation engine, and metadata instrument for showing reports Ad hoc, and a customer GUI that consents Ad hoc reports.

2.2.4 OLAP Mondrian

OLAP is the acronym for On-Line Analytical Processing. It is an application implemented in the arena of (BI) whose aim is to expedite the consultation of large amounts of data. It uses many dimensional structures (or OLAP Cubes) that contain summary data from huge Databases or Transactional Systems (OLTP). Implemented in company reports for transactions, advertising, administration reports, and data mining and comparable ranges.

The cause for implementing OLAP for queries is the rapidity of reply [10]. The relational database stores entities in distinct tables if they are normalized. This arrangement is good in an OLTP system nonetheless for difficult multi-table queries, it is moderately slow. An improved search method, although inferior from the operative viewpoint, is a multidimensional database. In order to obtain the online analytical processing (OLAP) functionality, two other applications are used: the Mondrian OLAP server, which combined with Jpivot, allows queries to

Datamarts, the results are presented through a browser and the user can drill down and the rest of the typical navigations. Mondrian, now renamed Pentaho Analysis Services, is the integrated OLAP engine in the Open Source Business Intelligence suite Pentaho. Mondrian is an ROLAP engine with cache, which places it near the concept of Hybrid OLAP. ROLAP means that in Mondrian there are no data (except in the cache) nevertheless they reside in an external Database Management System. It is in this database that the tables make up the multidimensional information with which Mondrian works and (the star models of our data marts for example) reside in this database. MOLAP is the name given to the OLAP engines in which the data exists in a dimensional structure [15]. Mondrian is responsible for receiving dimensional queries (MDX language) and returning the data of a cube, only this cube is not a physical thing nonetheless a set of metadata that denes how to map these queries that deal with dimensional concepts to SQL statements that already deals with relational concepts that obtain necessary information to satisfy the dimensional query.

Some of the advantages of this model are:

- Not having to generate static cubes saves from costs, and also unnecessarily occupies the memory.
- The possibility of using the resident data in the database, thus working with updated data is beneficial. This is useful in an operational BI environment.
- Although MOLAP systems traditionally has a certain performance advantage, Mondrians hybrid approach, is the use of cache and aggregate tables, that makes it possible to obtain very good performance, without losing the advantages of the classic ROLAP model. It is very important to take advantage of the database where the tables reside.

Viewer OLAP Pentaho Analyser: OLAP Viewer that contains the Enterprise version [16] is newer and simpler to custom than JPivot as shown in Fig. 4.

AJAX delivers a GUI that consents big malleability when generating the OLAP interfaces.

2.2.5 Dashboards

Dashboard panels are formed by a series of graphic controls that allows to be seen in a quick and intuitive way in the main markers. The performances of the business are measured so that users can get the operation of the company a screen. Currently, there is not much support for the implementation of these controls so there is lot of work done by hand. It also expects a series of wizards that allow them to develop them visually as the rest of the Pentaho applications [15]. Dashboards are a development of Pentaho, as they collect information on all the components of the platform including external applications, RSS feeds and web pages. They include content management and filtering, role-based security and drill down. They can also be integrated into third-party applications, portals or within the Pentaho platform.

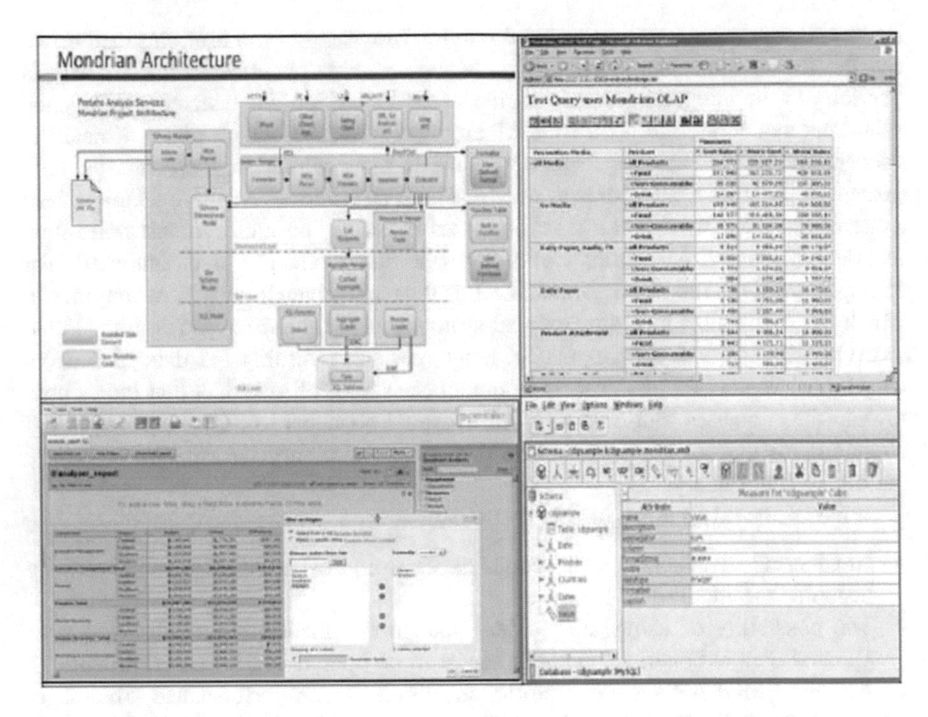

Fig. 4 Pentaho analyser interface to generate OLAP views that offer a complete reporting solution. Enhancing the whole features required in any reporting situation

To generate graphs, they rely on JFreeChart, a library to generate the most common graphs (2D, 3D, bars, time series, Gantt, etc.), interfaces to access different data sources, export to PNG, JPEG and PDF and Support for servlets, JSPs, applets and client applications.

All components of the Pentaho Reporting and Pentaho Analysis module can be a part of a Dashboard. In Pentaho Dashboards, it is very easy to incorporate a wide variety of types of graphs, tables and speedometers (dashboard widgets) and integrate them with the JSP Portlets, where you can view reports, charts and OLAP analysis. Pentaho offers the option of constructing dashboards [14] over and done with the web interface making use of the dashboard designer as illustrated in Fig. 5.

2.2.6 Jaspersoft

Jaspersoft is TIBCOs multi-client business intelligence platform. An enterprise class BI self-service software that allows generating embedded analysis of boards for large organizations. This BI tool offers pixel-enhanced visualization, Web and Mashup standard reporting, Bi self-service, Multitenancy for cloud and SaaS-based applications, Terabyte scalable memory engine for big data analysis, ETL and real-time integration.

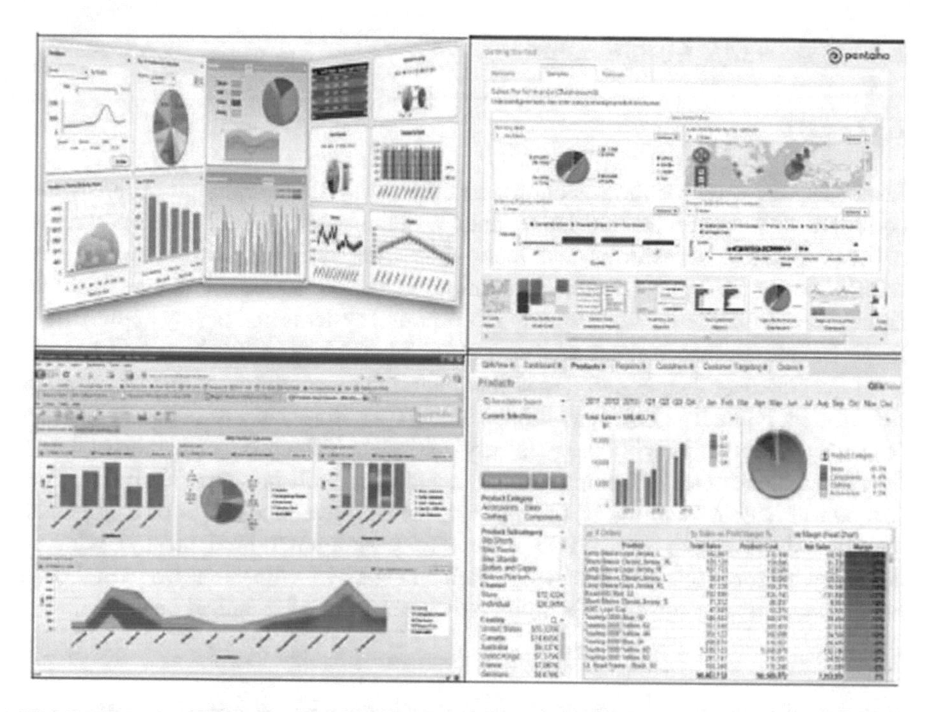

Fig. 5 Pentaho dashboards interface that offers the option of constructing dashboards over and done with the web interface making use of the dashboard designer

In addition, TIBCO Jaspersoft provides interactive reports to senior management, dashboards to executives, analytics, data exploration and discovery for analysis, and integration for data architects [11]. JasperSofts framework allows the user to easily integrate various data sources available in the company and through multidimensional analysis techniques that is presented in, control panels, dynamic reports, and provides the sensitive information to top management. This Open Source reporting solution is one of the most desired tools by many developers to implant in any Java application that needs a reporting scheme. The Reporting engine is the heart of Jaspersoft B.I solution [17], which has a different approach from Pentaho. Jaspersoft has unified ones projects well resolves current and associated projects; nevertheless, it is not engaged yet. Jasper can access the Mondrian code to adjust it and continues its developments with Mondrian.

2.3 Jaspersoft Components

2.3.1 ETL

Jasper ETL is used for the selection and processing of multiple data sources that configure and power the corporate data warehouse. Jaspersoft data integration software extracts, transforms and loads data from different sources into a data warehouse for the creation of analysis reports. This software uses more than 400 connectors to integrate a wide variety of corporate systems and legacy applications. JasperETL is, in fact, Talend Studio. Talend, is different to Kettle, which has not been integrated by Jasper and it is still an autonomous corporation that delivers its products individually [17]. Talend has similarly a native and instinctive user interface even though the methodology is very dissimilar. Talend is a code creator that is the consequence of an ETL implementation and it is a built-in Java or Perl code. This component is able to similarly assemble and create Java procedures or commands. This module is more adapted to a kind of programmer implemented with an advanced level of technical knowledge than it needs by Kettle as shown in Fig. 6. In conclusion, this method offers a higher advanced level of technical operation.

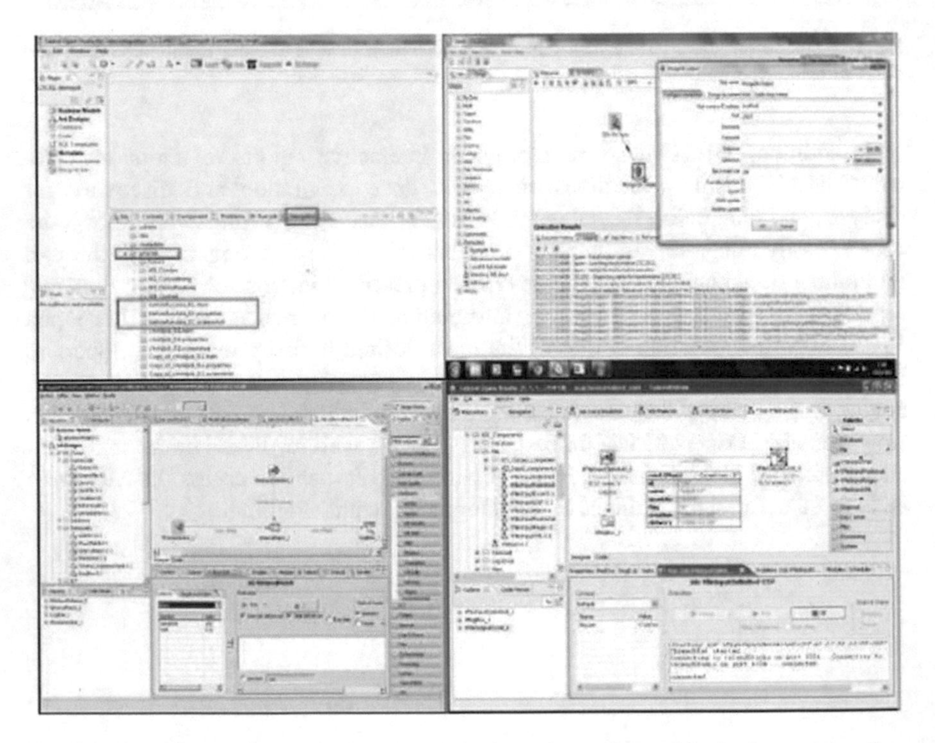

Fig. 6 JasperETL interface is, in fact, Talend Studio, it is a native and instinctive user interface, and a code creator as well

2.3.2 Web Application JasperServer

Jasperserver configured as a stand-alone application container that contains all of the elements described above adding security and resource accessibility capabilities. JasperServer is a hundred percent Java2EE that permits the management of the completely BI assets [18]. The general appearance of the web application is a little simple without giving up the power as it is illustrated in Fig. 7. Nevertheless, having all resources accessible on the top button bar creates a hundred percent useful application and contains all the essential assets for BI.

Its main features:

- It is the suite's report server
- Can be used stand-alone or from other applications
- Provides the entry point for reporting and data analysis
- User interface very easy to use and customizable
- It can execute, program and distribute reports to users and groups
- Can store issued reports
- Manage shared resources.

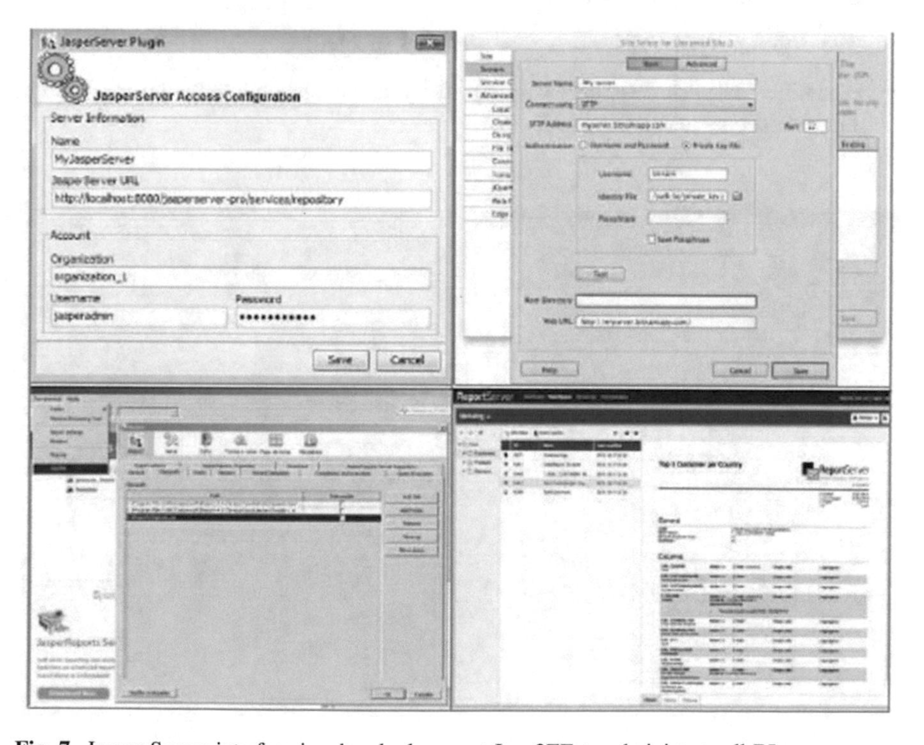

Fig. 7 Jasper-Server interface is a hundred percent Java2EE to administrate all BI resources

2.3.3 Reports

It is used for the design and presentation of dashboards and reports that configure dashboards with indicators required by the Management of the organization. This software allows presenting data from one or more sources in 10 highly interactive format types, for business users [11]. As said previously, the report engine is the solution of Jaspersoft and it is shown in Fig. 8. This module offers characteristics, such as, (i) Report development environment: Ireport is a structure that focuses on background NetBeans, (ii) System of metadata. It, accompanied by ad hoc views, the strongest point of the component, (iii) Web Interface for ad hoc views actually ne committed, (iv) The run-time JasperReports extensively was recognized and implemented in numerous, and (v) The reports are able to be transferred into PDF, HTML, XML, CSV, RTF, XLS and TXT.

The main features are:

- It is the basic library of the project
- It is the engine that runs the reports
- Widely used in the open source world
- Integrable in desktop and web applications java
- Great functionality and stability

Two different types of reports can be set or created:

- Predefined Ireport: IReport is an operational background that permits a huge amount of characteristics [19].

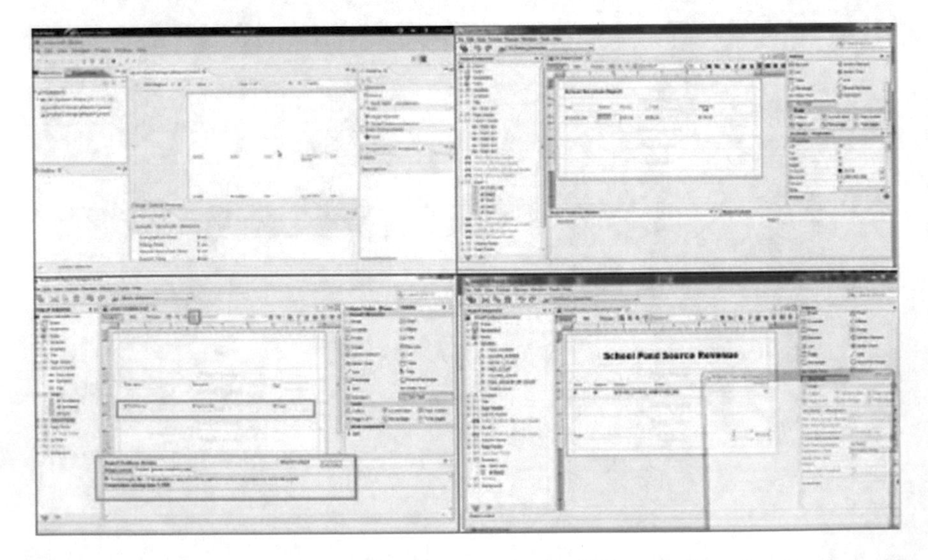

Fig. 8 Jaspersoft reporting interfaces to design and present different dashboards and reports with diverse indicators. Moreover, to present data from many resources in highly interactive formats for business users

- Ad hoc: it is the proper strong point of Jasper solutions. The corrector of ad hoc views is well organized and well contained tool for analysing [19].

It gives: (i) Choice of diverse sorts of prototypes and layouts, (ii) Variety of diverse data references, (iii) Authentication verification on the fly, (iv) Making of reports by dragging fields to the wanted place, and (v) Correction of whole features of the reports.

2.3.4 OLAP

Jaspersoft data analysis application is implemented to design, manage and display every type of data through OLAP analysis or in memory in order to detect problems, categorise tendencies and make faster and more accurate decisions. JasperAnalysis is used to design and support OLAP cubes that complement the structure of the scorecards that provide information research and analysis tools online. Mondrian is the OLAP engine that is implemented by JasperServer, furthermore, it works along with Viewnder-JasperAnalysis [20]. It is not JPivot but then again with a layer of makeups as illustrated in Fig. 9. Previously cited in Pentaho description.

Its main characteristics are

- It is the ROLAP user and server application.
- Allows users to scan data well beyond the usual reporting capabilities.

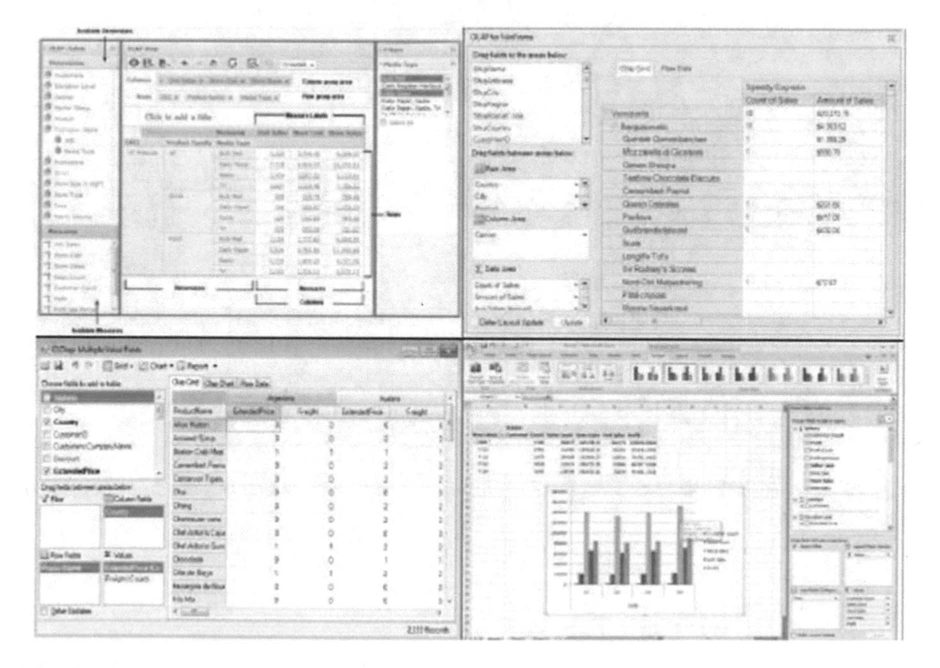

Fig. 9 The OLAP engine implemented by JasperServer and working with Viewfinder-JasperAnalysis, which contains a layer of makeup

- The server does all the heavy work, so the presentation layer is very light.
- It can be linked with reports, both as origin or destination of the same.

2.3.5 Dashboards

Jaspersoft dashboard software combines data and metrics to provide a summary graphical view of information for users to react quickly. Thanks to embedded dashboard solutions, it is possible to increase the competitiveness of applications with advanced reports and analysis. These dashboards extend their applicability through:

Dashboards:

- Supports iFrame with URL for peripheral content such as pictures, RSS feeds, etc.
- United dashboard interfaces mixes internal company info with peripheral data sources.
- The internal and peripheral data of the dashboard report are controlled by global parameters.

Fig. 10 Jaspersoft Dashboards Designers equipped with the option to choice predefined or ad-hoc dashboards

Embedded dashboard solutions:

- Delivers a profounder vision into data through reports and embedded dash-boards.
- Simply mix collaborating reports, diagrams and peripheral Web contention its own display.
- Provides interactive dashboards.

Two different type of Dashboard can be created through Dashboard Designer. Illustrated in Fig. 10.

- Predefined: Despite the designer boards do not bring with much logic [21], this tool is able to include its own developments, as it is a Java platform.
- Ad hoc: Dashboard Designer: As a web editor, this dashboard has a very basic and straightforward functionality.

3 Computer Algebra Systems

3.1 Sagemath

SageMath is a computer algebra system (CAS) assembled on mathematical sets and compared as NumPy, Sympy, PARI/GP or Maxima, as it gets a mutual power over and done with a common language which is built in Python.

The interface code mixes segments with graphics, texts or formulas improved with LaTeX rendered. Furthermore, SageMath is separated into a kernel that performs different calculation operations, in addition to an interface that visualizes the processes and interacts with the user. Furthermore, the CAS has a text based on command lines that uses a Python code, which empowers itself to perform interactive control calculations [22]. Python is a powerful programming language that supports object-oriented expressions and efficient programming. The core of SageMath is constructed in Python in addition to an adapted form of Pyrex called Cython. These features allow SageMath to perform parallel processing [23] and making use of both multi-core processors along with symmetric multiprocessors. Additionally, it is able to offer interfaces to other licensed tools such as Mathematica, Magma and Maple; this important characteristic permits users to compare software and contrast outcomes and performance. In addition, Sage is an environment of mathematical calculations of open code that allows carrying out algebraic, symbolic and numerical calculations. The goal of Sage is to create a free and viable option to Magma, Maple, Mathematica and Matlab, all of them powerful commercial programs. Sage serves as a symbolic calculator of arbitrary precision, nonetheless it can also perform calculations and solve problems using numerical methods. These calculations are performed through built-in and other external algorithms such as Maxima, NTL, GAP, Pari/gp, R and Singular [23]. Sage not only consists of the program itself, but also does the calculations, which can communicate with the terminal, that incorporates a graphical user interface through any web browser. Any and every set covers most functions, for instance, (i) Libraries of fundamental and singular functions, (ii) 2D and 3D charts of together purposes and data, (iii) Data management implements and responsibilities, (iv) A set of tools for aggregating GUIs to controls, (v) Implements for image treating via Python and Pylab, (vi) Implements to display and examine diagrams, (vii) Filters for introducing and transferring data, pictures, audio visual, CAD, and GIS, (viii) Sage implanted in documents LaTeX6 [24].

3.2 Matlab

Matlab is a computer algebra system (CAS) that offers an incorporated back ground that promises to develop and deliver a specific feature, for instance algorithm execution, data representation, and functions. Even though, it compromises interaction with other programming languages and other hardware mechanisms [25], among other are innovated. The Matlab suite has two additions that spread out of these functionalities: first, Simulink is a scenario that permits multi domain simulation and GUIDE, which is a graphical user interface (GUI). Second, Matlab toolboxes and Simulink blocks that enhance the blocks arrangements. Matlab has an interpreted programming language, which allows it to be executed in two interactive environments that makes use of scripts (*.m extension les). This feature permits vector and matrix type operations to work properly, in addition to calculations and object-oriented programming. The capability to work in .NET or Java environments has been possible through the release of new Matlab Builder tool, which in turn contains an embedded Application Deployment, which makes it feasible to use and manage Matlabs own functions such as library les. It is important to say that, Matlab Component Runtime (MCR) must be implemented/installed on the same computer where the core program is running, to ensure the correct functionality of Matlab [26]. Conducting various measurements and providing an interface for interacting with other programming languages is one of the great versatility that is provided by this CAS. In this way, Matlab can invoke subroutines, processes or functions that have been elaborated in C or FORTRAN [27]. In the dimension and time that this process is carried out, a wrapper function is created that allows them to be translated and returned by their Matlab data types. MATLAB also provides a number of specific solutions called TOOL-BOXES. These are very important to most MATLAB users and are MATLAB function sets that encompass the MATLAB background to re solve specific types of issues, for instance:

- Signal processing
- Design of control systems
- Simulation of dynamic systems
- Identification of systems
- Neural networks and others.

Probably the most important feature of MATLAB is its ability to grow. In summary, the most important features of MATLAB are:

- Writing the program in a mathematical language.
- Implementation of the matrices as a basic element of the language, which allows a great reduction of the code, since it does not need to implement the matrix calculation.
- Implementation of complex arithmetic.
- A great content of specific orders, grouped in TOOLBOXES.
- Ability to extend and adapt the language, using script files and .m function.

4 Materials and Methods

The action of measuring the execution times of an application is not an easy or an insignificant work. The results obtained can be significantly different in each measurement, from one computer to another. The factors that influence these times can be, among others, the algorithm used, the type of operating system, the number of processors and their speeds, the set of instructions of each processor, the amount and speed of the memorics (RAM and cache), and the mathematical coprocessor. In addition, the same algorithm used on the same computer could show different results in the measurement of the times; this could be due to, the times that other applications are use or to determine whether, there is enough RAM or not for the execution of the application or algorithm. The goal of this study is to make a comparison exclusively of the ETL and Reporting processes of both BI tools, trying to obtain independent conclusions from one computer to another. In the same way, perform a measurement of execution times as a function of the dimensions of the input data. To accomplish these purposes, two methodologies will be used: Calculating the processing time used by BI tools with the same input data proportions and calculate the number of instructions carried out by each tool.

4.1 ELT Measurement

The execution times and performance of the ETL processes of both BI tools were measured with Sage. For this, an algorithm in code C was implemented and it is illustrated as follow:

Algorithm 1 : ELT Measurement

```
int rbi, rdb, rsage;
cout "Run BI Tool"
cin rbi;
endl;
rbi=0
cout "Run Database"
cin rdb;
rdb=0
while rdb ≤ 6 do
   rdb=rdb+1
   cout "Run Sage"
   cin rsage;
   endl;
   while rbi = 1|rbi = 2 do

      if rbi =1 then
         count "Show Statistics"
         endl;
         return 0;
      end if

      if rbi =2 then
         count "Run Database Again"
         endl;
         return rdb;
      end if
      return 0;
   end while
end while
```

In order to measure CPU time, Sage uses the notions of CPU time, plus Wall time [23] that are intervals that the PC uses only for BI tool. The CPU time is used to the computations, and the Wall time clock flanked by the initiation and the completion of the calculations. The two calculations are subjected to unexpected changes. The easiest method to acquire the execution time is to write the word time in the command as illustrated in Fig. 11.

```
Sage: #To small data size
Sage. Time is_prime  (factorial (500) + 1)
False
Time: 0.09 s CPU, Wall: 0.009 s


Sage: #To higher data size, takes longer (in general)
Sage: time is_prime (factorial (100) + 1)
False
Time: 0.72 s CPU, Wall: 0.76 s
```

Fig. 11 The Sage algorithm is to calculate CPU time of ETL procedures in both BI tools for short and large data magnitudes

The time command is not properly adaptable and requires additional parameters to be able to calculate the exact times dedicated to only one type of program or application. For this case then, the CPUtime and Walltime commands are used. CPUtime performs the progressive measurements of the calculations made accurately which computes the times that were dedicated to Sage by the CPU. Walltime is the traditional UNIX clock. In addition, the previous and later times used in the implementation of the algorithms in Sage were documented and measured, and the differences are shown in Fig. 12.

The times used in the execution of the factorial function, together with the data of different size dimensions, that are stored in the list of CPU times as illustrated by the algorithm in Fig. 13.

4.2 Reporting Measurements

The execution times and performances of the Reporting processes of both Business Intelligence Tools (Pentaho and Jaspersoft) were measured with the Computer Algebra System Matlab. In order to carry out this process, an algorithm in C code was implemented and it is illustrated as follow:

```
Sage: #cputime only advances when the cpu runs
Sage: #(a taximeter cpu)
Sage: #runs this function several times to see as time increases
Sage: #if you want. Executes commands in between
Sage: cputime ()
2.0600000000000001

Sage: #walltime inexorable advance (it is an ordinary clock)
Sage: walltime ()
1298369972.163182
```

Fig. 12 The Sage algorithm with CPUtime and Walltime parameters to measure the ETL procedure in both BI tools

```
Sage: numbers = [2 ^ j for j in range (8,20)]

Sage: time = []

Sage: for number in numbers:

...Tcpu0 = cputime ()

...11 = factorial (number)

...Times.append (cputime () –tcpu0))
```

Fig. 13 The Sage algorithm to store the lists of the CPU times implemented in the execution of the factorial function, together with the data of different size dimensions

Algorithm 2 : Reporting Measurement

```
int rbi, rdb, rmatlab;
cout "Run BI Tool"
cin rbi;
endl;
rbi=0
cout "Run Database"
cin rdb;
endl;
rdb=0
while rdb ≤ 6 do
   rdb=rdb+1
   cout "Run Matlab"
   cin rmatlab;
   endl;
   while rbi = 1|rbi = 2 do

      if rbi =1 then
         count "Build Analytics"
         endl;
         return 0;
      end if

      if rbi =2 then
         count "Show Error"
         endl;
         return rdb;
      end if
      return 0;
   end while
end while
```

Additionally, to carry out this process, a function written in C language was used through that had a High-Resolution Performance Counter was applied. The purpose is to be able to calculate the execution times of the Re porting processes in both BI tools as shown in Fig. 14.

In that instance, Query Performance Counter functions as a clock () and Query Performance Frequency as CLOCKS PER SEC. Then, the first instruction generates the counter value and the next generates the frequency (in cycles per second, hertz). In this case, an LARGE INTEGER is the clearest method to characterize a 64-bit integer generated by a union.

```
/* returns "a - b" in seconds */
double performancecounter_diff(LARGE_INTEGER *a, LARGE_INTEGER *b)
{
  LARGE_INTEGER freq;
  QueryPerformanceFrequency(&freq);
  return (double)(a->QuadPart - b->QuadPart) / (double)freq.QuadPart;
}

int main(int argc, char *argv[])
{
  LARGE_INTEGER t_ini, t_fin;
  double secs;

  QueryPerformanceCounter(&t_ini);
  /* ...Reporting... */
  QueryPerformanceCounter(&t_fin);

  secs = performancecounter_diff(&t_fin, &t_ini);
  printf("%.16g milliseconds\n", secs * 1000.0);
  return 0;
}
```

Fig. 14 The Algorithm to calculate the run time of the reporting process in both BI tools

4.3 Computer System

The aim of carrying out this examination and the corresponding analysis, BI tools (Pentaho and Jaspersoft), CAS (Sage and Matlab), and databases are installed and configured on a computer with the characteristics which are described below: (i) Operating system: x64-based, (ii) Operating system version: 10.0.10240 N/D (iii) Compilation 10240, (iv) Number of processors: 1, (v) Processor: Intel (R) Core (TM) i5-3317U (vi) Processor speed: 1.7 GHz, (vii) Instructions: CISC, (viii) RAM: 12 Gb, (ix) RAM speed: 1600 MHz, (x) Cache: SSD express 24 Gb, (xi) Mathematical coprocessor: 80387, (xii) GPU: HD 400 on board.

4.4 Databases Analysis

Table 1 illustrates and describes the key characteristic of the Excel databases that are used to carry out this examination. The six dissimilar databases were obtained from UCI Machine Learning Repository [28].

Table 1 Illustration of the excel databases used for this examination

DataBase	Number of attributes	Number of instances	Size in Mb
DB1	21	65.055	0.009
DB2	26	118.439	0.017
DB3	35	609.287	0.134
DB4	40	999.231	1.321
DB5	51	1458.723	35.278
DB6	62	2686.655	144.195

5 Results

In this examination, the data resulting from CPU time measurements (using two different Computational Algebra Systems and in relation to the size of the different databases) was acquired from the input registers applied to the ETL and Reporting processes for both tools of BI. The considerable fluctuations in the computational time measurement resulted in each of the processes are directly linked to factors such as the implemented algorithms, the type of operating system installed on the test computer, the number and speed of its processors, set of instructions comprising the processor, the amount and speed of RAM and cache, and the mathematical coprocessor, among others.

Similarly, on the same computer, running in the same algorithm at different times can produce different results, caused by allocating the amounts of memory to other programs and their execution speeds. The data obtained from the CPU times taken by the computer during the execution of the ETL and Reporting processes, processing the six databases, are detailed in Tables 2 and 3. Similarly, the increase of the processing data in the tools of BI can be taken into account as a process of difference between such applications. The first evaluation (Table 2) shows that the computation times dedicated by the Pentaho ETL process and which were calculated by Sage were: 8; 12.01; 21; 32.01; 39.06 and 48.01 min. In contrast, the computation times employed by Jaspersoft, calculated by Sage as well were: 9.54, 19.32, 31.88, 44.73, 55 and 67.69 min, processing 0.009 Mb of DB1; 0.017 Mb of DB2; 0.134 Mb of DB3; 1.321 Mb of DB4; 35.278 Mb of DB5 and 144.195 Mb of DB6, correspondingly for both BI applications. The results obtained from the

Table 2 Outcomes of the CPU time of the ETL process used by both BI tools and the percentage increased after the processing of the databases

Time in minutes							
Tool	Process	DB1	DB2	DB3	DB4	DB5	DB6
Jaspersoft	ETL	8	12.01	21	32.01	39.06	48.01
	ETL	9.54	19.32	31.88	44.73	55	67.69
Increment in the process of data							
Jaspersoft	ETL	19.22%	60.85%	51.79%	39.75%	40.77%	40.99%

Table 3 Outcomes of the CPU time of the reporting process used by both Bi tools and the percentage increased after the processing of the databases

Time in minutes							
Tool	Process	DB1	DB2	DB3	DB4	DB5	DB6
Pentaho	Reporting	3.75	5.35	8.47	12.03	17.07	22.60
Jaspersoft	Reporting	3	4.02	6.05	8.13	11.16	14.15
Increment in the process of data							
Pentaho	Reporting	25%	32.99%	40%	48%	53%	59.75%

CPU time used during the ETL process, processing the data of the different databases by both BI tools, are illustrated in Table 2.

Oppositely, the second segment of the analysis, regarding the measurement of the Reporting process (Table 3), it can be seen that Matlab calculated the times used by Pentaho tool are as follows: 3.75, 5.35, 8.47, 12.03, 17.07 and 22.60 min. In contrast, the times used in the same process by the Jaspersoft tool were: 3, 4.02, 6.05, 8.13, 11.16 and 14.15 min, processing 0.009 Mb from DB1; 0.017 Mb from DB2; 0.134 Mb from DB3; 1.321 Mb from DB4; 35.278 Mb from DB5 and 144.195 Mb from DB6, correspondingly for both BI tools. The results obtained from the CPU time used during the Reporting process, processing the data of the different databases by both BI tools, are illustrated in Table 3.

The data results obtained showed that, the execution times of the ETL and Reporting processes of both BI tools, after having processed the databases, are shown and compared graphically. Thus, it is observed in Fig. 15 that the Jaspersoft tool had a significant increase in CPU times for the ETL process, represented by 19.22, 60.85, 51.79, 39.75, 40, 77 and 40.99% for processing DB1, DB2, DB3,

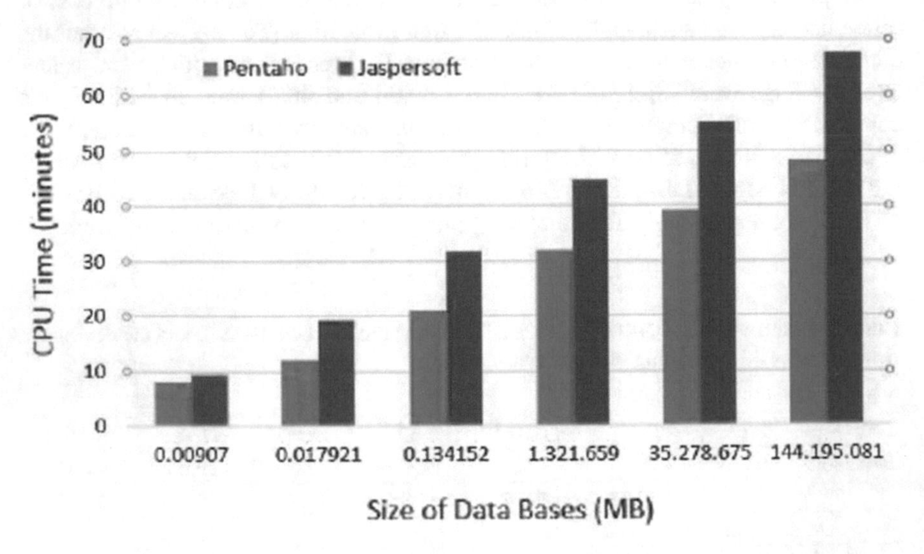

Fig. 15 CPU Time of the ETL process taken by both BI tools after processing the databases

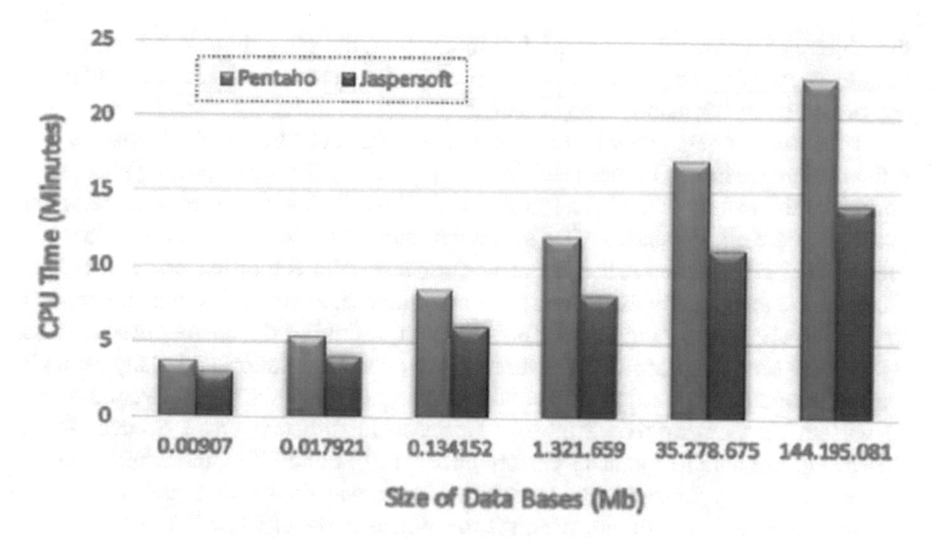

Fig. 16 CPU Time of the Reporting process taken by both BI tools after processing the databases

DB4, DB5 and DB6, correspondingly. This signifies that the Pentaho tool performs better, compared to Jaspersoft.

Then, as seen in Fig. 16, the results obtained from the Reporting process had a notable increase after being executed by the Pentaho tool; expressed by 25, 32.99, 40, 48, 53 and 59.75% processing DB1, DB2, DB3, DB4, DB5 and DB6, respectively. In this situation, Jaspersoft had an improved performance over Pentaho.

The results correspondingly exposed that the CPU times for both the ETL and Reporting processes are in a straight line associated with the data dimension of the databases. Finally, it can be established that Pentaho observed better performance for the ETL process and Jaspersoft superior performance for the Reporting process.

6 Discussion

Clearly, after the experiment and analysis of the data resulting from the ETL process, the Jaspersoft BI tool showed a marked increase in CPU times for data processing; this is denoted with an average of 42.28% over Pentaho when showing its performance in the six databases. The Pentaho tool exposed its data integration capabilities and ETL process competences as presented in "Pentaho Business Analytics: An Open Source Alternative to Business Intelligence" [12] and "Pentaho and Jaspersoft: A Comparative Study of Business Intelligence Open Source Tools Processing big data to Evaluate Performances" [29], with better performance.

Undoubtedly, for this piece of research, this study can demonstrate that the Pentaho tool showed a greater ETL processing capability; simultaneously it covered the objectives and requirements of data integration, the same time with big data. Its parallel processing engine provided a great performance and these characteristics are presented in "Pentaho Reports Review" [13].

In the second segment of this research, the BI tool "Pentaho" showed a significant increase in CPU times used for data processing for the Reporting process, it compared the results of the "Jaspersoft" tool. The difference showed 43.12% when processing the six databases. The evaluation part of the study has shown that the "Jaspersoft" BI tool has had a better performance with Reporting processes. This disclosure is specifically associated with other investigative results, which supports the Jaspersoft's increasing BI capabilities, circumscribing documents based on its operational output, interactive end-user search, data integration and analysis, as it was mentioned in the articles: "Pentaho Reports Review" [13] and "Pentaho and Jaspersoft: A Comparative Study of Business Intelligence Open Source Tools Processing big data to Evaluate Performances" [29]. In addition, the exploration of numerous security features could be an attractive way to examine and implement big data projects in the future, which is mentioned in the articles: "Threat Analysis of Portable Hack Tools from USB Storage Devices and Protection Solutions" [30] and "Optimizing windows Security features to block malware and hack tools on USB storage devices" [31].

7 Conclusion

Two of the best BI applications on the marketplace have been tested in this study: Pentaho and Jaspersoft. Both tools exhibit important features in their modules. On the one hand, Pentaho presents an ETL component easy-to-use and maintain, as well as great flexibility to perform transformations; It groups all information components only on a functional platform; Web application with Java J2EE application hundred percent extensible, flexible and configurable; OLAP Mondrian with a unified engine considerably used in JAVA environments; Configuration management is made up of almost all environments, which are intercommunicated with other apps using web services; Reports with an instinctive tool that makes it easy for users to generate reports easily; Dashboard Designer for the generation of Ad hoc dashboards, dashboards founded on SQL queries or metadata, and countless autonomy by providing an extensive variety of devices and choices.

On the other hand, Jaspersoft owns JasperETL (Talend) with native Java/Perl; The service fixes are well-defined, most of them supported by the same web application; Web application with a Java component j2EE hundred percent expandable, flexible and configurable; The reports and commands of the Ad hoc editor of the command boxes are solved in an outstanding way; Reports are expedited; Ad hoc reports have an interesting, flexible, powerful, instinctive and easy-to-use interface. The key focus of this experimental study is the evaluation of the ETL and Reporting processes of BI tools, measuring their performance through two computational algebra systems, Sage and Matlab. The evaluation of the ETL process produced noticeable results, showing marked increments used by Jaspersoft in CPU times, over those used by Pentaho. This made it factual as that Jaspersoft tool used 42.28% more time on the performance metrics when processing the data of the six databases. Pentaho meanwhile, has evidenced during measurement tests, the use of more CPU time in the Reporting process, compared to those used by Jaspersoft. In this way, it was also established that the notable increase in these execution times used by Pentaho, amounted to 43.12% on performance metrics at the time of processing the same databases.

Finally, this experimental analysis is a rather convenient reference document for many researchers, in addition to those who are supporting big data's processing decisions and the implementation of Open Source Business Intelligence tools, founded on the process perspectives. For the author, future research could be involved with the development of new experiments in the space of BI and Data Warehouse to support of organizational decision-making, taking this research as a reference.

References

1. List, B., Bruckner, R. M., Machaczek, K., & Schiefer, J. (2002). A comparison of data warehouse development methodologies case study of the process warehouse. *Database and Expert Systems Applications DEXA, France, 2453,* 203–215.
2. Dresner, H. (1993). Business intelligence: Competing against time. In Paper presented a Twelfth Annual Office Information System Conference, Garther Group. London: Earl, M.J., 1989.
3. Atre, S., & Moss, L. T. (2003). *Business intelligence roadmap: The complete project lifecycle for decision-support applications*. Boston: Addison Wesley.
4. Gonzalez, J. F. (2011). Critical success factors of a business intelligence project. *Novtica, 211,* 20–25.
5. Sallam, R. L., Hostmann, B., Schegel, K., Tapadinhas, J. Parenteau, T., & Oestreich, W. (2015). *Magic quadrant for business intelligence and analytics platforms*. Retrieved from http://www.gartner.com/doc/2989518/magic-quadrant-business-intelligence-analytics. Accessed November 9, 2016.
6. Gartner, Inc. (2016). *IT glossary*. Retrieved from http://www.gartner.com/it-glossary/business-intelligence-bi/. Accessed November 9, 2016.
7. Hazlewood, W. R., & Coyle, L. (2009). *On ambient information systems: Challenges of design and evaluation*. Retrieved from http://www.igi-global.com/article/ambient-information-systems/3873
8. Baldwin, H. (2014). *A Match made somewhere: big data and the internet of things*. http://www.forbes.com/sites/howardbaldwin/2014/11/24a-match-made-somewhere-big-data-and-the-internet-of.things/#5dda8d8f6513
9. Kune, R., Konugurthi, P. K., Agarwal, A., & Chillarige, R. R. (2016). The anatomy of big data computing. *Software: Practice and Experience, 46*(1), 79–105.
10. Pentaho A Hitachi Group Company. (2005–2016). *Pentaho: Data integration, business analytics and bid data leaders*. Pentaho Corporation. http://www.pentaho.com. Accessed 10 Nov 2016.
11. TIBCO Jaspersoft. (2016). *Jaspersoft Business Intelligence Software*. Available via TIBCO Software. http://www.jaspersoft.com. Accessed November 15, 2016.

12. Tarnaveanu, D. (2012). Pentaho business analytics: A business intelligence open source alternative. *Database System Journal, 3,* 13.
13. Innovent Solutions. (2016). *Pentaho reports review*. Retrieved from http://www.innoventsolutions.com/pentaho-review.html. Accessed: December 12, 2016.
14. Kapila, T. (2014). *Pentaho BI & integration with a custom Java web application*. Retrieved from http://www.neevtech.com/blog/2014/08/13/pentaho-bi-integration-with-a-custom-java-web-application-2/. Accessed November 11, 2016.
15. Pozzani, G. (2014). *OLAP solutions using Pentaho analysis services*. Retrieved from http://www.profs.sci.univr.it/˜pozzani/attachments/pentaho_lect4.pdf. Accessed: December 12, 2016.
16. Sanket. (2015). *Fusion charts integration in Pentaho BI dashboards*. Retrieved from http://www.fusioncharts.com/blog/2011/05/free-plugin-integrate-fusioncharts-in-pentaho-bi-dashboards/. Accessed November 13, 2016.
17. Vidhya, S., Sarumathi, S., & Shanthi, N. (2014). Comparative analysis of diverse collection of big data analytics tools. *International Journal of Computer, Electrical, Automation, Control and Information Engineering, 9,* 7.
18. Olavsrud, T. (2014). *Jaspersoft aims to simplify embedding analytics and visualizations*. Retrieved from http://www.cio.com/article/2375611/business-intelligence/jaspersoft-aims-to-simplify-embedding-analytics-andvisualizations.html. Accessed December 16, 2016.
19. Pochampalli, S. (2014). *Jaspersoft BI suite tutorials*. Retrieved from http://www.jasper-bi-suite.blogspot.com.au/. Accessed December 17, 2016.
20. Vinay, J. (2013). *OLAP cubes in Jasper Server*. Retrieved from http://www.hadoopheadon.blogspot.com.au/.2013/07/setting-up-olap-cubes-injasper.html. Accessed: November 19, 2016.
21. Informatica Data Quality Unit. (2013). *Data quality: Dashboards and reporting*. Retrieved from http://www.Markerplace.informatica.com/solution/dataqualitydashBoardsandreporting-961. Accessed December 21, 2016.
22. Sagemath. (2016). *Sagemath—Open-source mathematical software system*. Available via Sage. Retrieved from http://www.sagemath.org. Accessed November 21, 2016.
23. AIMS Team. (2016). *Sage*. Retrieved from http://www.launchpad.net/˜aims/+archive/ubuntu/sagemath. Accessed December 21, 2016.
24. Stein, W. (2016). *The origins of SageMath*. Retrieved from http://www.wstein.org/talks/2016-06-sage-bp/bp.pdf. Accessed: November 28, 2016.
25. MathWorks. (2016). *MATLAB—MathWorks—MathWorks Australia*. Available via MathWorks. http://www.au.mathworks.com. Accessed December 28, 2016.
26. Gockenbach, M. S. (1999). *A practical introduction to Matlab*. Retrieved from http://www.math.mtu.edu/msgocken/intro/introtml. Accessed: December 28, 2016.
27. Black, K. (2016). *Matlab tutorials*. http://www.cyclismo.org/tutorial/matlab/. Accessed: November 29, 2016.
28. Lichman, M. (2013). *UCI machine learning repository*. Retrieved form http://www.archive.ics.uci.edu/ml. Accessed December 10, 2016.
29. Parra, V. M., Syed, A., Mohammad, A., & Halgamuge, M. N. (2016). Pentaho and Jaspersoft: A comparative study of business intelligence open source tools processing big data to evaluate performances. *International Journal of Advanced Computer Science and Applications, 10* (14569), 1–10.
30. Pham, D. V., Syed, A., Mohammad, A., & Halgamuge, M. N. (2010). Threat analysis of portable hack tools from USB storage devices and protection solutions. In International Conference on Information and Emerging Technologies (pp. 1–5).
31. Pham, D. V., Halgamuge, M. N., Syed, A, & Mendis, P. (2010). Optimizing windows security features to block malware and hack tools on USB storage devices. Progress in Electromagnetics Research Symposium (pp. 350–355).

Part III
Internet of Things Based Smart Life

IoT Gateway for Smart Devices

Nirali Shah, Chintan Bhatt and Divyesh Patel

Abstract In the world, the revolution of any industry is to connect their product and appliances to the Internet and make it autonomous and remotely connected so any one can operate and watch it from anywhere at any-time. This idea is known as industrial Internet of things (IIoTs). For achieving this goal very first and compulsory needed thing is connectivity. For any IIoTs enabled world the most frequency model architecture is build once in a world and that connectivity is one time disruption which requires new technology to interrupt the old one and manage it with new features. For this challenging task we try to develop the small but open for all system as a part of IIoT for the world. In this we are using the devices which is already define in open standard unified point of sale devices (UPoS) in which they included all physical devices like sensors printer and scanner. Firstly use the method and device defined by the UPoS system for the particular device. Connect that device to the server and lastly connect with it to the cloud. This flow will help us for connecting different branches of the company ubiquitously. By this you can connect remote device anytime anywhere with the right access. To the beneficial this will help us in decision making to the administration of the industry for the industry growth. Furthermore it will connect the one common employee to the chairman of the industry.

Keywords Industrial IoT · UPoS · Sensors · Gateway · Autonomous system · Connectivity

N. Shah (✉) · C. Bhatt · D. Patel
Computer Department, Charusat University, Changa, India
e-mail: niralishah1992@gmail.com

C. Bhatt
e-mail: chintanbhatt.ce@charusat.edu.in

D. Patel
e-mail: divyeshpatel.ce@charusat.edu.in

N. Dey et al. (eds.), *Internet of Things and Big Data Analytics Toward Next-Generation Intelligence*, Studies in Big Data 30,
DOI 10.1007/978-3-319-60435-0_7

1 Internet of Things: Introduction

Internet of things (IoTs) is network of physical objects embedded with software using Internet network connectivity. In IoT the connection between to devices are carried out using different techniques and it is dependent on the location of the two devices. The techniques are Bluetooth, RFID, GPRS, using gateway like Raspberry Pi, Arduino connected through wired and wireless Internet connection [5].

IoT started at 1980 at Carnegie Mellon University, the first application was Internet appliance coke machine such that employees can connect to the machine to check the cold drink is there or not. The main idea behind IoT is any application or system which is working uniquely try to get data from the physical device and after processing it will deliver fruit full result for the further process. Internet of things refers to the concept that the Internet is no longer just a global network for people to communicate with one another using computers, but it is also a platform for devices to communicate electronically with the world around them [1].

–Center for Data and Innovation

1.1 IoT Provide Services Like

- *QoS*, *latency*, *security* using wireless and Bluetooth,
- *Data capturing* using microsystems, sensors and actuators,
- *Security*, *privacy* using cloud technologies,
- *Network* using IPv6, lower latencies.

1.2 IoT Applications in Real World

- Global manufacturers like 'Bosch, FCA, Harley-Davidson, Cisco, general electric (GE)' are working on smart systems and making their product to work with industrial IoT (IIoT) environment.
- Company like Siemens is good example of IIoT. There 75% of work I converted into machine to machine communication. in which more than 1000 of automation controllers in operation communicate with each other and tell the machine which product will process and also about the raw material to be added step by step. It will also take care of the errors to be occurs and at that time which action should taken [10].

- Industry's like GE has more than 10,000 sensors t be operated automatically and provide the data at specific places. Also consider the power consumption and provide it or inform it when low power notification on.
- Harley Davidson are able to build real time performance management system which will develop each and every port of the current product by step given by the system ad also track the process and notify to the particular person or process to communicate further [11].
- Cisco can benefit to the any IIoT system by developing the real time data analysis and it will be successful to develop virtual platform which use for the IoT, big data, cloud. Provide them virtual platform and create range of linking products and gather all data from different–different places for further use.

Embedded system is a core part of IoT so it is also consider as a part of IIoT because it is known as electronic/electro mechanical system designed to perform a specific function and it is a combination of software and hardware.

2 Different Application of IIoT

1. *Automation* automation of different devices like copier, fax machine, printer, scanner, point of sale (PoS) terminals, multifunctional peripherals, devices for storage, smart cards.
2. *Electronics consumption* electronics devices like music instruments, cameras, CD–DVD players, TV-setup boxes, PDA can operate remotely without physical lookout and this will benefit to major issues like security and distance management.
3. *Medical electronics* the operations like patient monitoring surgical system diagnostic system will operate and managed by the doctors and appropriate person without or with physical appearance [7]. It will be present the patient report to patient and also the adviser doctor team at their place. So using IoT it will make easy and fast medical system.
4. *Military/aerospace control* using IoT concept military system can operate satellite system for security purpose. Radar which is placed for some spying operation. For aerospace weather information at particular time and the system for aircraft management also manage and successfully implement using IoT concept.
5. *Industrial system* IoTs can manage and develop the system or application which is help for industry and also city as well by making the smart system by implementing sensors and special purpose controllers using networking concept and process controls.

3 Fundamental Components of IoT

IoT is dependent technology. It is being established by different independent technology and devices which make fundamental components of IoT.

Components are:

- *Hardware tools* making physical objects responsive and giving them capability to retrieve data and respond to instructions.
- *Software technology* enabling the data collection, storage, processing, manipulating and instructing.
- *Communication interface* most important of this entire communication infrastructure which is involving protocols and technologies helps to devices to talk and communicate with each other and transfer the data.
- Smart system for IoT is driven by the combination of sensors + connectivity + people and processes. The interactions between these things are creating new type of application which is smart enough for different functions [5].

Devices used in IIoT application is classified into two categories:

1. *Actuators* controlling mechanism can apply through motion for this type of devices.
2. *Node system* the second part of the whole IoT system which is used for decision making at the end of the task [3].

Gateway can classified in three types of functionality which is necessary for the basic gateway to work appropriately

1. *Gateway* works on request response architecture between client and server.
2. *Forward proxy* task defines for forwarding of request and data to one source to destination including hopes.
3. *Reverse proxy* act as single PC which work on functionality of gateway and also performs task like load balancing, authentication and encryption–decryption.

- Relation between IoT and big data analytics,
  - Privacy is implemented design by design.
  - Big data will consider about which data is collected how data will be processed and that data is used for which partied for what purpose.
  - Help to find purpose of collection at the time of collection and, at all times, limit use of the data to the defined purpose.
  - Collect and store only the amount of data necessary for the intended lawful purpose.
  - Allow individuals access to data maintained about them, information on the source of the data, key inputs into their profile, and any algorithms used to develop their profile.
  - Allow all individuals to correct and control their information.
  - Conduct a privacy impact assessment for every user.

- Consider data anonymization at the top to bottom level to make synchronization between them.
- Limit and carefully control access to personal data.
- Conduct regular reviews to verify if results from profiling are "responsible, fair and ethical and compatible with and proportionate to the purpose for which the profiles are being used."
- Allow for manual assessments of any algorithmic profiling outcomes with "significant effects to individuals" at the right time with right result.

4 IoT Protocols

Need of protocol in IoTs is to follow some frame work like device to device communication where device is located 30,000 ft. above the ground or it may be possible it will located below the ground. Here data can be sent from one device to another followed by the server [8]. Server infrastructure is established as one device connected too many other devices and servers.

Protocols of IoT can be defined as:

MQTT it is work on concept of device to server communication where collecting data from device and send it to the server. Message query telemetry transport only target on data collection on remote monitoring. Main goal of MQTT is to get the data from the device and transport it to the any information technology (IT) industry [10]. Same as large network of some physical devices and controlled and monitored from the cloud.

MQTT is not practically fast but generally the data taken in real time, secure and reliable data in MQTT it will placed on top of the TCP.

XMPP extensible messaging and presence protocol is as named suggested used when presence of people are involved. It will work on TCP and HTTP on top of TCP. Main benefit over MQTT is its addressing scheme. This will help to connect as a needle over the whole Internet bucket.

XMPP is not really fast but it will very as per need. Just like person to person case. XMPP use polling or checking for updates on demands that's why it will measured the real time data in seconds.

DDS data distribution services directly collect data from one device and send it to the other devices. So the function of the DDS is totally different from the MQTT and XMPP. As per the use of DDS it will compatible for use in industry and embedded application. DDS can deliver lots of the messages in compare to the other protocols in each second to the multiple server and devices.

Here use of TCP is not relevant because devices communicate with other devices in micro seconds so it is not possible to work properly in TCP. So that DDS can easily figure out the, location and destination of data though it has multiple destination. Mainly DDS is used in high performance integrated device systems.

Automotive test and safety, hotel integration, medical streaming, military usage, wind form operative management is the definition of the DDS.

AMQP business messages are always send by the advanced message queuing protocol the flow of AMQP is to connect the system, feed the message find the reliable transform receiver and transfer the data. AMQP is connecting over different organization. It can reliable over poor network or operate over distance. It is possible that the system network is not available simultaneously still system cannot lost. AMQP is reliable on different platforms [11].

Widely different message patterns are acceptable by AMQP. It gives the facility of generated message delivery at the level of at least and exact once. It has feature like message oriented communication and flow control. The message type followed by AMQP is bare message. Bare message will send by the sending application. Message divided between two pairs begin frame and end frame. The actual message written between open frames.

5 Approach on IIoT

The IIoTs is connecting the physical world of sensors, devices and machines with the Internet and, by applying deep analytics through software, is turning massive data into powerful new insight and intelligence. We are entering an era of profound transformation as the digital world is maximizing the efficiency of our most critical physical assets. To complete the IIoT transformation journey, it is important to conduct a holistic and strategic review that enables the organization to tackle the operational and customer engagement aspects, while also preparing the business for disruptive industry change that will go beyond anything we've seen in our lifetimes, and will demand a deep and broad transformation of the enterprise (Fig. 1).

Here in IIoT if you considering any scenario in which one company is manufacturing product this is used to measure a temperature in house. At the first site sensor just sense the temperature and display on it. But it is connected to the house members mobile then no need to check the temperature by going nearer to that. They will get message at periodic time the third component says that if this system is also connected to any fire system than if any problem is occur or temperature is goes beyond the limit than it will send the help to that house so it's called value added services. This whole system is known as platform as a service in IIoT [2].

In networking, an edge device is a device which provides an entry point into enterprise or service provider core networks. Examples include routers, routing switches, integrated access devices, multiplexers, and a variety of local area network and wide area network access devices. Edge devices also provide connections into carrier and service provider networks. Network providers and others have been pushing intelligence—compute power and the ability to run applications and analytics—to these edge devices for some time.

But the growth of the IIoTs extends the 'edge' beyond the network devices, into industrial and commercial devices, machines, and sensors which connect to the

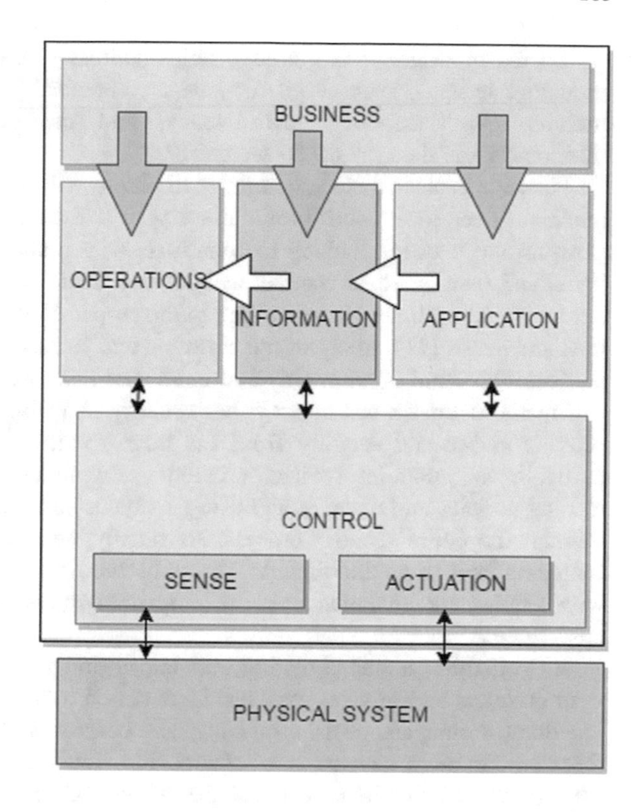

Fig. 1 IIoT architecture

network. Edge computing and analytics can, often should be, and increasingly is close to the machines and data sources. As the digitization of industrial systems proceeds, we expect that analysis, decision-making, and control will be physically distributed among edge devices, the network, the cloud, and connected systems, as appropriate. These functions will end up where it makes most sense for them to be.

IIoT will change the way industrial organizations generate, collect, and analyze data. Data will be generated faster and in greater volume than ever before. This will require today's plant information infrastructure to evolve. One part of this new infrastructure will be intelligent edge devices, which will include the latest generation of controllers, such as DCS's, PLC's and PACs. Besides providing control, these edge devices will securely collect, aggregate, filter, and relay data, leveraging their close proximity to industrial processes or production assets [5]. They will also be capable of collaborating with powerful analytics tools, detecting anomalies in real time, and raising alarms so that operators can take appropriate actions.

With edge computing and analytics, data is processed near the source, in sensors, controllers, machines, gateways, and the like. These systems may not send all data back to the cloud, but the data can be used to inform local machine behaviors as it is filtered and integrated. The edge systems may decide what gets sent, where it gets sent and when it gets sent.

Placing intelligence at the edge helps address problems often encountered in industrial settings, such as oil rigs, mines, chemical plants, and factories. These include low bandwidth, low latency, and the perceived need to keeps mission-critical data on site to protect IP.

The edge is also where autonomous machines will be found. These 'self-driving' machines need local control to interface with and direct mechanical systems, local analysis and decision making to coordinate with other machines, things, and systems, and usually some remote analysis and decision making to ensure that the autonomous machines are all pulling in the proper direction to support the dynamic business needs [11]. Many of the concepts that surface in Industry 4.0 discussions envision this kind of autonomy and local coordination.

Finally, a brief word about cyber security. Any implementation of IIoT must provide end-to-end security from the edge to the cloud. This should include hardening of endpoint devices, protecting communications, managing and controlling policies and updates, and using analytics and remote access to manage and monitor the entire security process. Eventually, security will be designed-in, not bolted-on later as an afterthought. But until then, existing plants will have to work with equipment manufacturers and other partners to create a more secure environment.

As you think about digitizing and transforming your industrial operations or your products and services, pay special attention to the edge. Consider the optimal location for analysis, decision-making, and control, and the best way to distribute these among edge devices, the network, the cloud, and other connected systems. The growth of the IIoTs extends the 'edge' beyond the network devices, into industrial and commercial devices, machines, and sensors which connect to the network.

6 Benefit of IIoT to the World

IIoT system can perform

- data collection,
- processing,
- helps to production decisions,
- allows quick activation,
- proactive monitoring system,
- troubleshooting and automatic alerts.

Develop single pane of glass:

It is an IT phrase used to describe a management tool—such as unified console or dashboard that integrates information from varied sources across multiple applications and environments into a single display.

The IIoT connects physical device like sensor and actuator to get the device data and it will improve the performance of the system by efficiency, safety, reliability and at last performance.

1. Data collected from the sensor are more cost effective because sensor doesn't use battery power like other devices.
2. Using big data, data analytics will be done. It will generate actionable result. These are used in decision making.
3. Generated result will be send to the appropriate person for the further processing on time.
4. When whole procedure take place on time it will improve result and it will benefit to the whole system [10]. It improves the efficiency of the system.

Major benefits to the world are:

- It will improved outcome economy for the real world.
- Improve of operational efficiency.
- If any system will introduced newly than easy to understand by everyone and also easy to configure because of it is divided into many sub-parts [2].

Big data analytics importance to IIoT:

The main purpose of the big data for IIoT is to handle company's data storage which is continuous stream of data hitting from more than one branches of company. Now in this IIoT world most of the company's are changing their platform towards platform as a service which is based on cloud so that the continuous expansion of their data storage can be avoided. Suppose the data are available on storage still cannot be useful for the use of that first you have to analysis.

Opportunities in big data analytics related to IoT

1. *Cyber physical cloud computing* connection between data storage and the front end communication channel are as important as input to any machine. So here any physical devices are connected to the local storage and from there it will connect to cloud. Data on cloud are every second updated by the connected device so that it will provide the right information and here the incoming and outgoing message buffer are different. So that in any emergency case any buffer can react immediately.
2. *Mobile crowd sensing* different mobile stations connected to the database using different road way like directly connected to cloud, connection from proxy server and passing from two servers. In these three types of classes like first class is of that type of data coming from any environment sensors like humidity and temperature. The second class of data which is of same type of data like scanner or fingerprint scanner. The last third class is the data which is needed in emergency case like the message will send to all when fire emergency occurred.

7 Background of UPoS

Unified PoS (UPoS) devices services is an architecture specification for application interface of PoS devices. The simple use of UPoS in industry where this kind of devices is made and used UPoS standard is operating system independent and language neutral [9].

7.1 *Inclusion in UPoS*

- Containing architectural overview.
- Text description of function for the interface for particular device.
- Description of relationship of classes' interfaces and objects using UML terminals.
- UML design will be implemented basis of C++, java, OO terminology and IDL.

7.2 *UPoS Standard will Not Include*

- API specification for specific language.
- Information about complete software components, hardware providers, software providers or third party providers will not provide by this document.
- No detail information about certification mechanism.

7.3 *Goal for UPoS*

- Provide common device architecture for international and across vendors, platforms and retail format interface in operating system.
- Application standard develop for device interface in operating system independent and language neutral.
- It will support multiple windows/com, windows/.NET and java platforms and many for because of same architecture for all this will benefit to reduce implementation cost for vendors.
- It will avoid completion between standard and encourage among implementation.

7.4 Dependencies

- UPoS success is depends upon some committees like OLE for retail PoS and java PoS technical committees which includes advance architecture into platform specific documentation [5], API definition and implementation.
- Technical implementation have two type of file

  1. *Definition file* detail description of interface and class file.
  2. *Example file* it has set of sample control classes, for illustrating the interface which is presented in application.

- Depends on input output device sharing for Microsoft COM model for the operation of the interface, explain into definition, goals and deliverables for OPOS.

7.5 Device Behavior and Sharing Model

Platform specific APIs are used by the applications. This used three types of common elements of UPoS.

Properties it is just a device characteristic which mostly held in binary or Boolean and string still many more types are there among this the property value is writable. So it can change or modify.

Methods are predefined with or without some parameters application has to call the particular method by passing the value or the parameters and in result it will benefit to the application with particular output.

Events in UPoS in like attributes which has some property declared which is held by the benefit of the application with predefined value and also targeted by the application to perform some particular task.

Device initialization and finalization for the particular application for that some rules are followed by the particular event.

For the initialization of the particular device firstly the instance of the any device is created which is followed by any unique ID or device name. Which already in the document mapped with it?

After the creation of the instance, it will go to the control event so the control is open and it will send proper service to pacified device and read the properties of the device for further process implementation.

Any device to be in used first it has to go into enable property which will state true and then the device can convert into operation mode. If this will not true than the property of the device is set to the disable mode and the device can not follow the instruction. For example if the scanner it will be able to display name into the available list. But if it will be selected to perform any task than error occurred because its enable property is set to false.

Algorithm for initialization and finalization

- Set the control reference.
- Call open method.
- Call the claim to device property.
- Perform the operable command
- Perform task to enabled device.
- Use the background task for the device.
- Task to be perform to disable the device.
- Call release to release the selected device.
- Run the close method.

7.5.1 Physical Device

Always deliver and operate as independent and it depends on the device standard.

The provision of customer-oriented IT at the POS implies integrating content in different formats and from different supply chain members in various devices [9]. Therefore, we will try to develop the system in which we used the UPoS devices and their methods which is declared openly. Using this you can make a system open in use for anyone who is implementing this type of IIoT system.

7.5.2 Overview of Technology

IoTs Application Architecture

The development of digital and network technology makes the controller node, sensor node, radio frequency identification and home appliances such as TV, air conditioning equipment can make use of the Internet protocol, wireless communication protocol to achieve information transmission and exchange [4] to integrate here the loose coupling and large scale deployment two characteristics are ensuring time with distributed system to complete the goal of difficulty in IIoT development two characteristics are ensuring time with distributed system to complete the goal of difficulty in IoT development. The structure of IoTs is divided into three parts:

1. Application layer,
2. Perception layer,
3. Network layer.

Application layer the user interaction of any system or application is depended on this layer how this software is developed the presentation of each system is depends on that. It will help mostly on management level the application used like remote medical treatment, green agriculture, and urban management.

Network layer used as bridge between other two layers. When the system is ubiquitous or the branches of the system is based everywhere than this type of bridge is necessary. Accurately and efficiencies is main goal of this layers.

Perception layer always connected to the development side. If we talk about any programming technology than this layer is work on back end side, this information collected from any device like GPS, temperature sensor, QR code label this layer always work at the bottom of the system to transfer between two or more devices [4].

In IoTs the developer develop engine which help the customer to complete the task without been abstracted using the current system. It will give benefit of searching individually as per their requirement with the already available data. The operation like search select and apply can help user to communicate with the other system and full fill the requirement of the particular result. Here in this paper the author explain the purpose using system of checking humidity of the soybeans.

Compass is nothing but the architecture of anything search engine. It will divide into mainly three parts logic device, API of the system and middleware. It will also provide easy compare to other engine and also predefined architecture to easily understand by any user [8].

The architecture work based on the requirements of application just like request response model, computing and processing of coming data, solving issues occurs during communication process, also database handling. Architecture always followed by the service oriented architecture using this it will help to organize widely spread industry program and also small unit of any industry. Starting from the API of the system data will go through the corresponding middleware and using this middleware it will go through the logical device. This chain will maintain this flow whenever data will transfer from one device to another. Here below the steps are describe in which the flow of data transfer from one device to another.

1. The application or API of the system consider some parameter to collect the data from the different device. Parameters are always define in earlier stage in the data gathering process.
2. The respective middleware cycles create between API and logical; device by the middleware. It is also possible that there are more than one middleware between end devices.
3. The subscribe–un subscribe mechanism is used between logical devices and system so that security mechanism is applied.
4. Data messages containing data requested by the logical device. It will follow some proper format for each message which will contain sender's address, receiver's address and data.
5. The task of the middleware is to follow register cycle each time and also generate the report to record and send it to the user.
6. Every time after getting the result and data. The user must unsubscribe and also request for the other data and be part to another cycle.

IoT Approach for Motion Detection

The IoTs has one limitation that is devices have limited computing power. Because of the problem you cannot fulfill the feature of security so to overcome this limitation IoT will revolutionize everyday life and help in situations like managing different sensor data, smart homes, and fire systems, health care. Also when one IoT project is going to be placed at that time we have considered the IoT challenges. IoT has main three challenges which is to be consider

1. Every device wants unique address to communicate.
2. Database management at every level.
3. Security police to be implemented correctly and also privacy also one of the important factor.

Raspberry Pi operating systems based on Linux runs by the Raspberry Pi and kernel known as Raspbian is the specialized kernel used by the Raspberry Pi. Benefit of the kernel is it is compatible to run all type of Linux based programs [3].

Raspberry Pi features:

32-bit ARM1176JZFS processor,
Clocked at 700 MHz,
256 MB RAM,
5V micro USB AC charger,
4AA batteries,
2 USB ports,
10/100 Ethernet port.

Uniqueness compare to other machine used in industry is it has low price in comparison to all. It can serve as a server like NTP server and DNS server for web traffic. TI is also used or run on low power and used batteries when low power detects also it will notify to the appropriate person for the external power source on power cut. So that the work not bends and the person is able to send power at a particular time.

It has some limitations due to its hardware. Firstly, it cannot run xS6 operating systems such as Windows and some Linux distributions. Secondly, it cannot run applications, which require high CPU utilization.

Here the main difference between Arduino and Raspberry Pi is Arduino is just a controller and Raspberry Pi is processor so you can run the program on Raspberry Pi. Raspberry Pi can work independently instead Arduino is work with any integrated large machine.

Evaluation of IoTs

WSNs are characterized by high heterogeneity because they are compliant with different proprietary and non-proprietary solutions. This wide range of solutions is currently delaying a large-scale deployment of these technologies in order to obtain a virtual wide sensor network able to integrate all existing sensor networks [6]. There usually is the need to use gateways with application specific knowledge to export WSN data to other devices connected to the Internet there is no direct communication between different standards unless complex application-specific conversions are implemented in gateways or proxies.

IoT scenarios and challenges:

1. Healthcare and wellness,
2. Home and building automation,
3. Improved energy efficiency,
4. Industrial automation,
5. Smart metering and smart grid infrastructures,
6. Environmental monitoring and forecasting,
7. More flexible RFID infrastructures,
8. Asset management and logistics,
9. Vehicular automation and smart transport,
10. Agriculture,
11. Smart shopping.

The data collected by the sensor are less in comparison of system but overall amount of the data collected by each and every sensor is in large amount so handle that data at one time and also fault tolerance are the biggest challenge in WSN.

In non IP solutions ZIGBEE, INSTEON, WAVENIS are worked as wireless protocol stack which defines task as physical link and network layers in this any of them can play role of sender, receiver or realer. The devices which are not in same range are achieved the functionality by multihop approach that relies on a time slot synchronization scheme.

IP based solution in this first IPv4 configuration is used but as the use of the device is increasing now it will shift to the IPv6 and as we know IPv6 is the auto configurable [6]. Also better the compare to IPv4. Another protocol used is 6LOWPAN. It is made up of low power which also connects to another IP networks and gateway devices. While handling 6LOWPAN comparison and neighbour discovery the edge router play important role. As is routes traffic in and out of the low pan. In IP based the transformation of data to one end to another end is transparent and efficient and also stateless.

High level and middleware solutions one working group of IETF suggested protocol like COAP, MQTT, AMQP, HTTP for IIoT to better performance which has main idea of request/response. Which has the main functionalities of low power and energy consumption using small embedded services? This all protocol mainly built upon the user datagram protocol the main concept of this type of protocol has three classes one is wrapper which containing all the functions and logics the second class is processor class which take the input and process on it and the third is descriptor which has the data file in XML format.

COAP–HTTP Proxy for IIoT

Here to main reason is there why HTTP can't use into constrained network those are:

1. The human readable, but less-efficient, representation of ASCII data.
2. Three way handshaking for the establishment of the connection is required for the HTTP on TCP connection for exchanging the data from source to destination.

COAP is the light weight protocol so it is used in consideration with any IoTs application here the TCP and HTTP is being replaced by COAP. COAP has more features than this two and it is used on network layer. Also it is able to provide security mechanism same as the other two. Here message is divide into different chunks and server always broadcast the message and the related client accept the broadcast message and again send the request to joint that particular group. If the broadcast messages is not relevant to the particular user that it simply discard the message [8].

Here transferring one protocol to the another is known as translating proxy. The proxy builds connection between the web and web of things by allowing client of HTTP to request resources from servers of COAP and same reverse procedure also possible where the client is from COAP and HTTP server handle that request using proxy we can also avoid unnecessary transmission of equivalent resources.

The main use of things in machine to machine communication here for constrained network COAP is presented as lightweight HTTP alternative for the use of some service.

Proxy as a gateway:

In sensor network the gateway must be used to the specific protocol here the work done by proxy to communicate with two different devices and it is not purely transparent one partner must have knowledge of the existence of the proxy.

1. Forward proxy

In this type of proxy client directly not connect to the server it just connect to the proxy and proxy will do the further task. In this proxy is work as a client for the server so server did not know the actual address of the client. These will benefit.

In this if proxy has more than one request from different client than it will send just one request so that it will reduce the network traffic.

2. Reverse proxy

In this type of proxy server directly not connect to the client it just connect to the proxy and proxy will do the further task. In this proxy is work as a client for the server so server did not know the actual address of the client. These will benefit.

In this if proxy has more than one response from different server than it will send just one request so that it will reduce the network traffic. If client is not able to work with server at a particular time than proxy will buffer that request or response and further respond to that service.

DZONE Industrial Internet System

According to DZONE IIoT is the connection between machine to machine for collecting data from the one end and send it to the other node using Internet and after that also make proper use of that data [2].

It will divide IIoT into mainly two parts

1. Operational technology (OT),
2. Information technology (IT).

OT describes the system communicating with the outer world and this is it will firstly connected to the physical objects like sensors and actuators after connecting with the device it start collecting the data. The major task of OT is to connection establishment with the device and also communicates with them continuously without breaking the connection [2]. After the data arrived it will start to process on them and also generate report to send forward and handle their own database by the OT the work of the Internet is being easy and act like human brain. Like it can sense and hear and see easily and also simultaneously.

Information technology (IT):

The main usage of the it over the OT is to maintain and control the system which is not physically observed if the system is being operable from the distance than it is really helpful to expand any industry to any region in that first of all the task is to make it ubiquitous and for that IT is used to put it to the software configure and operable simply it can divided into category like operation based and software based. Where physical devices can communicate with software and can send data as per the command run on the software application.

Control, operations, applications, business, information.

These five operations are major domains for any it based industry. These will control all functionalities and it will make better result to the current development. And improve it. Operations perform in background and it will generate the output and configure them. Also it has failure detection system for the sensors and other detection system.

Control system:

When establishing for the IIoTs, consider the following level of the system:

Device level

- Store the data and also encrypt them.
- Device authentication.
- M2M connections are compulsory to end to end devices.
- Security configuration to each device.

Gateway level

- Authenticated devices have to process all the way when it will connected.
- Cloud connectivity.

8 Design Approach

8.1 Hardware Design

In the above design there are three modules. This is connecting to each other it will generate a system. Where physical device is connected to the proxy server and that proxy server is connected to the local server. This chain will contain the flow and after generating data from the physical device like sensor (Fig. 2).

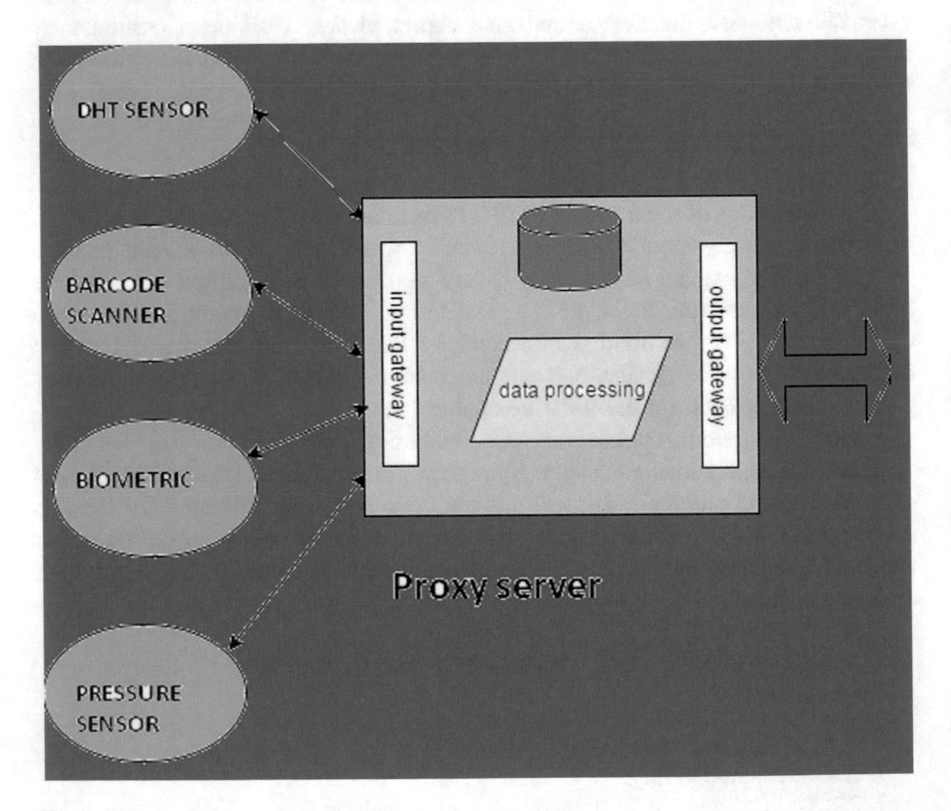

Fig. 2 System design

8.1.1 Module 1: Physical Device

Here we are trying to connecting three sensors for three systems.

Fire alert system: in this system the sensor used is DTH11 sensor which is used to experiment the system. That sensor is connected to the Raspberry Pi and that Raspberry Pi is connected to the proxy server.

8.1.2 Module 2: Proxy Server

Proxy server is used as a gateway between physical device and local server. Functionalities of proxy server are listed below.

1. Authentication of physical device.
2. Data storage at server:
   - Check the redundant data,
   - Time to receive date,
   - Time to send data,
3. Forward data,
4. Connectivity,
5. Immediate reaction,
6. Security.

8.1.3 Module 3: Cloud Connectivity

Proxy server is connected to local server and that server is directly connected to the cloud. In cloud data coming from the local server and all local server data are coming from different application and location and stored at one place.

The three phase mention above shown in the figure where starting from the sensor device and ending at the cloud all things are connected with other which shown the IoT things to things connection. Here both type of connection are used wire and wireless where the physical device is remotely place still it is under continuous monitoring.

9 Conclusion

Machine-to-machine (M2M) concept is one of the main features of IIoTs. So in this system developing one system which is a part of IIoT and also M2M communication. To bridge the gap, between wireless sensor network and cloud we try to develop this system. In this system, we propose the open source M2M connectivity with cloud based connectivity. In which the sensors and scanners are type of UPoS system so it proposes a modular architecture running as open standard, making it highly extensible via plugging. It enables multiple communication protocols binding, reuse of existing remote devices management mechanisms, and interworking with existing legacy devices. The proposed system will be helpful for connecting the remotely located sensors. This system would be helpful for the industry where the units are located at different location. So using this system one can make all the branches of the industry connected geographically. Also at any time anyone can operate or see the company progress or anything they want to know about the industry. Using this technique you can also alert to the people at particular place of any disaster.

Acknowledgements With immense pleasure, I would like to present this report as the part of my Dissertation work of IoT gateway for smart devices. I express my gratitude towards this college for providing me with proper resources and knowledge for the completion of my Dissertation work for this semester. I have received a good support, motivation and encouragement from many people. First, I would like to thank Professors Chintan Bhatt and Divyesh Patel for constant guiding me through each and every step of the process with knowledge and support. Thank you for your advice, guidance and assistance.

References

1. Lin, N., Shi, W., & Zhengzhou University Software School. (2014). *The research on internet of things application architecture based on web*. IEEE.
2. *Predix: The industrial Internet platform*. GE Digital.
3. Han, Y., Chen, Y., Wang, B., & Ray Liu, K. J. (2015). *Enabling heterogeneous connectivity in Internet of things: A time-reversal approach*. IEEE.
4. Ben Alaya, M., Banouar, Y., Monteil, T., Chassot, C., & Drira, K. (2014). OM2M: Extensible ETSI-compliant M2M service platform with self-configuration capability science direct. *Procedia Computer Science, 32*, 1079–1086.
5. Lunardi, W. T., de Mato Ramao Tiburski, E., Amaral, L. A., Marczak, S., & Hessel, F. *Context-` IoT: Discovery, search, selection, and usage of devices*. Porto Alegre: Pontificia Universidade Catolica do Rio Grande do Sul (PUCRS).
6. White Paper on Machine-to-Machine Communication M2M.
7. *Wireless Personal Communication, 69*, 999–1012 (2013). doi:10.1007/s11277-013-1060-9.
8. Connecting the web with the web of things: Lessons learned from implementing a COAP–HTTP proxy.
9. Unified POS international standard, for implementation of point of service peripherals, version 1.14.1 October 23, 2014.
10. Bhayani, M., Patel, M., & Bhatt, C. (2016). Internet of things (IoT): In a way of smart world. In *Proceedings of the international congress on information and communication technology* (pp. 343–350).
11. Bhatt, Y., & Bhatt, C. (2017). *Internet of things in healthcare—Internet of things and big data technologies for next generation healthcare* (pp. 13–33).

Smart Manufacturing in the Internet of Things Era

Th. Ochs and U. Riemann

Abstract The purpose of this chapter is to synthesize opportunities and challenges related to the IoT within the manufacturing environment based on a project at Villeroy & Boch (V&B). The chapter seeks to visualize the impact of IoT methodologies, big data [2, 3] and Predictive Analytics towards the ceramics production. Key findings and challenges are related to both to the technical and organizational dimension tended to overshadow optimism. Organizations, such as V&B have been working with big data sets but emerging business models are now looking to gain deep insights from the so-called "data exhaust" that's now become a "data gold mine" for business optimization embracing the move toward revolutionizing their production through big data and Predictive Analytics.

Keywords New production models · Advanced manufacturing · Big data and Predictive Analytics

1 Introduction

The world as we know and experience it today has been shaped by three major technological revolutions. Especially in the last 2–3 years the IT technology showed an increasing development which leads to new chances and opportunities. Technology is usually defined as artifacts, processes and machines and the knowledge—based on technical or engineering knowledge—used to design and operate them [30]. Technology from the sociological perspective is always a relational object because its creation, use and diffusion is based on social processes

Th.Ochs (✉)
Villeroy & Boch AG, Mettlach, Germany
e-mail: Ochs.Thomas@villeroy-boch.com

U. Riemann
BTS, SAP SE, Walldorf, Germany
e-mail: Ute.Riemann@sap.com
URL: http://www.sap.com

N. Dey et al. (eds.), *Internet of Things and Big Data Analytics Toward Next-Generation Intelligence*, Studies in Big Data 30,
DOI 10.1007/978-3-319-60435-0_8

relating things, signs and meaning, humans and institutions [26]. These opportunities will have a significant impact on our company and our value chain. This can be detected especially though the topic of "Industry 4.0", i.e. with the interconnectedness and the computerization of machines, assets and products which are summarized as the "Fourth Industrial Revolution". Based on the close linkage of computer technology with electronics and mechanics (Embedded Systems) and the online communication ability via (mobile) internet (IoT = Internet of Things) new potentials and opportunities in production, logistics, sales and marketing up to new concepts such as (smart) products and (smart) services emerge. A further technology push will be achieved with the processing capability of mass data in various formats, sources and levels of structure in real time (= big data). This, in a bundle with complex statistical methodologies and techniques (= Predictive Analytics), will generate new options to gain more insights. The emergence of this technology will push the boundary beyond traditional reporting. Companies such as Google clearly show: "the one that understands the data has the main key for success". As of today, this statement do no longer count only for companies that have built their business model on the internet [14] but becomes relevant for any company in any industry [15]. The rapidness in the usage of IT functionality increases dramatically as well. Cloud-based applications raise the bar in regards to the fast provisioning of an IT-solution, flexibility of usage, scalability, global accessibility and operating costs. "Smart manufacturing" becomes the new normal in a world where intelligent sensor-based machines, systems and networks are capable of independently exchanging and responding to information to manage industrial production processes.

While having smart manufacturing as one target, the "market of one" is becoming a reality as customer's desire highly customized and high quality products delivered at their choice of time and place and at competitive prices.

This is huge challenge not only for sales but as well for manufacturing. With multiple systems being deployed across multiple plants and warehouses, organizations are facing more complexity and more touches by employees, leading to a high cost of ownership, an inability to respond adequately, and poorly informed decision making due to old, incomplete, or inaccurate information.

A ceramics industry production may not be the first image that comes to mind when contemplating the Internet of Things. But Villeroy & Boch (V&B)—a German family owned ceramics manufacturer—becomes one of the savviest data crunchers. In our chapter we will outline, based on the yield data base project, how the company takes advantage of the new capabilities driven by the IoT. For years, the manufacturing production was driven by experiences, using systems such as Enterprise Resource Planning (ERP)/Production Planning and Steering (PPS) systems and methodologies such as Six Sigma or traditional analytics. Manufacturing insights have been made available with the support from IT department using analytical tools. As of now the production still relies mainly on experiences and predictions from heuristically data. Although V&B have been using proven modelling, statistical process control and data analysis methodologies to optimize production for years, the extreme composition and availability of big data Analytics [25] provide the chance of new approaches to further improve the production in its core elements which was not possible before. In that sense, big data and Predictive Analytics represents an extraordinary opportunity [29] as it promises new and valuable insights to not just improve competitive results but to improve sustainability to compete in today's economy.

Now that a new crop of tools raised the bar manufacturers such as V&B are enabled to crunch the massive amount of data collected though sensors and to predict all relevant data for a rich root-cause analysis of yield and quality data, combined with customer rejects/returns, supplier's quality data, and other critical measures. In other word: manufacturers such as V&B are ready to embrace the use of big data supported by computing methodologies, statistical modelling and the appliance of extended research results.

As sensor costs and other infrastructural elements become available and improved applications, tools and analytic methodologies are accessible, better insights into the root cause of the production and a day-to-day optimization of the manufacturing process becomes realistic. V&B have invested in these tools that allow the collection of manufacturing and yield data from the production equipment and input this information into a database that, when aggregated with multiple sources of data and analysed by statistical methodologies, lead to fact-driven actions within the production for enhanced quality and reduced overall cost. The volume, variety and velocity of data [23, 24] being produced in the plants is growing exponentially and gains opportunities which lead to a competitive advantage in a changing and dynamic market environment. With a modern equipment across the manufacturing floors data of different types can be generated, stored and analysed.

For V&B the ability to access, analyse, and manage vast volumes of data becomes important to successfully continue the operation of their production. Taking the sentence of "the more we know, the better we perform" into real business, V&B launches a big data and Predictive Analytics initiative. With the launch of these initiatives V&B aims to shift towards a fact-based and analytical steering of the entire production processes. The target is a comprehensive penetration of the entire production process to seize the chance of increasing the "Right at First time" factor—meaning a production with a high percentage of first quality without post processing. The overall goal of this initiative is to open the door into smart manufacturing enabling V&B to remain competitive in their markets.

2 Villeroy & Boch

V&B is an innovative company in the ceramics industry with a tradition for more than 265 years [29, 35]. The company focuses its business activities in the Company Divisions 'Bathroom and Wellness' and 'Tableware' [29, 35]. In times of digitalization and expanding competition, a sustainable strategy is not sufficient any more [13]. It requires as well a comprehensive approach to operationalize it to gain the ability to make decisions faster and smarter. As strategy needs to always come first, V&B decided to take benefits of the emergence of digitalization and its evolving opportunities by defining their own digitalization strategy.

While V&Bs digitalization strategy gave a clear direction for implementing and systematically executing big data and Predictive Analytics, the next step was to define the path forward to get benefit of the combination of available technologies [22] while gaining additional value from new technologies to further strengthen the companies competiveness [5]. To realize the advantage of these optimization potentials V&B has started a digitalization initiative. Within this initiative V&B focused mainly on the aspects of

- *Production operation* such as basic production execution, production control, repetitive manufacturing and manufacturing execution
- *Quality management* such as quality engineering, quality inspection and quality improvement with the overall target to improve the yield

to gain the ability to access, analyze, and manage vast volumes of data—this is especially because V&B expects an extraordinary opportunity from big data and Predictive Analytics [31] as these technologies promises new and valuable insights. The realization of these opportunities is challenging and per Thomas Ochs, the CIO of V&B, increasingly critical to successfully continue the operation of the production, focus on production efficiency which leads to an improved productivity as well as production quality [19] and overall improved operational efficiency.

3 Big Data, Predictive Analytics and Advanced Manufacturing

Big Data refer to large-scale information management and analysis technologies [17, 34] that exceed the capability of traditional data processing technologies [4, 6]. The most important step ahead for V&B is now the improvement of product quality and yield (the amount of output per unit of input) [19]. This requires the ability to make process data accessible as a new qualitative data source besides the existing pool of operating data (e.g. Post Processing (PP), Betriebsdatenerfassung (BDE)) delivering the necessary granularity for an enhanced gaining of knowledge.

3.1 Advanced Manufacturing System Requirements

Historically, instrumenting and connecting equipment and resources to improve the manufacturing of ceramics was always mainly conducted by human experiences. Such an approach did not scale well and—within a globally acting company, failed to reap the investment benefits. IoT technologies are now central in regards to a transformation of the manufacturing as it brings together people, processes, data, and things within a single networked connection. Sensors and data-acquisition systems allow the collection and aggregation of equipment operation information. Key to this are the systems and the infrastructure itself, which must be robust and resilient and run smoothly. To accelerate the journey through the phases of IoT-enabled manufacturing transformation, V&B focused on main elements that do address the various system requirements:

- Connected manufacturing based on the internet of assets. The availability of a cyber-physical platform that comprises smart machines, storage systems, and production facilities capable of autonomously exchanging information and, triggering actions and controlling each other independently provides a huge benefit: the short-term benefit is the need of less personnel required to enter data. The long-term benefit will be the ability to run autonomic or processes.
- Inside-driven manufacturing focusing on internal operations and processes and externally on delivering new revenue streams and potentially new business models. The powerful combination of challenging market conditions and the opportunities provided by smart manufacturing are pushing V&B to a shift in their strategic focus.

For V&B it was obvious, that the IoT technologies only make sense if they are integrated in an already stable running business. The basis for successful implementation is consequently therefore the horizontal integration of business processes. Covering the processes all the way through to the entire manufacturing the technology is fully covering the manufacturing as a "raw material to end product"

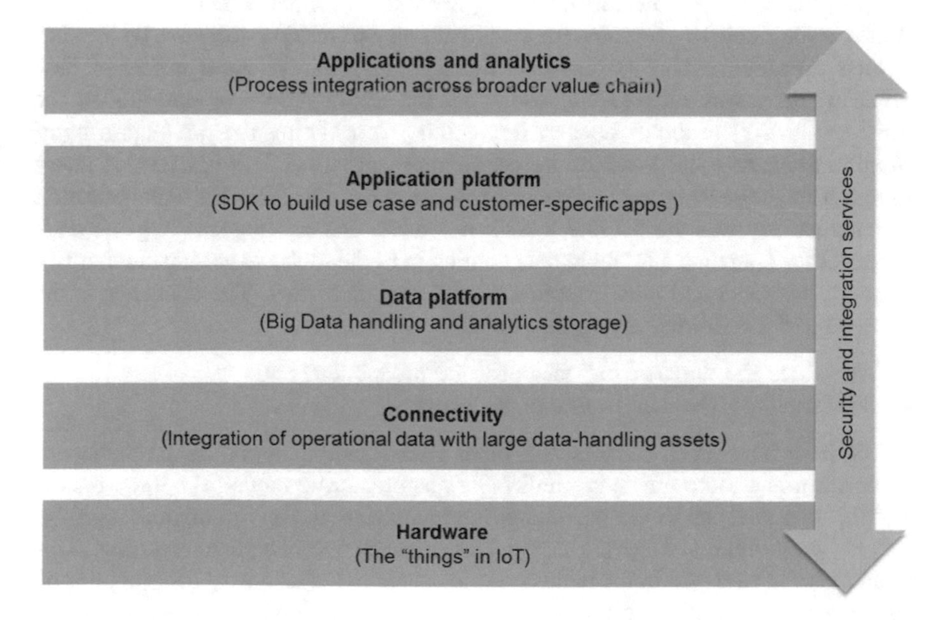

Fig. 1 Generic IoT stack

view. One of the prerequisites for this horizontal integration is a common data model and physical representation in the sense of an IT platform. This platform should be smart in the sense that it allows fast and simple correlation, analysis and visualization of data. More and more this common data platform is built on a cloud infrastructure to allow access for constituents in the extended eco system. Up to this point, you could still look at this as "simple" Computer Integrated Manufacturing (CIM from the eighties of the last century). What is new in IoT is the additional vertical integration from the business process down to the machine level. In the IoT terminology this is enabled by so called "Cyber Physical System". In manufacturing by connecting machines among themselves and to V&Bs enterprise system to enable a high connected manufacturing (lot size one, flexibility, configurability) and predictive quality (Fig. 1).

3.2 Challenges on Advanced Manufacturing for V&B

One major issue per Thomas Ochs is that "manufacturers are a cantankerous bunch" and rely "on a hodgepodge of old equipment with their traditional experience rather than trust in new technologies and scientific experiences with a strong link to the IT". This makes it difficult to get an end-to-end view of quality due to distributed systems. Changing a local or departmental quality management focus to a V&B-wide focus was one of the key targets. Preventing non-conformance within all stages of production was the overarching goal.

Now, V&B opens their manufacturing to capture new and valuable information hidden in their data to overcome the traditional way of making decisions by looking in their rearview mirrors at historical financial information. As a matter of fact, collecting data from all relevant machinery for a comprehensive snapshot of the entire production becomes next to impossible. Considering the production complexity influencing the yield in V&Bs ceramics production, V&B need a smart approach to diagnosing and correcting process flaws. So far a lot of information within a factory was treated like a sunken treasure and the target is to get it back into real life. There are a lot more data coming in as the tools are getting smarter to a much greater level and take machines tracking much further. The challenge is the answering of the following two questions:

- What does the "smart data" mean for the production? and
- What are the right data to further investigate?

The now introduced sensors and other data-gathering tools are propelling to move towards a more efficient, predictable manufacturing model. As V&B is now pushing data analysis to the manufacturing technician and environmental level the data are now employed by the technicians engaged in daily manufacturing. This means that the staff within the plants need to access to the data to quickly adjust and optimize the manufacturing process.

Big Data delivers huge data sets consisting of structured, semi-structured and unstructured data unmatched and in real-time. The real business need for V&Bs manufacturing comes from the key characteristics of a ceramics manufacturing process: the production process has a high complexity and variability stemming from various factors such as the raw material, the type of equipment used, the humidity within the plant, machine operator differences—just to name a few creating high volumes of various types if data. At the end, the quality of the production process can be stated in one key KPI: the production yield.

3.3 Production Yield

The production of ceramics uses various ingredients, and the manufacturing team must monitor these ingredients in addition to environmental factors related to the production flow to ensure the high quality of the final product. This unexplained variability can create issues within the production process and product quality which can negatively affect the production costs.

Product yield is a measure of output used as an indicator of productivity [27]. It can be computed for the entire production process (or for one stage in the process) as follows:

$$\text{Yield} = (\text{total input})\,(\%\text{good units}) + (\text{total input})\,(1 - \%\text{G})\,(\%\text{R})$$

or

$$\text{Y} = (\text{l})\,(\%\text{G}) + (\text{l})(1 - \%\text{G})(\%\text{R})$$

where l = planned number of units of product started in the production process, % G = percentage of good units produced, %R = percentage of defective units that are successfully reworked.

In this formula, yield is the sum of the percentage of products started in the process (or at a stage) that will turn out to be good quality plus the percentage of the defective (rejected) products that are reworked. Any increase in the percentage of good products through improved quality will increase product yield [27].

4 Yield Database Project

With the launch of the yield database project V&Bs has introduced a multi-step initiative. These steps are:

- Step 1: Big Data and Predictive Analytics to reduce product quality issues mainly in the production of bathroom & wellness ceramics in all ceramic plants starting with the plant in Mettlach (headquarter).

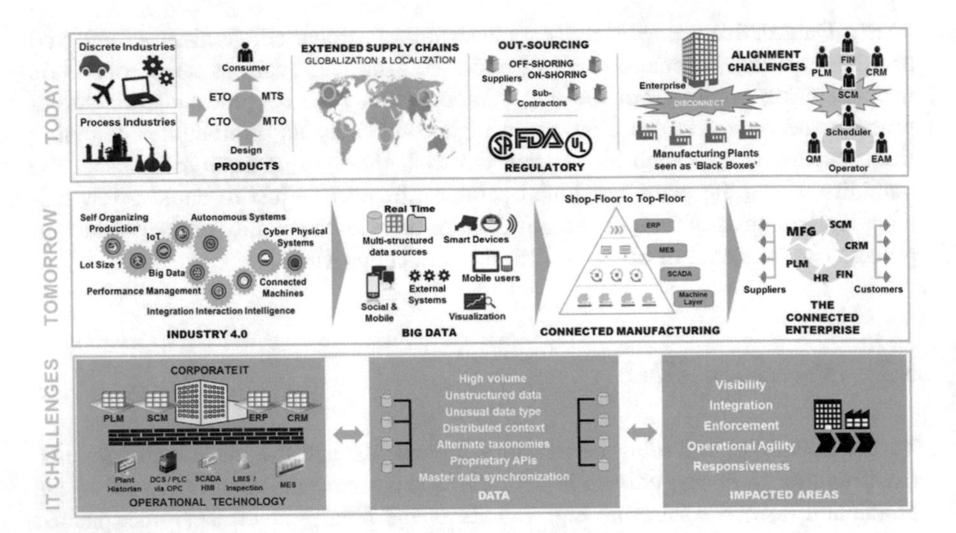

Fig. 2 Overview on path forward towards the new IoT-driven manufacturing

- Step 2: Smart manufacturing to further improve efficiency and increased manufacturing throughput leading to a shorter planned maintenance, a better identification of defective units and though a better and faster time-to-market [32] (Fig. 2).

4.1 Private/Hybrid Cloud Infrastructure Requirements

Now let's focus on cloud. On the topic of cloud models—especially the different flavors of cloud—there is an ongoing discussion about what the best model could be: public and private. There are many definitions and we think a little simplification is in order, at least for this chapter (this is where some of the cloud purists will jump up and down):

- Private cloud—hosted solutions, often managed by a partner/vendor.
- Public cloud—multi tenant solutions, managed by the vendor.

Keeping in mind that an integrated solution is successful when necessary data is available at the right place at the right moment. Besides smooth working interfaces, this also requires a more holistic approach to data maintenance. Therefore, it was a key importance that the data integrity was maintained:

- Only data which are absolutely needed is maintained.
- Data that only serves local needs is kept locally.
- Double data entry is reduced to the absolute minimum.
- Completed data is a pre-requisite for next steps.

Even though we have made a significant step forward in the use of a cloud environment, cloud manufacturing is still emerging as a novel business paradigm for the manufacturing industry, in which dynamically scalable and virtualized resources are provided as consumable services over the internet. One of the challenges is that few of the resources can adapt to changes in the business environment. There is limited support on migrating to different cloud modes in existing solutions. Furthermore, it is currently still difficult to set self-defined access rules for each resource so that unauthorized companies will not have access. In parallel a unified ontology needs to be developed to enhance semantic interoperability throughout the whole process of service provision in the clouds. The feasibility of this approach needs to be implemented throughout a global company, each of which has complex access requirements. Having stated briefly the requirements and challenges, V&B decided towards the solution of a hybrid cloud to manage the manufacturing requirements with some technical insights for the V&B IT:

- Configuration of cloud content and processes by discovering security, migration, cloud only and hybrid scenarios for faster, cheaper and flexible innovation cycles.
- Faster and cheaper consumption of innovation with the flexibility to fulfill the growing requirements of business process owners and business users with easier and simpler process and data integration.
- Easy and end-to-end data/process integration across on premise and cloud by "discovering" the integration platform, standard/custom packages and identifying stakeholders with the target of a TCO reduction.
- Validation of business process quality and consistency using Go-Live checks.
- Reduced cost of operations using proactive EarlyWatch checks, end user performance/experience monitors.
- Ensure business continuity by using end to end incident and escalation management process.

To best realize the value of this investment and harnessing the need of data security lead to a hybrid cloud solution. Both private and public cloud offerings should leverage the same platform. With this Platform as a Service (PaaS) offering, you can enable the eco-system and the IT departments of customers to do what in the cloud also needs to be done—manage the last mile, configure or extend to your needs. Regardless of whether you want to configure solutions to your needs, adapt a workflow or enhance a solution—the platform will support this. In addition, integration and data integrity is handled by the platform, regardless of a cloud to cloud or cloud to on-premise. With such a hybrid environment V&B can accelerate existing projects by bringing innovation and responsiveness to the business—while you can work on new projects in parallel. This will allow you to truly transform your business and bring cloud benefits like minimized infrastructure costs, risk reduction and the freeing up of internal resources to focus on essentials.

4.2 Big Data and Predictive Analytics

The overall target was to ensure a production flow to achieve “Right at first Time”, meaning a final yield (= production without mistakes and any post processing) as high as possible. For taking advantage of big data and Predictive Analytics to significantly increase its yield V&B has to establish a modernized IT platform in addition to a sophisticated network inside their plants to capture machine data as the main outpoint of the connected production. This step helped to overcome the lack of visibility into the production processes, status, and performance in manufacturing systems, which formerly lead to inaccurate and late decision making towards the product quality. In addition, V&B segments the entire production process into groups of linked production steps. For each group of production steps, real-time data are collected about the process steps, materials used and other environmental data and gathered them in a central database [19]. The next step was to specific mathematical and statistical methodologies to determine interdependencies among the different process parameters [1] (upstream and downstream) and their impact on yield. The target for V&B was and is to allow focused process changes on the detailed parameters level to enable the increase of yield [27]. As so far, the impact of a singular production parameter as well as the correlation and interdependencies between the various parameters was only limited transparent, the key success factor was the establishment of a central process database and the use of a statistical analysis software which proposes a significant value. Even though several data have already been captured during the entire V&B production process, the novelty was to intensify this data capturing and the integration of further data/information. This means, that in addition to the already existing capturing of production data additional sensor-generated data need are now seamlessly added to the data capturing and analysis process. By this, the new IT platform that collects data from underlying system databases, manuals, sensors from the machines, maintenance and environmental factors V&B creates a valuable knowledge repository for the phase 2 of the project. The new agile IT platform offers an integrated and simplified solution utilizing quality parameters to allow the quality technician, quality engineer, quality planner and plant manager to calibrate the production accordingly. With the adequate exposure to the problem solving domain the now implemented IT solution can learn continuously from every incident to apply previously gathered knowledge and assist the manufacturing engineers to solve problems faster. To enable this V&B invests in excellent numerical analytics and is now able to predict outcomes within highly confident bounds in some areas. The appliance of Predictive Analytics provide the depths to understand the collection of information of relevant parameters and patterns that are critical to production process and allow to have a steering of the processes on a detailed level. Having big data in place, Predictive Analytics helps to identify and understand the value of previously unused and to identify hidden patterns [21]. It allows a detailed analysis of process data as well as the identification of patterns and relationships among process steps to provide transparency on factors and effects on yield and to allow work-in-process decisions.

If you want to see it like this: Predictive Analytics is a logical next step after introducing big data developing into an advanced area of analytics designed to provide more intelligent, forward-looking (anticipatory), and more actionable analysis for making better decisions, and to derive better insight and optimize a decision-making process. Quality inspections providing inspection results to enable timely, informed release decisions and the support of product compliance and quality requirements are much easier enabled with a much better result. Enhanced efficiency in product quality assurance becomes easily and efficient. Integrated, simplified quality management with built in IoT connectivity and Predictive Analytics, improved quality measurement results through recording of defects, planning and flexible, harmonized action management with real-time monitoring and exception management of quality deviations is now reality. The benefits of Predictive Analytics for V&B can be summarized as:

Understand while getting from transactional to analytic data the IT enables the business to access all data, ensure self-serve, governed data and gain deeper, real-time insights.
Flexible and simple with the shift from transactional to analytical data the V&B IT landscape provides solutions that are easy to use, simple to understand and to manage and business-driven. In addition, the analytics platform leads to a cleaner, simpler and more flexible architecture.
Transform the future with business scenarios to visualize new outcomes, predict new opportunities and plan new strategies based open a scalable platform with lower total cost of ownership (TCO).

The meaningful and proactive use of big data that has now being introduced at V&B allows to answer difficult questions that business intelligence has not equipped for previously. Predictive Analytics unleashes the value of big data as it is now possible for V&B to leverage native, in-database predictive algorithms for high-performing analytics.

What does this mean for V&B? While having now an (big) data volume of real-time shop floor data (and still have the historical data available) and the capability to apply statistical assessments [16] towards the previously identified production areas another level of transparency is achieved: the knowledge embedded within the manufacturing process and their human experts in the plants can be now unpacked in two ways:

- *Semantics* establishing evidence-based relationships between entities and observations when the manufacturing process is under control.
- *Data metrics* sensing and measuring the Data to understand the evidence and enable knowledge capturing instead of just data.

V&B has built an IT solution to encapsulate the knowledge of the entire production process and provide sophisticated data analytics to facilitate the engineers do their work and thus to reduce the yield.

Where previously servers are managed at plant level as a standalone system that was not fully integrated to other subsystems on the shop floor or ERP system, V&B has now an IT environment (data center to cloud), where appropriate systems are integrated [10] with synchronized data. This connected approach allows the deployment of integrated applications, with one code base to fit any "mobile" device (workstations, laptops, tablets and phones) across all roles in the manufacturing process.

This newly established IT platform and the support of big data and Predictive Analytics form the foundation to deliver these new insights, including identifying correlations between production data, raw materials, and other environmental data to be turned into valuable actions, with real-time controls and measurements, based on a set of intelligent production rules.

The now connected manufacturing allow real time visibility to the current production status, ability to analyze, predict "What if" scenarios and empower proactive decision making. The link to dedicated real-time production data immediate enable real-time decision-making which leads ultimately to an improvement of the production quality. With this link of previously isolated data sets of historical data and real-time data V&B gains the capability to conduct sophisticated assessments revealing important insights in real life and not in any laboratory environment. The integration between all aspects of the manufacturing process and data points was the door-opener for big data Predictive Analytics as it serves as a basis for the analysis core production activities synchronized across devices, and the entire supply chain and product lifecycle management. A data- and fact-driven production has many advantages. With Predictive Analytics V&B can take the step forward into a detailed and focused analysis not only into (historical) process data. They can identify patterns and correlations between process steps, input factors and environmental factors which enables V&B to either optimize the entire production process focused on the factors with greatest effect on e.g. yield or decide on the entire production process at a very early stage e.g. to stop the entire production line.

4.3 Advanced Manufacturing

Hearing and seeing what happens within the production at the very moment it occurs, is just the first step. The merging of the virtual and the physical worlds through cyber-physical systems and the resulting fusion of technical processes and business processes are leading the way to the smart manufacturing approach of V&B. Smart factory products, resources and processes are characterized by cyber-physical systems; providing significant real-time quality, time, resource, and cost advantages in comparison with classic production systems [11, 12]. The smart manufacturing at V&B is designed according to sustainable and service-oriented business practices. These insist upon adaptability, flexibility, self-adaptability and learning characteristics, fault tolerance, and risk management. After having introduced a higher level of automation as a new standard in their plants and a flexible network of system-based production systems which, oversee the production processes a flexible production systems, which will be able to respond in almost real-time conditions is on its way allow to allow an in-house production processes that is radically optimized.

In addition additional value is being added with Advanced Analytics. An Internet of Things-driven smart manufacturing uses sensors placed in different parts of the machines to provide continuous visibility into the machine, the entire production process and its operating conditions. These data can be analyzed in real time while the noise—the irrelevant information—are filtered. All relevant data are put into a context, so that business can react upon as soon as they fell outside standard operating limits. These context data can be many things: production machine breakdown, walking around on the shop floor, getting feedback from quality inspections regarding the used raw materials, weather conditions. The real time machine data and the context data need to be merged to have the full context, understanding the production environment entirely and know what the best to act in these production circumstances. As already stated in our chapter, this means that V&B has to get advantage from a combination of the large scale Analytics provided by big data and prediction Analytics in combination with the IoT [25]. After gaining value of the analysis of big data and identifying useful patterns additional benefits can be realized with the use of IoT:

- Increasing uptime of the manufacturing with less cost of unplanned downtime and reduced maintenance windows.
- Accelerated factory output with a better time-to-market.

This leads to a continuous analysis of data across the entire IT network using data gathered throughout V&Bs manufacturing network. To increase the uptime the monitoring of yield, cycle time, maintenance schedules and process health got introduced. For this, V&B now collects not only the relevant data for those KPIs and transform them into actionable activities, the information goes into a report that contains the indication with the capability to drill down into the detailed levels of information. This allows an interactive analytics with the possibility to ultimate decide on appropriate actions. In addition issues within the production flow can be identified in real-time providing the ability to find the root cause much faster than before and thus to solve the problem at short-term. The ability to drill down from the issue warning to the detailed root-cause enables an immediate and precise call for action and has a flow back into the knowledge database to further analyze more persistent ones.

5 Solution

For achieving the benefits proposed by big data and Predictive Analytics to survive in a highly competitive world [9]. V&B has to establish a comprehensive IT platform. The building blocks of V&Bs manufacturing platform and network from floor-to-floor to the data center in their headquarter are composed of specific architectural components:

- A private cloud infrastructure is used where the data are stored and ingested.
- Analytic tools are available to enable analytics, drill down and enhanced reporting.
  The BI results and visualization tools link reported results, predictions and enable a drill down to manufacturing operations.
- The link to the production and the connected machines is established to extract the data throughout the corporate intranet.

The most important step for V&B was—and it is still ongoing—the collection of necessary internal information and, making these data accessible for Predictive Analytics. Key data categories that have been identified and data sources are mainly:

- Excel files for those data that are manually generated but with high relevance e.g. laboratory data.
- Technical data collected from production systems and machines (e.g. from sensors generated such as temperature)

This new infrastructure allows continuously pulling information from different sources within the production and linking them together. On the shop floor, automation assets are distributed over the manufacturing network. In parallel, V&B establishes a rigorous process to ensure a necessary data quality level to allow a consistent measurement and fact-driven results. This, by consequence requires a high process stability to gather the necessary data based on a stable environment.

After having established the basic production and IT infrastructure, the creation of a robust data set that encompasses all necessary inputs was the next step on V&Bs path forward: the IoT gateway acquires machine Data in real-time, sending these data to the data center where various data preparation tasks such as filtering the noise of data are performed before they are formatted and ready-made for further Analytics. This means as information is typically generated and stored in different formats, and much of the data are "noisy" [33]—they cannot be understood and interpreted correctly by machines, because they include unstructured text or have other limitations. To ensure robust data and facilitate analytics, V&B has to extract and validate the gathered information before creating any models. While this sounds simple, it was challenging for V&B since all variables per process step, representing diverse inputs as e.g. resistance, temperature, pressure, and machine location, needs to be captured. In addition, the required data for each model and analysis differ, depending on what equipment or sensor are in use and is needed. V&B had to ensure that they examine complete information of high quality rather than relying on aggregate data sets that capture averages or on a sampling of inputs, since such methods can lead to false positives or missed patterns. This means for instance that V&B has to compile all sensor, process, inspection, and environmental data. With the use of the established yield database V&B ensures that the data are effectively gathered and stored to keep the best data quality for the analytics and for a continuous optimization. These data can be accessed for ad hoc reporting, workflows and predictions. The successful way forward is not just to collect as much information as possible, but also to make sure that the data fulfil a certain level of quality and are easy-to-understand for the end-users e.g. with a role-based dash-board enabling actions on monitored and predictive actions.

6 Further Perspectives

Since the cloud environment has still some unanswered questions (e.g. security questions) and the concept of big data and Predictive Analytics are still about to emerge the path along will move while the technology emerges. The question for V&B is how to balance the new technologies with the existing manufacturing process. The problems with the core manufacturing do not stop after having introduced RFID chips and sensors to the production plant; the production still may encounter unexpected issues where experience is still the key to solve it and manual intervention is required. One question is also what shall move into the fact-based knowledge/pattern recognition and what portion of manufacturing will rely intuition and experience?

Another question is that the new design and manufacturing techniques lead more complex inspection, testing, and validation procedures also create delays and higher costs. The question here is business-driven: what is the business case and what is the optimum of automization and tool support?

7 Conclusion

Smart manufacturing is about harnessing the power of the "intelligent production". The key is increasing agility and awareness of manufacturing environments from plant maintenance & production to machine & equipment data and sales to finance. In complex manufacturing environments, this can mean an improvement in overall equipment. The now implemented solution enables V&B to predict the failure points and hot spots or flashpoint in the plant as different production modes are entered.

The production at V&B in terms of volume and demand for quality grows aggressively in the past and in the meantime, the company has to compete in a market with decreasing margins. Reliability and safety is the key for battery manufacturing. The biggest challenge here is to provide stable and improved battery manufacturing quality. So far, the manufacturing processes are still highly isolated with a lot of manual intervention. So there was a huge need for change:

- Requirement for a unified, real-time manufacturing execution platform to support the high volume and high speed complex production.
- Necessity to standardize the manufacturing process and master data definition to support significant business growth.
- Need to visualize the production line, provide timely information to management for them to make timely decision to deal with the fast changing market.
- Target to build an end-to-end traceability system to control the product quality with non-disputable data records.

"The promise of these new technologies are that, by putting all this data together, running the proper analytics and making use of computing technologies, we can do a better job of manufacturing," said Ochs, "But what holds up the promise from being fully realized is the accessibility of data—the capability of bringing all that information together into one place. This is now possible. "V&B decided for a new way of manufacturing, a clear return on investment of technologies and solutions requires infrastructures and capabilities that can scale as manufacturing operations become more complex. A key consideration is the ability to statistically use production data—this is where big data is a game changer. Additionally V&B demystified big data analytics with the implementation of a robust scalable platform [7] that empowers the organization to be proactive instead of reactive by:

- Accelerating big data processing across flexible data management options to support greater data availability at lower cost.
- Acquiring data from a variety of sources and applications and combining structured, unstructured, machine, and human data for comprehensive insights.
- Analyzing big data with solutions that uncover new insights across the organization to leverage issues and realize new capabilities.
- Achieving tangible results from your big data initiatives with services that apply advanced data science to your business.

big data and Predictive Analytics helps V&B to make better decisions by using accurate, reliable, and scientific information to analyze risk, optimize processes, and predict failure [36]. The use of Predictive Analytics to improve yield was important have a clear understanding what data are available and which data are of relevance for future analytics. One major challenge was to invest in the systems, the infrastructure and skill sets that allow V&B the optimization of existing process information e.g. by collecting data from various sources and indexing them so that an analysis becomes more easily to apply. Additionally, the building of an adequate skill level within the teams was important to focus on those identified production patterns and turn them into action. Among the factors already examined e.g. temperature, quantity etc. the analysis revealed a number of previously unseen sensitivities—for instance, degree of humidity significant effects in yield. By resetting its parameters accordingly, V&B was able to reduce its waste and its production costs, thereby improving overall yield—even though the Data were less than complete.

There are several reasons why smart manufacturing is now obtainable for V&Bs manufacturing—most of which directly correlates with the dramatically improved availability of data. Today it's possible to monitor anything you want on your production line, without having to spend a fortune. As a result, the V&B has now affordable access to granular data. Additionally with smart manufacturing, collating resource usage across the entire company makes it possible to improve strategy and better understand the total organizational footprint. What truly drives V&Bs smart manufacturing is the ability to seamlessly connect these data sources into one contextual source, analyze the data and take strategic action. Simply gathering mountains of data has little value if the organization doesn't have the ability to put it in perspective, and ultimately take meaningful action on the provided insights. This seamless connectivity between all the machinery, devices and sensors—all of which are producing and yielding big data prime for Analytics—ultimately creates what many refer to as the IoT. Not only do the sensors provide information needed to optimize speeds, feeds and process approaches [37, 38], it also enables the machine producer to alert the customer when to take precautionary action, such as regular maintenance. Actions that will enable V&B to use smart grids to schedule time consuming activities during low demand periods and avoid to slow production down during peak energy demands. This will lead to a greater product customization and to the achievement of the production target of "lot size 1" at an affordable price while having at the same time more efficient processes within the production—previously opposite targets.

V&B uses Predictive Analytics to conduct detailed analysis of historical and real time production data with performance rankings and key insights related to production decisions, expected quality results associated with cost/budget. It gives V&B greater insight into the production across the entire plant network, the production output can be monitored and adjusted closely and immediately, and smarter decisions can be made regarding production schedules—consequently more economical business models, such as just-in-time or lean manufacturing strategies, can be adopted. The project enables real-time insight into the shop-floor environment, resulting in increased utilization and productivity and decreased waste from the top floor to the shop floor. The entire project—so far—has produced tangible results:

- The digitalized manufacturing platform at V&B has built the integration for production design to manufacturing, this has significantly accelerated the idea to production process.
- With the digitalization transformation, the average production volume per line has been increased with stable quality. And it is still being continuously improved.
- The production quality has significantly increased, highly strengthen the competitive advantage of the product in the market. The yield is declining continuously.

References

1. Auschitzky, E., Hammer, M., & Rajagopaul, A. (2014). *How big data can improve manufacturing*. New York: McKinsey & Company.
2. Bergoli, E. Horeyr, J. (2012). In *Design Principles for Effective Knowledge Discovery from Big Data, Software Architecture (WICSA) and European Conference on Software Architecture (ECSA) Joint Working IEEE/IFIP Conference on, Helsinki, August 2012.*
3. Bertolucci, J. (2014). Big Data Fans: Don't Boil The Ocean, In *Information Week*, 12/2014.
4. Borkar V., Carey, M.J., & Li, C. (2012). In *Inside Big Data Management, Ogres, Onions, or Parfaits?, EDBT/ICDT 2012 Joint Conference Berlin German.*
5. Brown, B., Chui, M., & Manyika, J. (2011). *Are you Ready for the era of 'Big Data'?, McKinsey Quarterly, McKinsey Global Institute, October 2011.*
6. Davenport, Th., Barth, P., & Bean, R. (2012). How 'Big Data' is different. In *Opinion & Analysis, Fall 2012*, July 30, 2012.
7. Dingli, A., Attard, D., & Mamo, R. (2012). Turning homes into low-cost ambient assisted living environments. *International Journal of Ambient Computing and Intelligence (IJACI), 4* (2), 1–23.
8. Eaton, C., Deroos, D., Deutsch, T., Lapis, G., & Zikopoulos, P. C. (2012). *Understanding Big Data: Analytics for enterprise class Hadoop and streaming data*. New York: Mc Graw-Hill Companies. ISBN 978-0-07-179053-6.
9. Eaton, C., & Zikopoulos, P. (2011). *Understanding Big Data: Analytics for enterprise class Hadoop and streaming data* (1st ed.). New York: McGraw-Hill Osborne Media.
10. Gerhardt, B., Griffin, K., & Klemann, R. (2012) *Unlocking value in the fragmented world of big data analytics*. Cisco Internet Business Solutions Group, June 2012. http://www.cisco.com/web/about/ac79/docs/sp/Information-Infomediaries.pdf
11. Heesen, B. (2015). Big Data Management, Stand der Innovationsadoption, In *ERP Management, GITO mbH Verlag für Industrielle Informationstechnik und Organisation*, 2015.
12. Heesen, B. (2016). *Effective strategy execution, management for professionals* (p. 2016). Berlin, Heidelberg: Springer Verlag.
13. Hu, H., Wen, Y., Chua, T. S., & Li, X. (2014). Toward scalable systems for big data analytics: A technology tutorial. *IEEE Access, 2*, 652–687.
14. Juneja, D., Singh, A., Singh, R., & Mukherjee, S. (2017). A thorough insight into theoretical and practical developments in multiagent systems. *International Journal of Ambient Computing and Intelligence (IJACI), 8*(1), 23–49.
15. Kimbahune, V. V., Deshpande, A. V., & Mahalle, P. N. (2017). Lightweight key management for adaptive addressing in next generation Internet. *International Journal of Ambient Computing and Intelligence (IJACI), 8*(1), 50–69.
16. Kimball, R., & Ross, M. (2011). *The data warehouse toolkit: The complete guide to dimensional modeling*. Indianapolis: Wiley.
17. Laney, D. (2001). *3-D data management: Controlling data volume, velocity and variety, META Group Research Note*, February 6. http://goo.gl/Bo3GS
18. Leibenstein, H. (1966). Allocative efficiency vs. "X-efficiency". *The American Economic Review, 56*(3), 392–415.
19. Manyika, J. et al. (2011). *Big Data: The next frontier for innovation, competition, and productivity*. McKinsey Global Institute, http://www.mckinsey.com/~/media/McKinsey/dotcom/Insights%20and%20pubs/MGI/Research/Technology%20and%20Innovation/Big%20Data/MGI_Big_Data_full_report.ashx
20. McAfee, A., & Brynjolfsson, E. (2012). Big Data: The Management Revolution. *Harvard Business Review, Harvard business review, 90*(10), 60–66.
21. Mohanty, S., Jagadeesh, M., & Srivatsa, H. (2013). *Big data imperatives: Enterprise 'Big Data'warehouse, 'BI'implementations and analytics*. New York: Apress.

22. Morabito, V. (2015). Big data and analytics. In *Big data governance* (pp. 83–104). Springer, Switzerland.
23. Ochs, T., & Riemann, U. (2016a). Industry 4.0—How to manage transformation as the new normal. In Palgrave handbook of managing continuous business transformation.
24. Ochs, T., & Riemann, U. (2016b). Big data and knowledge management. In *Proceedings of the international conference on Internet of things and big data*. Scitepress.
25. Ochs, T., & Riemann, U. (2017). Industry 4.0: How to manage transformation as the new normal. In *The Palgrave handbook of managing continuous business transformation* (pp. 245–272). Palgrave Macmillan, UK.
26. Odella, F. (2017). Technology studies and the sociological debate on monitoring of social interactions. In *Biometrics: Concepts, methodologies, tools, and applications* (pp. 529–558). IGI Global.
27. Rogers, M. (1998). *The definition and measurement of productivity*. Melbourne Institute of Applied Economic and Social Research.
28. Russom, P. (2011). *Big data analytics, TDWI Best Practices Report, TDWI Research, Fourth Quarter 2011*. http://tdwi.org/research/2011/09/best-practices-report-q4-Big-Data-Analytics/asset.aspx
29. Sagiroglu, S., & Sinanc, D. (2013). Big data: A review. In *Collaboration Technologies and Systems (CTS), 2013 IEEExplore Digital Library*.
30. Schaller, A., & Mueller, K. (2009). Motorola's experiences in designing the Internet of things. *International Journal of Ambient Computing and Intelligence (IJACI), 1*(1), 75–85.
31. Singh, S., & Singh, N. (2011). Big data analytics, 2012 international conference on communication, information & computing technology, Mumbai, India. IEEE, October 2011.
32. Smart Manufacturing Leadership Coalition. (2011, November). Implementing 21st century smart manufacturing. In *Workshop Summary Report*.
33. Tybout, J. R. (1992). Making noisy data sing: Estimating production technologies in developing countries. *Journal of Econometrics, 53*(1–3), 25–44.
34. Vailaya, A. (2012). What's all the buzz around "big data?". *IEEE Women in Engineering Magazine, 6*(2), 24–31.
35. Villeroy & Boch homepage. https://www.villeroy&boch.com
36. Ward, J. S., & Barker, A. (2013). Undefined by data: A survey of big data definitions. In *Proceedings of IEEE CloudCom 2013* (pp. 647–654). IEEE Computer Society, December 2013.
37. Warwick, K., & Harrison, I. (2014). Feelings of a cyborg. *International Journal of Synthetic Emotions (IJSE), 5*(2), 1–6.
38. Warwick, K., & Shah, H. (2014). Outwitted by the hidden: unsure emotions. *International Journal of Synthetic Emotions (IJSE), 5*(1), 46–59.

Home Automation Using IoT

Nidhi Barodawala, Barkha Makwana, Yash Punjabi and Chintan Bhatt

Abstract The main agenda of IoT is to enable us monitoring and controlling physical environment by collecting, processing and analysing the data generated by smart objects, which is an advancement of automation technologies, making our life simpler and easier. Nowadays Internet of Medical Things also became popular in which medical devices are connected and provides integration for taking care of patients and other aspects related to healthcare. Before the technologies like microcontrollers and smart phones are introduced, establishing home automation was a real burden with interference of electricians, installer and monthly maintenance costing. IoT is providing us home automation system using smart devices to get over this hindrance, which allows us to easily control the home appliances. The presented chapter introduces the home automation system using BASCOM, which also includes the components, flow of communication, implementation and limitations.

Keywords Smart devices · Automation · Microcontroller · Home appliances

N. Barodawala (✉) · B. Makwana · Y. Punjabi · C. Bhatt
U & PU Patel Department of Computer Engineering,
CSPIT, Charotar University of Science and Technology, Changa, Anand, India
e-mail: nidhidbarodawala@gmail.com

B. Makwana
e-mail: barkha.makwana007@yahoo.com

Y. Punjabi
e-mail: 14pgce035@charusat.edu.in

C. Bhatt
e-mail: chintanbhatt.ce@charusat.ac.in

N. Dey et al. (eds.), *Internet of Things and Big Data Analytics Toward Next-Generation Intelligence*, Studies in Big Data 30,
DOI 10.1007/978-3-319-60435-0_9

1 Introduction

IoT [1, 2] is an acronym for "Internet of Things" which is the combination of economic significance, technical, social and provides dynamic global infrastructure which has capabilities of self-configuration [3]. In IoT objects obtain intelligence by enabling context related decision and also by making themselves recognizable. The fact of IoT objects is that they can communicate and exchange information about themselves [4], whether it is wireless or wired. The main aim of IoT is to make the things smart by getting them connected at anytime, anyplace with anyone and anything which we call ubiquitous in nature. IoT is the next era of computing which is outside the domain of traditional desktop [5]. Sensors and actuator plays an important role in connecting the IoT objects. IoT provides several benefits. Some of them are mentioned here:

- Simplifies daily life
- Assist people with personal safety
- Security
- Health
- Provide ubiquity

In healthcare, IoT devices are connected and data is collected from various sources like temperature monitors, glucose levels in blood, health tracking, monitors and these are essential for the patients. In addition, these measures sometimes need an interaction with the health professionals which in turn requires the data to be delivered to the smart devices. This platform makes the patient and health professional interaction possible. As a part of IoT some hospitals are implementing the concept of "Smart Beds" which can detect the patient movement and can help patient to get up without help of nurses [6]. Healthcare can also be applied at home for example, by making smart toilets which can automatically takes the urine samples and send the data to the doctor if some disease found in that sample, through this preventive care can be taken at early stages. Healthcare and IoT are two major concerns nowadays which are in trend and giving benefits to more and more people adopting it.

So why there is an emergent need of home automation? There are several reasons for it, which are mentioned here: It is a fact that the main power to control, monitor and simplify everything within our daily life and homes, is in the hands of home automation system. There are so many devices and activities happening around us in day to day life and so many devices are used. With the help of home automation homeowners can fuse together their activities like, entertainment, lighting, temperature control and security. Home automation is one way to manage everything. It is easy to use and main idea behind home automation is to be able to control everything from afar and cut your energy costs. Therefore, the requirement of home automation becomes very important nowadays.

Home Automation is a system that simplifies our lives which enables us to control home appliances automatically and electronically. It transforms our normal house into smart house and it is like "Giving your Home a Brain" which can reduce the power consumption [7]. Before this advancement interfaces like buttons and switches were used in the normal houses [8], now home automation replaces all these things with a central control system including the user friendly GUI on smart phones and is economically feasible.

Main aim of home automation or smart home is to monitor the activities of an individual within its own environment and it also monitors how a person interacts with devices. Based upon these observations and interactions, physical environment can be controlled using IoT [9, 10] technology and an improved living experience can be provided to an individual. Controls of all the day to day life aspects are included in it. Humans generally interact with the devices like bulbs, fan, AC, etc. With the help of IoT, environment settings can be made such that, it can respond to the human activities automatically. If such controlled environment can be made, it would be very beneficial. Even it becomes more useful when user itself can control these devices from their mobile phones and get the information. Nowadays demand of controlling devices through mobile phones is growing rapidly. Home automation is the most popular topic while talking about controlling devices through mobile phones.

Home automation system provides features [7] like:

- Security and fire
- Energy management
- Home appliances controlling
- Entertainment

Here we are introducing the concept of home appliance controlling using smart device in which previously Bluetooth was used to access and control the appliances. Due to some following drawback of Bluetooth the system is not sustainable.

- Less secure
- Low bandwidth
- Limited range (5–30 m)
- Only up to 7 devices can be connected at a time.

To overcome this inconvenience, we provide a solution using Wi-Fi which is having much larger range of 32 m indoors and 95-m outdoors [11]. By using this Wi-Fi module, we can have access to anything if we have a suitable microcontroller for interfacing that Wi-Fi module, for that we implemented the home automation system which can handle electrical components using the Android device through Wi-Fi. With advancement of wireless technology various different connections are introduced like GSM, Wi-Fi, ZIGBEE and Bluetooth. Each one of these has their own unique features and applications. Among these wireless connections, Wi-Fi has been chosen because of its suitable capability.

Wi-Fi capabilities are more than enough to be implemented in this system. Nowadays all devices like laptops, mobiles have facilities to connect to Wi-Fi. In recent years' home automation is becoming so much popular. It provides interoperability: some limit can be set for temperature according to weather condition; lights can be switched on or off based on daylight. Home automation can also give facility to access and monitor the house with the help of your laptop or even mobile phones. It should also be able to extend and reduce if required, so it also provides benefits like saving energy and it is also expandable. Media is also concerned about topics related to energy conservation.

House equipped with these kind of system can provide comfort, flexibility, elegance, security, and the main benefit it provides is reduction in cost with the help of optimization of electricity consumption and heat. Example: turning on the sprinkler at the middle of night if it detects thieves. Home automation also employs sensors which can detect the presence of individual in a room, and it can also control the volume level of music based on different factors. Different technologies and standards are used in current home automation systems like Bluetooth, ZigBee, X10. We use Bluetooth only where there exists hardly any infrastructure to connect appliances. ZigBee can be a better solution than Bluetooth. In future ZigBee chips may exists in normal house. These things can make mornings easier as alarm clocks can communicate with coffee maker and order it to make strong coffee. GSM network is preferred due to its wide spread coverage for communication between the home appliances and people. These things make everything online for all time. There are many projects are available on the Internet that are Based on Bluetooth, Wi-Fi and ZigBee. And we can combine those projects and make one large project to control our Home.

The main thing in our project is how to make our home smart and better than the traditional houses. Smart House is an environment in which technology in enabled which allows device and system to be controlled automatically. Many Features that we can include making our Home Smart.

Lighting: We can make our lighting system smart by controlling it through our phone. To control it we have to create one Android or IOS application that control our home system. We can control it and change it according to our mood. If we want deemed light by giving the command through our phone we can deemed it. Change the traditional switching method to on or off the lights.

TV: We can change the channel according to our mood there is no need to do it by traditional way.

Refrigerator: To maintain the cooling automatically inside the refrigerator, if refrigerator wants to store some items the list of items will be generated automatically and it will send to super market stores computers and it will also send to your phone to get connected with your refrigerator

Structure of chapter is arranged as following: Sect. 2 introduces modules of the system, Sect. 3 presents the architecture of system, Sect. 4 include implementation flow along with output, Sect. 5 contains system limitation and future work followed by Sects. 6 and 7 which includes conclusion and references respectively.

2 Modules of the System

Here in this section you are going to learn about each of units that can be used to construct the home automation system. Following are the modules used in constructing the system:

1. Serial bot
2. Wi-Fi module
3. Bascom controller
4. Relay board
5. USB ASP loader
6. Bascom AVR-IDE

2.1 Serial Bot

It is an application made by Simon Hope Communication which gives access to any embedded device using Bluctooth or Wi-Fi. This application is easily available in play store. It is a Telnet/SSH client with integrated support for RS232 serial connections via Wi-Fi to Serial or Bluetooth to serial [12].

Serial bot can be used to connect to serial ports such as industrial equipment, router console parts and many other devices which have RS232 connections, from our android devices. It is a terminal app specifically designed to support serial ports via Wi-Fi or Bluetooth to serial adapters. Its size is 778 K and current version available is 1.8.1. In connection type options are given like local, telnet, SSH, Serial-Wi-Fi, serial-bot.

Select telnet as shown in Fig. 1 and write the IP address and port address of particular host. Here it is 192.168.16.254:8080. Through this you can easily send commands to the device connected to it.

2.2 Wi-Fi Module

Robodu UART Wi-Fi module by Hi-Link was used which is shown in Fig. 2. With the help of Wi-Fi module, the serial device does not need to change configuration and we can easily transmit data over Internet [12]. It supports network standards of wireless and wired like for wireless it is IEEE 802.11n, IEEE 802.11g, IEEE 802.11b and for wired it is: IEEE 802.3, IEEE 802.34. Wireless transmission rates are as follows:

For 11n: up to 150 Mbps
For 11g: up to 54 Mbps
For 11b: up to 11 Mbps

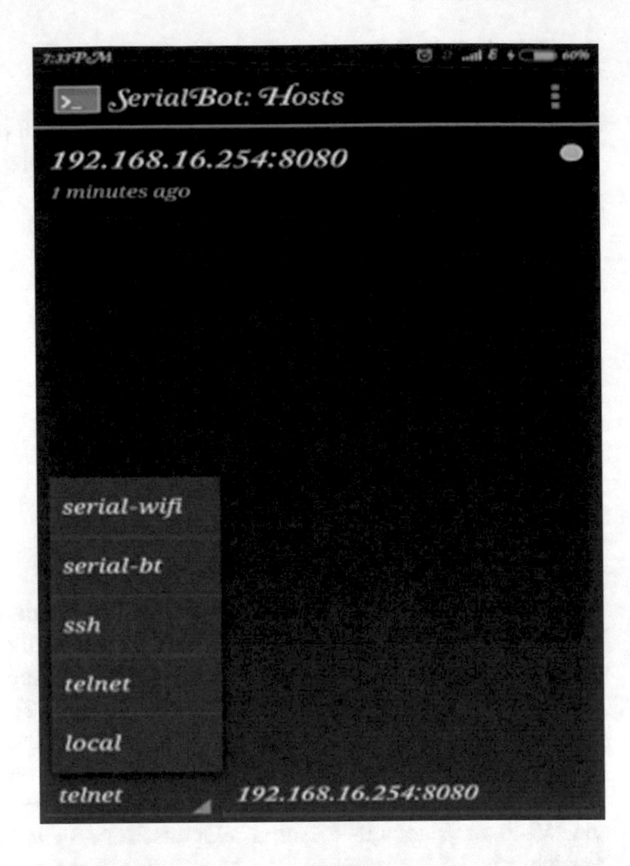

Fig. 1 Serial bot

Fig. 2 Wi-Fi module

Fig. 3 Bascom controller

It provides 2.4–2.4835 GHZ frequency range. This Wi-Fi module provides good range and if we make some changes in it, it can work as a client router.

2.3 *Bascom Controller*

Bascom includes following components which are shown in Fig. 3.

1. ATmega 8 microcontroller—Atmel
2. Opto coupler (OC)
3. DC–DC relays
4. Relay driver-Toshiba ULN2803APG

2.4 *Atmega8 Microcontroller-Atmel*

Atmega8 microcontroller is a low-powered, 8-bit microcontroller which is based on AVR RISC structure. Designer can optimize the power consumption and increase

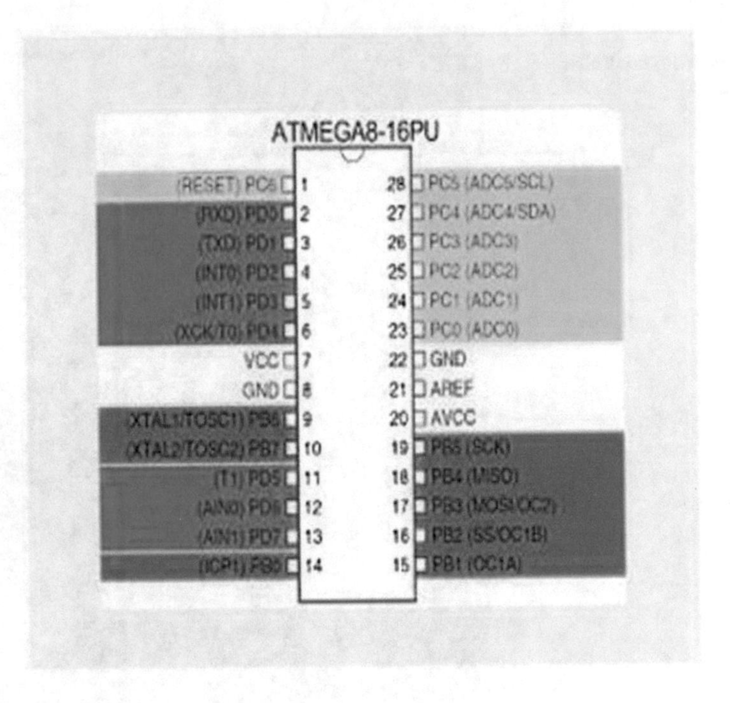

Fig. 4 Pin diagram

the processing speed with help of this microcontroller [13]. This microcontroller can achieve throughput of 1MIPS per MHZ and can also execute instruction in a single clock cycle.

Atmega 8 microcontroller provides following features:

1. High performance, low-power
2. Advanced RISC architecture
3. Special microcontroller features
4. 512 bytes of EEPROM
5. 1 Kbyte of SRAM

Pin diagram of Atmega8 microcontroller is shown in Fig. 4.

2.4.1 Memory

The Memory of ATmega8 microcontroller is 8 kb of flash program, EEPROM of 512 bytes and internal 1 KB SRAM.

2.4.2 I/O Ports

There are three ports and 23I/O lines can be obtained from it, PORT B, PORT C and PORT D are the names of three ports.

2.4.3 Interrupts

At PORT D there are two external interrupt source are located and internal peripherals generates 19 events that supports 19 different kind of interrupt vectors.

2.4.4 Timer/Counter

There are 3 internal timers are available, 2 timers are 8 bit and 1 timer is 16 bit, which supports internal or external clocking and offers different operating modes.

2.4.5 Serial Peripheral Interface/SPI

ATMega8 integrated with 3 communication devices one is serial peripheral interface.

2.4.6 USART

Most powerful communication solutions are USART and ATMega8, synchronous and Asynchronous data transfer scheme are supported by them. 3 pins are assigned for that.

2.4.7 Two Wire Interface/TWI

ATMega8 has another communication device named two wired interface. Designers can setup a communication between two different devices with the help of just two wires having a common ground connection. By means of open collector outputs TWI outputs are created, thus to makeup the circuit external resistors are required.

2.4.8 Analog Comparators

In the IC one comparator module is integrated which gives facility of two inputs of the analog comparators to get connected with the two voltages via external pins that are in turn attached to the microcontroller.

2.4.9 Analog to Digital Convertor

Input signal can be converted and analogue into digital data of 10-bit resolution using analog to digital convert.

2.5 Opto Coupler (OC)

It is a type of Optical Coupler. Opto Coupler is also known as photo coupler, optical isolator. With the help of light, it transfers electrical signals between two isolated circuits [14]. The main function is to protect the circuit from high voltage. There are different types of Opto Couplers. The most common size with maximum output we are using is 30, 70 and 80 V and the maximum output of the Opto Coupler is up to 80 kV.

Four general types of Opto Couplers are available; they have an infra-red LED source having different photo-sensitive device.

These four OCs are named as: Photo-transistor, Photo-Darlington, Photo-SCR and Photo-triac.

2.6 DC–DC Relays

Relays are like switches and because of their simplicity they are used in many applications. DC–DC contains sensing unit, the electronic coil, which is powered by DC current [15]. It is one type of electromagnetic device which used to isolate two circuits electronically and also connect magnetically [15]. It is separate circuit. Following are applications where DC–DC relays can be used:

1. Appliances
2. Security
3. Telecommunication
4. Motors
5. Industries

2.7 Relay Drivers

Relay driver is one type of electrically operated switch. The main aim of the Relay drivers is to directly control electric motor which requires handling high power and different loads [16]. In this project, Relay drivers can handle up to 240 V power supply. Relay drivers are comprised of eight NPN Darlington pairs with high voltage and high current. They contain following features:

- Input compatible with various types of logic.
- Output current 500 mA (maximum).
- APG-DIP-18pin.

Some applications are hammer, relay, lamp driver, display (LED) driver.

3 Relay Board

As shown in Fig. 5, relay board consists of 8 relay switches. It is an 8 channel relay interface board which can be controlled by Bascom AVR. The main functionality of relay board is to trigger the DC voltage to AC voltage, which is done through switches. Port for power supply and equipment for controlling the appliances are also there on the relay board. It can be directly connected with the microcontroller [17].

Red LED lights are also attached with it. It triggers 12 VDC–240 VAC. Internal circuit is working on DC to AC triggering and basically used for triggering AC voltage from PC. There are many Relay Boards is available. This boards are ideal for switching appliances, controlling the lighting switches and much more. This relay board uses 12 V relays to switch or control heavy loads, with each relay suitable for switching up to 10 V. Each relay output is controlled with +5 V TTL Signal.

Features of 8 Relay Board:

- 8 Relays for large connection
- High degree of isolation between control signal and output signal with the use of op-to isolator for isolation.
- Each relay has a LED to indicate the relay state.
- For cable connection PCB power terminals are included on relay board.
- Standard '0.1' headers are provided for easy connection to other control system or our development board.
- 12 V power supply is needed for relay oils.
- Ideal for use in industrial and commercial system.
- Board dimension: 205 × 66 mm.

4 USB-ASP Loader

For AVR microcontroller it is a USB boot loader. If a certain hardware condition is met, loader loads the data on USB. Mainly it is used for connection of micro-controller to pc and to dump a code for assembly language or hex file [18]. USB ASP loader is an one type of USB boot loader for AVR microcontroller. Each AVR has at least 2 KB boot loader section.

Fig. 5 Relay board

USB ASP loader is shown in Fig. 6.

5 Bascom AVR IDE

It is a window BASIC COMPILER IDE for AVR family. There are many microcontrollers available to write the code but we used Bascom AVR IDE which is very powerful and AVR-Eclipse is a plug-in for eclipse IDE which adds tool chain support, through CDT for the gcc compiler [19]. Here we used BASCOM AVR IDE for writing assembly language code for the microcontroller. For AVR family, Bascom is windows Basic compiler, which runs on WIN 7/WIN 8, VISTA, XP. Variables used here can be long as 32 characters. It is constructed with BASIC labels.

Instead of using interpreted code BASCOM AVR IDE uses fast machine code. It is a testing simulator. There are some key advantages, which are as follows:

- It is structured BASIC with labels.
- Contains structured programming with IF-THEN-ELSE-END-If, Do-Loop, WHILE-END, and SELECT-CASE.
- Fast machine code instead of interpreted code.
- Labels and variables can be as long as 32 characters.

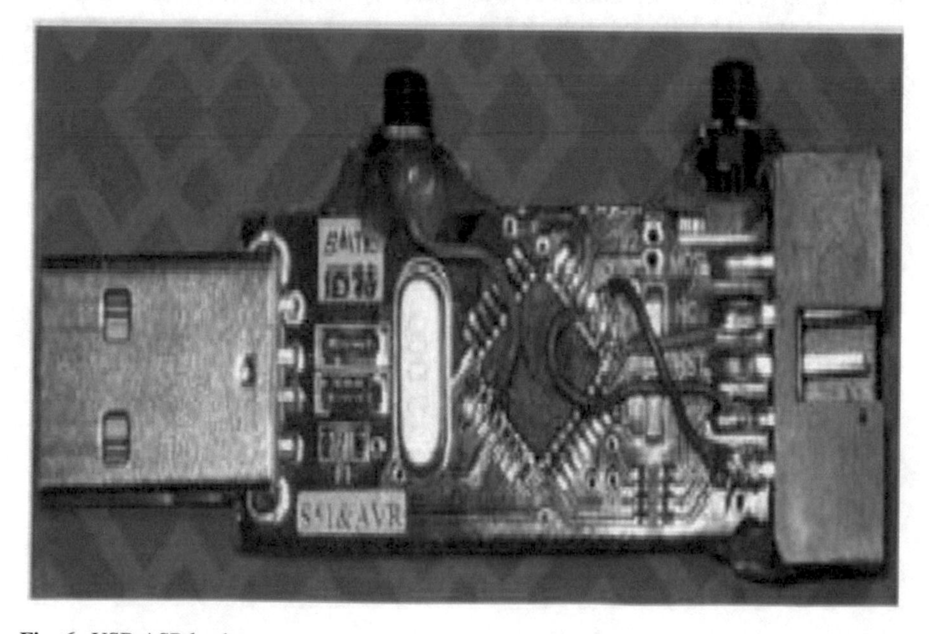

Fig. 6 USB-ASP loader

- All the compiled programs can work with all AVR microprocessors that have internal memory.
- Integrated simulator for testing.
- Local variables, user functions, library support.

In Fig. 7 the screenshot of the editor is shown, in which you can observe the coding written in code explorer.

Following are the steps by which you can make a program in BASCOM-AVR-IDE:

1. Write a program with BASIC.
2. Compile it to fast machine binary code.
3. With integrated simulator, test the results.
4. Now you can program the chip or hardware with this integrated programming.

6 System Architecture

6.1 Architecture

Figure 8 describes the overall system architecture of Home Automation. It starts with the serial bot application which was described in Sect. 2.1. When power supply is given through the adapter to the circuit the Wi-Fi module provides the Wi-Fi signals, and we can connect the serial bot application with it. Then after we need to enter valid IP 192.168.16.254 and port address 8080 as discussed in Sect. 2.1, it will check, if it is not connected it will give the message "Enter Valid IP and PORT", but if it is connected the output is provided to the relay board.

Fig. 7 Bascom AVR IDE

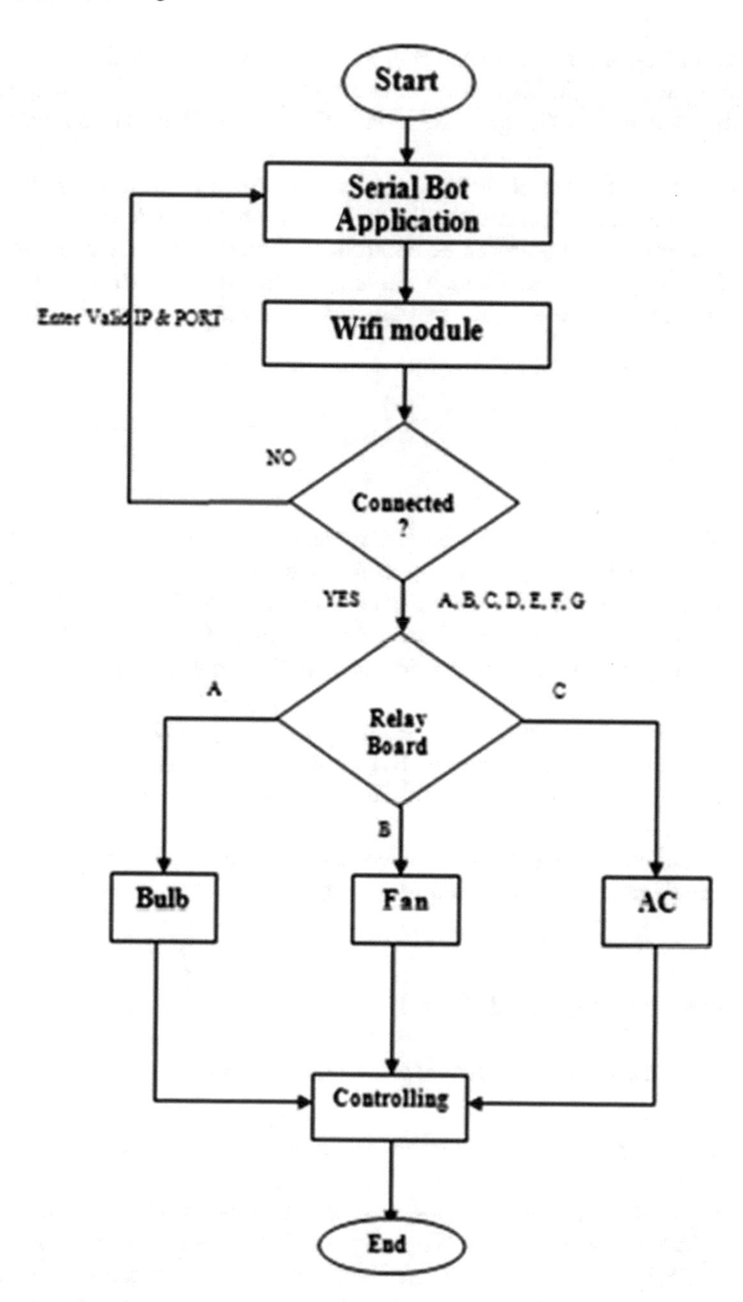

Fig. 8 Architecture

In this project, we have attached 8 switches on relay board and through this we can control various eight home components up to the 240 V. Serial bot application is already available on Google play store. We have to give the command using telnet.

On relay board 8 channel switches are there to which A, B, C, D, E, F, G ports are given, if we trigger 1st port it will send the signals to port A to which a bulb is attached and this way a bulb can be controlled. Similarly, if we trigger 2nd port it will send the signal to B to which a fan is attached, we can control a fan.

By using the same procedure, we can control all the appliances attached to the relays.

6.2 Controlling Circuit

After all these implementations we need to understand the final controlling circuit, through which controlling of home appliances can be done. To understand this, we need to focus on the small blue part of the relay board. At the bottom No, C and X are written, from which No and C are made common. C part of each is going to connect to the phase of each of the electrical appliances which is indicated as P in the figure.

As you can see in the Fig. 9 from each C part a connection is given to positive. No part is connected to the negative. Here it is neutral for the electrical appliances which are indicated as N in the figure. The electrical components such as bulb, fan are connected to P and N.

So the final controlling of these devices are done through these implementation, in which appliances gets connected to the circuit.

7 Implementation and Result

7.1 Implementation Flow Along with Output

Now let us understand how actual implementation is done. In figure implementation flow is given.

First we need to provide the power supply to Bascom circuit, which enables Wi-Fi module that provides the Wi-Fi signals. Now we can connect our android device with Wi-Fi and we can use serial bot application for sending the command. In serial bot application we need to select telnet option and write particular IP and Port address. Here it is 192.168.16.254:8080 where 192.168.16.254 is IP address and 8080 is port address, all this is shown in Fig. 10a.

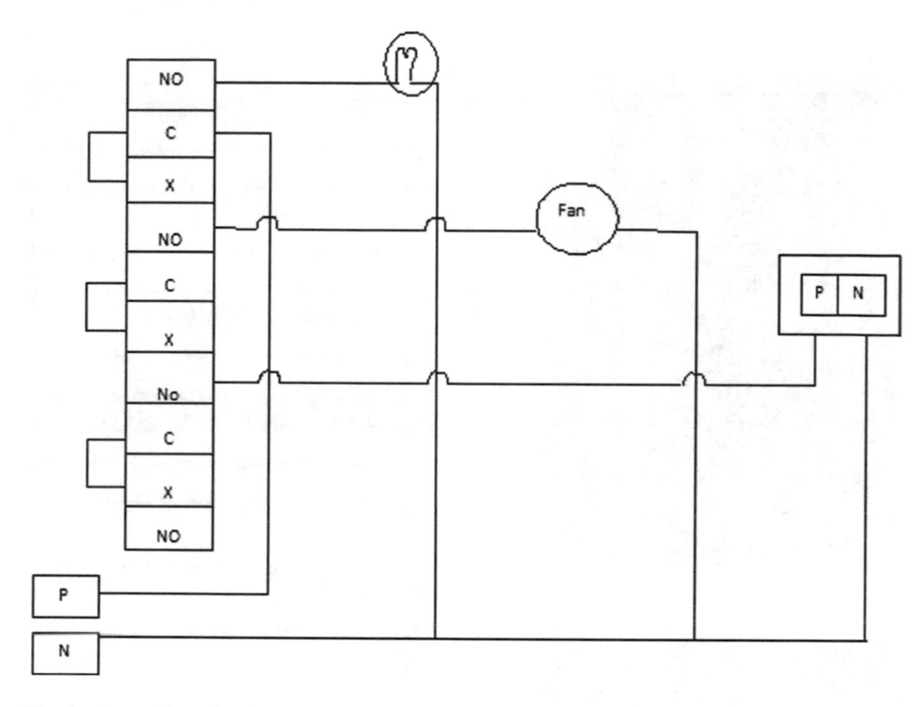

Fig. 9 Controlling circuit

Then after in Bascom controller the code that we have written in Bascom IDE is going to be implemented. With the help of USB ASP loader, we can dump this code into hardware. Now let us see the implementation flow and how it is working.

As you can see in the Fig. 10c, for each command port number is implemented which is in binary form. For X = "A" binary is given as 0100, for X = "B" binary port is 0110. Similarly, for X = "C" and X = "D" it is 0101 and 1000 respectively. This way for each port up to "G" binary value is given. From the Bascom controller the command which we have triggered is sent as an output of Bascom to relay board. In relay board as we have discussed earlier 8 channel switches are there. The current provided as input to the relay board is DC current and home appliances are running on AC current so we need to convert this DC current into AC current. For this DC–AC relays are attached. Relay is one small switch is given, its main function is to trigger the AC current.

Whenever DC current is provided this switch is triggered which finally enables AC current. Here we have used (12 VDC–240 VAC) relay which means it can take up to 12 V DC and can trigger up to 240 V. So finally the devices connected to the controlling circuit which is explained in previous section can be controlled. If A is given as input then it will control the device connected to the first relay, here bulb is attached, so we can control bulb. For B fan is attached so we can control fan. Similarly, for each relay and port, one device is connected. Here 240 VAC is given

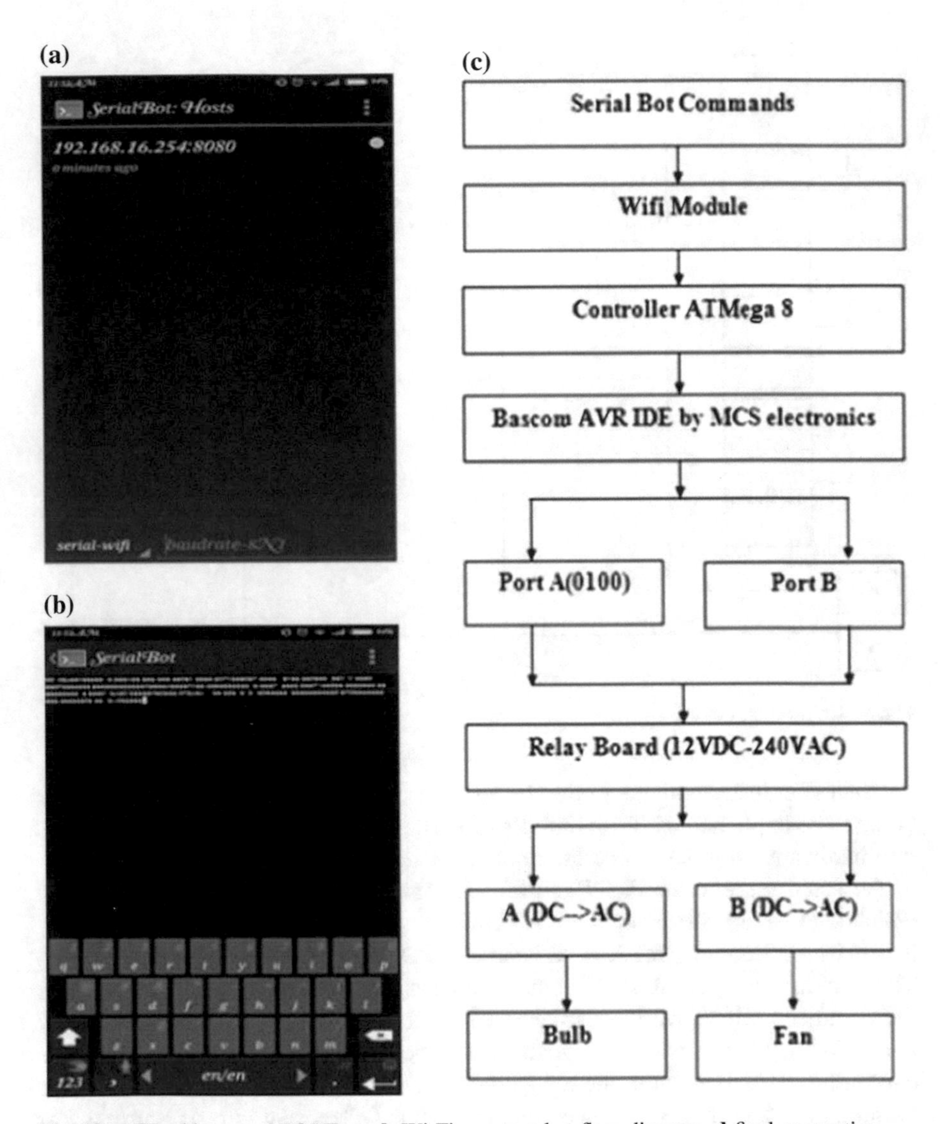

Fig. 10 **a** IP address and PORT no. **b** Wi-Fi command, **c** flow diagram, **d** final connection

Fig. 10 (continued)

which indicates that we can control all the home appliances which are running on 240 V.

As a result, the implementation of the home automation system is done which is show in Fig. 10d.

In this project we make one circuit in which we used capacitor, resistors, LED, microcontroller, Wi-Fi, ROBODU, 4 switches and I/O ports. To connect all these devices, we used single core cable. On relay board we used seven relays so that we can control different Home equipment.

8 System Limitation and Future Work

There are many projects available on Home Automation using Bluetooth. But in this project range is provided and it is basically work in specified Bluetooth range. To overcome this limitation and Bluetooth range problem we use Wi-Fi in our project which provide better range than Bluetooth.

Every system has its pros and cons. There are some limitations of the system. In this section we are going to learn the limitations of the system and how the system can be extended to the next level in near future.

8.1 Limitation

Following are the limitations of the system:

- The system runs on Wi-Fi signals so if the Wi-Fi strength is weak it might cause a problem of connecting the serial bot with it.
- If the signal strength is weak or some hardware problem is there.
- There are chances of delay in the output, so we can say it might restrict the speed of implementation.
- Another problem with the system is that instead of going into microcontroller, signals are sometimes bounced back to the serial bot application.
- The system can only control the appliances which are running on 240 V AC.
- It will work only in Wi-Fi connected range.

8.2 Future Enhancement

- In future this system can be extended up to controlling all the devices of the house by adding more relays using an android device.
- High voltage appliances which are running beyond 240 VAC will be controlled using this system.
- Full controlling of the electrical components of home will be implemented in very near future.

This implementation will make our lives easy and makes it easy to access and control the home appliances.

8.3 Future Ideas

Health Care and IoT: Growing awareness about the health among people nowadays has led to the disciplines like "Health Advisor" and "Self- Healthcare"

[20], which made realized people to live a healthy lifestyle and plays an important role in bringing the care point from hospitals to patient's home. Enabling technology related to health care requires capturing data related to individual's health and various other parameters.

IOT (Internet of Things) provide such facilities to not even capture but also to analyse and share with different connected environment. This will help to make the cost of care low. This innovation in healthcare will increase the efficiency and can make healthcare affordable.

How IOT will help in Healthcare: Some of the examples can easily explain how IoT will help in healthcare field. They are as follows:

- **Disease Management**: Patient's physiological conditions like glucose level in blood for diabetics and blood pressure for hypertensive patients can be monitored and observed by the IoT devices and these devices can send periodic data and through which various important analysis can be done.
- **Preventive Care**: Health supervision is necessary to observe and take preventive actions on time. For this reason, IoT wearable technology is been introduced which made supervision easier.
- **Clinical Monitoring Remotely and Assisted Living**: IoT enables medical providers to monitor and transfer routine assets on personal level while saving cost.

IoT Prospects in Healthcare:

IoT changes the direction to where health care can be forwarded, how quality outcomes can be produced, increment in efficiency and how these new technologies can be made affordable to the individuals. Analysts can predict the health responsibilities which a patient can take for himself. Healthcare related technologies, systems, services, applications, software, medical devices in terms of the IoT are listed as follows [21]:

Technologies:

- Cellular
- NFC
- BLE
- Wi-Fi
- ZigBee

Healthcare related services:

- Maintenance
- Support
- Integration of systems
- Professional services

Applications related to healthcare:

- Inpatient monitoring
- Management of medication
- Telemedicine
- Workflow management
- Connected imaging
- Clinical operations
- Others

Healthcare on the base of users:

- Research laboratories
- Diagnostic labs
- Defence and government institutions
- Hospitals and clinics
- Surgical centres

Medical devices required for healthcare:

- Stationary devices
- External wearable devices
- Implanted devices

Today's healthcare environment requires focusing on individual's health, patient protection, availability of medicines at lower rates, and personalized care to individuals. With the help of IoT all these things can be made possible and can also provide ultimate facilities to the people who actually needs to focus on their health and confused about their health issues.

Patients and providers both will get benefit from IoT's presence in healthcare environment. Some other use of IoT is medical applications, wearable devices. Hospitals use IoT technology to keep track on location of medicines and patients.

Smart Lighting: Using only one button that turn on/off all the lights of a room it and by double tap it can switch off all the lights. If there is a dark outside the room the light will automatically turn on there is no need to turn it on by switch. When there is no one in the room and the room has been empty for some amount of time, motion sensors in particular areas can signal the automation system to perform switch on or off operations. This automated system can also provide facility to turn off all the lights by pressing one single button.

At sunset landscape lightning can turn on and off again at sunrise. If natural lights is present in the room the lights will automatic get deemed. When there are shades or on cloudy day in the middle of summer this will help.

Smart Music: In Our generation music is important in our daily life and we can make it smart or better. In each room some dedicated keypads are placed which can help playing stream music in every room. We can play music according to our mood, and also can play our favourite playlist.

To activate particular zone of music we can tap one single button accordingly, it can change the LED lights and assign the colour for each family member with their own playlist setting. Music will automatically start when we on the shower and double tap on volume button to increase and decrease the volume. To control music streaming in the shower, we can install a touch screen behind the waterproof wall mount having a shower. You can program your own setting and control it also with your smart phone application.

Nowadays TV is in home of every individual, after coming from work we can automatically tur on favourite channel. At night the lights in each bedroom flashes a warning an after 10 min TV will switch off and remote control and keypads gets disabled.

Smart Lock: We can also make our home locks smart, we can control it with our phone and we can also lock and unlock it with our smart phone application. It will also helpful to protect our homes from thieves.

When the bell rings, we can see who's at door by automated bedroom system. When someone press the doorbell the identity of that person will play through speaker, and the security camera rotates to the front door. When some door is open it will notify in your phone and pop-up will be displayed on your and gives you alert message to close the door. From anywhere in the world, we are able to access security videos.

When any unwanted motion is happened with your home's lock or some other person try to open it, the message will automatically send to your phone and gives you an alert message. You can also create one temporary access code for your guest, which they can punch into your home through smart locks when you are not at home and want to give them access, notifying you of their arrival.

If you forget to lock your door then set some predefined time after that the door will automatically locked, so there is no need to take tension when you forgot to lock your door.

9 Conclusion

This was our first attempt to develop a home appliance controlling application on Bascom and from which we have concluded that, it gives us basic understanding about IoT's latest research and project. Results of the project is achieved by creating a hardware which can be used to control the basic home appliances like fan, bulb, etc. with use of serial bot application through which a user can easily access or control the devices of home. Through this system we can easily control the home appliances using Wi-Fi which overcomes the range problem of Bluetooth in previous work. The goal of making a handy system with easiest control is achieved and it satisfies the true meaning of the tag "Brain of Home" because it actually works as a brain which has all the ability to handle home appliances. Here we have used Bascom. You can also use Arduino or Raspberry Pi to implement this system or similar kind of projects and can build your own hardware using simplicity of IoT.

References

1. Internet of Things (IoT). (2016). In a way of smart world. In M. Bhayani, M. Patel & C. Bhatt (Eds.), *Proceedings of the International Congress on Information and Communication Technology* (pp. 343–350).
2. Shah, T., & Bhatt, C. M. (2014). The internet of things: Technologies, communications and computing. *CSI Communications,* 7–9.
3. Vermesan, O., & Friess, P. (2014). *Internet of things-from research and innovation to market deployment*. River Publishers, IERC_cluster_Book.
4. Rose, K., Eldridge, S., & Chapin, L. (2015). The internet of things: An overview understanding the issues and challenges of a more connected world. *Internet Society*.
5. Buyya, R., Gubbi, J., Marusic, S., & Palaniswami, M. (2013). Internet of Things(IOT): A vision, architectural elements, and future directions. *Future Generation Computer System, 29,* 1645–1660.
6. Chouffani, R. (2005–2017). Can we expect the internet of things in healthcare.
7. Leviton Manufacturing Co. (2013). Home Automation: Your introduction to the simplicity of control.
8. Bruni, R., Matteucci, G., & Marchetti Spaccamela, A. (2015). Dipartmento di Integneria informatica, automatica e gesrionale Antonio Ruberti, Research report.
9. Bhatt, Y., & Bhatt, C. (2017). Internet of things in healthcare. In *Internet of things and big data technologies for next generation HealthCare* (pp. 13–33).
10. Dey, N., Ashour, A. S., & Bhatt, C. (2017). Internet of things driven connected healthcare. In *Internet of things and big data technologies for next generation healthcare* (pp. 3–12).
11. Web Reference. http://www.diffen.com/difference/Bluetooth_vs_wifi
12. Web Reference. http://www.dx.com/p/hi-link-hlk-rm04-serial-port-ethernet-wi-fi-adapter-module-blue-black-214540
13. Atmel- ATmega8. (2013). Atmega8L, Atmel-24868-bit-avr microcontroller-atmega8-L_datasheet.
14. Web Reference. https://en.wikipedia.org/wiki/Opto-isolator. Accessed 8 Feb 2017.
15. Web Reference. https://nphheaters.com/products/solid-state-relays/what%20are%20relays%20&%20how%20do%20they%20work.pdf?/pisphpreq=1
16. Toshiba, Toshiba bipolar digital integrated circuit silicon monolithic, TD62783APG, TD62783AFWG, Datasheet, 2012-11-12.
17. Web Reference. http://www.ebay.com/itm/8-Channel-5V-Relay-Module-Board-Shield-With-Optocouple-For-PIC-AVR-DSP-Arduino-/121995275964
18. Web Reference. https://www.obdev.at/products/VUSH/prjdetail.php?pid=118
19. MCS Electronics. (2016). Embedded systems basic compliers development, BASCOM AVR 8051.
20. Venkatramanan, P., & Rathina, I. (2014). Leveraging internet of things to revolutionize healthcare and wellness. *RFID Journal.*
21. Web Reference. https://www.marketsandmarket.com/Market-Reports/iot-healthcare-market-160082804.html

A Prototype of IoT-Based Real Time Smart Street Parking System for Smart Cities

Pradeep Tomar, Gurjit Kaur and Prabhjot Singh

Abstract Smart city is a vision that aims to integrate multiple information and communication solutions to residents with essential services like smart parking inside the all streets. Today, the parking systems has been changed by new advances that are empowering urban communities to diminish levels of congestions altogether. Internet of Things (IoT) is also new advancement which helps in detection of vehicle occupancy and congestion by basic intelligence and computational capability to make a smart parking system. The main motivation of using IoT for parking is to collect the data easily for free parking slots. This work presents the prototype of IoT-based Real Time Smart Street Parking System (IoT-based RTSSPS) with accessibility of data to make it simpler for residents and drivers to locate a free parking slot at the streets. Firstly, this work presents the introduction of IoT for smart parking with technology backgrounds, challenges of accessing IoT and database. Secondly, this work presents the prototype design of IoT-based RTSSPS with architecture and algorithm. IoT-based RTSSPS architecture is divided into three parts IoT-based WSN centric smart street parking module, IoT-based data centric smart street parking module and IoT-based cloud centric smart street parking module with street parking algorithm, evaluation and future directions.

Keywords Internet of things · Smart cities · Sensors · Data-centric · Cloud-centric and WSN-centric

P. Tomar · G. Kaur (✉)
School of ICT, Gautam Buddha University, Greater Noida, Uttar Pradesh, India
e-mail: gurjeet_kaur@rediffmail.com

P. Tomar
e-mail: parry.tomar@gmail.com

P. Singh
Salesforce Inc., San Francisco, USA
e-mail: prabhjot27@gmail.com

N. Dey et al. (eds.), *Internet of Things and Big Data Analytics Toward Next-Generation Intelligence*, Studies in Big Data 30,
DOI 10.1007/978-3-319-60435-0_10

1 Introduction

These days, one of the real outcomes of the vital advancement in the field of communications is the improvement of the parking systems. This is expected basically to current needs as far as accessibility and access to information whenever and from wherever. For sure, it means to make systems more secure, more effective, more solid and all the more ecologically well disposed without fundamentally having to physically modify existing foundation [1]. The unstable development of smart city makes numerous logical and designing difficulties that call for quick research endeavors from both scholarly community and industry, particularly for the advancement of proficient, adaptable, and solid smart city. IoT is perhaps the most widely discussed technology for landscape challenges in smart cities by using protocols, architectures, and IoT services. Ahmed and Anguluri [2] present the concept of smart city for the inclusive development of the cities to counter the issues of infrastructure, governance and make ourselves capable and competitive enough to address the issues of sustainable development and bring people together. It also includes industry and institutions from different parts of the country to make cities smart, eco-friendly, energy efficient and better place to live in through tailored innovative solutions. Smart cities are the foundation of smart things like smart education, smart health, smart parking inside the city. This work gives main focus on the smart parking through IoT due to the congestion of vehicles in the street. The main motivation of using IoT for parking is to collect the data easily for free parking slots. IoT objects will communicate with sensors, actuators, bluetooth associations, radio recurrence IDs, and so forth. Through IoT, this work can make this parking system even more real where objects can communicate with each other without human obstruction. Mechanized parking in the streets this is unrealistic.

These days, many vehicles are now guided by Global Positioning System (GPS) and sensors can be easily put in parking slots, so that monitoring of parking slot data can be done through drone and IoT to make a database for street. Sensors are best in light of the fact for vehicle detection. So this study identifies the availability of free slots in the streets and creates the database of streets. The extended Mobile Ad-hoc Network (MANET) are the most emerging research domains in contemporary decade. MANET can be visualized as a group of network nodes that can perform a collaborative task. In such case, the network nodes may be homogeneous or heterogeneous type where the different nodes may use the same or diverse network protocol or routing methodology. In addition, the network node can have different kind of nature based on the node movement. Based on the physical location of the network, the velocity of the nodes can be the same or different. Typically, the vehicular Ad-hoc network is a subset of MANET that has been deployed within a group of the ground vehicle. The majority of such network cases consist of a series of vehicles that can generate a connectivity to establish an Internet infrastructure [3].

The proposed prototype of IoT-based real time smart street parking system is combination of WSN-centric, cloud-centric and data-centric IoTs with the sensor hubs. Sensor hubs installed nearby the roadside, each hub is placed at the center point of parking spot and card. At the point when a hub recognizes entry of a vehicle, it communicates with the client through switches. The switch advances the parcel to the main server and the main server gather the information from hubs and displays parking direction through display board.

1.1 Technology Background of IoT

In the context of Next Generation Internet (NGI) everyday devices are globally connected and managing an increasing number of devices requires scalable and efficient addressing mechanism. Due to the economics of scale in NGI, energy, ubiquitous access and secure interaction increases the complexity of operation. Distributed Address Assignment (DAA) proposed by the Zigbee Alliance does not ensure that the device may fail to access and available addresses from its neighbor which is referred as addressing failure. Mobility is another interesting challenge which needs to be addressed essentially in the context of NGI [4]. IoT are as of now the most prevalent information and communication innovation for smart cities. IoT is an idea that imagines all objects like advanced cells, tablets, computerized cameras, drone, sensors, and so forth. When every one of these objects is associated with each other, they empower increasingly keen procedures and administrations that bolster our fundamental needs, economies, environment and wellbeing. Such huge number of objects associated with web gives numerous sorts of administrations and creates gigantic measure of information and data. The IoT will enable connectivity for virtually any physical object that potentially offers a message, and will affect every aspect of life and business [5]. IoT furnishes us with loads of sensor information. Constant sensor information investigation and basic leadership is frequently done physically however to make it versatile, it is ideally computerized. Computerized reasoning gives us the system and instruments to go past unimportant continuous choice and robotization utilize instances of IoT. As per [6] the idea of the IoT originates from Massachusetts Institute of Technology, which is devoted to making the IoT by using radio frequency and sensor networks. As per Uckelmann et al. [7] IoT is an establishment for associating things, sensors, and other brilliant advances. Data and communication technologies can get data from anyplace by growing altogether new systems, which shapes the IoT. Radio-Frequency Identification (RFID) and related recognizable proof advances will be the foundation of the up and coming IoT. IoT advancements is possible because of 4G Long Term Evolution (LTE), Wi-Fi, ZigBee and Bluetooth Low Energy (BLE) technologies which are being utilized for most recent application to make a city smart.

Methodological perspectives adopted in the analysis of social implications of scientific and technical developments and in particular the current sociological debate involving monitoring technologies and social research. A particular attention is devoted to network analysis, an emergent area of research that focuses on the relational implications of technologies in organizations, small groups and other contexts of participation. Using case studies and examples of technologies implementation, this study describes the advancements in this field of enquiry and highlights the main elements of the structure of interactions in virtual and technology mediated communications. It also pays attention to the implications of human interaction monitoring technologies such as IoT [8]. Things might be basic sensors (e.g. temperature sensor in a room), more perplexing sensors (e.g. electrical power measuring gadget), actuators (e.g. HVAC room controller, engine), or complex gadgets (e.g. mechanical electrical switch, PLC giving home, building or modern robotization). The IoT application may extend from a straightforward checking application, for example, gagging the temperature in a working, to an unpredictable application, for example, giving complete vitality mechanization of grounds. IoT correspondences might be required disconnected, where data is traded each day or on request, or internet taking into consideration ongoing control. Building control applications can give effective utilization of the vitality in a building while at the same time protecting solace (warm, power, and so on.) to building tenants. Customary arrangements are utilizing complex Building Administration Systems (BMS) interconnected with Programmable Intelligent Controllers (PLCs) that send requests to actuators in view of sensor information.

The system must consider numerous parameters, for example, climate figures or constant vitality costs. Substitute, developing IoT arrangements depend on sensors and actuators collaborating together in a self-sufficient way. So as to give the continuous building control locally (e.g. in a room, in a zone and so forth.), this nearby gathering of gadgets trades non ongoing data (e.g. temperature set focuses) with different gatherings and with larger amount administrations or applications to fabricate the worldwide building control application. This requires more wise gadgets (e.g. more insight in neighborhood actuators), however gives a similar application without the intricate structure of BMS and PLCs, interconnected with ongoing imperatives. Additionally, the neighborhood self-governance guarantees the system is stronger and dependable, while discharging the correspondence imperatives [9].

1.2 Architecture of IoT Layers

The architecture of IoT layers is dynamic in nature. The architecture of IoT has four layers i.e. sensors connectivity and network layer, gateway and network layer, management service layer and after that comes the application layer as shown in Fig. 1. This architecture provides communication stack for communication through IoT.

1.2.1 Sensors Connectivity and Network Layer

This layer has a group of smart devices and sensors as shown in Figs. 1 and 2 and grouped according to their purpose and data types such as environmental, magnetic,

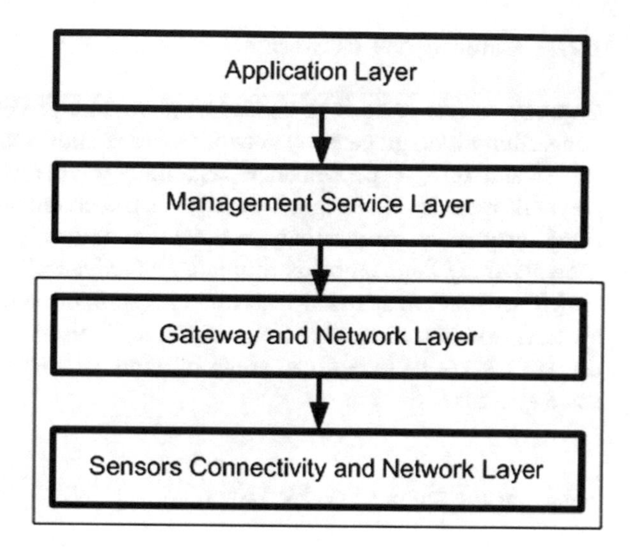

Fig. 1 Layers of IoT architecture

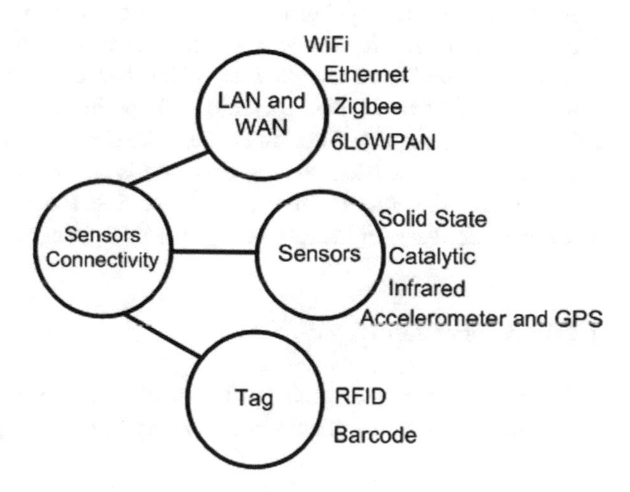

Fig. 2 Sensors connectivity and network layer

obstacle and surveillance sensors etc. WSN formation is made and the information is delivered to a targeted location for further processing. Real-time information is collected and processed to get a result or generate some output. The sensor network is then made to communicate with each other via sensor gateways. They can be connected using the Local Area Network (LAN) (ethernet or Wi-Fi), Personal Area Network (PAN) (6LoWPAN, Zigbee, bluetooth).

1.2.2 Gateway and Network Layer

Capacity of this layer is to bolster huge volume of database produced by sensor connectivity through gateway network layer as shown in Figs. 1 and 3. It requires a robust and reliable performance, regarding private and public network models. Network models are designed to support the communication, Quality of Service (QoS) necessities for inactivity, adaptability, transmission capacity, security while accomplishing large amounts of vitality effectiveness.

IoT sensors are amassed with different sorts of conventions and heterogeneous systems utilizing distinctive innovations. IoT systems should be versatile to productively serve an extensive variety of administrations and applications over vast scale systems.

1.2.3 Management Service Layer

Information analytics, security control, process modeling and device control are done by the management service layer. It is also responsible for an operational support system, security, business rule management, business process management. It has to provide service analytics platform such as statistical analytics, data mining, text mining, predictive analytics etc. The data management manages information flow and it is of two types: periodic and aperiodic. In periodic data management, IoT sensor data requires filtering because the data is collected periodically and some data may not be needed so this data needs to be filtered out. In aperiodic data management, the data is an event triggered IoT sensor data which may require immediate delivery and response e.g. medical emergency sensor data.

1.2.4 Application Layer

This layer at the highest point of the stack is incharge of conveyance of different applications to various clients in IoT as appeared in Fig. 4. This application layer

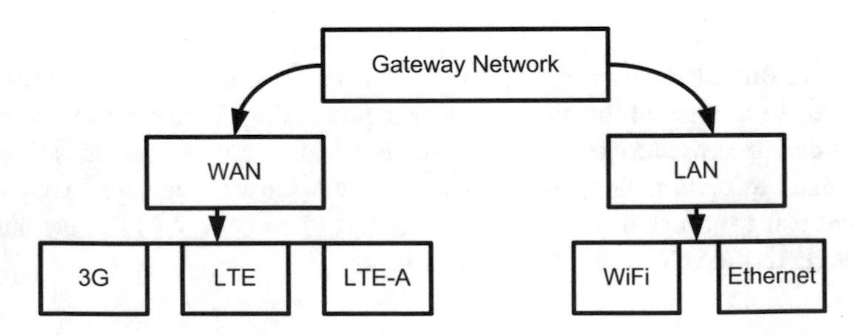

Fig. 3 Gateway and network layer

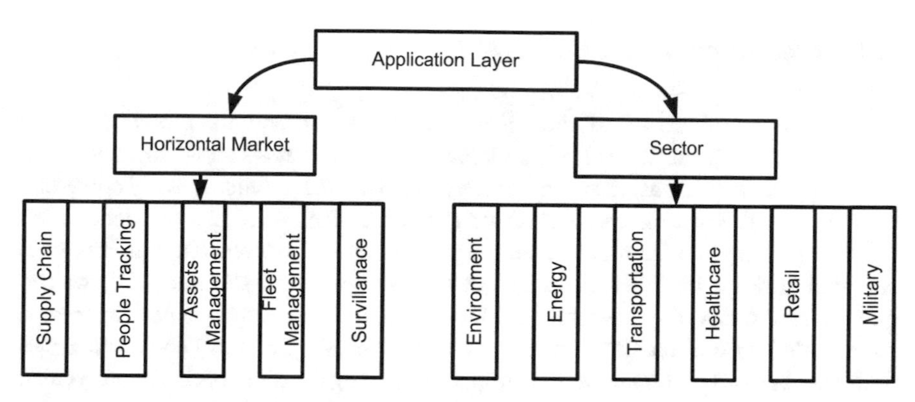

Fig. 4 Application layer

serves to the client of assembling, coordinations, retail, environment, open security, human services, nourishment, medication and so forth. Different applications from industry divisions can utilize IoT for administration improvement.

2 Parking Systems

The traditional parking system is unable to provide some real help to the residents and drivers in the streets. Parking congestion in cities is a major problem mainly in developing countries; to encounter this, many models of parking system have been proposed by different scholars. Different ways have been proposed to make the parking system smarter, reliable, and robust [10]. Smart parking systems are environments facilitated with technology that act in a protective and proactive function to assist an inhabitant in managing their daily lives specific to their individual needs. A typical smart parking implementation would include sensors and actuators to detect changes in status and to initiate beneficial interventions [11].

2.1 Vision Based Parking System

Vision based parking system have two classifications. First one is to search the quantity of empty spaces for parking of vehicles. The second one is to check the individual parking slots from the individual resident property.

2.2 Traditional and Multilevel Parking Systems

The infrastructural growth of cities is unable to cope up with the growing demand for the parking space. It becomes a major issue in old streets and near educational and religious institutions, near market area and also during fairs. A study conducted by the Central Road Research Institute (CRRI) clearly says, that the demand for efficient parking is quite higher than the parking supply offered. Though this work can find parking facilities close-by but the poor management and long waiting queues lead to driver's exasperation. Due to this, they are forced to park in front of someone's house or on streets adding to already existing chaos. People often park their cars in front of the house and shop, to buy things and get back to their vehicle aggravating the traffic congestion. Traditional parking facilities are typically an inefficient, aesthetically unappealing, and even environmentally unfriendly. Therefore resident require a smart parking system for their streets. The Centre for Science and Environment check the high rise building with multi-level parking facilities and according to them 20–40% of the capacity of the multilevel parking was in use, even as the surrounding area remained gridlocked with cars. Hence, it is clear that multilevel parking requires a large amount of capital but is unable to encourage the public to use the facility. The need of the hour is a low cost and an effective method of a parking facility that offers good benefits to the users because multiple parking is not possible in streets.

2.3 Intelligent and Smart Parking Systems

The intelligent and smart parking systems proposed till now are mostly based on webcam which are used to monitor the parking space. The system then uses image processing techniques to check the occupancy of that slot. Tang et al. [12] propose a WSN-based smart street parking system where remote sensors are put into a vehicle for parking slot, with every parking area furnished with one sensor hub, which recognizes and control the parking. The position of parking area is identified by sensor hubs and maintains data of parking through the remote sensor system. The data helps in parking through different administration functionality, for example, finding empty parking areas, vehicle-toll, security administration, and measurement report. They executed a model of the system utilizing crossbow bits. Kumar and Siddarth [13] presents the WSN-based smart parking system in light of remote sensor is used to observe and observing and oversee the person parking slots. It also gives giving computerized direction and booking ahead of time administrations too.

But every system have their own challenge like less information for drivers on parking availability and price is a big challenge for parking and drivers spend a lot of time in searching for a parking space. Inefficient use of existing parking capacity in street creates a big problem for users. Excessive use of automobile use in the small or large family is also a big issue. Automobile dependency imposes many costs on society. Inconvenience parking spaces that are nearby residents and businesses create a problem due to less information and managed database. To overcome all the challenges this study proposes the IoT-based parking system by going through the following related work for parking systems.

3 Related Work for Smart Parking Systems

Deshpande et al. [14], proposes a webcam based parking system that uses IRIS infrastructure with different sensing and organizing component. Sensing components gather the information while organizing components gives offices to questioning later and verifiable sensor information. This system requires a lot of data transfer over a wireless network which consumes a lot of bandwidths. Lee et al. [15] presents the ultrasonic sensors for recognition and identification of vehicle at the beginning of parking. Identification of vehicles is based on their attributes like the shape and design. One could make a database of these qualities for every vehicle and make the identification calculation coordinate their examples, therefore, having the capacity to recognize and distinguish particular sorts of vehicles. Hsich and Ozguner [16] proposed a parking algorithm gives a viable strategy for vehicle parking. No costly calculation is required, and demonstrated that the calculation can stop with any underlying positions and proper calculation of parking slots which can control an vehicle to stop in a restricted space. According to Kumar and Siddarth [13] Smart Parking Management System (SPMS) in light of WSN innovation which gives propelled highlights like remote parking checking, computerized direction, and parking reservation instrument. Initially, the system would ready to graphically show ongoing data and help client to save parking slot from remote area. So, this system helps clients to effectively find empty parking spots rapidly and securely. Chen and Chang [17] developed Parking Guidance Information System (PGIS) in which there are three sorts of hubs like screens hubs, directing hubs and sink hub. These hubs speak with the remote channel and self-arrange into a specially appointed system. Nawaz et al. [18] display ParkSense, a cell phone based system that identifies if a resident has removed the vehicle from parking space. ParkSense influences the universal Wi-Fi guides in urban ranges for detecting unparking occasions. It uses a vigorous Wi-Fi signature coordinating a way to deal with identifying driver's arrival to the stopped vehicle. Also, it utilizes another approach in view of the rate of progress of Wi-Fi guides to detect if the client has begun driving. They demonstrate that the rate of progress of the watched cases are much related with genuine client speed and is a decent pointer to whether a client is in a vehicle. Pereira et al. [19] presents an all-encompassing system engineering comprising of heterogeneous gadgets. According to Reddy et al. [20] a parking direction and data system that uses image processing technique to detects only car at a particular slot. If there is anything else than a car, it will not consider it as booked. They used ARM9 microcontroller, webcam, and global system for mobile communication (GSM) to monitor parking space and a touch screen LCD to display the results. As mentioned earlier using webcam can create a lot of information which can be exceptionally hard to transfer over the WSN. According to Addo et al. [21] despite the fact that tending to key security and protection concerns comprehensively minimizes end-client selection obstructions, recognitions identified with the dependability of an IoT application hangs fundamentally on the execution of security and security best practices in the promptly noticeable IoT gadget. According to Barone et al. [22] parking is turning into a costly asset in any significant city on the planet, and its restricted accessibility is a simultaneous reason for urban activity clog, and air contamination. In old urban areas, the structure of the general population parking spot is unbendingly composed and frequently as on-road opens parking spaces. Lamentably, these open parking spaces can't be held in advance amid the pre-trip stage, and that regularly prompt to a weakness of the nature of urban portability. Tending to the issue of overseeing open parking spaces is in this manner essential to get naturally friendlier and more advantageous urban communities. Marquez et al. [23] show an open air parking model utilizing WSN as a conceivable answer for taking care of the issue of parking and can enhance the personal satisfaction in smart urban communities, on the grounds that the ideal opportunity for finding a free parking spot could be diminished, and also enhance vehicular activity. Rhodes et al. [24], used community wayfinding to enhance the effectiveness of smart parking system (SPS) and in this way lessen movement blockage in metropolitan situations, while expanding proficiency and gainfulness of parking structures. A huge bit of movement in urban zones is represented by drivers scanning for an accessible parking street. According to Sharma et al. [25] when the car goes into the parking zone, the stockpiling process begins, with the utilization of lifts vehicle gets stopped without getting hindered from different vehicles. They executed the capacity and recovery handle utilizing JAVA dialect. Utilizing A*, D* Lite, Unblocking calculations, it gets actualized. They can look for the utilized and unused spaces, utilizing D* Lite they arranged a way to retrieve the vehicle and utilizing unblocking calculations they put away and recovered vehicles without obstructing the way of another vehicle. They additionally utilized the Manhattan Distance to choose the closest way for the vehicle from the lift. According to Zhanlin et al. [26] cloud-based keen vehicle parking administrations in brilliant urban areas are imperative for IoT worldview. This sort of administrations will turn into a fundamental part of a non-specific IoT operational stage for keen urban areas because of its unadulterated business-situated components. Since the above related work and solutions are not IoT-based, this work aims to provide a smart solution based on IoT i.e. the next generation solution that will enable people to know before-hand about the availability of free parking space.

The IoT addresses the purchaser hardware showcase and also the B2B business. Contingent upon investigators, the estimation of IoT is evaluated to be from $300 B to near $2000 B in 2020 (contingent upon what is represented in IoT advertise). Gartner has a characterized concentrate on IoT included esteem. Ten billion IoT protest shipments are normal every year beginning in 2020, which will be utilized as a part of a wide assortment of uses. Associated LED lights are relied upon to be the biggest by a wide margin. In term of innovations, Gartner states sensors will assume a basic part inside IoT engineering. However IoT design stays questionable; there won't be one single engineering for all business sectors portions. More than 80% of the $300 B incremental income evaluated for 2020 will be gotten from administrations.

The incremental cost of equipment and implanted programming is moderately little, though the administration and examination opportunity is substantially bigger. At first, a significant part of the provider center in the IoT markets will be on equipment and programming, as organizations attempt distinctive methodologies and capabilities with an end goal to construct consciousness of their items. As plans of action develop, in any case, the market will progressively be driven by administrations (counting information examination administrations). There will be administration openings related with organizing and dealing with the various things those common purchasers will communicate with in a normal day, both in their home and in their day by day ventures. For sure, the esteem chain for IoT gadgets and administrations will be multilayered, with use information utilized utilizing examination programming that is intended to haul out patterns helpful for further item and administration advertising activities. IoT creates "Enormous Data" that examination must deliver to change immense volumes of information into a little (promptly integrated by a human personality) amount of usable/noteworthy data [9].

4 IoT-Based Real Time Smart Street Parking System

The motivation behind this work is to give benefits to the citizens through smart street parking system using IoT because IoT can collect the data for free parking slots through drone. Drone images acquired by a UAV system which is the latest geo-spatial data acquisition technique are used in this project for mapping free parking features. This reduces the project cost and time substantially. Geo clipping video output of the drone are stored in a database and these data can be viewed together with the Google 2D map of the captured areas. Generated frames from the captured video which contain XY coordinates for individual frame are used to convert those raster into vector maps. Developed vector maps are stored in both ArcGIS server and specific map engine. Each individual free parking is stored with a unique ID. The government has been struggling very hard since last 15 years to reduce the parking problem in the unmanaged street. Vehicle population is eating up its street space at an alarming pace. The concept of multi-level parking failed in streets. The proposed prototype of IoT-based RTSSPS is divided into three modules in which each segment has the four layers as per architecture of IoT. A prototype of IoT-based RTSSPS is designed by using IR sensors on each parking slots to detect the vehicle by matching the token numbers. When a vehicle is in front entry gate of street, vehicle operator will generate a token with unique id and this unique id goes to the parking sensor server by using first module of architecture and takes database from cloud module through the data centric module. The sensor readings are received and analyzed by Arduino UNO, the microcontroller. If the sensor reading will change from low to high for more than 5 s then the slot is considered as occupied and is its low than slot is vacant. It forwards the result to the application which displays the parking space. If a vacant parking slot is found the result is shown on the 16X2 LCD screen on the passageway of the parking facility. A motor controlled by an IC L283D opens the gate through which the car enters the parking street and with the help of token number.

4.1 IoT-Based Real Time Smart Street Parking System Architecture

IoT-based RTSSPS architecture is a better approach as it overcomes the disadvantages of other traditional and smart parking system. It basically involves sensors to detect the presence of vehicle on the street parking. The sensor data is transmitted to the management system which facilitates the administrators and managers with real-time information. In this IoT-based RTSSPS, sensor nodes, master nodes, sink nodes are used to set up their sensor and networking layer. IoT-based RTSSPS have focused on optimizing the sensor node placement so that least number of sensor hubs is utilized to get the greatest scope of the zone. Sensors are best to detect the vehicle, and place at the parking slots because vehicles have huge amount of metals, so this study can utilize a sensor for vehicles parking.

The IoT-based RTSSPS have two main things smart street parking architecture and smart street parking algorithm. Smart street parking architecture utilize WSN driven, cloud driven and information drove IoT with base stations, switches, sensor hubs, and a remote server. Sensor hubs are conveyed close by the roadside, every hub is planted in the center of a parking spot and one on parking card. Every sensor hub recognizes the world's attractive field, Wi-Fi, daylight and road light intermittently. At the point when a hub recognized a vehicle, it transfers the SMS to the portable of the client through the switch to main station. This main station gather the info from different hubs and transmits to the display board which can be mounting in the starting and leave purpose of the road, remote server furthermore on the portable of the considerable number of individuals living here.

To lessen the number of sensors and to give full scope, this study will attempt to utilize the street parking calculation through hereditary qualities or insatiable approach which helps in inquiry streamlining of sensors. IoT-based RTSSPS architecture is divided into three parts IoT-based WSN centric smart street parking module, IoT-based data centric smart street parking module and IoT-based cloud centric smart street parking module (Fig. 5).

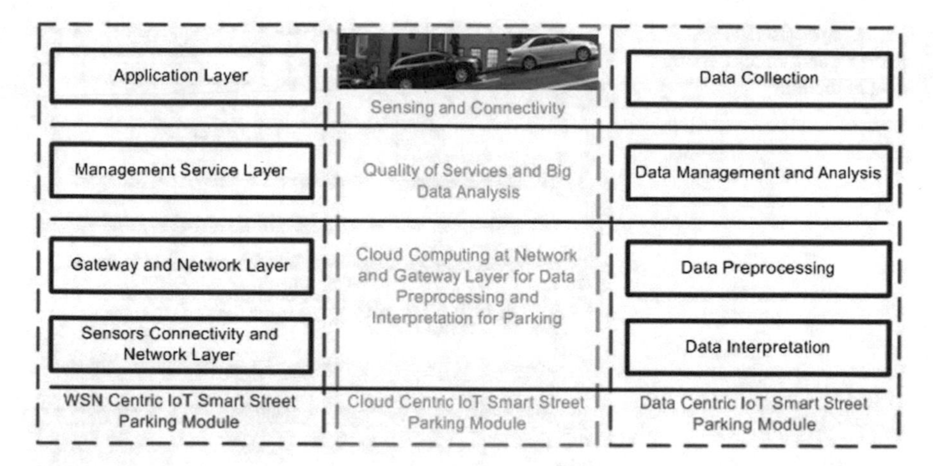

Fig. 5 IoT-based real time smart street parking system architecture

4.1.1 IoT-Based WSN Centric Smart Street Parking Module

IoT-based WSN centric smart street parking module are a combination of the application layer, management service layer, gateway and network layer, sensors connectivity and network layer which officially portrayed in the design of IoT. WSN driven IoT helps in the simple establishment of sensor hubs. The parking slots are exhibited to clients through objects. In this architecture, application layer deals with the sensing and connectivity of vehicle when the driver wants to park into the parking slot and collect the data of free parking slots and owner information. Management service layer provides the QoS, information analytics, security control, process modeling and device control through big data analysis (Fig. 6).

Gateway and network layer and sensor connectivity layer must be able to support massive volume of IoT data preprocessing and interpretation produced by wireless sensors and smart devices through cloud computing and gateway network. This layer grouped with smart devices and sensors, according to their purpose and data types such as environmental, magnetic, obstacle and surveillance sensors etc. WSN centric IoT formation is made in this stack and the information is delivered to a targeted location for further processing. Real-time information is collected and processed to get a result or generate some output from data-centric IoT smart street parking architecture. WSN as a conceivable answer for take care of the issue of stopping and can enhance the personal satisfaction in smart urban communities, on the grounds that the ideal opportunity for finding a free parking spot could be diminished, and also decrease contamination and to enhance vehicular activity.

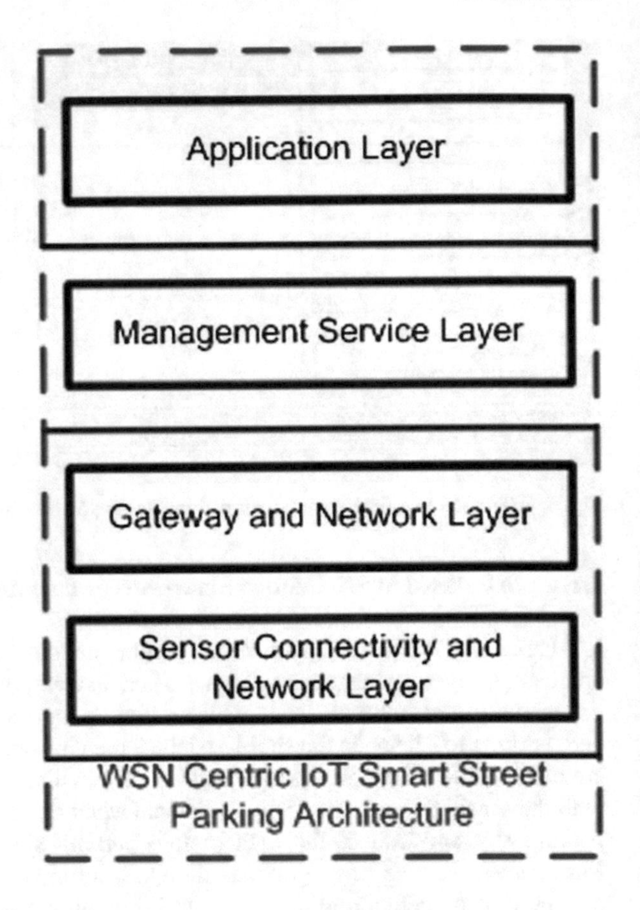

Fig. 6 IoT-based WSN centric smart street parking system module

4.1.2 IoT-Based Data Centric Smart Street Parking Module

From last consecutive years this study have seen an unrivalled explosion in the amount of data which is transmitted via many digital devices and for reducing the traffic, there is a need of strong communication technique so that large amount of data and information can be transmitted. The most principal issue in world today is the large amount of valuable big data that is flowing among various networks and swap information with more compression and security in both the time and space for data transmission for data storage [27]. A big amount of data will be collected and generated through data-centric IoT which emphasizes on data collection, processing, storage and visualization. A generalized system for data collection is required to effectively exploit spatial temporal characteristics of data which is prepared with the help of drone. Extraction of meaning full information from raw data is non trivial. It usually involve data pre-processing because in smart cities, adaptability and robustness of algorithm to compare data at large scales of time and space is essential and for smart cities applications, visualization is important for data representation in users. The information gathered through application layer and WSN drove IoT, preprocessed and transmitted to the database system which is further utilized by the client through upper layer applications with LCD, LED and AMOLED (Fig. 7).

The data-centric IoT architecture maintains the following modules to communicate with the WSN and cloud-centric database:

- Parking management and data module which manage the free parking slots.
- Security management module through cloud computing which alerts the illegal parking.
- Sensor management module which manages the location and searches optimization through updated database.
- Server management module to control the servers which are linked with database and notify the situation through SMS.

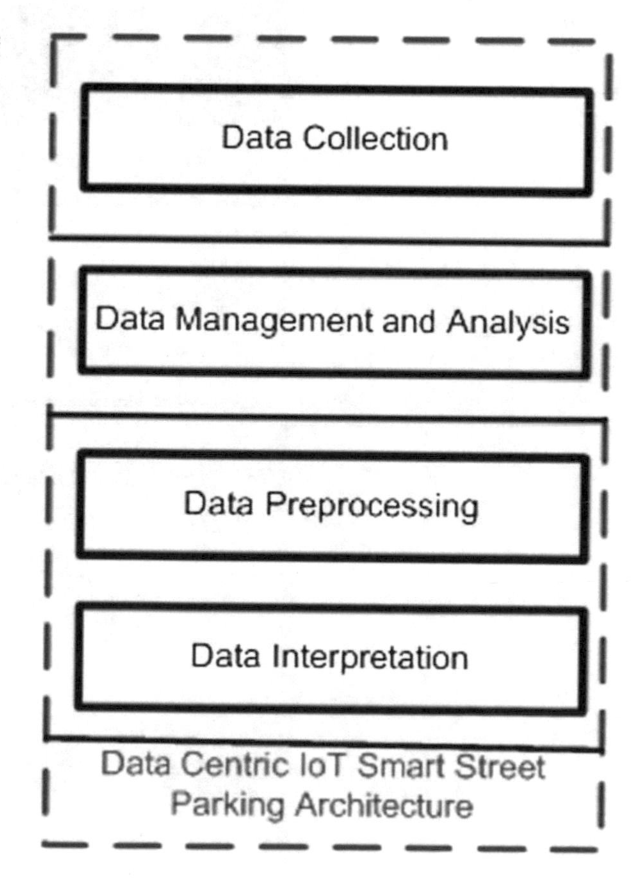

Fig. 7 IoT-based data centric smart street parking system module

4.1.3 IoT-Based Cloud Centric Smart Street Parking Module

IoT-based cloud-centric smart street parking module is the middle part of the whole architecture and makes communication between WSN centric IoT and data centric IoT through cloud database by using analytic tools. Cloud computing provides the software to collect the information of vehicle parking and to manage the system, services, and overall behaviors with changing physical environment of the street through proposed architecture (Fig. 8).

This cloud computing environment also helps in optimal parking space with the proposed algorithm according to the availability of space, and search and display the path to the parking space on the LED screen at the entrance. So that the drivers can enter the parking slots and spends less time for the parking space.

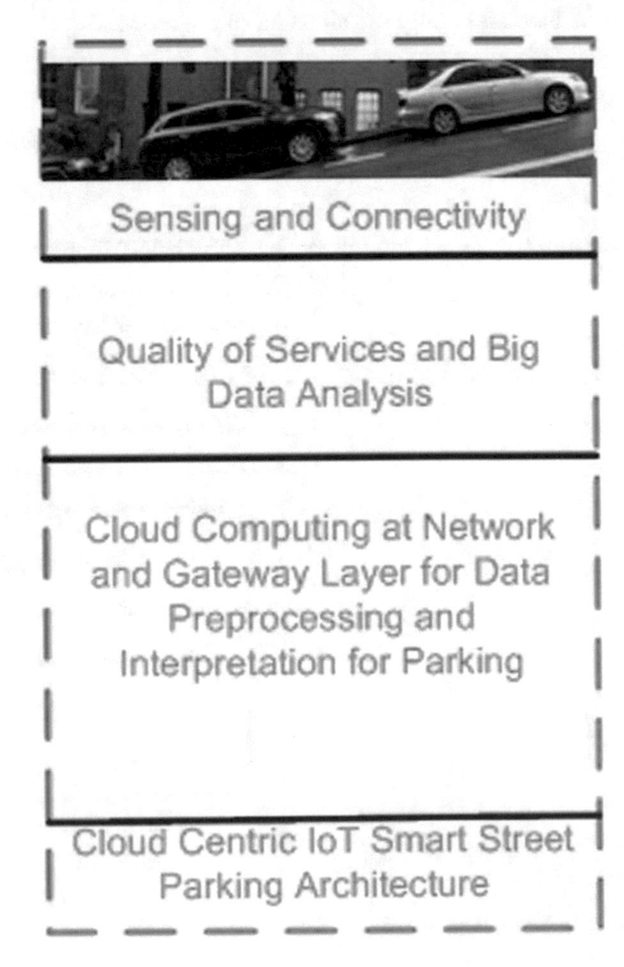

Fig. 8 IoT-based cloud centric smart street parking system module

4.2 IoT-Based Real Time Smart Street Parking System Algorithms

Numerous urban communities have embraced a parking direction and data system to attempt to reduce this activity clog. In our proposed system, this work creates the database of the streets by using drone and gives every resident an id number to their parking slots. When a resident or driver enters inside the street for parking, this system generates an id number which is combination of id number of parking slot, last two digit of car registration number and last two digit of resident house number. Suppose id of parking slot is 76A, owner car number is 1234 and house number is 1976 than unique id is 76A3476 and submits this unique id at the sensor location of real parking space. When a resident enter inside the street with nique id it automatically help to search the parking slot and help to park the vehicle. Reservation authority sends a sms to the user as soon as user reserves the parking space. At the parking slot host identifies the user by scanning the code with a parking ticket or IoT, generated by authority. But if the vehicle is not owner own vehicle than authority sent the car number to owner via sms and owner give permission for this number than authority generate the unique id number. This algorithm has been proposed for the car in the event and car out event to resolve the issue of street parking for outside and owner of parking. Sensor hubs have restricted figuring force and memory.

4.2.1 Implementation of Street Parking Algorithm

RTSSPS (PL, N)

Here RTSSPS is smart street parking system, PL is a linear array with N parking slots.

Input: S = C, CL, Db, PS, Dmin, Dmax, C_i, C_j, C_{next}

Here S denote the entry at Street Node. C denotes the Nodes and CL denote the current location of C and C_i denotes the next node, Dmin and Dmax denotes the maximum and minimum distance, PS denotes the parking slots and Db denotes the Database of streets.

Output: Signal for L293D

Step 1 **Initialize**: CL → Current Location of Car,
Car enters into parking street and search for empty parking space, authority generate the unique id for the resident or drivers

Step 2 **If** Parking Slot (PS) = 0, then L293D = 0 and
Car will not allow entering the street
Else if

If Parking Slot (PS) = 1 or more than 1, Than L293D = 1 or more than 1
Car will allow entering the street after getting the unique id

Step 3 **for** all C-nodes there is 2 neighbor tables of Ci do
Calculate the Dmin and Dmax for parking slot with IoT/Parking Ticket
Here w is weight/distance of the link between node Ci and node Cj
if w < Dmin for owner parking slot then

$$\begin{aligned} \text{Dmin} &= \text{w} \\ \text{update} : \text{C}_{\text{next}} &= \text{Cj} \end{aligned}$$

end if
Else if w < Dmax for owner parking slot then

$$\begin{aligned} \text{Dmax} &= \text{w} \\ \text{update} : \text{C}_{\text{next}} &= \text{Cj} \end{aligned}$$

end for
return C_{next}

Step 4 **Vehicle-In Event**

- *On arrival of the vehicle at the gate, the sensor will detect its presence and will report the same to the microcontroller.*
- *The microcontroller will check for the availability of the space in the parking lot.*
- *It communicates the result to the web application which displays it.*
- *If space is available, a parking ticket is issue with unique id, the gate opens and the vehicle enters.*
- *When the user parks the vehicle according to the ticket number, the sensor detects the presence and reports to the microcontroller.*
- *The microcontroller updates the database.*
- *If the parking space is not available then the gate will not open.*

Step 5 **Car-out Event**

- *At the point when the client escapes the parking spot, the sensor distinguishes the free opening and reports to the microcontroller.*
- *The microcontroller communicates the same to the web application. The database is updated. The slot status is now free from occupied.*
- *The result is displayed on the web application and user moves towards the gate.*
- *The entryway sensor matches the ticket number and gives the sms to the owner and is owner say yes than gate opens for vehicle.*

Step 6 **End**

5 Evaluation of IoT-Based RTSSPS

Performance evaluation: A prototype of the RTSSPS is built in order to test the working. The evaluation of the IoT-based RTSSPS was done based on the observations made while testing the prototype. It was watched that an ideal opportunity to look for a free parking space was lessened significantly when IoT-based RTSSPS was deployed. According to a survey conducted by the Telegraph, the average time taken by the drivers to find parking space is 20 min. Using IoT-based RTSSPS it was observed that this time was reduced to 1–2 min since the drivers will be able to locate a free parking spot using a smart token number. Hence, there is no need to wander around and time is only consumed in entering and parking the automobile in the parking facility.

The aim of the IoT-based RTSSPS is to help the drivers to easily spot a vacant parking space. The web application is able to fulfill this requirement. It shows real-time accessibility of the parking and facility which are authorized to manage the application. The clients can simply see the status and position for the parking space. The position of the parking opening is dictated by the accompanying condition:

$$\begin{aligned} &\text{If } ((S = \text{ High})\ \&\&\ (T > 5)) \text{ then,} \\ &\quad P_N! = \text{ EMPTY} \\ &\text{Else} \quad P_N = \text{ EMPTY} \end{aligned}$$

where S is sensor signal, T is time and P_N is nth parking slot. When the sensor signal detects a presence it becomes High and when it remains High for 5 s then the spot is considered to be occupied else the status remains to be empty. Since the time to find a parking spot is reduced, the traffic congestion caused due to the drivers roaming around in search of free parking space will also be reduced. Hence, congestion density of various markets that use traditional parking system were compared against the IoT-based RTSSPS. Congestion density at a street place can be expressed as follows:

$$\text{Congestion Density} = \frac{(Tsearch + Tparking) * Naverage_{cars}}{ECs}$$

where ECS is equivalent car space which is a designated area for comfortable circulation space for vehicles and easy retrieval. ECS has been taken to be equal to 23 m^2.

6 Conclusion

The work presents review of the existing research done in field and tries to develop a IoT-Based RTSSPS suitable for developing countries, that improves performance by reducing the number of users that fail to find a parking space. This work presents the prototype design of IoT-based RTSSPS with architecture and algorithm. IoT-based RTSSPS architecture is divided into three parts IoT-based WSN centric smart street parking module, IoT-based data centric smart street parking module and IoT-based cloud centric smart street parking module with street parking algorithm. Our proposed architecture and algorithm has been successfully implemented in a real situation by the proper communication of each module like IoT-based WSN centric smart street parking module, IoT-based data centric smart street parking module and IoT-based cloud centric smart street parking module. The sensors are to be fitted on the side of the street and connected to the controller at the intersection and communicate with the WSN centric IoT smart street parking module. These are some hectic jobs which are to be dealt before implementing the system, but once implemented, it will make our parking system more convenient and smarter. During implementation, this study finds that the algorithm significantly reduces the average waiting time of users to search the parking space. In our future work, this study considers the security aspects of our system as well as implements our proposed system in large scales in the real world.

References

1. Benadda, M., Bouamrane, K., & Belalem, G. (2017). How to manage persons taken malaise at the steering wheel using HAaaS in a vehicular cloud computing environment. *International Journal of Ambient Computing and Intelligence (IJACI), 8*(2), 18.
2. Ahmed, D. M., & Anguluri, R. (2014). Conceptual understanding of smart cities. *International Journal of Science and Research, 3*(12), 1470–1473.
3. Mukherjee, A., Dey, N., Kausar, N., Ashour, A. S., Taiar, R., & Hassanien, A. E. (2016). A disaster management specific mobility model for flying Ad-hoc network. *International Journal of Rough Sets and Data Analysis, 3*(3), 72–103.
4. Kimbahune, V. V., Deshpande, A. V., & Mahalle, P. N. (2017). Lightweight key management for adaptive addressing in next generation internet. *International Journal of Ambient Computing and Intelligence, 8*(1), 20.
5. Schaller, A., & Mueller, K. (2009). Motorola's experiences in designing the internet of things. *International Journal of Ambient Computing and Intelligence, 1*(1), 11.
6. Sweeney, P. J., II. (2005). *RFID for dummies*. New York: Wiley Publishing Inc.
7. Uckelmann, D., Harrison, M., & Michahelles, F. (2011). *Architecting the internet of things*. Berlin: Springer.
8. Odella, F. (2016). Technology studies and the sociological debate on monitoring of social interactions. *International Journal of Ambient Computing and Intelligence, 7*(1), 26.
9. Jammes, F. (2016). *Internet of things in energy efficiency: The internet of things* (Vol. 16, pp. 1–8). New York: ACM Digital Library.
10. Biswas, S. P., Roy, P., Patra, N., Mukherjee, A., & Dey, N. (2016). Intelligent traffic monitoring system. *Advances in Intelligent Systems and Computing, 380*, 535–545.
11. Poland, M., Nugent, C. D., Wang, H., & Chen, L. (2009). Smart home research: Projects and issues. *International Journal of Ambient Computing and Intelligence, 1*(4), 32–45.
12. Tang VW, Zheng Y, & Cao, J. (2006). An intelligent car park management system based on wireless sensor networks. 1st International Symposium on Pervasive Computing and Applications. http://citeseerx.ist.psu.edu/viewdoc/download?doi=10.1.1.460.6553&rep=rep1&type=pdf

13. Kumar, P. V., & Siddarth, T. S. (2010). A prototype parking system using wireless sensor networks. *International Journal of Computer Communication and Information System, 2*(1), 276–280.
14. Deshpandey, A., Nath, S., Gibbons, P. B., & Seshan, S. (2003). IRIS: Internet-scale resource-intensive sensor services. In *ACM SIGMOD International Conference on management of data* (pp. 664–667).
15. Lee, S., Yoon, D., & Ghosh, A. (2008). Intelligent parking lot application using wireless sensor networks. In *International Symposium on Collaborative Technologies and Systems, Collaborative Technologies and Systems* (pp. 48–57).
16. Hsieh, M. F., & Ozguner, U. (2008). A parking algorithm for an autonomous vehicle. In *IEEE intelligent vehicles symposium* (pp. 1155–1160).
17. Chen, M., & Chang, T. (2011). A parking guidance and information system based on wireless sensor network. In *IEEE International Conference Information and Automation* (pp. 601–605).
18. Nawaz, S., Efstratiou, C., & Mascolo, C. (2013). ParkSense: A smartphone-based sensing system for on-street parking. In *Proceedings of the 19th Annual International Conference on Mobile Computing and Networking* (pp. 75–86). ACM. https://www.cl.cam.ac.uk/~cm542/papers/mobicom2013.pdf
19. Pereira, P. P., Eliasson, J., Kyusakov, R. Delsing, J., Raayatinezhad, A., & Johansson, M. (2013). Enabling cloud connectivity for mobile internet of things applications. In *Seventh International Symposium on Service-Oriented System Engineering* (pp. 518–526).
20. Reddy, P. D., Rao, A. R., & Ahmed, S. M. (2013). An intelligent parking guidance and information system by using image processing technique. *International Journal of Advanced Research in Computer and Communication Engineering, 2*(10), 4044–4048.
21. Addo, I., Ahamed, S. I., Yau, S. S., & Buduru, A. (2014). A reference architecture for improving security and privacy in internet of things applications. In *IEEE International Conference on Mobile Services* (pp. 108–115).
22. Barone, R. E., Giuffrè, T., Siniscalchi, S. M., Morgano, M. A., & Tesoriere, G. (2014). Architecture for parking management in smart cities. *IET Intelligent Transport Systems, 8*(5), 445–452.
23. Marquez, M. D., Lara, R. A., & Gordillo, R. X. (2014). A new prototype of smart parking using wireless sensor networks. In *Colombian communications and computing (COLCOM)* (pp. 1–6).
24. Rhodes, C., Blewitt, W., Sharp, C., Ushaw, G., & Morgan, G. (2014). Smart routing: A novel application of collaborative path-finding to smart parking systems. In *2014 IEEE 16th Conference on Business Informatics* (Vol. 1, pp. 119–126).
25. Sharma, S., & Chhatarpal, R. H. (2014). Survey on internet of things and design for a smart parking area. *International Journal of Inventive Engineering and Sciences, 2*(9), 11–16.
26. Zhanlin, J., Ganchev, I., O'Droma, M., Zhao, L., & Zhang, X. (2014). A cloud-based car parking middleware for IoT-based smart cities: Design and implementation. *Sensors, 14*, 22372–22393.
27. Sawlikar, A. P., Khan, Z. J., & Akojwar, S. G. (2016). Efficient energy saving cryptographic techniques with software solution in wireless network. *International Journal of Synthetic Emotions (IJSE), 7*(2), 19.

Smart Irrigation: Towards Next Generation Agriculture

A. Rabadiya Kinjal, B. Shivangi Patel and C. Chintan Bhatt

Abstract Watering to plants or on whole field of crop is most important and pain taking task on daily basis. Quantity of water required by plant is one of the effective parameter for greenhouse effect on plants. To make this challenging work to informal, some analytical and historical data is required so that irrigation cycle of crop may become easy task for farmers. This project is made with the use of Node MCU board; consist of ESP8266 Microcontroller with in-built Wi-Fi module. Soil Moisture sensor is set in the field, which keeps track of moisture level in field soil. That collected data are sent over cloud to make people's nurturing activity pleased and tranquil. Data from the cloud is collected and irrigation related graph report for future use for farmer is made to take decision about which crop is to be sown. "Smart Irrigation Analysis" is an IoT application which provides remote analysis of irrigation on the field to the end user which is better than traditional irrigation of crop on field. Smart irrigation application has an automated recurring watering schedule; sensing and analysis of water used for crop and also senses moisture level giving real time data.

Keywords Analysis · Internet of things · Moisture · Node MCU · Sensing

A. Rabadiya Kinjal · B. Shivangi Patel (✉)
CHARUSAT, Changa, Nadiad, India
e-mail: shivangihpatel@gmail.com

A. Rabadiya Kinjal
e-mail: kinjalrabadiya20@gmail.com

C. Chintan Bhatt
U & P U Patel Department of Computer Engineering,
CHARUSAT, Changa, Nadiad, India
e-mail: chintanbhatt.ce@charusat.ac.in

N. Dey et al. (eds.), *Internet of Things and Big Data Analytics Toward Next-Generation Intelligence*, Studies in Big Data 30,
DOI 10.1007/978-3-319-60435-0_11

1 Introduction

"The Internet of Things (IoT) is the interconnection of uniquely identifiable embedded computing devices within the existing Internet infrastructure. Internet of Things or IoT basically is connecting Embedded System to internet." IoT [1, 2] is mainly concerned with making devices work with connected data over an internet. The data can be from different types of sensors which are embedded with real environments. These data are then carried over the internet to their respective controllers. The controllers then process the data and send relative commands to the actuators. The sensors and the actuators used both work on different environments. So they need to communicate with each other.

Basically all devices in an IoT environment communicate with each other. The to-and-fro communication between sensor and actuators uses different protocols such as Message Queuing Transport Telemetry (MQTT), Constrained Application Protocol (CoAP), XMPP, REST, 6LowPAN etc. These devices are connected to the internet through capillary networks. The above data concludes that Internet of Things consist of multiple specialized hardware and software that perform specific functionalities through web APIs which uses various protocols to create seamless connection to internet so that sensory data system sense that data and control system can take action on actuators set on specific environment (Fig. 1).

The data sensed by the respective sensors can also be stored virtually in cloud from where it can be used by either the user remotely or analytics to derive the results. IoT is mainly developed into two technologies: wearable and embedded.

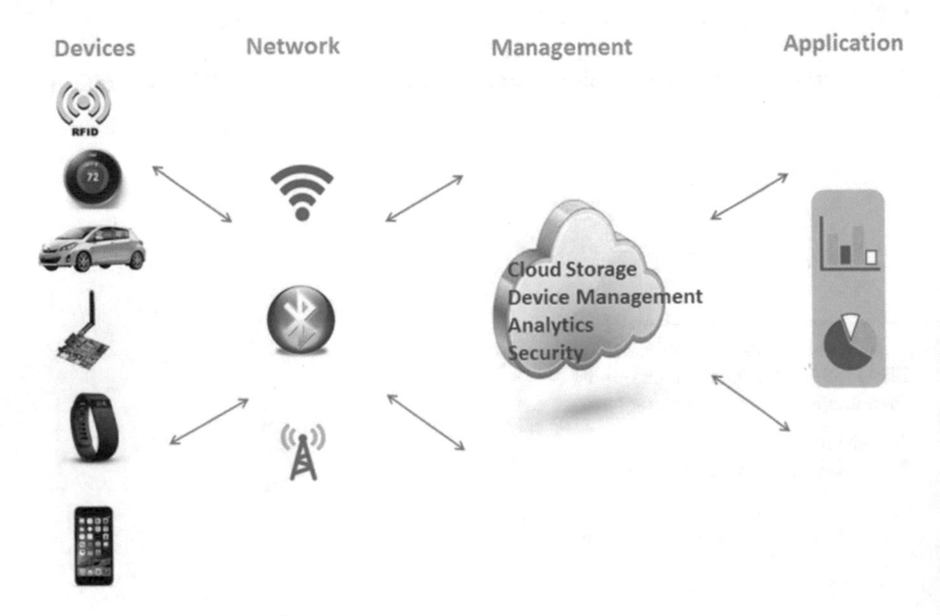

Fig. 1 IoT architecture [10]

The wearable devices are mainly developed using their own platforms depending on the developer. The embedded platforms mainly use Arduino, Raspberry Pi, Intel Galileo, Intel Edison Netduino etc. India is known as land of agriculture with many traditions and large variety of cultures. Traditionally agriculture is practiced by performing series of tasks such as tilling the land, managing it, sowing the seeds, irrigating it periodically, maintaining the growth, harvesting on perfect time and finally transporting it to the traders. India being a developing country faces many challenges towards its all sectors. In agriculture main challenge is the increase in growth of population, which leads to need for more food output to 50% by 2025. "By collecting real-time data on weather, soil, air quality, crop maturity and even equipment and labor costs, predictive analytics can be used to make smarter decisions" [3]. This is known as smart agriculture. Different Smart Agriculture solutions may be categorized as (Table 1).

Remote equipment monitoring includes different types of equipment used in the field to monitor the proper growth and maintenance of plants. Different equipment may include smart drones, smart tractors etc. Drones carry a camera which is piloted by the farmer on the ground to survey the land. They fly up to height of 100 ft. from ground level and captures images of different areas of land. The images are analyzed and mapped under number of variables from moisture level of soil to chlorophyll content of leaves and also distribution of weeds and infections to the crops if present. The images captured are in the RGB (Red, Green, Blue) format

Table 1 Agriculture sensor solutions

Different agriculture solutions	Sensors/tools	Purpose of sensor
Remote equipment monitoring	UAV-unnamed aerial vehicles/drones, smart tractors, smart water pumps and power meters [12]	Monitoring plant growth and maintaining different equipments used in farm
Sensor-base field and resource mapping, climate monitoring and forecasting [13]	Climate sensors: anemometer, wind vane, pluviometer Yield monitors: solar radiation Soil sensors: soil temperature, moisture and humidity Leaf and plant sensors: leaf wetness, fruit/trunk diameter	To monitor air pressure, temperature, humidity, monitoring climate variations to prevent growth of fungus and microbial contaminants Wind speed/direction
Remote crop monitoring	Image based sensor: smart drones, water sensors, soil sensors	Frequent analysis of field areas, automation and control system
Livestock tracking and geo-fencing	GPS (global positioning system) tracker collar, GNSS (global navigation satellite system)	Tracking livestock in the field, spatial proximity analysis
Smart logistics and warehousing	GPS tracker	Tracking logistics which are used for transportation to avoid food loss

where infected parameters will be categorized distinctively. Main equipment used in the field is tractor. Smart tractors are laden with tech and sensors. The steering is GPS controlled and is programmed with optimized route planning to recommends the shortest route across the field. It also has different sensors which are controlled and connected to mobile devices for their respective uses. Next equipment used is for watering the field which includes smart water pumps with smart irrigation system. Water pumps are attached with smart power meters which can automatically regulate the pump remotely according to the requirements. These water pumps are attached to the irrigation systems across the field which works smartly according to the moisture level of the soil.

Sensor based field and resource mapping measures different parameters with the help of their respective sensors. Parameters like air temperature, air humidity, soil temperature and humidity, leaf wetness, atmospheric pressure, solar radiation, trunk/stem/fruit diameter; wind speed and direction, rainfall etc. can be measured using their respective sensors. The sensors which are used mainly for agriculture purposes are ultra-low power (0.7 μA). They support different radio technologies which may fall under long range (3G/GPRS/LoRaWAN/900 MHz), medium range (ZigBee/802.15.4/WiFi) and short range (RFID [Radio Frequency Identification]/ NFC (Near Field Communication)/Bluetooth 4.0) [4]. They support Over the Air (OTA) programming. Also they support different encryption libraries such as AES (Advanced Encryption Standard), RSA etc. to protect the data which is exchanged between the sensor and the actuator. The sensors are mainly robust water proof.

Remote crop monitoring helps to identify the crops affected by the environmental conditions like too dry or wet, infected by insect, weeds or weather related damage. If a crop is infected by disease or if there is growth of weeds in any area, then the crop emits IR waves which can be detected showing crop stress or crop damage. Not only is the growth of crops affected by the weather but also logistics of harvesting and transportation. Knowing the climatic conditions well in advance can help a farmer to decide when to irrigate the field, when to medicate it etc. Temperature and humidity of soil and surrounding air respectively can be measured by placing sensors in field. All the details related to the climatic conditions according to the geography of the area can be received on the mobile devices and accordingly the sensors placed in the field can be controlled.

Agriculture is incomplete without Livestock. Livestock may include different type of cattle used by the farmers. These need to be tracked continuously by the trencher and should be placed in a geo-fenced area. Geo-fencing is the use of GPS satellite networks or RFID to create virtual boundaries around a location. It is used in livestock industry where a herd of cattle would be equipped with GPS units and if the herd moves out of boundaries set by rancher then he would receive an alert. Finally when the crop is ready it is harvested and the logistics of harvesting and transporting food to distribution centres is crucial. During the distribution process, wastage of food happens on very large scale. It happens due to unusual rainfall, high temperature. That's why transportation is managed in such way that food is not hold for longer.

Having advance knowledge of weather conditions and the routes which can be affected, helps farmers to take better decision for faster transportation. Crop production management can be varied using Smart Agriculture in the field. This requires farmers to use knowledge of marketing, Information and Technology and Decision support in order to increase economic returns. The agricultural specialists help farmers to implement smart agriculture programs. Agriculture being a very broad concept is very difficult to analyse its each module. Considering a small irrigation module, smart irrigation facilitates watering of the field by controlling it remotely. The physical presence of farmer is not necessary to control the pumps or sprinklers placed in the field. The soil moisture sensor placed in the field senses the moisture level of the soil and sends the data to the mobile device over an internet. Mobile devices is then remotely controls the functioning of the system according to the moisture level of the soil. Also the data send over the internet to the mobile device can be used by analytics to derive their results.

The main aim to develop this application is to provide information about irrigation analysis on field for particular season of crop. This application is related to IoT application. So in this application there is a combination of hardware, software as well as soil moisture sensor. In this application output is in form of graph as well as serial monitor (of Node MCU). It is possible because of in-built Wi-Fi module on Node MCU. Using Wi-Fi data will get uploaded on cloud. From that captured data graph will be generated and that can be used for further analysis. In this project, thingspeak.com is used for storage of data and it is open source solution for IoT. For communication with cloud storage it uses HTTP protocol with method POST, which provides better security then GET method. Using this data collected from sensor to cloud and represented in form for .csv or JSON form.

Moisture level of soil during growth of plant is precise significant factor in life of plant. Enough watering is necessary in soil which may vary due to temperature and ground water level of that area. When soil is moist then growth of root is better than expected. The process of transpiration is important to preserve water in plant body. If moisture is at excessive level then soil pathogens may destroy whole plant body. The main purpose of the application is to capture sensor data which is send by sensor to cloud storage. Users can see graph generated based on those stored data, so they can easily use this. Suppose farmer wants to decide the crop on field according to available water then he can use this application based data as well as can control the pump for irrigation on field. Graph also gives .csv file data with time stamp.

2 Smart Agriculture Solutions by Different Service Providers

Different companies are carrying out research work in agriculture domain to make it smarter using various IoT solutions. They use different sensors to measure all parameters related to agriculture. Different parameters that are measured using sensors are:

- Atmospheric pressure
- Leaf wetness
- Humidity
- Temperature
- Luminosity
- Soil moisture and temperature
- Trunk, stem and fruit diameter
- Solar and ultraviolet radiation
- Different climatic conditions like rainfall, wind speed etc.

The sensors that are used to measure these parameters are embedded with the circuit known as sensor board. Different sensor boards are available that help to measure all these parameters at a time. Atmospheric pressure is an important parameter that needs to be monitored continuously. Sudden change in atmospheric pressure may damage the crop. Atmospheric pressure is usually measured in units of kPa, within range of 15–115 kPa. This value is converted to analog voltage value in the sensor board and again with the help of library functions it is again converted to corresponding pressure value while displaying. Another parameter that needs to be monitored is humidity. There are certain crops that need to be cultured in dry environment. Presence of humidity may damage the crop. The sensor provides the analog value which is proportional to the relative humidity level in the atmosphere. The library function used in programming return the value in relative humidity percentage. If these sensors are connected with other sensors on the board then it can control excess humidity level by turning on the drier. One major issue concerned in implementing this system is vast field area. The fields that are very large in area and continuous monitoring all the parameters become very difficult. That's why these sensors need to be connected to each other so that they can perform in required way. Another important parameter is the measurement of temperature. Appropriate temperature is required for proper growth. If a plant is cultivated in greenhouse environment then temperature can be effectively controlled. Temperature in the range of −40 to +125 °C can be measured with the help of sensors. Temperature sensors measure the present climatic temperature.

There are others types of sensors called resistive sensors which give more accurate results then other sensors. These types of sensors are mostly used to measure the small parameters which need accuracy in measurement. Parameters like soil moisture level, temperature of soil, leaf wetness, luminosity measurement etc. need to measure accurately. Resistive sensors work mostly on the principle of difference of resistance of metals used in the sensors. Even a small change in the value of the parameter can be captured by using these sensors. Wetness of leaves is measured using sensor to determine the level of water content in the plant. Plants like tobacco which is mainly cultivated for their leaves, their wetness level and moisture level needs to be monitored continuously. Harvesting at the right time can be more beneficial in cultivating such crops. Also level of transpiration from leaves can be monitored using these sensors. When the leaves are condensed then resistors show low resistance value which may increase in absence of condensation, and is low when totally flooded in water. The output voltage value is inversely proportional to the level of humidity at the surface of sensor. The value returned by the library function is the amount of the percentage of condensation on the surface of sensor which is placed on the leaves.

Luminosity is the amount of light present at that time or amount of sunlight present at that time. For plant to carry out photosynthesis, fix amount of sunlight is required to carry out the process. Excessive sunlight may burn the cells and can damage the plants. This luminosity can be measured using its sensors. These resistive sensors give higher value in darkness in range of 20 MΩ and gives value in range of 20 kΩ in presence of light. This sensor can share the power supply with the dendrometer, digital humidity and temperature sensor and humidity, temperature, leaf wetness and solar radiation sensors, which can be inhibited centrally. Value is measured through analog to digital converter. Value is outputted in spectrum form of human visible range. Along with amount of light to be measured, its amount of radiation to be measured and controlled is necessary. Amount of solar radiation measured in PAR (photosynthetically active radiation). It is the as voltage proportional to the intensity of the light in the visible range of the spectrum. PAR is key parameter in photosynthesis process. Also the amount of ultraviolet radiations is also measured in the same manner. Excess amount of UV radiation can damage the crop. Leaf and dendrites may get burned in presence of more amounts of UV and solar radiation. So they need to be monitored and controlled accordingly. Fruit diameter dendrometer, trunk diameter dendrometer is a type of sensor that works on the variation of internal resistance that pressure causes due to stem growth or trunk growth or growth of fruit or stem etc. the circuit reads the value of resistance variation and then library function returns the corresponding growth in mm.

Finally the most important parameter that affects all types of crops and that is required to be monitored continuously is soil moisture level and soil temperature level. Maintaining soil parameters in required amount is very important as all the vital functions are carried out through the soil only. Water and other required nutritional things are absorbed from the soil through roots. These can function at right temperature and at right moisture level. Also the quantity of water required to irrigate the fields need to be monitored. Soil moisture sensor is used to measure the moisture level of the soil. If the moisture level is low then automatically it can regulate the pump or the sprinkler to water the field. After watering if the moisture level and the temperature of the soil reach the required value according to soil type, then it can send command to pump or the sprinkler to stop automatically. Soil water tension is a Parameter that depends on moisture that reflects the pressure needed to extract the water from the ground. The resistance of the sensor is proportional to the soil water tension. The library function returns the output in frequency (Hz) of sensor adaption to circuit. Moisture sensors are placed at different depths so that moisture at different depths can be measured and accordingly decision regarding watering of soil can be taken.

The main climatic changes that affects almost all crops and vegetation are wind speed, rainfall and direction of wind. Wind speed is with the help of anemometer. When the arms of anemometer completes 180°, the switch closes for short duration of time and outputs value in digital form whose frequency is proportional to speed of wind. The library functions returns this value in units of km/h. Direction of wind is measured with the help of wind vane. The wind vane comprises of basement that turns uninhibitedly on a base with various arrangement of resistances associated with relating switches that are regularly and are shut (maybe a couple) when a magnet in the cellar follows up on them which distinguishes different positions. The equivalent resistance of the wind vane form a voltage divider, whose output can be measured in the analog input. The library stores value of the variable that corresponds to an identifier of the pointing direction. Amount of rainfall is measured using instrument called pluviometer. The pluviometer consists of a small bucket that, once completely filled closes a switch, emptying automatically afterwards. The sensor is connected directly to the sensor board through a pull-up resistance which triggers the interruption of the controller when the rainfall stimulates a bucket emptying event. Different functions are defined in the library to measure from the pluviometer. Measurement of rainfall and calculation of precipitation in mm can be done at either current time or can be read for before 1 h or after 24 h.

All the sensors that are described here are the basic sensors required to measure all the necessary parameters related to agriculture to make it smarter. Researches are being carried out to develop and promote other sensors that can measure other minute parameters. Also future work can be carried out to develop such boards in which all the sensors can be embedded one at a time. If all the sensors and device are connected together then it can develop a full smart agriculture system that can make current agricultural system better.

3 Programming the Sensor Boards and Storage in Database

There are many devices that support Over the Air Programming. In this process first setup all the sensors on field for observation. Then using device like laptop and gateway, all the sensors can programmed at single point of time.

When sensors collect data then for future use application will transfer data for storage. From storage unit users can generate many reports related to crop health and if face any problem then users can take advanced step to stop damage of crop. There are mainly two types of storage that smart agriculture use:

- Local Data Base
  - When network is not available or any local information user need then storage only used at local level and it can’t be access via World Wide Web.
- External Data Base
  - When users want that data must be available throughout the world and remain on cloud then user can use external database and there are many devices are available that can upload field data to cloud and can be access from anywhere in the world.

4 Right Time to Irrigate

For more growth of any crop, it is vital decision that to determine perfect time to irrigate soil and quantity of water given to crop. Various physical parameters also affect irrigation and growth of crop. People mainly follow one method that shows when to irrigate soil and check soil moisture depletion. When plant starts drying it means there is strong need of watering otherwise crop is going to die because plants start use the water remain themselves. Soil has different types of profiles: clay, sand and loam soil. Clay particles have highest capacity to hold water, sand has lowest capacity and loam soil has moderate capacity to hold water. Figure 2 shows three typical curves for sand, clay and loam soils. As Fig. 2 shows, the plants will use the water in the soil until the moisture level goes to the permanent wilting point (PWP). Once the soil dries down to the PWP, plants can no longer extract water from the soil and the plants die (Fig. 3).

5 Related Work

5.1 Technology in Detail

This application is a combination of hardware and software. In this application Node MCU is used. NodeMCU is IoT hardware platform. It utilizes the Lua scripting dialect. It depends on the eLua project, and based on the ESP8266 SDK 1.4. It utilizes many open source ventures, for example, lua-cjson, and spiffs. It incorporates firmware which keeps running on the ESP8266 Wi-Fi SoC, and

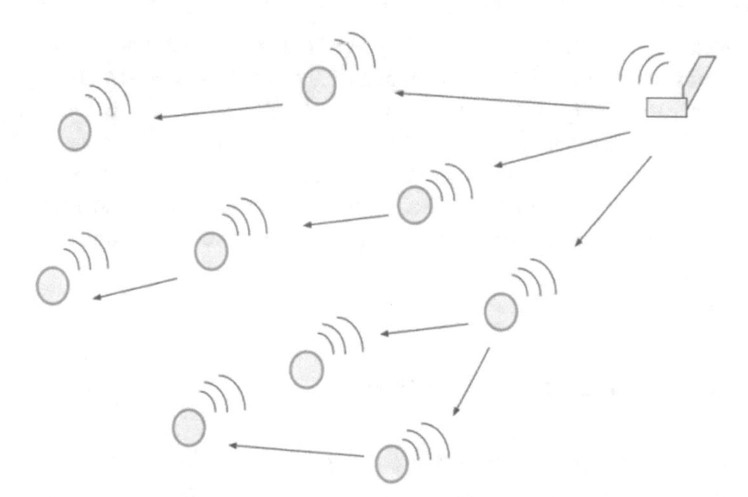

Fig. 2 Over the air programming process

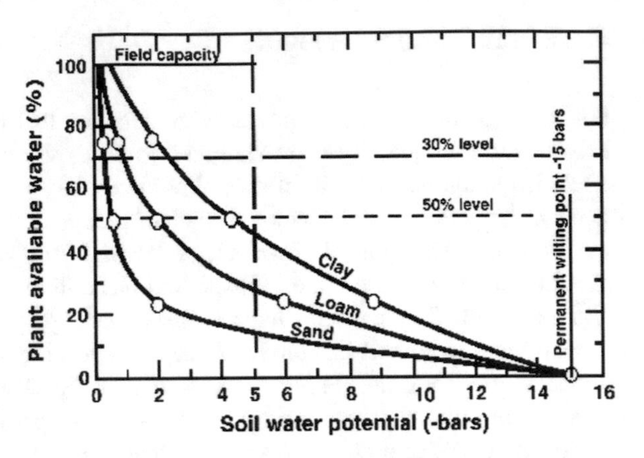

Fig. 3 A diagram of typical tension and water amount for sand, clay [11]

equipment which depends on the ESP-12 module. The ESP8266 is a Wi-Fi SoC coordinated with a Tensilica Xtensa LX106 center, generally utilized as a part of IoT applications. In this application, soil moisture sensor is used for sensing the moisture on field [5].

The soil moisture sensor is used to measure the water level in particular soil. It widely used in soil science, agricultural science, environmental science, horticulture, botany [6] and many other small scale agricultural experiments. Soil Moisture sensor is mostly used to:

1. Measure the water loss in form of vapour due to evaporation and plant uptake and to evaluate optimal soil moisture level for particular species of plant.
2. Measure the moisture in greenhouse to control the irrigation in greenhouse which in not easy to check soil by seeing.
3. Expand small scale biological experiments with small scale investment.

The horizontal position of the sensor gives such assurance that it gives result for one particular depth of field. If sensor position is vertical then it may happen that soil moisture may vary according to depth of soil. From above two cases it is advisable to put sensor horizontal. A trenching shovel is used to make hole in the soil so that soil moisture sensor can be put appropriate to collect data. If probes of soil moisture sensor are fully covered by soil then it will throw accurate result of moisture. Soil moisture sensor positioning is one of the most vital factors to get accurate result.

5.2 Communication Technologies

All sensors used in agriculture area are going to deploy in remote area of field to collect the information about whole field and for that wireless in most suitable way for deployment. In wireless there are three limitations other than cost are selection of node or sensors communication technique, Range of Communication, battery life and data Integrity/Security.

There are four standards available which can be used in agriculture based areas:

- ZigBee
- Bluetooth
- Wibree
- Wi-Fi.

All above are for ISM band which have frequencies of 2.4 GHz. Here focus must be on vital specifications those are transmission range, power consumption, cost and data security. ZigBee is used in most of applications and with 250 Kbps data rate is acceptable for frequent data transmission. In agriculture most of data is not in real time or in stream so it can be used as communication technology. It has low power consumption and with high transmission range with 128 bit security key processing power. Due to 128 bit encryption key and its low power consumption it gives 250 kbps data rate still there is no better technology than ZigBee.

5.3 *System Requirement*

1. **User Characteristics**

This application is useful for farmer or crop analytics that make important analytical result throughout for one or many crop season. They can estimate require water quantity during growth of crop and take decision that which crop farmer has to sow seed according to their soil type (Table 2).

2. **Hardware and Software Requirements**

- **Hardware requirement:**

Node MCU:

NodeMCU is IoT hardware platform. It utilizes the Lua scripting dialect. It depends on the eLua project, and based on the ESP8266 SDK 1.4. It utilizes many open source ventures, for example, lua-cjson, and spiffs. It incorporates firmware which keeps running on the ESP8266 Wi-Fi SoC, and equipment which depends on the ESP-12 module. The ESP8266 is a Wi-Fi SoC coordinated with a Tensilica Xtensa LX106 center, generally utilized as a part of IoT applications [5].

Table 2 Hardware requirement

No.	Hardware component
1	Node MCU
2	Jumper wire
3	Soil moisture sensor
4	Bread board

Jumper wires:

A jumper wire is just a connection wire which is used to connect two on more electrical components.

There are three types:

1. Male–Female
2. Male–Male
3. Female–Female.

- **Software requirement:**

In this project Arduino IDE tool is used which is installed in PC/Laptop. With the help of this tool code can be prepared for this hardware and dump into the processor for functionality.

Arduino Integrated Development Environment Tool:

Arduino, open-source Arduino Software (IDE) [7] which gives easy GUI in which developer can customize the code, then upload the code to run on board. It is compatible with Windows, Mac OS X as well as Linux. Basically this tool is written in Java and using many other open source software.

5.4 System Design

1. **System Hardware Design**

In this project, there are many hardware modules. The steps are given below for how to design circuit.

- First, prepare the circuit for this Wi-Fi module on the breadboard.

  Node MCU Configuration Description:
  3V3 pin on the Node MCU connected to VCC on HL-01
  GND pin on the NodeMCU connected to GND on HL-01
  A0 pin on the NodeMCU connected to A0 on HL-01.

- Then, after assemble the circuit as shown below

  Cable Configuration Description:
  + pin to one probe of soil moisture sensor HL-69
  − pin to another probe of soil moisture sensor HL-69.

2. **System Software Design**

When Hardware design is complete, software design comes in focus. Development tool Arduino IDE must be installed in PC/Laptop then after coding can be done for circuit. After that connection of Arduino on a Laptop/PC using a serial cable is necessary.

- Once installed, run the program on Arduino IDE.
- Click the menu "Tools—Board—NodeMCU 1.0 (ESP-12E Module)"
- Click the menu "Tools—port—(select Port Arduino detected on your computer except COM 1)"
- Then enter Sketch below, and at last click upload.

3. **Input/Output and Interface Design**

- **Input Design**

In the input design, Soil Moisture Sensor is used; with the help of this IC user get current moisture level of soil (Fig. 4).

6 Implementation Details

1. **Implementation Environment**

The project starts with use of 100 cm^2 box which contains clay with water spread on n continues it. The result is captured from 8:00 a.m.–11:00 p.m. with 1 h gap between result where place was Nadiad, Gujarat with temperature 41 °C. Arduino IDE tool is used for the implementation. There are mainly three part of system:

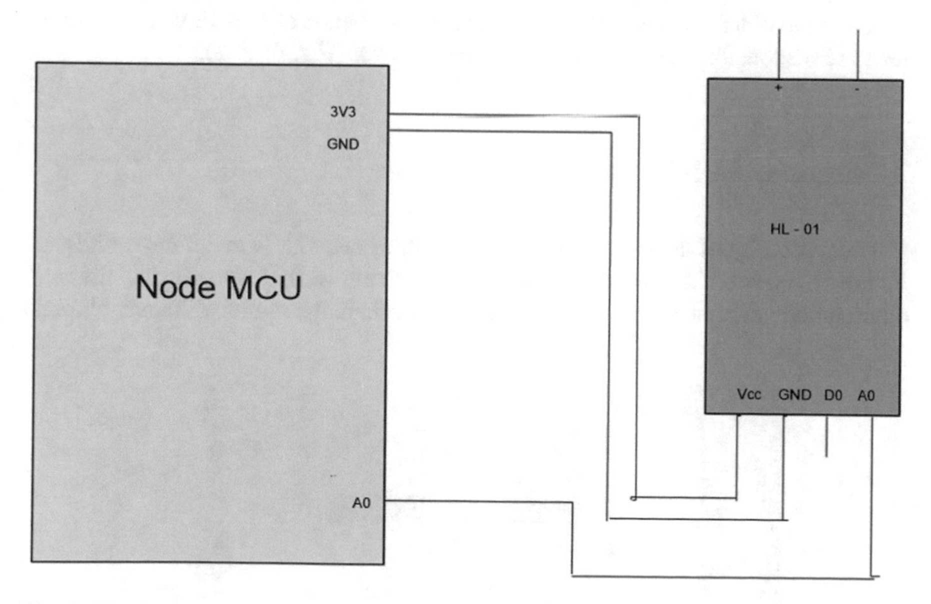

Fig. 4 Circuit diagram

1. Sensor data
2. Cloud data
3. Controlling of pump.

During the time of implementation, Wi-Fi module is used for connectivity between sensor output and Cloud data sensor. Based on the cloud data graph is generated and according to graph pump can be control. There may be situation when snow may fall in field and sometimes temperature may rise so range from −20 to 100 °C temperature can be barred by Node MCU. If network connection is lost and it has captured data of field then 4 MB Memory in built on Node MCU can store data in it and transfer data when network connection is established. Here the box is considered as field as actual field is very large so single sensor is not enough to take result of whole filed. More sensors need to be planted so that they can capture result and finally combining those results, final graph can be plotted.

For soil moisture sensor surrounding medium should be wet soil. It has capacity to pass the electricity as well as it also contains various positive or negative ions. To check dielectric permittivity of soil, wet soil must be able to produce capacitance. In wet soil dielectric permittivity can be measured in terms of amount water the sample soil holds. It is proportional to the voltage induced by soil in surrounding medium. Sensors are placed vertically or horizontally however the water content is calculated as an average over the length of sensor. If sensor pin is 2 cm long then it induces electromagnetic field as shown in figure but at the extreme point of pin have little or no sensitivity. Figure 5 shows influence zone and cross section of the sensor placed in soil which create electromagnetic field. This electromagnetic field cannot be avoided by any technique as there is presence of ions in wet soil. These ions are responsible for sensing Soil Moisture level (Fig. 6, 7, 8).

7 Security in IoT

Physical security of devices used in any IoT [8, 9] network is major research issue. Different types of sensors are used in the agriculture field for sensing different environment parameters like soil moisture level, temperature, different climatic

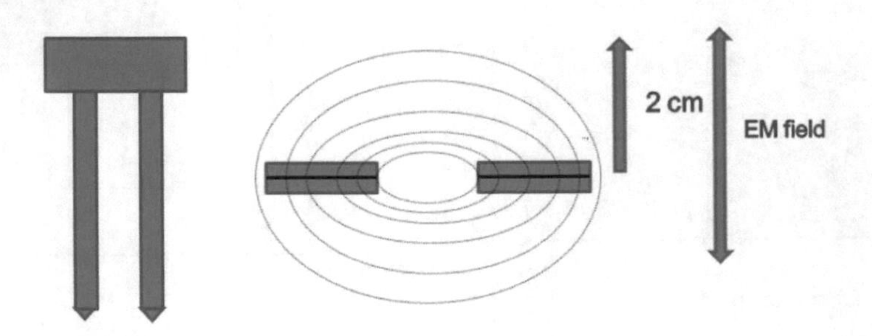

Fig. 5 Electromagnetic field around sensor

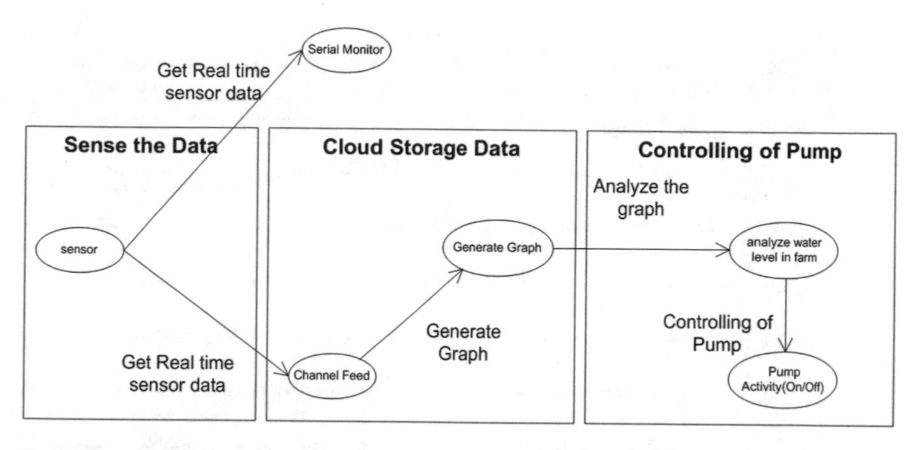

Fig. 6 Proposed architecture

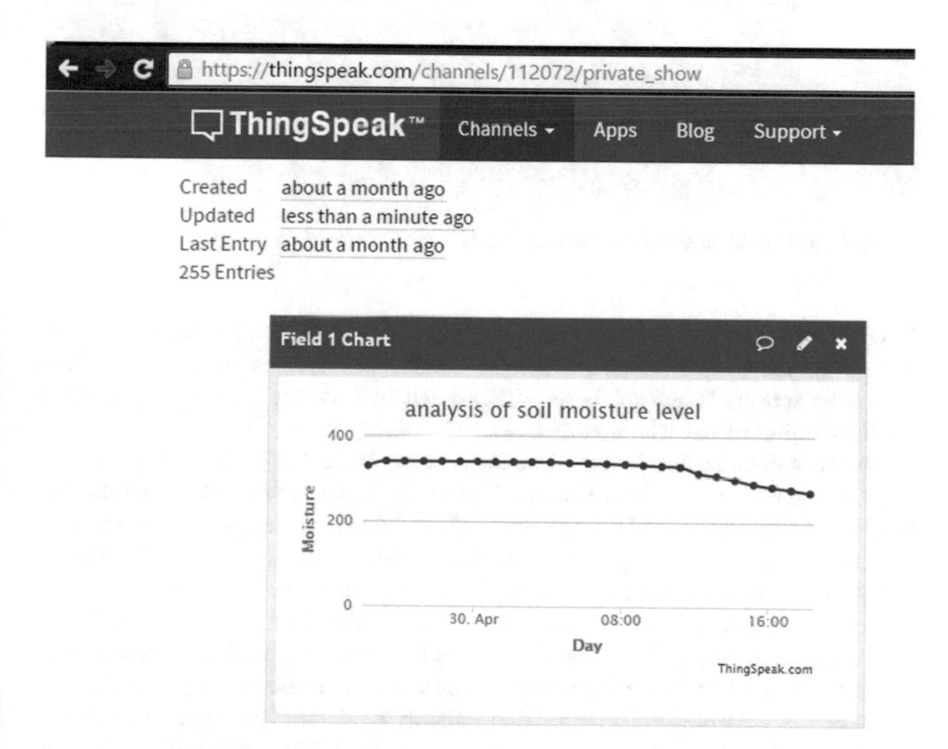

Fig. 7 Analytical graph

conditions etc. these sensors are laid in the field at different distances. There are chances of physical damage to these sensors. If any of the sensors in the network gets damaged then the output may get affected. Also other systems that are further connected with these systems may not get proper signals. Soil moisture sensor may

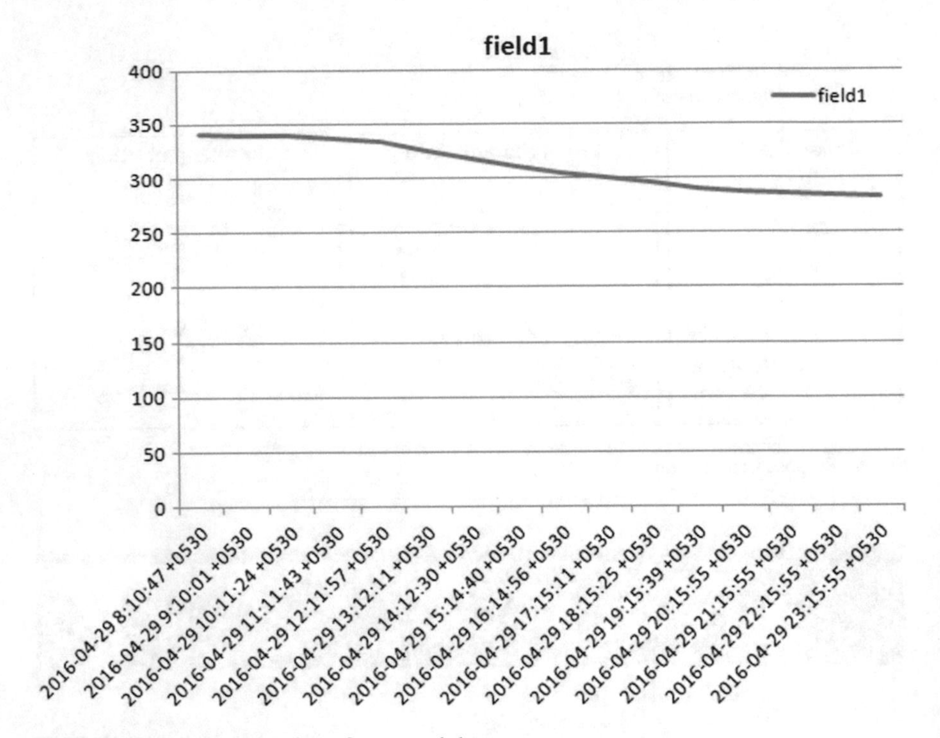

Fig. 8 Graphical representation of measured data

be connected with the irrigation systems, temperature sensor may be connected with the green house system, and different climatic sensors are connected to their respective sensor. If any of these breaks down in mid-way then it may result in malfunctioning of the whole system.

Device authentication is also a major concern in field of IoT. End devices are small enough but need authentication from the main system to ensure that the data received is from the right device. Each small end device authentication and to keep track of each device is a major problem. Agricultural land is a very large in area and to lay sensor and their connected devices in the field and to maintain them regularly requires more attention. When new devices are introduced in the network, they should automatically authenticate themselves to the system. Different protocols are developed according to the requirement. These protocols are employed at different layers of IoT network. IoT in agriculture mainly works in three layers categorized as sensor layer or collection layer where all types of data are sensed by various sensors and those data must reach to some server for further processing or analysis so here transport layer or network layer comes and play vital role to transfer data and application layer where actuators are placed for specific purpose or here as result analytical report may be generated.

Sensor layer: Main task of sensor layer is to collect or sense the real world physical parameters. The sensed data is in digital form. These data are further processed through different means. Humidity, temperature, pressure, gas levels etc. are different parameters that are sensed. Also location of sensors, their information related to features, parameters are also stored at this layer.

Transport layer: The main task of this layer is to collect all the sensed data and use them for further processing. Transport layer is the main central layer of this network. Network layer includes integration and communication of all the intermediate devices that are in the network.

Application layer: This is the main layer that is responsible for the communication between agricultural market analysis and the farm. It analyses and process the information collected from the field and derives necessary results in order to give better techniques to the farmers.

8 Conclusion

This application is used when analysis of used water and irrigation cycle is required. According to analytical result farmer can take decision about required water for particular crop in particular season. Analytical result can be useful when sowing of seed is started. According to this historical data, analysis says that what kind of crop is suitable for this soil.

Acknowledgements We have worked on this project for over 4 months. While working on this project we have come to realize the actual set of skills necessary to work in professional development environment. Making a product which includes both hardware and software is no small task considering the numerous details that have to be taken care of. We have put our best in the making of this project. It is effect of many hands and uncountable hours from many people. Our acknowledgements go to those who helped through their suggestion, comments, guidance, and feedback. We express a deep sense of gratitude to the Head of the Computer Department, Prof. Amit Ganatra for providing such environment where we can implement our work. Moreover, we would like to thank our guide Prof. Chintan M. Bhatt and Prof. Ritesh P. Patel who has helped us throughout our project development.

We also convey our thanks to Prof. Jignesh Patoliya and Prof. Hitesh Patel (EC Department) as they gave us knowledge of different embedded devices and their functionality and utility in Internet of Things.

Last but not the least, we would also like to thank our friends and classmates, who have co-Operated during the preparation of our report and without them this project has not been possible. Their ideas helped us a lot to improve our project report.

References

1. Bhayani, M., Patel, M., & Bhatt, C. (2016). Internet of Things (IoT): In a way of smart world. In *Proceedings of the International Congress on Information and Communication Technology* (pp. 343–350).
2. Shah, T., & Bhatt, C. M. (2014). The internet of things: technologies, communications and computing. *CSI Communications*, pp. 7–9.

3. Precision_agriculture. http://www.research.ibm.com. [Online]. IBM Research. Cited April 29, 2016. http://www.research.ibm.com/articles/precision_agriculture.shtml
4. Plug-sense. http://www.libelium.com. [Online]. Libelium Comunicaciones Distribuidas S.L. Cited April 30, 2016. http://www.libelium.com/products/plug-sense/
5. Wikipedia. NodeMCU. https://en.wikipedia.org. [Online]. Cited May 1, 2016. https://en.wikipedia.org/wiki/NodeMCU
6. Soil_moisture. http://www.vernier.com. [Online]. Vernier Software & Technology. (2016). Cited May 1, 2016. http://www.vernier.com/experiments/awv/11/soil_moisture/
7. Arduino. https://www.hackster.io. [Online]. Hackster Inc. (2016). Cited April 29, 2016. https://www.hackster.io/arduino
8. Bhatt, Y., & Bhatt, C. (2017). Internet of things in healthcare. In *Internet of things and big data technologies for next generation healthcare* (pp. 13–33). Berlin: Springer.
9. Dey, N., Ashour A. S., & Bhatt, C. (2017). Internet of things driven connected healthcare. In *Internet of things and big data technologies for next generation healthcare* (pp. 3–12). Berlin: Springer.
10. Breakfast-with-ecs-the-internet-of-things-iot-part-2-disrupt-or-be-disrupted. http://emergingtechblog.emc.com. [Online]. EMC Corporation. (2015). Cited April 29, 2016. http://emergingtechblog.emc.com/breakfast-with-ecs-the-internet-of-things-iot-part-2-disrupt-or-be-disrupted/
11. Martin, E. C. (2001, January). *Methods of determining when to irrigate*. [Online]. Available http://cals.arizona.edu/pubs/water/az1220/. Accessed August 17, 2016.
12. Norris, J. (2015, October 9). *Precision-agriculture*. Retrieved March 21, 2016 from http://www.nesta.org.uk/; http://www.nesta.org.uk/blog/precision-agriculture-separating-wheat-chaff
13. https://ccafs.cgiar.org/climate-smart-agriculture
14. Agriculture Board Technical Guide. http://www.libelium.com/downloads/documentation/agriculture_sensor_board_2.0.pdf
15. http://www.iosrjournals.org

Greening the Future: Green Internet of Things (G-IoT) as a Key Technological Enabler of Sustainable Development

M. Maksimovic

Abstract New technologies and the revolution of Internet of Things (IoT) fuel innovation in every area of science and human life, providing anytime and anywhere access to information in novel ways and contexts and brings people, processes, data and things as well as places, organizations and facilities together in unprecedented ways. Despite the numerous benefits IoT offers, manufacturing, distribution, and utilization of IoT products and systems are the resource and energy intensive and accompanied by escalating volumes of solid and toxic waste. In order to minimize the potentially negative influence of technological development on human and environment, it is necessary to successfully deal with challenges such as increased energy usage, waste and greenhouse gas emissions, and the consumption of natural and non-renewable raw materials. This is the reason for moving towards a greener future, where technology, IoT and the economy will be substituted with green technology, green IoT and the green economy, respectively, what implies a whole world of potentially remarkable improvements of human well-being and hence contributes to the sustainable smart world. This chapter presents an analysis of the significance of greening technologies' processes in sustainable development, exploring the principles and roles of G-IoT in the progress of the society through the examination of its potential to improve the quality of life, environment, economic growth and green global modernization. It has been shown that the G-IoT holds the potential to transform and bring numerous benefits (among environment protection, customer satisfaction, and increased profit are the most significant) in diverse sectors using the latest technology approaches and solutions alongside eliminated or minimized the negative impact on the human health and the environment.

Keywords Green Internet of Things (G-IoT) · Green economy · Health · Environment · Sustainable development

M. Maksimovic (✉)
Faculty of Electrical Engineering, University of East Sarajevo,
Vuka Karadzica 30, 71123 East Sarajevo, Bosnia and Herzegovina
e-mail: mirjana@etf.unssa.rs.ba

N. Dey et al. (eds.), *Internet of Things and Big Data Analytics Toward Next-Generation Intelligence*, Studies in Big Data 30,
DOI 10.1007/978-3-319-60435-0_12

1 Introduction

The rapid development and utilization of Information and Communication Technologies (ICTs) are causing a direct and dramatic impact on all aspects of life, transforming an "industrial" to a global "information" society. New ICT solutions, the Internet of Things (IoT) particularly, are changing the way in which people organize their lives, interact with each other and take part in the various domains of society. The IoT as a worldwide network of intercommunicating physical objects/ "things", through interconnecting people and things anytime, with anything and anyone, anyway and anywhere, dramatically reshapes and transforms a life of individuals, businesses, and society in general. Hence, thanks to the Internet, mobile technology and the IoT, people, places, organizations, and facilities are connected nowadays in unprecedented ways. Despite tremendous benefits technology development offers and the "smartness" brought by IoT to different aspects of the society, their rapid growth influences more waste, greenhouse gas (GHG) emissions and/or the natural and non-renewable raw materials consumption. The resource-intensive manufacturing and distribution of ICT products and systems, increased energy demands during their utilization and rising volumes of diverse waste, hold the potential to influence negative to human health and the environment [1], and as such represent major challenges towards sustainable development and sustainable place for living.

Society is proceeding towards a greener future to minimize the negative impacts of technology applications on human and environment, by developing methods and techniques without harming effects on the environment and human health, conserving natural resources and generating methods of sustainable development—to account for people, planet, and profit. Technology utilization without damaging, over-exploiting or depleting natural resources, reduction of energy consumption, the producement of completely reused or reclaimed products, decreasing the amount of waste and emissions during production and usage, and alternatives to technologies which do not cause a negative impact on health and the natural environment, are the leading goals of green, environmental or clean technology (Fig. 1). In other words, the green technology involves environmental sustainable designing, manufacturing, using and disposing, with no or significantly decreased impact to the environment [2].

The policy makers of the ICT sector and ICT sector itself have started to realize the opportunities based on an ICT role in supporting the green development and the green economy growth. A green economy can be defined as one of the effects of improved quality of life and social equity accompanied with preserved and enhanced environmental quality [3]. Hence, green ICTs will make a future greener, in which human will be more aware of technology impact on the environment and human health. Still, there are certain obstacles to the introduction, developing and implementing of green technologies such as: technical obstructions (e.g., deficiencies of a fast-broadband infrastructure and e-services' deployment), commercial

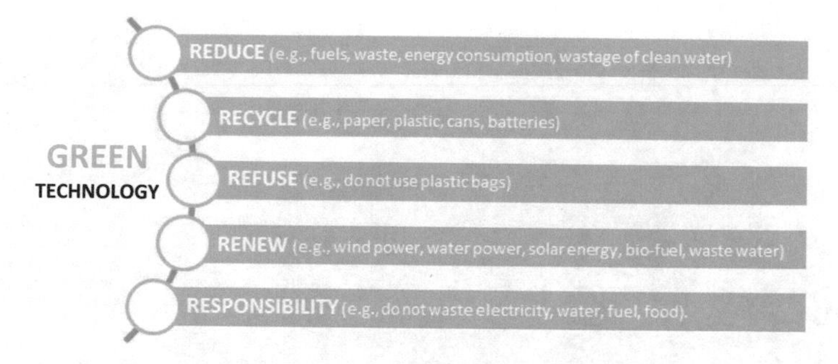

Fig. 1 The goals of green technology

barriers (e.g., deficiency of effective competition) and regulatory obstacles (standards and regulations development and implementation lag behind technological development) [4].

Considering the environmental impact from the beginning and involving it in IoT development and appliance, lead to the evolution of green IoT (G-IoT) opening a whole universe of potentially remarkable improvements of human well-being and developing the world's economic system. Even G-IoT is still early in the maturity, it is more energy-efficient, reduces carbon emissions and pollutions, and poses a great potential to strengthen economic and environmental sustainability. For that reason, companies, and the general public are engaged in IoT innovations, to develop green and sustainable products and services and explore opportunities for G-IoT business worldwide [5]. To get as much as possible insights into the G-IoT principles and its role in sustainable development, this chapter examines the ideas, significance and current impacts of G-IoT innovations in a way towards the sustainable smart world. Relying on the ideas, principles and potential benefits of G-IoT, a comprehensive analysis of the G-IoT impacts to key sectors in the green economy (agriculture, forestry, fisheries, water, energy, industry, transport, waste, buildings and tourism) has been given. With the G-IoT vision, developing products and services which have minimum or no impact to human health and do not harm the environment, minimizing energy consumption, pollution, and emissions and preserving natural resources, the world has the potential to become more environmentally aware and sustainable place for living. Therefore, smart systems and G-IoT can be considered as a key technological enabler of sustainable growth of society.

2 Greening Technology and IoT: Maximizing Benefits and Minimizing Harm

The numerous IoT applications (smart home and assisted living, healthcare and wellness, public safety, factory automation/smart manufacturing, sustainable energy, mobility, food production and distribution, environmental monitoring,

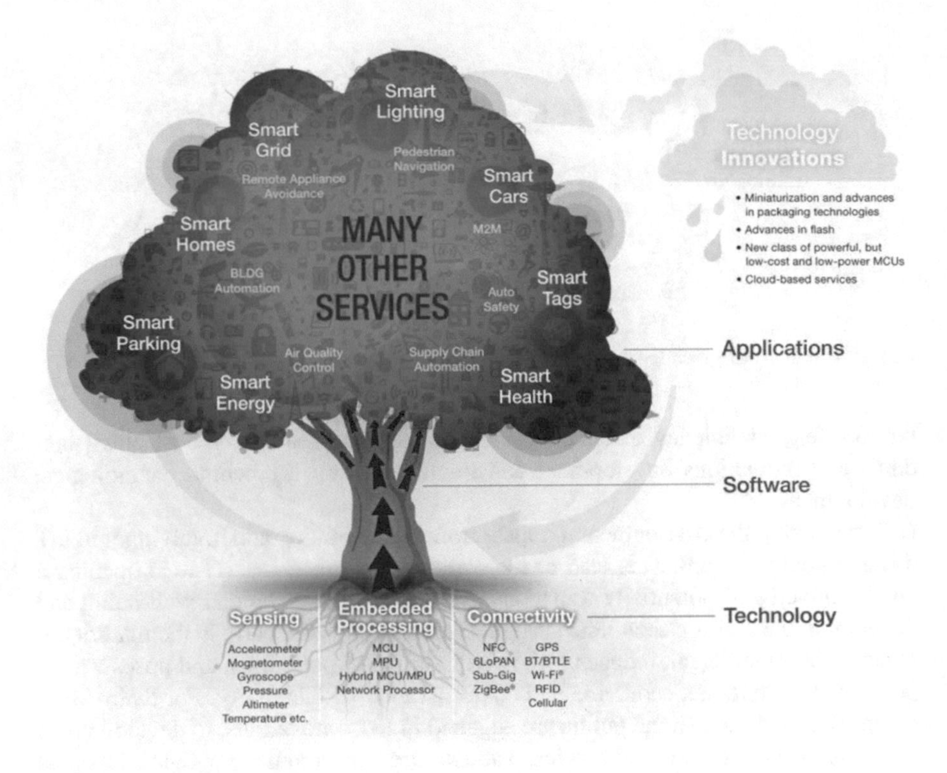

Fig. 2 Internet of Things (IoT) [6]

buildings, transportation, wearables, smart cities, etc.) show why IoT is one of the current and future strategic trends (Fig. 2).

It is anticipated that the creation of smart things/devices and environments for numerous applications, such as health, energy, food, climate etc., will be achieved in years to come. It is expected that trillions in value will be achieved through reduced costs in numerous applications in healthcare, industry, buildings and many other diverse areas [7]. All of this will be the consequence of greater investment efficiency in aspects of operational efficiency, customer service, organizational cooperation, and strategic decision-making processes [8].

On the other side, the fast growth of ICT and IoT in business and daily life will result in a significant growth of the total number of ICT and IoT devices and the total electricity consumption. Various estimates state that 19–40 billion smart things/devices will be in use by 2019 [7] what will consequently lead to substantial economic and social benefits. At the same time this will cause more electronic waste (e-waste) (if just 10% of IoT devices are trashed per year beginning in 2021, this could mean 2+ billion new e-waste items annually [9]) and significantly increase the amount of energy applied to perform different tasks. In other words, ICTs affect the environment from two different perspectives.

Along one side, ICTs generate environmental problems at each point of their full life cycle from initial production, through distribution, to consumption and disposition. Throughout the manufacturing process of ICT equipment, a significant amount of natural and raw materials, chemicals and electricity, is being used. Alongside, dangerous waste related to the use of hardware and software, significantly influence the environment. The total use of electrical energy by ICT sector (personal computers, peripherals, telecommunication networks and devices, and data centers) is a permanently increasing trend [10]. There are predictions that electricity consumption compared to the level in 2010, will be doubled by 2022 and tripled by 2030 – 1700 TW. Increased utilization of electrical energy cause the growing GHG emissions. ICT equipment (computers, monitors, servers, data communication equipment, cooling systems) is responsible for approximately 2% of global CO_2 emissions [11]. ICT's share of global emissions is projected to double to 4% by 2020. It is anticipated that global Internet use, which relies on ICTs, will rise from 30 to 40% per year, what may lead to 60% of global energy resources utilization by ICTs. Clearly, it will be impossible to deal with such burden [12]. Despite these worrying facts, the environmental impact of ICTs and IoT didn't receive the same degree of attention as the application, benefits, and potential of rapid growth and adoption of novel ICTs and IoT solutions.

In order to have a sustainable place for living, there is a need to put a lot of efforts to reduce toxic pollutions by reducing carbon production and the energy consumption. This is the reason for moving towards green technologies, especially to G-IoT. Therefore, ICTs and IoT, from the other side, can be considered as an instrument in addressing the environmental problems. Modeling, developing, utilizing and disposing of computer, servers and associated subsystems, in an economically responsible way, applying dematerialization, with improved energy proficiency and decreased waste, insignificant or no impact on the environment and human health, improving recycling and reuse of ICTs, is recognized as a greening ICT process, which has potential to maximize benefits and minimize harm at the same time.

The greening ICT process includes greening communication technology, greening computing technology, using smart grid and applications [13, 14].

Having in mind that the most energy is spent during communication and that IoT devices are energy-limited, the deployment of green communication and system model in IoT has been a core challenging issue [15]. Hence, the primary focus on green communication technology is in decreasing energy consumption and CO_2 transmission in communication and networking device. Creation of online collaboration environments through teleconferencing, virtual meetings and participating from home and other remote locations, consequently leads to cost-savings, ease of management and decreased energy consumption. Evolving communications architectures, green wireless communication, energy efficient routing, relay selection strategies for green communication, energy efficient packet forwarding and networking games are the main research focuses on greening communication technology.

Green computing refers to the environmentally-aware utilization of ICT equipment (computers, servers, and peripherals) and related resources. Hence, research in green computing technology involves:

- Storage consolidations and virtualization of server resources (reduce the total number of devices running in the server room and decreasing the energy needed for their operation, improve associated cost-efficiency and reduce GHG emissions, reduce time required for maintenance and management, simplify storage space);
- Desktop virtualization and replacing personal computers with thin clients (reduce energy consumption, decrease the cost of maintenance and management);
- Printer consolidation (decrease consumables (paper, toner, ink, and energy), sharing printers, maximize printing substitution);
- Measuring energy usage of ICT equipment (measure server room consumption, manage energy costs); and
- Responsible recycling, reuse, and disposal of ICT equipment.

Smart grids are energy networks, built on advanced infrastructure, computer-based remote control and automation, that manage energy flow, supply and demand in a sustainable, reliable and economic manner. Smart grids, nanosolar, orbiting solar arrays, concentrated solar power, enhanced geothermal systems, carbon capture and storage and nuclear fusion power represent the future of green ICT.

Alongside decreased consumable use and electricity usage, decreased expenses or investments, enhanced features and functionality for business together with meeting customer's needs and realizing credits or rebates from public local utilities represent leading elements that push the adaption of green ICTs [2].

As can be seen, green ICT is a wide concept which encompasses many technologies like e-commerce, virtualization, IoT, telecommuting, smart grids, supercomputers, and Cloud computing [16]. IoT vision is founded on devices embedded with electronics, software, sensors, and network connectivity, which have power to collect information "sensed" from the surroundings and exchange them among each other, as well as with the environment, and react autonomously to the "real/physical world" with or without direct human intervention [17].

2.1 The G-IoT: Concept, Requirements, and Challenges

The G-IoT concept is established on the same computing devices, communications protocols, and networking architectures as IoT, but with more ecological, and energy efficient production paradigms for making products and services (significantly reduced energy usage, carbon emissions, and toxic pollutions). Hence, the G-IoT is enabled by the hot green ICTs such as: Radio-Frequency IDentification (RFID), Wireless Sensor Networks (WSN), cellular networks, Machine-to-Machine communications (M2M), energy harvesting devices and communications, cognitive radio, Edge/Fog/Cloud computing, and Big data analysis. G-IoT, same as IoT, regardless of use cases, consists of sensing nodes, local embedded processing nodes, connectivity nodes, remote network/Cloud and/or Fog/Edge-based embedded processing nodes, software to automate tasks and enable new classes of services, and full security across the signal path (Fig. 3) [6]. Constantly growing number of IoT/G-IoT devices implies large quantities of data generated. Volume, variety, velocity, value, and veracity represent the foundational characteristics of data generated in IoT/G-IoT vision. Extraction of useful information from such data requires utilization of diverse big data techniques. Artificial intelligence, machine learning, intelligent data mining, computer vision, big data analytics and other techniques are commonly carried out in order to reveal hidden patterns, anomaly detection, perform predictive modeling and make actionable decisions. Furthermore, the success of numerous IoT and G-IoT applications depends mainly on security and privacy issues. Hence, the security must be built into each element and the overall system [18]. The mandatory requirements that any IoT/G-IoT solution must meet in order to fulfill certain aspects of security and safety include confidentiality, data integrity, accountability, availability of services, and admission control [19]. Therefore, ensured privacy, security, and consumer trust are the keys to realizing the full potential of the IoT/G-IoT applications.

Despite tremendous efforts that have been made in advancing IoT technologies, G-IoT still faces significant challenges from many aspects. It requires a major technological breakthrough to efficiently connect millions of devices to the Internet with limited radio spectrum, to intelligently coordinate physical devices toward a complicated task, to realize the more realistic energy use by different parts of G-IoT systems, to effectively integrate big data analysis and Edge/Fog/Cloud computing

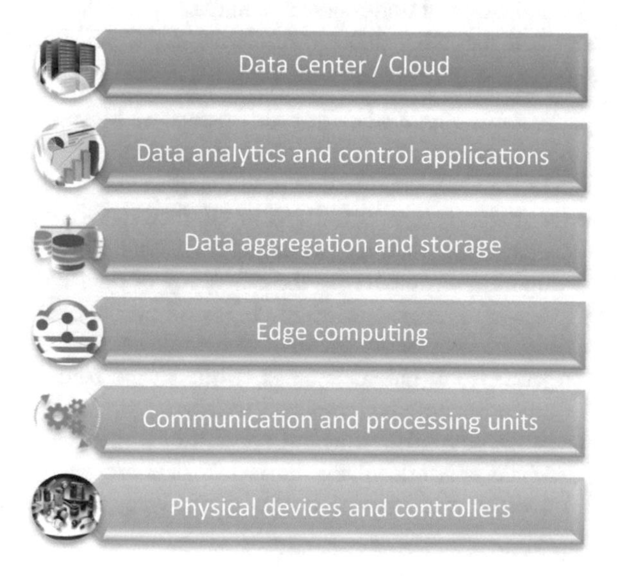

Fig. 3 G-IoT layer architecture

for smart operations, to flexibly implement and validate technological innovations in existing system, therefore satisfying service objectives in the sense of Quality of Service and Quality of experience. Better understandings of service requirements for different G-IoT applications and more realistic energy consumption models for G-IoT system components are the challenges that must be considered. Similarly, as for the green ICTs, the major challenges for the realization of G-IoT vision can be summarized in:

- Adoption and implementation of hardware and software that reduce energy consumption, development, convergence and interoperability of technologies for identification and authentication, security and privacy;
- Application of algorithmic and distributed computing techniques that will enable the creation of G-IoT system level architecture, through the extension of computing, networking and storage from Cloud to the Edge of the G-IoT network;
- Continual improvements of communication technologies, networks, and services to support dynamically changing environments.

To satisfy the G-IoT goal for energy efficiency, sensor nodes should work only when is necessary, use radio optimization techniques and energy-efficient routing techniques, as well as use miniature high-capacity energy storage technologies (wireless charging or harvesting energy from the environment). Energy can also be saved through Edge/Fog/Cloud computing, namely power-saving virtual machine techniques and diverse mechanism for energy-efficient resource allocation. Alternative computing and virtualization are one of the most effective instruments for more cost-effective, greener computing. By sharing storage capacity with other smaller devices, some purely mobile, are everything that is needed to deal with voluminous data sets [20]. Another goal, decreasing waste and contamination, and the utilization of non-renewable materials, implies the significance of materials' and devices' recycling and reuse. Hence, to satisfy general principles of G-IoT, it is required to perform the following tasks [21–23]:

- Eco-friendly design and the usage of bio-products in manufacturing G-IoT components (the future lies in production and utilization of biodegradable devices and nano-power processing units);
- Reduce energy consumption of facilities (e.g., through power-down G-IoT equipment when not in use and switching off by carrying out of sleep scheduling algorithms; improving the efficiency of data center cooling and power supply; designing energy efficient chips and disk drives, etc.);
- Utilize renewable green power sources such as solar, wind, water, oxygen, geothermal, biogas sources, while in the future can be expected the usage of biodegradable batteries and wireless power.
- Move powerful, decentralized and intelligent processing and storage to the edge of the network where the G-IoT data are generated, what enable successfully overcome of latency, bandwidth, privacy and cost challenges.

- Transmit information only when is needed/required (e.g., predictive data delivery or Edge/Fog/Cloud architecture that reduces greatly the quantity of data being transmitted by sending only the critical information);
- Minimize the length of the data path (e.g., implementation of energy-efficient routing techniques or network working mechanism);
- Minimize the wireless data path length (e.g., implementation of energy-efficient architectural design, cooperative relaying);
- Trade off processing for communications (e.g., data fusion, compressive sensing);
- Apply advanced communication techniques (e.g., MIMO (multiple-input-multiple-output) or cognitive radio utilization);
- Implement security mechanism and data privacy into each G-IoT system component and the overall G-IoT system.

In short-term, finding ways to change existing systems into "sleep" mode whenever it is possible and making their work more efficiently can be regarded as an appropriate solution. Most of the today's work focuses on energy savings in the data acquisition process and in multi-directional communications. In the mid-term, there is a need to include energy harvesting at each step in the Cloud's processes. Longer-term goals will require the system of systems redesign in order to make them more intelligent, adaptive, and optimized (no-power, battery-less devices, redesign of the backbone and data centers) [12, 24].

The practice of using green ICTs and G-IoT in a way that maximizes positive environmental benefit and minimizes environmental pollution makes them major key technological enablers of a clean or green economy. The introduction of billions of "smart" devices in coming decades, will contribute to the significant economic transformation process, a ton of employment opportunities, and trillions of USD in economic development and cost-savings and thus leading to the sustainable growth of society [7]. In other words, G-IoT promises to improve existing and create new business models and processes, and scale down costs and risks. Besides the creation of numerous benefits for consumers, G-IoT is expected to provide significant improvements in diverse areas and global economic growth as well. Therefore, the recent significant public and private investments are seen in renewable energy and energy efficient technologies, as well as in other environmental sectors, based on consumption of environmentally friendly goods and services and waste management and recycling [25]. ICTs can bolster up the development of the green economy by decreasing direct impacts on the environment and human health, increasing the enabling effects of ICTs on the development of the green economy and transforming the behavior and values of individuals, economic and social structures, and public utility processes [11, 26]. Thus, the green economy requires the cooperation of all business functions in order to achieve profit and long-term positive contribution to the environment. Even G-IoT attracts significant interest and it is taken for granted that it will be an important issue for several years to come, the advancements and effects of G-IoT will not depend only on financial resources. Its success and substantial impact on sustainable development will depend on take-up by individuals, businesses, and governments, together with legal and institutional frameworks.

3 G-IoT Applications Towards the Achievements of Sustainable Development

Technology can help in the realization of sustainable world vision, only if the "right" decisions on technological innovations are made at an early stage of their development. In this way, technologies become an indispensable tool for sustainable development because they can be used as a tool in ensuring that people have access to clean water and clean and affordable energy, to live in a less toxic environment, to manage natural resources in more effective and efficient way, and have effective environmental governance regimes. In recent years, many citizens have begun to accept the "living green" trend, which includes, among others, clever use of e- and m-services, teleconferencing, the usage of green informatics tools and services through the Internet, especially G-IoT, the extended use of eco-friendly products, green constructions, green roofs, renewable energy use, energy saving at home, awareness of recycling importance, and so on. Hence, greening technologies and moving towards G-IoT, enable the realization of benefits that new technologies offer, while at the same time influence on the environment and energy usage are minimized (Fig. 4).

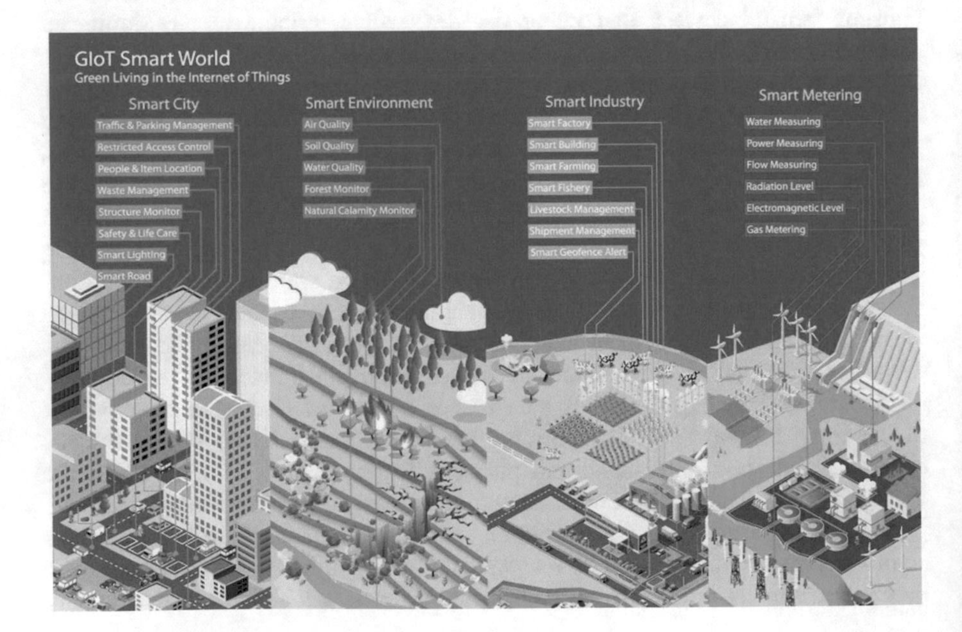

Fig. 4 G-IoT smart world vision [27]

All of this extends to a profit for the environment and for the society [5]. In other words, green ICTs offer novel manners to successfully implement green economy and achieve sustainable development. Green ICTs and G-IoT are seen as crucial factors for a sustainable economic development. The domains where they can show their benefits are numerous: reduction of GHG emissions and waste as well as conserving natural resources and maximizing the use of raw materials [28].

Utilization of the G-IoT concept, by designing products and services for sustainable environment, using energy-efficient computing, power management, server virtualization, green metrics, assessment tools and methodology and renewable energy sources, have potential to significantly transform following key sectors in the green economy [29]: agriculture, forestry, fisheries, water, energy, industry, transport, waste, buildings, and tourism.

3.1 Agriculture

Agriculture and forestry are very important categories that cover a scope of GHG sources (and some sinks) [30]. While modern farming methods increase agricultural output, they often harm the land, biodiversity, and local water sources. Furthermore, agricultural sector produces large quantities of CO_2 and methane when natural land uses are converted for other uses. Nevertheless, agriculture remains fundamental to economic development and environmental sustainability [31]. Realization of the green agriculture vision demands physical capital assets, financial investment, research, and capability building in managing soil fertility and plant and animal health, sustainable and effective water utility, crop and livestock diversification, and adequate farm automation system. Applying the concept of G-IoT can bring significant improvements in a whole agriculture sector and the food supply chain through automation of production, cultivation factors, and inputs, precision farming, logistics, traceability, remote monitoring, decision making, forecasts, etc. Technologies like sensor-based, satellite remote sensing, geographical information systems, advanced data analytics and smart information platforms significantly contribute to more productive, sustainable and precise agriculture practices. Access to high-speed Internet and affordable smart devices are prerequisites for making agriculture smart and sustainable (monitoring of plant/crop, soil, climate and insect-disease-weed, and livestock feed levels, precision and remotely controlled, automatic irrigation and precision fertilization, etc.) (Fig. 5).

Examples of IoT applications in realizing the concept of smart farming can be found in numerous works. Some of the presented solutions are: Rowbots, a small and self-driving robotic platform which performs certain actions regarding corn cultivation, and collects information in order to maximize growth and minimize waste [33]; "personal food computers" which are small indoor farming environments that reproduce climate variables, powered with an open source platform to gather and share data [34]; smart irrigation system for kiwi production improvement [35]; numerous platforms which bring new and powerful smart agriculture IoT

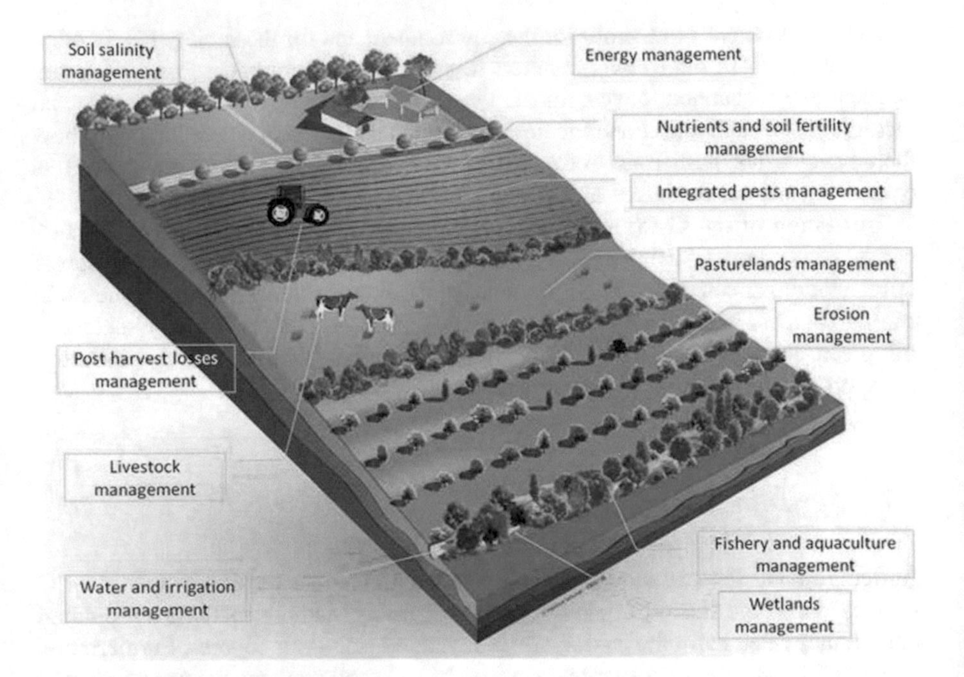

Fig. 5 Sustainable agriculture platform [32]

solutions developed by leading companies which explore and employ IoT in agriculture (Libelium, Link Labs, ThingWorx, Kaa), and many other solutions for monitoring climate and forecasting, remote monitoring crop, soil, livestock feeding as well as tracking livestock, smart logistics and predictive analytics for crops and livestock. Weather G-IoT concept is used in outdoor or indoor farming, livestock monitoring, storage monitoring, vehicle tracking, fish farming or forestry, all G-IoT solutions are sensor based (cheap and easy-to-install sensors), with software applications and data analytics solutions that enable remote monitoring and communications 24/7 in order to gain insight into problem of interest (e.g., soil health and crop production, livestock health, used machinery at farm, storage conditions, level of used energy). Such G-IoT solutions are able to made adequate decisions and perform certain actions, reduce the consumption of raw and non-renewable materials, decrease the negative influence on the environment and human health and finally lead to better output in the sense of improved production and increased profit.

Enhanced productivity and energy and resource efficiency, as the result of implementing novel technology solutions in the agriculture sector, lead to emissions reduction. It is forecasted that by 2030, 2.0 Gt CO_2 equivalent emissions per year could be averted by putting through a smart agriculture concept. In addition, smart agriculture provides substantial economic and social benefits. It is anticipated that by 2030 1 billion MW of energy will be saved, 897 average kg/Ha of land yield will increase, 16% of global hazardous emissions will be reduced along 2 billion USD of additional revenues to companies across the agriculture sector while cost-savings through decreased water use could grow up to 110 billion USD [36].

3.2 Forestry

Forestry industry, together with agriculture, is undergoing fundamental technological modifications and changes. At the same time, forests and agricultural land significantly influence climate changes because of the importance of their carbon stock and also for GHG exchange between the atmosphere and soils and vegetation that can go both ways. Hence, reducing deforestation and increasing reforestation, are essential [31].

The IoT-based forest wireless monitoring system can monitor the growth information and the surroundings of trees continuously and accurately as well as conduct reasonable and precise control over them. In other words, the applied sensors and intelligent systems detect and deliver high-resolution forest-based monitoring data wirelessly or through a cellular network to an online database which will provide real-time updates. Cloud-hosted data are accessible to staff from any device and at anyplace. Data analytics enable better insight into trees growth, the influence of climate conditions, pests, and diseases, and help in making intelligent predictions, early warning and prevention actions (e.g., in the case of forest fire detection the firefighters will receive real-time data about the velocity and direction of fire spread and therefore they will be capable of reacting in appropriate manners). Alongside forest resources and their management, ICT-enabled technologies participate significantly in the other forestry-wood chain sectors such as: harvesting, transportation, processing, production, and recycling (Fig. 6).

Compared to ICT and IoT, green ICT and G-IoT hold the potential to much more reduce carbon emissions, preserve natural resources and via predictions and intelligent decisions make forestry-wood chain sector sustainable. Therefore, the forest-based sector plays a central role in the transition towards a greener economy and a more sustainable society. Subjects that invest in the forest sector will speed the transition to a greener economy.

3.3 Fisheries

Modern aquaculture as intensive, extremely productive, automatically operated, ecological and safe along controlled facilities, require the support of adequate ICT solutions. Smart sensors for water quality, monitoring fish behavior and equipment, automatic control and information management, facility management and fault diagnosis, logistics and traceability of fishery product quality, lead to the smartness and sustainability of aquaculture [37]. Furthermore, keeping in mind that

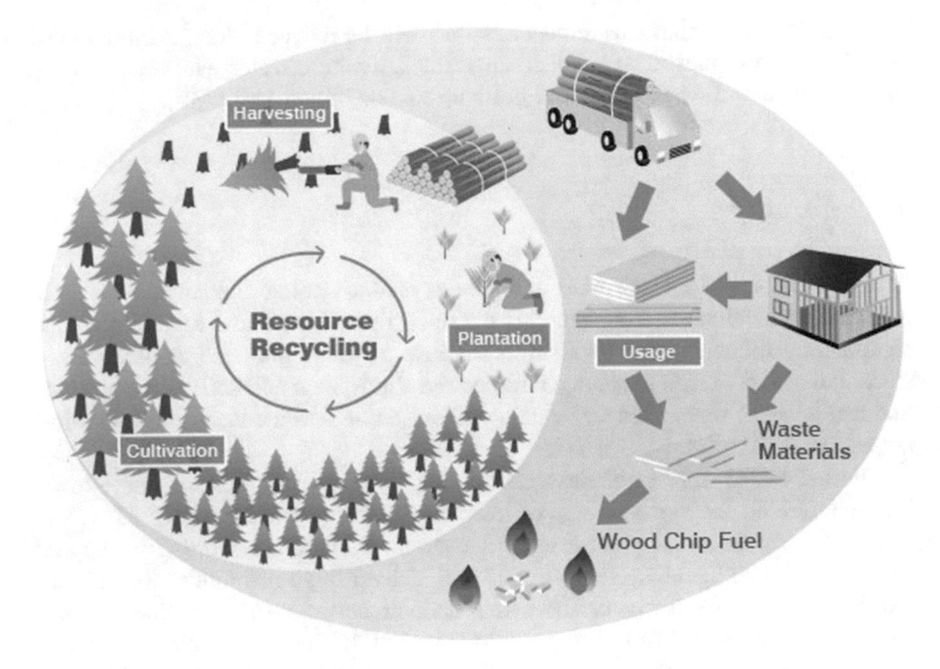

Fig. 6 Sustainable forestry-wood chain

production capacity in fisheries and aquaculture depend directly on the natural environment [38], G-IoT potential to protect and restore natural ecosystems, through surveillance systems, remote monitoring, sensor tele-detections and prevention actions [11], makes it the significant tool to improve this sector. In other words, G-IoT together with the large-scale, distributed, low-power, low-cost WSN for aquaculture and fishery, has the potential to realize the remote dynamic monitoring and controlling of water quality. In [39] is presented an IoT-enabled aquaculture fishery for reducing seafood deficit. The deployment of cellular-loaded buoys that float above the aquaculture farming units enable the data collection regarding water salinity, temperature and the growth rates of the seafood and consequently send data back in order to make adequate decisions and actions. With the help of sensors and IoT, the quantities and species of fish caught can be easily monitored, satisfying in this way the catch share regulations [40]. IoT and fly-fishing drones using an integrated remote camera enable easier fish targeting and catching [41]. Furthermore, G-IoT can assist very successfully in monitoring fish behavior and equipment, facility management, logistics and traceability of fish product, etc. [37]. Regardless of the application, intelligent predictions, early warning and forecasting models, precision feeding decisions and other automatic control methods can be made based on data collected and analyzed (Fig. 7) [37].

It is anticipated that G-IoT with cheaper products, extended service life, lower operating costs, reduced energy use, easy to use and widely spread with no or minimum impact on the environment will significantly improve aquaculture and

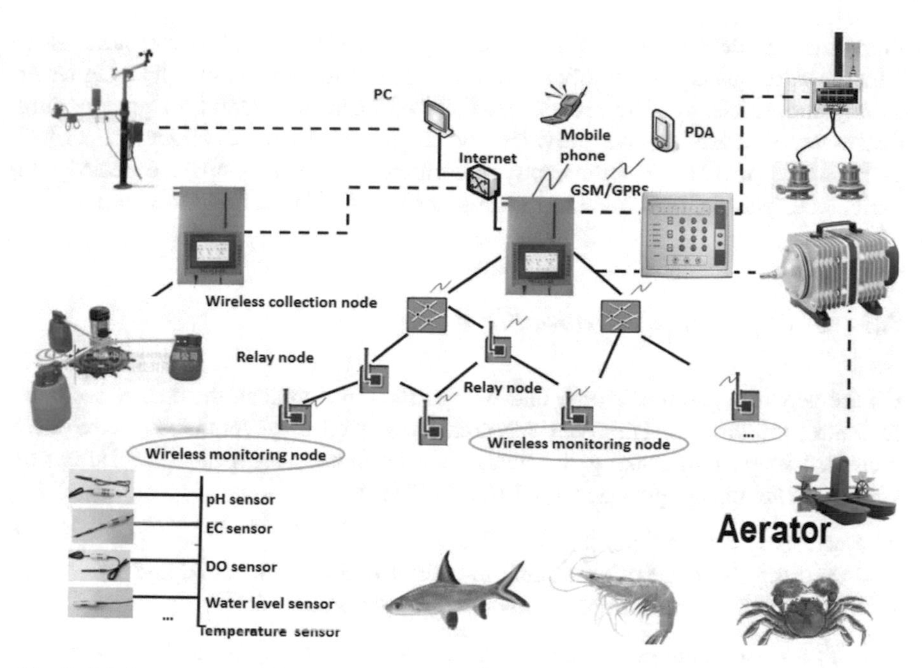

Fig. 7 IoT in aquaculture [37]

fisheries sectors, satisfying the world fish demands. However, an involvement of many innovative approaches in fisheries and aquaculture sectors and making them greener, demands public intervention, through government and private researches and development as well as reorienting public spending to strengthen fisheries management. Greening this sector will lead to numerous benefits.

3.4 Water

Reliable access to water supply and sanitation services are considered as one of the major challenges for sustainable development [42]. Growing world's water resource scarcity implies the need for the much more effective management of this valuable natural resource. ICT-enabled technologies can serve to improve the sustainable water management even water consumed by the ICT sector can be significant [43]. This can be overcome moving towards green ICTs and G-IoT concepts.

Technologies such as satellite imaging, geographic mapping, and sensor web based technologies have a huge potential to be applied in the water management sector. Real-time collected and analyzed data about current conditions enable individuals, businesses and water companies to predict, perform preventing actions and other intelligently made decisions in sustainable water resource management [42]. The G-IoT concept will play a major role in shifting from current practice to more sustainable one. Transitioning to a green economy in water requires sustainable water management policies in order to raise investment in enhancing water supply and efficiency. The proper water management is essential in safeguarding freshwater ecosystems and provides social and economic development. G-IoT vision has potential to significantly contribute to water savings, especially in agriculture, energy, building and industry sectors as well as in ICTs sector.

3.5 Energy Supply/Renewable Energy

On the way to a green economy and sustainable development, the energy sector is of central importance. Therefore, researchers, as well as governments, show their increased interest in making the energy sector more sustainable. The European Union has set challenging goals until 2020 [28]:

- Decreasing carbon emissions by 20%;
- Increasing the share of renewable in energy utilization to 20%; and
- Saving 20% of the European Union's energy consumption.

ICTs are an important enabler in creating a resilient, reliable and secure global energy system. With the help of ICTs, energy efficiency in grids and speeding the energy sector decarbonization can be improved. Smart grids utilization leads to a better balance between energy demand and supply. It is forecasted that smart grids in the future will be composed of micro-grid networks, connected to each other via the Cloud, and be able to monitor, run or disconnect themselves and heal, based on the data collected with smart metering devices (Fig. 8) [36, 44].

The success of smart grids is based on well-defined ICT solutions, and G-IoT role in future of smart grids can be considered as crucial. Alongside manners for energy savings during producing and operating G-IoT components, their usage contributes significantly to energy savings. For instance, replacing products with online services (dematerialization), putting business online by implementing new ways of working (teleworking and videoconferencing) and exploring the viability of using green suppliers and energy from renewable resources [1], make significant improvements in energy efficiency and therefore reduce costs and decrease negative impact on the environment. G-IoT components together with Cloud-based automated energy management system will improve energy efficiency and consequently reduce carbon footprint. With the help of smart devices, sensors and smart meters, and advanced big data analytics, it is possible to predict consumption, reduce outages, and monitor assets.

Next-generation power networks technologies will be highly automated and IoT-based, allowing changeable and decentralized generation, storage and distribution of energy, as well as facilitate distributed and renewable energy sources [36]. ICT-enabled solutions are able to better integrate renewable energy sources into the grid. With the assistance of accurate prediction algorithms, better renewable energy

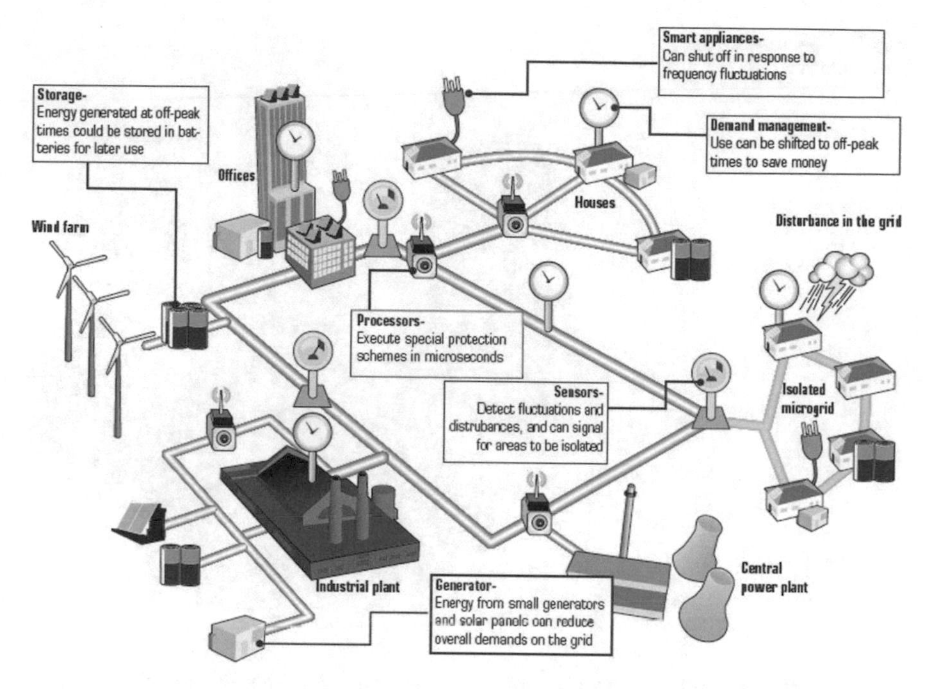

Fig. 8 Smart grid vision [44]

integration to the system and reduced amount of energy wasted can be achieved [45]. More efficient energy usage and better integration of alternative energy sources into the power grid consequently lead to economic growth without damaging the environment. Enhancing incentives for investing in renewable energy is an essential role for government policy, according to the green economy concept. Feed-in-tariffs, direct subsidies, and tax credits represent a scope of required time-bound incentives which make the risk/revenue profile of renewable energy investment more interesting. From a sustainability perspective, energy sector based on smart ICTs can lead to significant environmental benefits, saving up to 6.3 billion MW of energy. Additional benefits are: 700,000 km saved grid, 810 billion USD of additional revenues, universal access to energy, reduced costs and emissions, improved client service, lessen outage time and enhanced reliability [36].

3.6 Industry/Manufacturing

The IoT, as a synthesis of interconnected smart machines, materials and products, and optimal use of gathered, processed and analyzed data, have potential to enhance traditional factories, by increasing productivity, quality, and production flexibility, and at the same time, reduce energy and resources usage and GHG emissions [36].

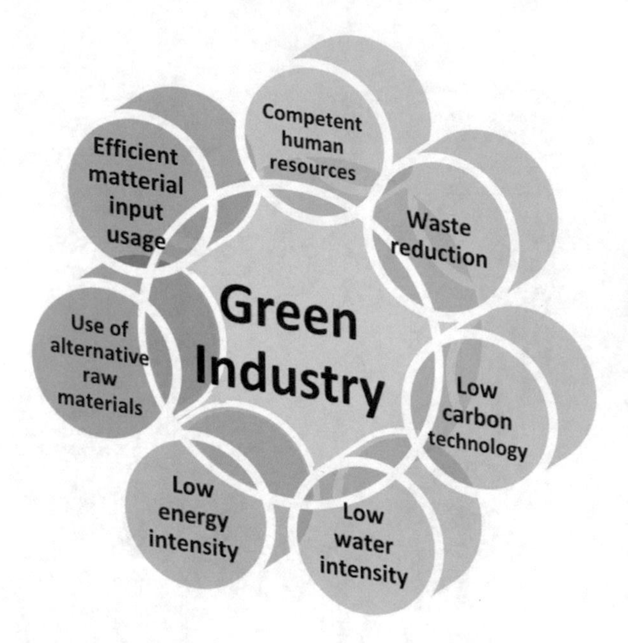

Fig. 9 Green industry characteristics

Knowing that manufacturing encompasses diverse emissions during direct energy transformation in the specified industries [30], greening the industry/manufacturing sector requires redesign, remanufacturing and recycling to increase productivity, prolong the useful life of manufactured goods and reduce emissions (Fig. 9).

Emerging green technologies, such as G-IoT, are capable of increasing productivity and growth, by realizing cleaner production, transforming factories from traditional cost centers into high-performance plants, efficient resource management and reductions in waste and pollution (Fig. 10) [46, 47]. For instance, authors of [48] adopted an energy-efficient IIoT (Industrial Internet of Things) architecture composed of sense entities domain (three-layer architecture: sense layer, gateway layer, and control layer), RESTful service hosted networks, Cloud server, and user applications. Through forbidding the direct communication between two sensor nodes, making gateway nodes as relay nodes, and implementing a sleep scheduling and wake up protocol, the proposed structure may save the energy and extend the overall system lifetime.

It is forecasted that by 2030, the 4th industrial revolution, namely the IIoT, will transform traditional factories into self-organizing, self-maintaining and flexible factories. Innovations like virtual manufacturing, user-centric production, 3D printing, virtual production networks, circular supply chain and smart services represent the future of industry/manufacturing sector [36]. G-IoT can help monitor, manage, optimize and coordinate both production and use of goods and services. However, the regulation made by government and pricing play a key role in making industries more sustainable. Green industry establishment and development require a support from all stakeholders, national and/or international [5].

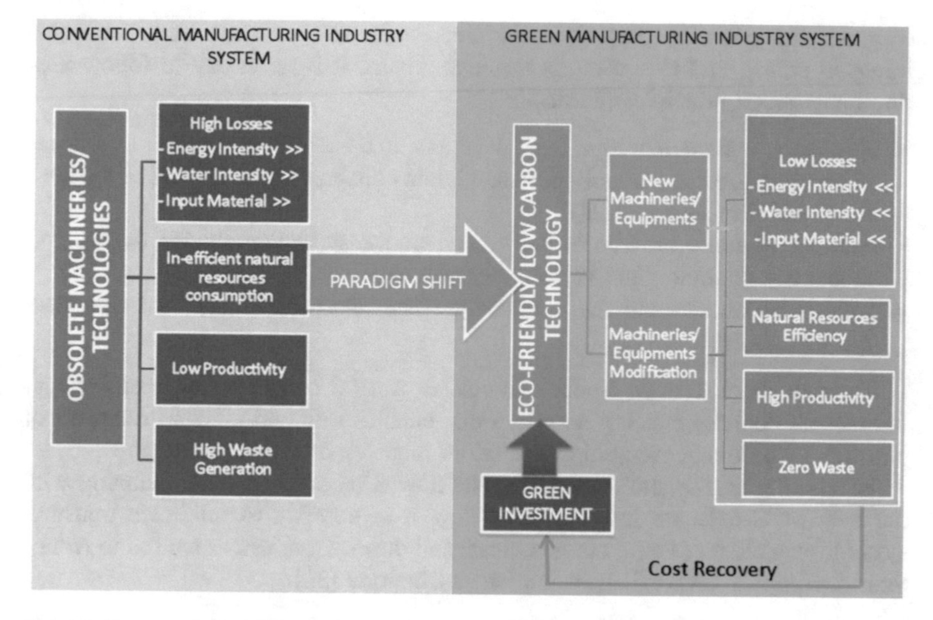

Fig. 10 Reasons for moving towards green industry/manufacturing [47]

Smart and green industry/manufacturing enhances production and quality while reduces costs, energy, and natural resource use along minimized environmental impact. It is predicted that by 2030, smart manufacturing solutions will create 4.2 billion MW energy saved and save 81 billion liters of water, accompanied with reduced 12.08 Gt CO_2 equivalents and significant economic benefits (11.9 billion USD per year in operating cost-savings) [36].

3.7 *Transport*

Around 95% of the world's transportation energy comes from gasoline and diesel what makes the transport sector responsible for about 25% of global CO_2 emissions. Emissions and pollutants from this sector increase faster compared to any other sector. Though road transportation (2 billion vehicles are expected to be on the road by 2030) represents the bulk of GHG emissions (about 75% worldwide), aviation and shipping emissions are growing rapidly. According to predictions, energy use in transport could double by 2050, what implies the necessity for significant reduction of associated emissions and pollution [36, 49].

With the technology development, numerous opportunities to create a smarter and much more sustainable transport sector have appeared. Various approaches for the smart, reliable and robust traffic system have been proposed to realize the vision of green transportation [50]. To make green or more commonly said sustainable transport sector, and to reduce its negative effects, it is necessary to follow next three interlinked principles (Table 1):

- Avoiding or reducing trips (e.g., ICT and IoT-based communication such as video conferencing technologies and remote collaboration tools reduce the need for physical business travel);
- Shifting to more environmentally efficient transportation modes (e.g., intermodal transportation and vehicle sharing); and
- Improving vehicle and fuel technology (e.g., electric vehicles and driverless transportation).

Through the IoT, any product or vehicle can be connected to another (e.g., Internet of Vehicle) creating a system that enables safe and efficient transport of products by reducing congestion, emissions, and resource consumption.

Monitoring traffic jams, optimizing the flow of goods and route planning with the help of G-IoT, are just examples how it is possible to minimize transport systems' negative influence on the climate and the environment as well as to reduce their dependence on fossil fuels and energy imports [36]:

- Traffic control and optimization is realized by connecting cars, roads, lights and control systems using smart sensors, cameras, location-based applications and intelligent infrastructures (Fig. 11). Collecting and processing real-time information on traffic conditions make traffic, driving and parking more efficient (e.g., recommending optimal driving speed and the best route to avoid congestion or to find the nearest available parking place) [51, 52].
- Connected private transportation includes connecting people and vehicles (e.g., via smartphone-enabled car-sharing or car-pool platforms) so that people may travel together. Car sharing significantly contributes to reducing overall fuel consumption and emissions as well as time and money savings. Like traffic control and optimization, which are projected to reduce 6% of global GHG emissions by 2030, connected private transportation also has potential to contribute to carbon footprint reduction (5%).
- Smart logistics—the connection of vehicles, products and load units using ICT-enabled solutions, leads to improved operational efficiency and planning, and reduced amount of waste. Connected devices or Internet of Vehicles can significantly contribute to route and load optimization. Consequently, smart logistics will help to reduce 10% of global GHG emissions by 2030.

Therefore, through innovative G-IoT solutions, transport systems have the potential to become seamless, inclusive, affordable, safe, secure, and robust. It is anticipated that with technology advancements in the transport sector, following benefits can be accomplished by 2030: 750 billion liters of fuel savings, 1 trillion USD of avoided costs and around 42 billion hours saved [36]. Additionally, attractive jobs and strengthen local economies may be the result of reduced emissions in the urban transport sector [53].

Table 1 The impacts of different types of transport policy/technology [49]

	CO_2 reduction potential	CO_2 reduction cost-effectiveness	Mobility/access/equity improvement	Air quality improvement	Traffic congestion reduction	Noise reduction	Safety improvement
Travel reduction/modal shift	Medium–high	Variable and uncertain	Medium–high	Medium–high	Medium–high	Variable	Medium–high
Fuel economy improvement	Medium–high	Medium–high	None	Low	None	None	None
Electric/fuel cell vehicles	Variable	Low–medium	None	High	None	High	None
Biofuels	Variable	Variable	None	Low	None	None	None

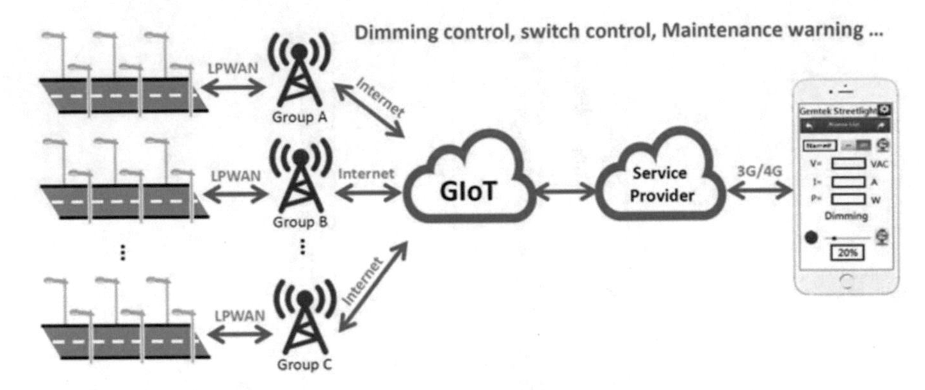

Fig. 11 Street lights' management [27]

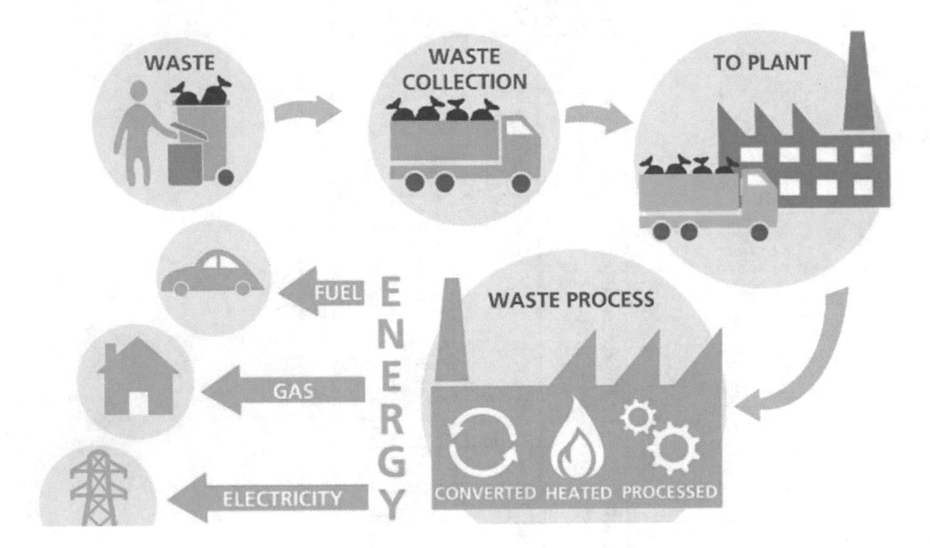

Fig. 12 Waste management and recycling [55]

3.8 Waste Management

E-waste (electrical and electronic equipment waste) represents the fastest growing waste stream (3–5% per year) [54] and the global concern, but at the same time, it is one of the most controllable parts of the waste stream. Greening this sector means the appliance of technologies or practices to reduce or remove emissions and pollutants made in operations and existed in the environment as well as gathering and reusing or recycling waste, turning it into valuable resources (Fig. 12).

Most crucial demands in the waste sector are for decoupling waste from economic development and addressing the fast growing challenges of e-waste and increasing their recycling rate. E-waste recycling consequently leads to the removing health and environmental hazards, resources preservation, reduced energy consumption, and economic development [56]. Alongside this, the numerous benefits obtained by waste management in the 21st century are also the result of implementing advanced ICT solutions in this sector.

Companies around the world are installing IIoT systems in order to more efficiently manage waste (e.g., Enevo, Compology, Bigbelly, Sutera, etc. [57]). Cloud-based solutions collect data from wireless sensors that measure the container filling level and other multiple sensors and information sources (e.g., GPS—Global Positioning System, smart devices, RFID tags) that provide streams on garbage truck location and traffic congestion, and transmit all captured information so that waste management companies can obtain the data they need to analyze, predict and optimize the important services they offer (e.g., optimize garbage truck routes and bin collection times) [58, 59]. Evidently, upgrading IoT solutions to G-IoT will lead to much more improved, sustainable, and cost-efficient waste management and benefits to many parties, directly and indirectly involved in trash disposal.

Some of the currently present and expected economic benefits of recycling e-waste are:

- Energy efficiency and conservation of precious, critical and rare natural materials (e.g., gold, silver, platinum as well as valuable copper, aluminum and iron), including plastics and glass. E-waste recycling can save up to 70% on the energy as the consequence of savings in costs and precious power resources [60]. These energy savings lead to many indirect benefits such as reduced reliance on finite oil reserves, reduced pollution and GHG emissions [56].
- Electronic products that are refurbished and repaired, appear as cost-effective and more convenient compared to completely new products.
- Companies for recycling e-waste creates more job opportunities and strengthen the economy. The potential revenue from the e-waste recycling is around 2 billion USD, anticipated to rise to almost 3.5 billion USD by 2020 [61].

3.9 Buildings

To realize smart and green building vision it is necessary to consider and integrate ICTs, architecture and urban planning. The main components of smart home and intelligent buildings include automation systems, sensors, and smart grids. The inclusion of the IoT technology into the vision of the smart and green building includes integration of all electrical devices and other smart devices (smart meters for electricity, gas and water, home automation gateway, home smart appliances, HVAC (heating, ventilating, and air conditioning), audio and video, lighting, fridge, washing machine, etc.), data analytics implementation, monitoring, controlling and alerting actions in order to provide safety, security, convenience, energy and cost savings (Fig. 13) [44, 62].

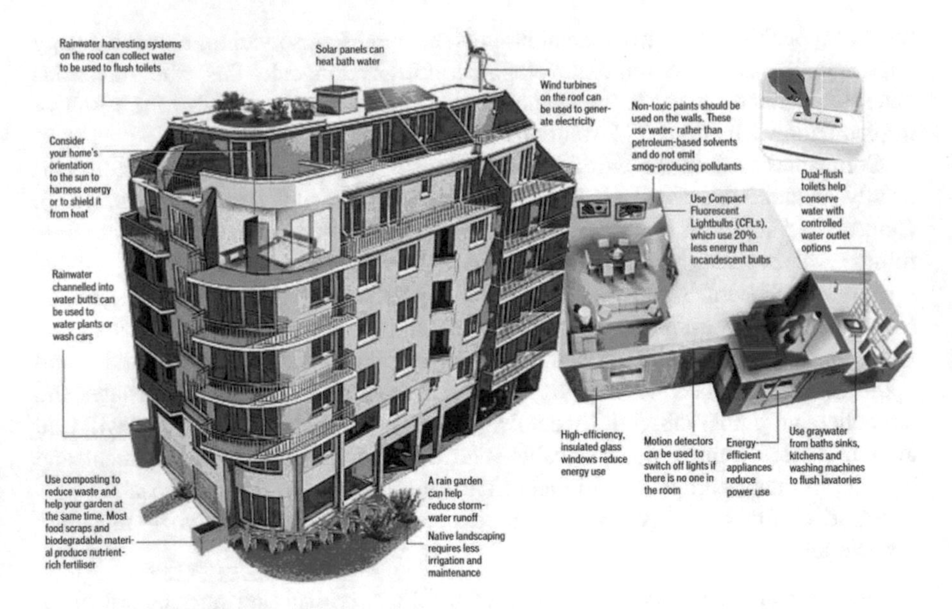

Fig. 13 Making buildings green [63]

In other words, users through their smart devices have access to data collected via smart devices and other smart home solutions. Based on data processed, users are able to remotely manage energy use, cooling or heating, lighting, surveillance, entertainment appliances and much earlier detect any faults or abnormalities. These solutions could be applied from smaller homes to the larger complexes, helping to a more efficiently usage of resources and energy and leading to the realization of green and sustainable concepts [36]. Green building is the practice of improving the complete building life cycle, from design, construction, and operation to maintenance and removal. Usage of renewable energy, heating by the sun, conserving water, utilization of local/natural resources and recycling materials are examples how people can contribute to green building.

Therefore, G-IoT and future technology approach in Internet communication, smart devices and appliances will make buildings smarter and more efficient in term of energy and resource usage, make them more sustainable, and user and environmentally friendly (Fig. 14).

Additional energy savings of around 1/3 could be achieved worldwide in building sector by 2050 with the right government policies [64]. Therefore, a range of policy instruments, such as sustainable building standards, economy stimulated, capacity building efforts, etc. already has proven particularly cost-effective and efficient for greening the building sector. Smart building solutions have potential to offer numerous benefits by 2030: 16% breakdown of GHG emissions, 300 billion liters of water saved, 5 billion MW of energy saved, 360 billion USD of cost-savings as well as notably improved living and working conditions [36]. All of this will consequently lead to a full accomplishment of green or sustainable cities

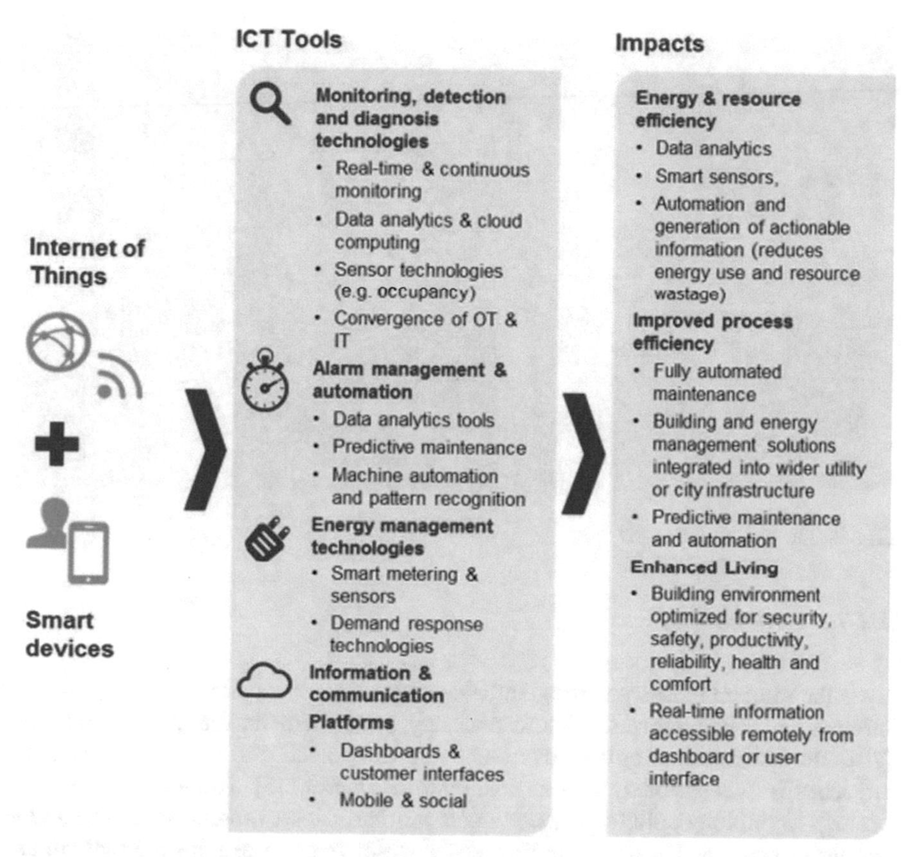

Fig. 14 Smart building technology vision for 2030 [36]

vision. According to Leung [21], 30 mega cities with 55% in developing countries will exist by 2023 while 60% of world population will live in urban cities by 2025. These facts justify the necessity of moving towards a green/sustainable place for living.

Green city is a city that consists of quality of environmental assets, efficient use of resources and climate change risks, alongside maximized economic and social co-benefits [65]. In order to measure sustainability of smart cities, the Green City Index series covers approximately 30 indicators across following categories: CO_2 emissions, energy, buildings, land use, transport, water and sanitation, waste management, air quality, and environmental governance (Fig. 15) [66, 67].

The measurement of quantitative and qualitative indicators together means that the evaluation of current environmental performance as well as the city's intentions to go greener are taken into account. The analysis of the Green City Index and results of comparative analysis of numerous cities can be found in various reports [65–69].

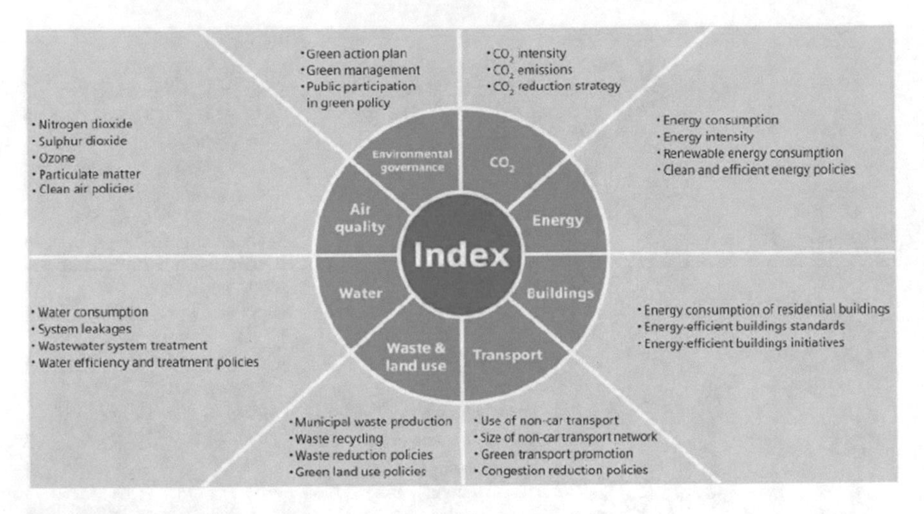

Fig. 15 Green City Index [67]

3.10 Tourism

Even the tourism industry is frequently seen as a slower adopter of technology, the Internet has revolutionized it more than any other factor in the last few decades [70]. The latest technology innovations have the potential to improve operational efficiency in this sector as well as customer satisfaction [5]. Furthermore, the 21st century technology solutions make smart tourism vision omnipresent. The foundations of smart and at the same time sustainable, green or responsible tourism are in the usage of sensors, cameras and smartphones, Cloud services and IoT, Big data analysis, a collaboration of the web and mobile services. The advent and application of mobile technologies, especially the usage of touch screens through mobile tours and apps, have further impacted the tourism and travel industry, contributing to the improvements in tourist experiences. Furthermore, with the help of IoT technologies, various sensors and gadgets, travelers are able to record and share data (e.g., time, speed/pace, distance, location, elevation), to instantly look for a help if they get into any trouble at any point of their journey while their health conditions can be monitored remotely and continuously 24/7 via teleconsultation [71]. The usage of G-IoT can significantly contribute in making tourism smart and sustainable, bringing the huge benefits to its stakeholders as well as the tourists. In order to achieve full potential G-IoT can offer in the smart tourism sector, there are certain implications which need to be addressed. The main concern is the privacy and security of customer's data. An additional concern is a deep system dependence on technology and network services what implies the requirement for trained and knowledgeable staff so that the whole business model can be revolutionized [72].

Making the fast growing tourism sector green and sustainable can be managed only by using recyclable or renewable technologies, protecting the environment, respecting local cultures, improving local communities, involving businesses, staffs and tourists in sustainable practices, minimizing energy usage and pollution, and conserving cultural and natural assets. As it is already known, travel and tourism industries are able to generate significant social, economic and cultural development of any region [70]. Hence, a great number of tourist places of interest are trying to bring the tourists in various ways [73]. While the direct contribution of travel and tourism was 2155.4 billion USD in 2013, it is predicted that it will rise by 4.2% annually in 2014–2024, to 3379.3 billion USD in 2024 [71]. G-IoT will evidently play a significant role in achieving economic and environmental benefits in the tourism sector.

4 Concluding Remarks

To have a sustainable place for living and development, conserving planet's natural resources appears as the crucial precondition. Despite tremendous benefits and society progress influenced by achievements in ICTs sector, the 21st century technology solutions and their growth imply the acceleration of energy and other resources usage accompanied by increased hazardous emissions and e-waste. Hence, the goal of modern society is to use the same modern technology solutions to save the planet, establish the sustainable living place for people and make the profit. Moving towards green technologies, their development and usage will require less energy and other resources, as well as minimize the negative influence on the human health and the environment. In this way, the green technologies, as less energy and resource consuming, are able to contribute significantly to sustainable development.

This paper demonstrates the G-IoT concept and its potential to transform and bring numerous benefits in various sectors, which are considered as crucial sectors in the green economy. The usage of G-IoT offers promising results in conserving and sustainable usage of natural resources in agriculture, forestry, fisheries and aquaculture sectors. However, the most benefits can be expected from the G-IoT utilization in realizing smart energy, buildings and industrial sectors, which are presently seen as sectors that consume the largest amounts of energy and natural resources, and as the largest sources of GHG emissions. The appliance of G-IoT vision in transport and tourism sector, and particularly in waste management, will also contribute significantly to socio-economic and environmental benefits.

Obviously, incorporation of the G-IoT concept in the ICT sector and driving sectors of green economy require the engagement of all participants (governments, management, employees, and consumers) as well as official legislation, standards and policies to enforce G-IoT practices. In order to achieve sustainable development, environmental and economic savings with the help of G-IoT, it is mandatory to consider organizational structures and consumers demands, the ways to mitigate negative factors (reduce energy usage and GHG emissions in processes, software, hardware, and infrastructure, utilization of less harmful materials, recycling and reusing possibilities) and boost the companies' competitive advantage.

To conclude, G-IoT can change the planet and people's lives for the better. A smarter, healthier and more prosperous world and other forecasted benefits justify the investment and collaboration in developing, innovating and utilizing the G-IoT as well as regulation approaches for promoting green manners and widespread adoption of G-IoT.

References

1. Vidas-Bubanja, M. (2014). Implementation of Green ICT for sustainable economic development. In *MIPRO 2014*. (pp. 1592–1597). Opatija, Croatia.
2. Green ICT. (n.d). *Green ICT publications*. [Online]. http://greenict.govmu.org/portal/sites/greenict/download.html
3. UNEP (United Nations Environment Programme). (n.d.). *Green economy initiative*. [Online]. www.unep.org/greeneconomy
4. ITU. (2014). *Green ICT technologies: How they can help mitigate the effects of climate change*. [Online]. http://www.itu.int/en/ITU-D/Regional-Presence/AsiaPacific/Documents/ICTCC_Session_7_Green%20ICT%20Technologies%20V4.pdf
5. Maksimovic, M., & Gavrilovic, Z. (2016). Connecting sciences in green: Internet of things and economy. In *ENTECH '16/IV. International Energy Technologies Conference* (pp. 173–182). Istanbul, Turkey.
6. Karimi, K. (n.d.). *The role of sensor fusion in the internet of things*. [Online]. http://eu.mouser.com/applications/sensor-fusion-iot/
7. Thierer, A., & Castillo, A. (2015). *Projecting the growth and economic impact of the internet of things*. [Online]. http://mercatus.org/publication/projecting-growth-and-economic-impact-internet-things
8. Lim, Y. (2015). *Answers to IoT, the latest trend in IT-talk service-oriented IoT (1)*. LG CNS. [Online]. http://www.lgcnsblog.com/features/answers-to-iot-the-latest-trend-in-it-talk-service-oriented-iot-1/
9. Vertatique (2014). *'Internet of Things' presents Green ICT challenges*. [Online]. http://www.vertatique.com/internet-things-will-rerquire-more-storage-and-bandwidth
10. Buchalcevova, A., & Gala, L. (2012). Green ICT adoption survey focused on ICT lifecycle from the consumer's perspective (SMEs). *Journal of Competitiveness, 4*(4), 109–122.
11. Andreopoulou, Z., Stiakakis, E., & Vlachopoulou, M. (2013). Green ICT applications towards the achievement of sustainable development. In *E-innovation for sustainable development of rural resources during global economic crisis*. Hershey: IGI Global.
12. Kantarci, B. (2016). *Cloud computing and the urgent mission of Green ICT*. [Online]. http://www.cloudcomputing-news.net/news/2016/mar/15/cloud-computing-and-urgent-mission-green-ict/
13. Ozturk, A., et al. (2011). Green ICT (Information and communication technologies): A review of academic and practitioner perspectives. *International Journal of eBusiness and eGovernment Studies, 3*(1), 1–16.
14. NCB. (n.d). *Green ICT guidelines for businesses*. National Computer Board. [Online]. http://www.ncb.mu/English/Documents/Downloads/Reports%20and%20Guidelines/Green%20ICT%20Guidelines%20for%20Businesses.pdf

15. Abedin, S. F., Alam, M. G. R., Haw, R., & Hong, C. S. (2015). A system model for energy efficient Green-IoT network. In *International conference on information networking (ICOIN)* (pp. 177–182). Cambodia.
16. Radu, L. D. (2016). Determinants of Green ICT adoption in organizations: A theoretical perspective. *Sustainability, 2016*(8), 731.
17. European Research Cluster on the Internet of Things. (2014). *Internet of things*. [Online]. http://www.internet-of-things-research.eu/about_iot.htm
18. Neuhaus, C., Polze, A., & Chowdhuryy, M. M. R. (2011). *Survey on healthcare IT systems: Standards, regulations and security*. Potsdam: Hasso-Plattner-Instituts für Soft-waresystemtechnik an der Universität Potsdam, Universitätsverlag Potsdam.
19. Maksimovic, M., & Vujovic, V. (2017). Internet of Things based e-health systems: ideas, expectations and concerns. In S. U. Khan, A. Y. Zomaya, & A. Abbas (Eds.), *Handbook of large-scale distributed computing in smart healthcare*. New York: Springer.
20. Schäfer, T. (2013). *Green IT: Sustainable communications and information technology*. [Online]. https://www.alumniportal-deutschland.org/en/sustainability/economy/green-it-sustainable-information-technology/
21. Leung, V. (2015). Green internet of things for smart cities. In *International workshop on smart cities and urban informatics*, Hong Kong.
22. Pazowski, P. (2015). Green computing: Latest practices and technologies for ICT sustainability. Management, knowledge and learning. In *Joint International Conference* (pp. 1853–1860). Bari, Italy.
23. Zhu, C., Leung, V., Shu, L., & Ngai, E. C. H. (2015). Green internet of things for smart world. *Access IEEE, 3*, 2151–2162.
24. Cavdar, C. (2016). *The increasingly concerning carbon footprint of information and telecommunication technologies*. [Online]. https://www.ecnmag.com/blog/2016/04/increasingly-concerning-carbon-footprint-information-and-telecommunication-technologies
25. Fulai, S. (2009). What does it take for a transition towards a green economy? In *The 3rd OECD World Forum on "Statistics, Knowledge and Policy" Charting Progress, Building Visions, Improving Life*. Busan, Korea.
26. APC. (2010). *Global Information Society Watch—Focus on ICTs and environmental sustainability*. Association for Progressive Communications. [Online]. https://www.giswatch.org/sites/default/files/gisw2010thematicictssustainability_en.pdf
27. Shoaee, A. (2016). Green IoT Solutions. *Smart city—Smart environment—Smart industry—Smart metering solutions*. [Online]. https://www.linkedin.com/pulse/green-iot-solutions-smart-city-environment-industry-ali-shoaee
28. Stollenmayer, P. (2011). *How the Earth can benefit from Green ICT*. Eurescom. [Online]. https://www.eurescom.eu/news-and-events/eurescommessage/eurescom-message-archive/eurescom-messge-2-2011/how-the-earth-can-benefit-from-green-ict.html
29. Mazza, L., & ten Brink, P. (2012). *Green economy in the European Union*. UNEP. [Online]. http://www.ieep.eu/assets/963/KNOSSOS_Green_Economy_Supporting_Briefing.pdf
30. Anderson, J., Fergusson, M., & Valsecchi, C. (2008). *An overview of global greenhouse gas emissions and emissions reduction scenarios for the future*. London: Institute for European Environmental Policy (IEEP).
31. Andreopoulou, Z. (2012). Green informatics: ICT for green and sustainability. *Agrárinformatika/Agricultural Informatics, 3*(2), 1–8.
32. FAO. (n.d.). *Towards a sustainable agriculture platform in partnership with farmers' cooperatives and organizations*. Food and Agriculture Organization of the United Nations. [Online]. http://www.fao.org/tc/exact/sustainable-agriculture-platform-pilot-website/en/
33. Rowbot. [Online]. http://rowbot.com/
34. Tharrington, M. (2015). *The future of smart farming with IoT and open source farming*. DZone/IoT Zone. [Online]. https://dzone.com/articles/the-future-of-smart-farming-with-iot-and-open-sour
35. Libelium (2017). *Smart irrigation system to improve kiwi production in Italy*. [Online]. http://www.libelium.com/smart-irrigation-system-to-improve-kiwi-production-in-italy/

36. GeSi. (2015). *#SMARTer2030-ICT solutions for 21st century challenges*. Brussels, Belgium: Global e-Sustainability Initiative (GeSI). [Online]. http://smarter2030.gesi.org/downloads/Full_report.pdf
37. Li, D. (2014). *Internet of Things in aquaculture*. Beijing Engineering Research Center for Internet of Things in Agriculture, China Agricultural University. [Online]. https://www.was.org/documents/MeetingPresentations/WA2014/WA2014_0609.pdf
38. OECD. (2015). *Green growth in fisheries and aquaculture. OECD green growth studies*. Paris: OECD Publishing.
39. Briodagh, K. (2015). *How to solve seafood with IoT technology*. IoT Evolution. [Online]. http://www.iotevolutionworld.com/iot/articles/407059-how-solve-seafood-with-iot-technology.htm
40. Darrow, B. (2016). *How connected sensors can help you park and put fish on the table*. Fortune. [Online]. http://fortune.com/2016/04/08/internet-of-things-commercial-fishing/
41. Bridgwater, A. (2016). *Fly-fishing drones, an IoT angle on angling*. Internet of Business. [Online]. https://internetofbusiness.com/fly-fishing-drones-iot-angle-angling/
42. ITU. (2010). *ICT as an enabler for smart water management. ITU-T technology watch report*. [Online]. http://www.itu.int/dms_pub/itu-t/oth/23/01/T23010000100003PDFE.pdf
43. OECD. (2009). *Towards Green ICT strategies: Assessing policies and programmes on ICT and the environment*. [Online]. https://www.oecd.org/sti/ieconomy/42825130.pdf
44. Vijayapriya, T., & Kothari, D. P. (2011). Smart grid: An overview. *Smart Grid and Renewable Energy, 2*(4), 305–311. [Online]. http://research.iaun.ac.ir/pd/bahador.fani/pdfs/UploadFile_1231.pdf
45. System Energy Efficiency Lab. (n.d). *Internet of things with applications to smart grid and green energy*. [Online]. http://seelab.ucsd.edu/greenenergy/overview.shtml
46. UNIDO. (2014). *Emerging green technologies for the manufacturing sector*. Fraunhofer. United Nations Industrial Development Organization. [Online]. http://www.gita.org.in/Attachments/Reports/Institute_Emerging_green_trends_Future_of_Manufacturing.pdf
47. Day, B. (2015). *Green transportation and ICT industries development in Indonesia. Indonesia green infrastructure summit: Green solutions in ICT and transport*. Jakarta. [Online]. http://slideplayer.com/slide/5307023/
48. Wang, K., Wang, Y., Sun, Y., Guo, S., & Wu, J. (2016). Green industrial internet of things architecture: An energy-efficient perspective. *IEEE Communications Magazine*.
49. IEA. (2009). *Transport, energy and CO_2*. Paris: International Energy Agency. [Online]. https://www.iea.org/publications/freepublications/publication/transport2009.pdf
50. Roy, P., et al. (2016). Intelligent traffic monitoring system through auto and manual controlling using PC and Android Application. In *Handbook of research on applied video processing and mining*. Hershey: IGI Global.
51. Biswas, S.P., Roy, P., Mukherjee, A., & Dey, N. (2015). Intelligent traffic monitoring system. In *International Conference on Computer and Communication Technologies—IC3T 2015*, Hyderabad, India.
52. Solanki, et al. (2015). Advanced automated module for smart and secure city. In *1st International Conference on Information Security And Privacy (ICISP 2015)*. Nagpur, India: Elsevier Procedia CS.
53. Lefevre, B., & Enriquez, A. (2014). *Transport sector key to closing the world's emissions gap*. World Resource Center. [Online]. http://www.wri.org/blog/2014/09/transport-sector-key-closing-world%E2%80%99s-emissions-gap
54. Cucchiella, F., D'Adamo, I., Koh, S. C. L., & Rosa, P. (2015). Recycling of WEEEs: An economic assessment of present and future e-waste streams. *Renewable and Sustainable Energy Reviews, 51*, 263–272.
55. Tausif, M. (2015). *"Waste Management" & recycling—Kingdom of Saudi Arabia*. [Online]. https://www.linkedin.com/pulse/moving-saudi-arabia-towards-sustainable-future-zero-wast-shaikh
56. Nayab, N. (2011). *Benefits of e-waste recycling*. [Online]. http://www.brighthub.com/environment/green-computing/articles/71375.aspx

57. Musulin, K. (2015). *Refuse revolution: 4 companies transforming the trash bin.* [Online]. http://www.wastedive.com/news/refuse-revolution-4-companies-transforming-the-trash-bin/405405/
58. Medvedev, A., et al. (2015). Waste management as an IoT enabled service in smart cities. *Internet of things, smart spaces, and next generation networks and systems. Lecture Notes in Computer Science*, vol. 9247, pp. 104–115.
59. Tracy, P. (2016). *How industrial IoT is revolutionizing waste management.* [Online]. http://industrialiot5g.com/20160728/internet-of-things/waste-management-industrial-iot-tag31-tag99
60. CFES. (2012). *The environmental and economic benefits of electronic waste recycling.* [Online]. http://cashforelectronicscrapusa.com/the-environmental-and-economic-benefits-of-electronic-waste-recycling/
61. Morgan, K. (2015). *Is there a future for e-waste recycling? Yes, and it's worth billions—A methodology to help organizations in e-waste management.* [Online]. https://www.elsevier.com/atlas/story/planet/is-there-a-future-for-e-waste-recycling-yes,-and-its-worth-billions
62. Vujovic, V., & Maksimovic, M. (2015). Raspberry Pi as a sensor web node for home automation. *Computers & Electrical Engineering, 44,* 153–171.
63. Infosight. (n.d.). *Green buildings/smart cities.* [Online]. http://www.info-sight.net/services/smartcity-green-buildings/
64. Soares Goncalves, J. C. (2011). Buildings and UNEP's green economy report. In *UNEP-SBCI 2011 Symposium on Sustainable Buildings.* Leverkusen, Germany.
65. OECD. (2016). *Green cities programme methodology.* [Online]. www.ebrd.com/documents/technical-cooporation/green-city-action-plan-in-tirana.pdf
66. Thorpe, D. K. (2014). *What is the best way to measure the sustainability of cities? SmartCitiesDive.* [Online]. http://www.smartcitiesdive.com/ex/sustainablecitiescollective/what-best-way-measure-sustainability-cities/243106/
67. Busch, R. (2012). *The Green City Index—A summary of the Green City Index research series.* Munich, Germany: Siemens AG.
68. European Commission. (2015). *In-depth report: Indicators for sustainable cities.* Brussels: European Union.
69. Batten, J. (2016). *Sustainable cities index 2016—Putting people at the heart of city sustainability.* Netherlands: Arcadis.
70. Iyer, V. R., Chakraborty, S., & Dey, N. (2015). Advent of information technology in the world of tourism. In *Emerging innovative marketing strategies in the tourism industry.* Hershey: IGI Global.
71. Mimos Berhad. (2015). *IoT idea book: Experiential travel and tourism.* [Online]. http://www.mimos.my/wp-content/uploads/2016/01/IoT-Idea-Book-Experiential-Travel-and-Tourism.pdf
72. Kaur, K., & Kaur, R. (2016). Internet of things to promote tourism: An insight into smart tourism. *International Journal of Recent Trends in Engineering & Research (IJRTER), 02* (04), 357–361.
73. Dey, N., Acharjee, S., & Chakraborty, S. (2015). Film induced tourism: Hunting of potential tourist spots by frame mosaicing. In *New Business opportunities in the growing e-tourism industry.* Hershey: IGI Global.

Design of Cloud-Based Green IoT Architecture for Smart Cities

Gurjit Kaur, Pradeep Tomar and Prabhjot Singh

Abstract In the smart cities, objects can smartly communicate with the people through Internet of Things (IoT). It will make smart cities a greener place by detecting pollution through IoT and environmental sensors. In order to maintain the sustainability of green place in smart cities, the emerging technology, i.e. Green IoT automatically and intelligently makes smart cities sustainable in a collaborative manner. Governments and a lot of organizations around the world are doing a lot of efforts to campaign the importance of the reduction of energy consumption and carbon production as well as emphasize on the Green IoT for smart cities. Various IoT related smart cities architectures are already presented in literature. But this work presents the concept of the "Green IoT" to create a green environment which will apprehend the idea of energy saving in smart cities. In this chapter, design of Green IoT architecture is proposed for smart cites with the focus to reduce energy consumption at each stage and ensure realization of IoT toward green. This proposed Green IoT architecture is based on the cloud based system which reduces the hardware consumption.

Keywords Green IoT · Green radio frequency identification · Green data center · Green cloud computing sensors · Green machine to machine · Green wireless sensor network

G. Kaur (✉) · P. Tomar
School of ICT, Gautam Buddha University, Greater Noida, Uttar Pradesh, India
e-mail: gurjeet_kaur@rediffmail.com

P. Tomar
e-mail: parry.tomar@gmail.com

P. Singh
Salesforce Inc., San Francisco, USA
e-mail: prabhjot27@gmail.com

N. Dey et al. (eds.), *Internet of Things and Big Data Analytics Toward Next-Generation Intelligence*, Studies in Big Data 30,
DOI 10.1007/978-3-319-60435-0_13

1 Introduction

In next generation, the internet will make the world where physical things would reliably be consolidated into information frameworks which would give savvy administrations to clients. These interconnected things, for instance, sensors or convenient contraptions would produce and accumulate volumes of data which can be further handled to find supportive information to reinforce savvy and universal administrations [1]. The ascent of the internet, the always expanding universality of information, and its low flag to-clamor proportion have added to the issue of data over-burden, whereby people have admittance to a greater number of information than they can absorb into important and significant data. A significant part of the accomplishment of Web 2.0 has been accomplished after a viable handling of this issue [2].

IoT alludes to a worldwide, distributed network of physical items that are equipped for detecting or communicating with other things, different machines or PCs. IoT is a main constituent of the upcoming smart world and can be characterized as a network device with self-configuring abilities derived from advanced protocols where virtual and physical objects can easily communicate. IoT empower network for all intents and purposes any physical protest that possibly offers a message, and will influence each part of life and business. This article takes a gander at the ideas and advancements in three application territories that Motorola is adding to now and in the coming years [3]. This will connect objects and people which can communicate through various sensors and actuators, radio frequency identifications via bluetooth connections, etc. Such "smart" objects arrive in an extensive variety of sizes and capacities, incorporating simple object with embedded sensors, household apparatus, mechanical robots, automobiles, trains, and wearable protests, for example, watches, arm ornaments or shirts [4]. Their esteem lies in the endless amounts of information they can catch and their ability for correspondence, supporting constant control or information examination that uncovers new experiences and prompts new activities. Through IoT, this world can become technically more advanced and intelligent where human beings can communicate with objects and the objects can communicate with other objects without any human interference. IoT object will communicate through sensors, actuators, bluetooth connections and Radio Frequency Identification (RFID) etc. [5]. Every object has intelligent interfaces, which provide desired information after incorporated into the intelligent information network. Ambient intelligence is a human interface allegory alluding to the earth of registering which knows and receptive to the nearness of human collaboration. The point is to place awesome accentuation on the part of being easy to understand and effective and offer help for human cooperation. We are as yet taking a stab at a future world where we will be encompassed by clever interfaces that are to be put in ordinary items. These items will then have the capacity to perceive additionally react imperceptibly to the nearness of individuals [6].

But IoT will influence our surroundings in a few diverse ways. Every object of IoT, from its creation, all through its utilization till disposal presents environmental problems. Manufacturing sensory objects, PCs, laptops, and their different segments consumes power, materials and chemicals, water and produces perilous waste. All these objects and systems will increment carbon dioxide outflows and affect the earth. Also, the aggregate energy consumption by servers, personal computers (PCs), screens, information exchange equipment and cooling systems for data centers is gradually increasing. Green house gas emission will increase rapidly by increase of energy consumption. Every PC being used creates about a huge amount of carbon dioxide consistently. All electronic objects, sensors and PC segments used in IoT will contain poisonous materials. Even customers dispose of a large number of old computers, electronic gadgets two to three years after purchase, and most of this end in landfills therefore polluting the earth environment. It is noted that about 20–50 million tons of computer parts and mobile phones are dumped every year which is the major chunk of waste. So there is huge pressure on IoT related industries, businesses and the individuals who design the IoT related objects and systems to make IoT systems environment friendly and green throughout its lifecycle i.e. from birth to death to rebirth [7].

As IoT is integrated with so many sensory elements to sense and communicate the data using advanced communication technologies which ultimately shoot up the energy consumption. But only Information and Communication Technology (ICT) accounts for about 2% of global CO_2 emissions. The entire world today is talking green and it's our social responsibility to save our environment. It's not green with envy, but green as in becoming more eco-friendly, energy saver, adherence to global standards, environment friendly, efficient usage of the computing resources like Energy Star, Restriction of Hazardous Substances, etc. so that the system could be managed more efficiently throughout its life and even at the time of disposal. Green Information Technology (IT) is the practice of environmentally sustainable computing. It aims to reduce the negative impact of IT system on the environment by designing, manufacturing, operating and disposing of computers, sensors, products etc. in an environmentally-friendly manner. The real objectives of Green IoT are to reduce the use of harmful toxic materials. It should have the capability to improve energy efficiency throughout its life and support the biodegradability and recyclability of obsolete products and the redundant plant waste. Green IoT follows energy efficient procedures using advanced hardware and software which will reduce the greenhouse effect even for the existing applications so the impact of the greenhouse effect will reduce on the IoT itself [8].

The constant advancement of Next Generation Internet (NGI) [9] enhances the interest for productive and secure correspondence fit for reacting viably to the difficulties postured by the rising applications. For secure correspondence between two sensor hubs, a mystery key is required. Cryptographic key administration is a testing errand in sensor organizes as the threatening condition of sensor systems makes it more inclined to assaults. Aside from asset limitations of the gadgets, obscure topology of the system, the higher danger of hub catch and absence of a settled foundation makes the key administration all the more difficult in Wireless Sensor Network (WSN).

Technically, Green IoT based smart cities focus on green design, green manufacturing, green operations, green deployment, green retention and even green reuse with a very small impact on the environment. These green smart cities includes green smart aerospace and aviation systems, green smart homes, green smart buildings, green smart e-health, green smart logistics, green smart retail, green supply chain management, green smart transportation, green smart recycling, green smart environment monitoring etc. with minimal utilization of energy. In green IoT smart cities objects like mobile phones, computers, laptops, cars, and electronic appliances can communicate with each other with distinctive addresses in energy saving mode. These sensory devices are able to communicate intelligently via internet protocol and can provide green support in managing different tasks for the users. It can support various technologies [10] i.e. identification, communication, data and signal processing technologies with reduced energy consumption. During the design and development of Green IoT based smart cities energy efficient procedures needed to adopt.

2 Green IoT-Based Smart Cities

One of the biggest targets of the digital era is to design these green smart cities. Green IoT is an important technology to build green smart cities. Over the next ten years, buildings will be the main energy consumer which will emit greenhouse gasses on the earth. The best way for a building to become smarter by empowering owners and managers to collect energy and operational metrics into a single, centralized location and apply enterprise-wide analytical and optimization capabilities to gain insights from that information. There are some hardware and software systems which can provide important information related to electric demand and energy usage. By using real-time data gathering building managers can do the analysis to solve problems related to service proactively before they occur. Even they can visualize energy consumption, environment and portfolio performance for floor space, etc. Also, there is a need to make sustainable green smart cities in an economical way. This could include ICT based smart networking to reduce vitality transmission costs and enhance the versatility of these utility systems [11]. In IoT based smart cities, objects or things are made uniquely addressable by using a unique way of identification where these smart objects transmit the information like location, state, type, context or other sensory measurements to the other devices. These things are also heterogeneous in some capabilities. A green smart city includes green smart homes, green smart heath, green smart farming, green smart education system, green smart water system, green smart transportation and green smart retail etc. working very energy efficiently and keep the environment clean and green [12]. The main features of a smart city are represented in Fig. 1.

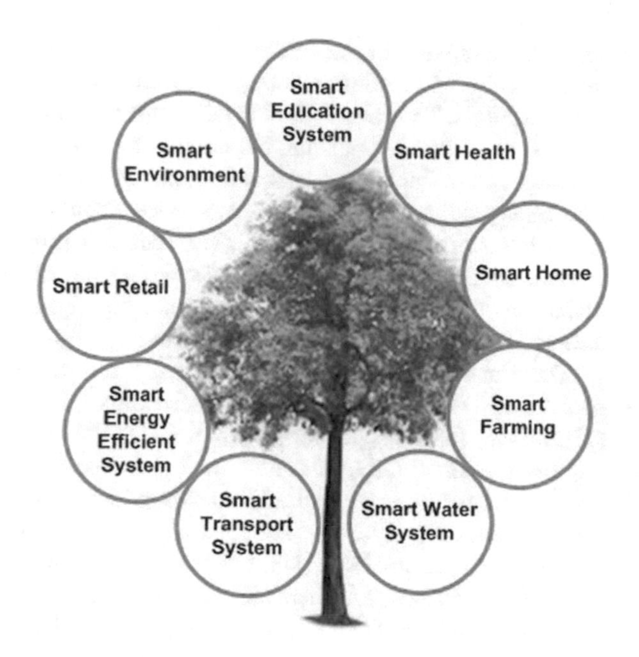

Fig. 1 Features of green smart city

There are certain standards developed for these smart cities. For designing these smart cities there are global city standards for life of the cities, e-health, leisure, safety, education, transportation, water, finance etc. Smart governance, smart environment, smart mobility, smart people, smart economy and smart living are the key characteristics of the smart cities. The green smart cities have other important features like reforestation zones, greenhouses, green floors and roofs, green bridges, green urban areas, green energy management, fish voyage channels, flood restoration facilities, farmland, forest areas, which have the advantages of nature-based solutions.

There are some smart cities in the world which are taking care of the environment. Denmark is the first country in the world to pass an environmental protection law and engage large number partners to save its environment. The city of Copenhagen is considered as one of the best smart city around the world with the vision to become the world's first carbon neutral capital. So as to accomplish this yearning objective, the city is implementing new and innovative solutions for managing waste, transport, water and utilizes alternate energy sources efficiently. Carbon neutrality will bring about a superior personal satisfaction, advancement work creation and saving. Songdo city in South Korea has already automatic climate control facilities along with computerized accessibility. Its roads, water systems, electricity systems and waste management systems have electronic sensors which help to track and respond to the movement of residents. Dubai together with its neighboring city Abu Dhabi is also deploying over 5000 Wi-Fi hotspots to offer free internet to cover all other areas of a smart city, including healthcare, industrial, education, safety, telecoms, tourism and utilities, where 250,000 smart meters are set to be deployed. In the safety arena, one of the city's government's plans is to introduce Google's Glass technology to the police to create the world's smartest police stations. The country is also taking urban landscape to an extreme with plans to build an artificial mountain, tall enough to make it rain more often in the desert nation. Even in India, the target is to build more than 100 smart cities. Barcelona is also rolling out more e-Government services, contactless services, more city apps, smart bus stops and more. Seattle has started a program to decrease vitality utilization by means of examination of ongoing information whereas San Francisco dispatches the I-80 Smart Corridor extend highlighting various cameras, sensors, and actuators, high innovation street signs.

There are certain standards developed for these smart cities. The application space skill for smart cities lies in International Organization for Standardization (ISO) TCs. Joint Technical Committee (JTC-1) commitments are imperative to empower a Smart City, with its norms about covering the data layer, application layer and support layer and additionally adding sensor layer. All types of communications and systems administration are controlled by the International Telecommunication Union (ITU), with commitment from JTC1/SC6 while security is added by JTC1/SC27. There are predominantly two foundations of city indicators i.e. ISO/TC268 and ITU-T Smart and Sustainable Cities Focus Group. ISO/TC268 is focused on sustainable development in society, where one of the working groups is focused on developing city indicators and the other in developing metrics for smart society infrastructures. These are indicators referred by the ISO/TR 37150 according to global city indicators, green city index arrangement and smart city acknowledged by ICT.

Worldwide city indicators cover up the general city life, for example, instruction, wellbeing, amusement, security, transportation, wastewater, water, back and so on. The global city indicators have now become an ISO Standard i.e. ISO 37120:2014 and ISO TC 8. This standard is being used by a number of smart cities. The green city index arrangement covers CO_2, water, transport and vitality etc. These are essentially focused on the indicators which are identified with ecological effect and, once again, are not specifically identified with ICT. This section provides a selection of the indicators used in the three sets reviewed in ISO/TR 37150:2014, and in the technical report on Smart and Sustainable City KPIs being developed by the ITU-T Smart and Sustainable Cities Focus Group [13]. A few indicators incorporated into the smart city acknowledged by ICT. To make the smart cities greener these are the vital indictors as represented in Fig. 2.

A city is not brilliant when the diverse systems which characterize it are not ready to impart and work together in systems. The communication among these things is mainly depend on emails, calls, video conferences, optical fiber links, social networks, broadband and distributed computing etc. Many smart devices in the city, which are the major parts of the city's physical infrastructure, depend on strong foundation the transmission carrying capacity i.e. bandwidth because this will help to transmit the important information throughout e.g. empty parking spots,

Fig. 2 Green index indicators for smart cities

vitality utilize, auto accidents, climate conditions etc. This data can be accessed by the city through the mobile phones or other electronic devices or sensors and the system will suggest other intelligent alternatives for moving around thc city. All together for the smart city system to work, all individuals and devices must have the capacity to communicate with each other. So ICT is a major part for the effective capacity of the above system [14].

The IoT will give them a typical domain where these heterogeneous things will have the capacity to communicate with each other utilizing standardized communication platform. Moreover, the computer and databases ought to have the capacity to impart and nourish into each other, empowering a smooth and effective data stream between the distinctive partners of the city. At the point when the physical foundation is included into the advanced systems the versatility of the city will be quite prominent. But the problem is how to communicate byte of data energy efficiently and hence this exponential traffic can choke the whole communication system. There is a need to rethink about the current architecture of the smart city. Even the sensors and devices of smart cities need to consume their own energy very carefully so that they not only to able to communicate for indefinitely long but also form extensive network even when infrastructure is weak or not available. So there is a need to develop a smart green IoT based architecture by using advanced communication and protocols which communicate the data energy efficiently in a more reliable way. The communication technologies and protocols need to be designed for the IoT platform where all objects or things are combined to analyze location, intent and even emotions over network energy efficiently. The integration of IoT and wireless sensor network can give fabulous solutions to establish communication services which will involve advanced communication protocols that connect smart devices in environment friendly way.

3 Features of Green IoT Smart Cities

In green IoT smart cities the main features includes are green smart home, green smart office, green smart healthcare system, green smart grids, green smart transport system, green smart farming, green smart waste management system and green smart environment.

3.1 Green Smart Home

Due to global warming, more and more people are getting energy-conscious and looking forward to the energy management solutions for their home. The smart home will enable new opportunities for architects, builders, technology providers and homeowners at the same time. These new sensor-based home products are controllable from an owner's smart sensory devices such as a smartphone or tablets for controlling the Air Conditioners (AC) system, lights, managing doors and for operating other domestic functions of his personal device. For instance, based on the weather forecast information, a smart home can automatically open and close the windows. Today many service providers have already launched residential applications that can be controlled via the set-top box or web by allowing their users to monitor their houses from anywhere via smart mobile phones.

Apple Inc. has developed a Home Kit so that people can control their home in a smart way. Not only Apple but Google also developed a product called Nest which provides the home automation solution and provide security and energy saving tool for the smart home. It has self-learning equipment which learns the daily habits and does programming itself for the schedule that matches up with the living style. Another major player iRobot Corp developed robotic technology-based solutions called Roomba which integrate adaptive navigation and visual localization along with app control. It has the capability to clean carpets, help to keep clean floors of the entire house etc. Even LG developed its own software which will allow people to connect and communicate with their smart homes. This natural language processing based App is known as HomeChat which will enable the users to send texts to its compatible LG appliances. Google has also announced home automation Brillo operating system.

For this Green smart home concept, the management of the power is very much essential i.e. to understand the peak and optimal times for usage of appliances. For these Green Smart home, there is a requirement of real-time energy monitoring tools which can give the feedback regarding usage of energy and its cost which will help to take necessary decisions accordingly. Also, a good power management system should be installed which can turn on and off the appliances automatically. Modernized indoor thermostats can limit carbon impression. One option is to control the temperature inside the home via cell phone. By chance somebody leaves an entryway open; a smart thermostat can switch off the air conditioner or heater automatically. Also an intelligent indoor system can ventilate the home while the owner is at work and turn it moves down before arrival back. Smart indoor regulators like the Nest can also take the temperature inclinations, making it much simpler to maximize energy efficiency. Smart home lighting arrangements can give more noteworthy control over the lighting at home thus makes the home more eco-accommodating. Indeed, even the smart window systems can lessen the Heating, Ventilation and Air Conditioning (HVAC) vitality utilization by keeping it cool and warm it when winter arrives. A smart water system can save gallons of water per year. Systems like Cyber Rain can be keenly adaptable to any size yard to screen and can be controlled by any internet associated gadget. Vitality administration systems like Savant watch out for the home's energy utilization so one can find which system, apparatus or hardware are utilizing the most vitality and after that control their utilization particularly amid pinnacle power value periods. So by utilizing these smart machines and innovations keen homes can consequently lessen the vitality utilization of the home and make it green.

Ambient Assisted Living (AAL) innovations incorporate the utilization of home inserted sensors and systems, body worn sensors, robots and inserts. As of late the business is demonstrating a developing enthusiasm for video and PC vision based arrangements and this due to the way that such items are continuously persuading less expensive to be created with cameras and sensors are being coordinated on the semiconductor itself [15].

3.2 Green Smart Office

The Green IoT is an advanced concept for designing the smart offices. Green Smart office makes life easier and improves the business. This Green Smart IoT-based office is connected to the internet, sensors and mobile devices which will enable the employees to manage time, resources and space in an efficient way. These types of smart offices result in a high return on investment and reduce the operational costs. These green smart office solutions also ensure the greater productivity along with better resources and space management in a better collaborative way. If there are any problems while communication amongst the devices, technologies or platform of smart office solutions can automatically go for another solution and contribute towards more efficiency. The processes get automated and the office activities get optimized with the smart office.

One way to reduce your office's energy usage is to install dimmer switches for the lighting. Most offices have huge windows that allow in a lot of natural light, so the artificial light can be adjusted throughout the day depending on how much natural light is available. For example, employees can turn the lights all the way on in the morning, dial the dimmer down as the mid-day sunshine comes through the windows, and then crank it back up as the sun begins to set. The second option is to install motion sensor lighting so the lights are never left on when no one is in the office even if the person who locks up at night forgets to turn them off. To use less energy on heating and cooling, smart offices can install programmable thermostats. These models should have the ability to preset temperatures so the office can be a comfortable temperature during business hours without wasting energy during non-business hours.

Adding plants throughout the office is also a smart way to go green. Plants can cool down the air around them, so one has not to rely on the air conditioner as much to keep the office at a comfortable temperature. There are smart tools which one can use with plants, like Plant Link, which is a device that texts the message when the plant needs water, sunlight or fertilizer.

3.3 Green Smart Healthcare System

The healthcare industry can be the biggest beneficiary of the IoT revolution. By building green IoT-based systems in hospitals and clinics, along with establishing IoT-based outpatient monitoring solutions, hospitals can improve access to patient care, while increasing care quality and reducing operating costs. The basic building blocks of an IoT-based system are sensors that collect patient data, Internet gateways for transmitting that data and cloud computing to store and process this data. The cloud platform is also used to analyze this data to generate valuable insights for doctors and medical staff. The final stage involves the creation of web and mobile applications which the medical staff can use to decide on the next course of action [16].

In patient-centric IoT systems, the patients are also given mobile applications and even wearable devices for monitoring their health. Adidas Smart Run is a Global Positioning System (GPS) enabled running watch that has a built-in heart rate monitor for the athletes who want to avoid wearing a chest strap to track heart rate. The continuous measurement of heart rates will only fuel the big data challenge that the medical community is facing. However, intelligent algorithms that can interpret heart rate information will provide insights that may ultimately transform health care and disease prevention strategies. Softweb Solutions developed the cross-platform app that communicates with the Pebble watch and enables the patient or caregiver to manage their emergency contacts, medicine alerts. The caregiver also receives notifications for medicine, activities, dementia fence and fall detection.

In green smart healthcare systems all the monitoring devices, systems, machines etc. should work energy efficiently. Indeed, even the hospitals ought to execute a framework that conveys exact ecological control while incorporating effective control systems. It ought to give ideal lighting conditions by utilizing a mix of lighting control strategies, including occasion based booking, inhabitance location, coordinated exchanging, sunshine detecting, and undertaking tuning. It additionally guarantees that the central plant systems operate at peak efficiency at peak efficiency by overseeing framework limit in light of building load requests and operational rules. It should have the component of interest controlled ventilation and and occupancy-based control in patient rooms, restorative workplaces, or exam rooms, to cut vitality utilization when these spaces are empty.

3.4 Green Smart Grids

The green smart grid is the combination of the traditional electrical power grid along with the recent IT technologies. The smarter control of these grid's is provided through high-speed, two-way communication, sensing the real-time coordination of all the devices via client meter or via end user gadget. So the smart grid is not characterized by a solitary innovation but rather it is a distributed, web based system which can provide better control of existing grid infrastructure. It integrates new assets with the existing operational systems and can engage these new devices to provide entirely new benefits to the grid. Such combination helps to get effective resource utilization which can optimize energy consumption. It has the features to exchange the generated power. It will allow dispatch the energy on the basis of demand. It will also empower clients to use energy more efficiently by accessing a complete data about their power utilize and having new management options. Due to these smart features the smart grid can operate energy efficiently [17].

3.5 Green Smart Transport System

For some transportation frameworks, the cost of growing the foundation is too high. Therefore, the concentration must move to enhancing the nature of transportation inside the current foundation. The second release of a hit, Intelligent Transport Systems: Smart and Green Infrastructure Design fundamentally inspects the victories and disappointments of transportation systems over the span of the previous decade. The new subtitle mirrors this current version's attention on meta-standards basic to pushing forward and effectively fabricating green smart transport systems that exploit keen/green advancements.

In green smart transport system for smart cities, IoT can solve the problem of toll charges, screening of travelers and their luggage, boarding business bearers and the merchandise moved via cargo etc. by supporting the security arrangements of the transportation. Observing automobile overloads through Personal Digital Assistances (PDAs) of the clients and sending of astute transport system will make the transportation of merchandise and individuals more effective. Transportation organizations would turn out to be more effective in pressing holders since the compartments can self-check and measure themselves. Utilization of IoT innovations for overseeing traveler baggage in airplane terminals and aircraft operations will empower robotized following and sorting, expanded per-sack read rates, and expanded security.

3.6 Green Smart Farming

In smart cities for smart farming procedures, cutting edge technologies and innovation are expected to enhance the generation and quality of crops. Smart farming includes applying the various inputs (water, pesticides, and manure), finishing pre and post harvest operations, and checking ecological effects. Different productive methodologies, for example, intelligent water system, smart underground sensors, and smart insect identification have been intended to perform undertakings for brilliant farming. Comparative methodologies can be connected to forest observing, where the real concentration is timberland fire checking since flames frequently result in critical harm to nature. So, by utilizing IoT empowered keen gadgets the characteristic assets can be utilized proficiently.

Green smart farming speaks to the utilization of present day ICT into agribusiness, prompting to what can be known as a third green revolution. Taking after the plant rearing and hereditary qualities upsets, this third green revolution is assuming control over the agrarian world based upon the consolidated utilization of ICT arrangements, for example, exactness gear, the IoT, sensors and actuators, Global Positioning Systems, Big Data, Unmanned Aerial Vehicles (UAVs, rambles), mechanical autonomy, and so on.

Green smart farming applications don't target just vast, traditional cultivating misuses, yet could likewise be new levers to help other normal or developing patterns in agrarian abuses, for example, family cultivating (little or complex spaces, particular societies or potentially cows, conservation of high caliber or specific assortments), natural cultivating, and upgrade an extremely regarded and straightforward cultivating as indicated by society and market awareness. Smart green farming can likewise give extraordinary advantages regarding natural issues, for instance, through more productive utilization of water, or enhancement of medications and data sources.

3.7 Green Smart Waste Management System

Green smart cities management system focus on reduction and segregation of waste at source, door-to-door collection, recycling and reuse of waste, generation of wealth from waste. The ideal smart green city attempts to produce a closed-loop system with zero emissions and zero waste. This means residents consume only what energy, water, and food inputs are necessary, recycling whatever waste is produced, and minimizing outputs of greenhouse gases (CO_2, Methane), air pollution and water pollution. Above all else, innovative waste management systems will be essential to the success of smart green cities. Currently, most of the produced things follow an unsustainable cradle to grave life cycle, which not only pollutes our environment but fails to recognize waste as a valuable resource. In order for a smart green city to use resources in a way that results in both cost and energy savings and minimal contributions to climate change, it must follow a cradle to cradle cycle where all materials that are produced are utilized to the fullest extent possible. With the right technology, just about everything which are thrown away has the potential to be composted, recycled or converted through waste-to-energy processes into energy, biofuel, or biochar. IoT services can solve the problem of waste collection and can also enable dynamic scheduling and routing in a smart city. IoT cloud based framework can help for association of waste accumulation process and applications for waste truck drivers and administrators [18].

3.8 Green Smart Environment

Expanding number of vehicles, urbanization and numerous modern exercises have increased air pollution extensively in the most recent couple of decades. Air pollution observing is a most tedious task. In advanced green smart cities the utilization of IoT innovation can make air pollution monitoring fewer complexes [8].

4 Approaches for Software and Algorithm Used in Green IoT Smart Cities

The main approaches like Green Data Center Design, Virtualization for going green, Green Power Management and Green Computing [19] for algorithm and software used in Green IoT smart cities are listed as follows:

4.1 Green Data Center Design

A green data center is used to store, manage and dissemination of data of green smart cities. Its design is such that it consumes minimal power resources for operation and maintenance. It saves energy even for computing and supporting electronic resources such as cooling, backup, and lighting etc. It can work with solar, wind or hydropower. The entire infrastructure of the green data center is designed for low power and for reduced carbon footprint.

4.2 Virtualization for Going Green

Through virtualization, multiple different operating systems can run simultaneously on the same physical hardware. So several physical systems for smart cities can be combined into the single powerful virtual machine. So it reduces the original hardware cost, power, and cooling consumption. Virtualization and going green are good for the environment but it can be great to design green smart cities.

4.3 Green Power Management

Due to the high energy consumption in peak hours, most systems for light industrial and commercial industrial market, consume more energy on peak hours, the price for electricity is high. Smart power management systems allow automatically to do power management functions for green smart cities.

4.4 Green Computing

According to [7] and [20], Green computing is an environment friendly technology. It is the field of outlining, assembling, utilizing and discarding PCs, portable workstations, servers and different subsystems in such a way that it has minimum effect on the environment. It incorporates appropriate reusing approaches, utilized gear transfer suggestions, government rules and proposals for obtaining green PC hardware. Green figuring fundamental component is to cover control use, least paper utilization alongside the proposals for new hardware and reusing old machines. As PCs have poison metals and toxins which can transmit harmful radiations so never dispose of PCs in a landfill. One needs to reuse them. So green computing can be a key feature for making smart cities green.

5 Cloud-Based Green IoT Architecture for Smart Cities

In this section, Green IoT Architecture for smart cities is proposed which will take care of communication, standardization, interoperability aspects. The main feature of this proposed architecture is that it is based on the cloud platform which automatically reduces the consumption energy for many systems and makes the environment clean.

The proposed architecture of Green IoT consists of five layers i.e. Presentation layer, Application layer, Big data analytic layer, Network layer, and Smart City Sensor layer and infrastructure as shown in Fig. 3. The architecture defines the main communication paradigms for the connecting entities. It provides a reference communication stack along with insight about the main interactions in the domain model. It describes how communication schemes can be applied to different types of Green IoT networks. It is important that heterogeneous networks of sensors in different types of networks are able to communicate with each other.

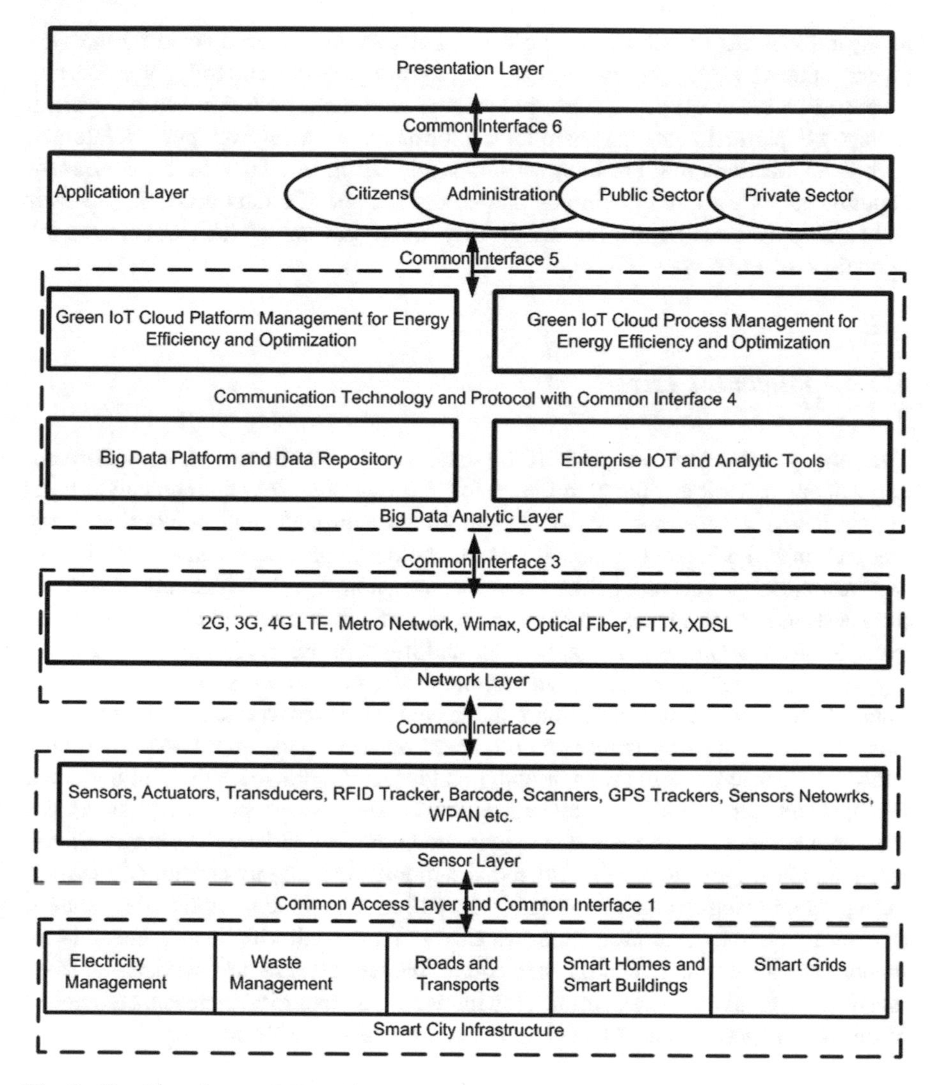

Fig. 3 Cloud-based green IoT architecture for smart cities

5.1 Presentation Layer

The presentation layer mainly receives the information from application layer. Information can be communicated in different formats via different sources. Thus, the presentation layer is responsible for integrating all formats into a standard format for efficient and effective communication. The presentation layer follows information programming structure schemes developed for different languages and provides the real-time syntax required for communication between two objects such as layers, systems or networks. The data format should be acceptable by the next layers; otherwise, the presentation layer may not perform correctly. Various city systems like water supply system, power supply system, pollution control system, transport department etc. can share their information by using web portals, internet, mobile applications that are built on this layer. Using this layer both government departments as well as individuals can access the specific data according to their rights. This data can be used to design even more services which can enhance the operations of the city.

5.2 Application Layer

This layer put at the highest point of the stack is incharge of conveyance of different applications to various clients in Green IoT through different communication techniques. It analyzes the massive data and information through cloud computing, fuzzy recognition and other technologies. The smart cities applications can be for the user, administration, private and public sectors as shown in Fig. 3. With application layer the latest information collected from big data analytic layer rapid responses can be given to physical phenomena i.e. garbage and its monitoring, digital video monitoring, noise and dust monitoring, underground, fire control, smart education, urban transportation public utilities, smart home, public transportation, smart business, smart buildings, railway transportation, smart business park, smart water, building energy, air transportation, urban drainage service, smart transportation, digital urban management, smart business centers, smart parking, smart properties, smart gas measurement etc. All the ecological environment correspondence like smart air pollution monitoring, water quality diagnostics monitoring, supply consumption monitoring, monitoring water resources, key pollution source, automobile exhaust, monitoring environment protection, water resources monitoring management, comprehensive monitoring of energy are a part of this layer. These new services based on real time physical world data, improving infrastructure integrity, increasing efficiency of urban management and addressing environmental degradation.

5.3 Big Data Analytic Layer

The data management and information flow layer is of two types: Periodic and Aperiodic [21]. In periodic data management IoT sensor data requires filtering because the data is collected periodically and some data may not be needed so this data needs to be filtered out. In aperiodic data management, the data is an event triggered IoT sensor data which may require immediate delivery and response for example medical emergency sensor data. In this proposed architecture big data analytic layer is used and it has big data platform and data repository as shown in Fig. 3 for periodic and aperiodic data which will help in saving and optimization of energy through enterprises IoT and analytic tools. The Green IoT communication technologies, networks and services approaches should be able to support dynamic environment through internet architecture evolution, protocols and wireless system access architectures with improved security and privacy. This layer provides the green IoT cloud platform and cloud process management for energy efficiency and optimization for the application layer. Even cloud can be segregated to fog to save more energy. It also control the management services like information analytics, security control, process modeling and device control to green IoT cloud platform and cloud process management. It is also responsible for an operational support system, security, business rule management, business process management. It has to provide service analytics platform such as statistical analytics, data mining, text mining, predictive analytics etc.

5.4 Network Layer

The lower communication layers are mostly specific for Wireless sensor network and the network and higher communication layer should preferably use common protocols in order to allow interoperability across networks. The most viable communication standard for wireless sensor network is Institute of Electrical and Electronics Engineers (IEEE 802.5.14) which defines the physical and link layer for short distance communication for smart cities with low power consumption and at low cost. It operates at the Industrial, Scientific and Medical (ISM) frequency bands i.e. 800/900 MHz and 2.4 GHz. The other communication technologies like ZigBee, WirelessHart, WIA-PA and ISA.100.11a depending upon their distances to communicate [22]. The proposed architecture is covering additional frequency bands e.g. TV white space, regional bands which operate at ultralow power for different applications like train control. Bluetooth is also a wireless short range protocol. Bluetooth 4.0 adopts a technology Bluetooth 4.0 is a low energy protocol and lightweight variant for low power applications.

The main requirements of these communication technologies is the power consumption and small computational footprints for wireless sensor network so IP protocol suite is the main candidate for these layers. Even the previously specific standards who defined their own protocol can be shifted to IP. So the WSN and IoT IPv6 is the feasible solution for smart cities applications. The IPv6 over Low-Rate Wireless Area Net- work (6LoWPAN) working group defines the mapping of IPv6 on IEEE 802.15.4 e.g. RFC 6282:

5.5 Smart City Sensor Layer and Infrastructure

This layer can supports different kinds of sensors operating with minimal power consumption and installed in different systems of the smart cities. There are three types of sensing models i.e. RFID, Wireless Sensors Networks (WSN) and crowd sourcing. RFID is an automatic identification technique to detect the tagged objects. These passive RFID tags are not battery operated rather they can take the power from the reader's interrogation signal to communicate the ID to the RFID reader. These types of systems can be useful for many applications e.g. retail and supply chain management for smart cities. WSN plays an important role in urban sensing applications. It is a feasible solution for the applications related to transportation and access control which will collect process and analyze the important information gathered from a variety of environments. The wireless sensors are smaller in size, cheaper, more intelligent and widespread (e.g., embedded camera).

As the social networking is booming a new type of sensing paradigm i.e. smart phone technology has evolved by encouraging the citizens of the smart cities to contribute towards the smart city management. It plays an important role in government citizen interaction. So, this layer must be able to support massive volume of IoT data produced by wireless sensors and smart devices. IoT sensors are aggregated with various types of protocols and heterogeneous networks using different technologies. IoT networks need to be scalable to efficiently serve a wide range of services and applications over large scale networks.

6 Conclusion

The telecommunication systems play a very important role in smart cities. It must be highly reliable and available as well as flexible, economical, and environmentally conscious. To satisfy these difficult requirements, a new architecture of smart cities is proposed which will use cloud—IoT smart systems energy efficiently and make the cities green. By adapting the proposed communication technologies gives provision of a wide variety of services in smart cities. The proposed cloud services and visual communication tools using high speed broadband communication networks in smart cities can improve business in corporate and government sectors also. Meanwhile, sensor networks utilizing variety of wireless technologies in green smart cities give access to information on the flow of goods and the status of equipment and the environment. They also facilitate the use of remote control. This makes possible the implementation of smart cities that are safe, secure, and environmentally conscious. Sensor layer provides the all green IoT through WSN to the users for using different application through cloud platform and process. In future, cooperation between communities can be encouraged as sensors and actuators, communication technologies and control systems are becoming more primitive and intelligent. Further, there is a need to develop advanced solutions for computing and communication technologies, dynamic networking and reliable software engineering with minimal energy usage. Another real research zone is to anticipate an improvement way along which most recent advancements which can be utilized to enlarge existing smart cities frameworks and applications by adding extra elements.

References

1. Mhetre, N. A., Deshpande, A. V., & Mahalle, P. N. (2016). Trust management model based on fuzzy approach for ubiquitous computing. *International Journal of Ambient Computing and Intelligence (IJACI), 7*(2), 14.
2. Hazlewood, W. R., & Coyle, L. (2009). On ambient information systems: Challenges of design and evaluation. *International Journal of Ambient Computing and Intelligence (IJACI), 1*(2), 12.
3. Schaller, A., & Mueller, K. (2009). Motorola's experiences in designing the internet of things. *International Journal of Ambient Computing and Intelligence (IJACI), 1*(1), 11.
4. Roberto, M., Abyi, B., & Domenico, R. (2015). Towards a definition of the internet of things (IoT). *IEEE Internet of Things, Issue, 1,* 1–86.
5. Tomar, P., & Kaur, G. (2016). Smart street parking and management system for smarts cities through internet of things. In *Proceedings of 2nd International Conference on Advances in Management and Decision Sciences, Organized by the School of Management* (pp. 49–55). Gautam Buddha University, Greater Noida, U.P., ISBN 978-81932836-5-3
6. Curran, K., McFadden, D., & Devlin, R. (2011). The role of augmented reality within ambient intelligence. *International Journal of Ambient Computing and Intelligence (IJACI), 3*(2), 19.
7. San, M., & Gangadharan, G. R. (2012). Green IT: An overview. In *Harnessing green IT principles and practices* (pp. 1–21).
8. Karim, S. F., Zeadally, S., & Exposito, E. (2015). Enabling technologies for green internet of things. *IEEE Systems Journal, PP*(99), 1–12.
9. Kimbahune, V. V. K., Deshpande, A. V., & Mahalle, P. N. (2017). Lightweight key management for adaptive addressing in next generation internet. *International Journal of Ambient Computing and Intelligence (IJACI), 8*(1), 20.
10. Xu, Wei, Wang, Ru Chuan, Huang, Hai Ping, & Sun, Li Juan. (2011). The research and application of vehicle identification using KPCA based on the car networking environment. *Applied Mechanics and Materials, 88–89,* 709–713.
11. Mohanty, S. P., Choppali, U., & Kougianos, E. (2016). Everything you wanted to know about smart cities: The internet of things is the backbone. *IEEE Consumer Electronics Magazine, 5*(3), 60–70.
12. Vermesan, O., & Peter, F. (2013). *Internet of things: Converging technologies for smart environments and integrated ecosystems* (pp. 1–363). Herning: Rivers Publications.
13. ISO/IEC JTC 1. (2014). Preliminary Report, Smart Cities.
14. Fantacci, R., & Marabissi, D. (2016). Cognitive spectrum sharing: An enabling wireless communication technology for a wide use of smart systems. *Future Internet, 8*(23), 1–17.
15. Dingli, A., Attard, D., & Mamo, R. (2012). Turning homes into low-cost ambient assisted living environments. *International Journal of Ambient Computing and Intelligence (IJACI), 4*(2), 23.
16. Nandyala, C. S., & Kim, H. K. (2016). Green IoT agriculture and healthcare application. *International Journal of Smart Home, 10*(4), 289–300.
17. Al-Ali, A. R., & Aburukba, R. (2015). Role of internet of things. In *The Smart Grid Technology in Journal of Computer and Communications.*
18. Medvedev, A., Fedchenkov, P., Zaslavsky, A., Anagnostopoulos, T., & Khoruzhnikov, S. (2015). Waste management as an IoT-enabled service in smart cities. In *Lecture notes in computer science* (Vol. 9247, pp. 104–115).
19. Saha, B. (2014). Green computing. *International Journal of Computer Trends and Technology, 14*(2), 46–50.
20. Roy, B. C. (2014). Green computing. *International Journal of Computer Trends and Technology (IJCTT), 14*(2), 46.
21. Kumar, M. S., Sahoo, P. K., & Wu, S.-L. (2016). Big data analytic architecture for iintruder detection in heterogeneous wireless sensor networks. *Journal of Network and Computer Applications, 66*(C), 236–249.
22. Yan, C., Han, F., Yang, Y.-H., Ma, H., Han, Y., Jiang, C., et al. (2014). Time-reversal wireless paradigm for green internet. *IEEE Internet of Things Journal, 1*(1), 81–89.

Internet of Things Shaping Smart Cities: A Survey

Arsalan Shahid, Bilal Khalid, Shahtaj Shaukat, Hashim Ali and Muhammad Yasir Qadri

Abstract Driven by the advances in hardware and software technologies, the term Internet of things has emerged as a worldwide framework of 'smart' internet-based interconnected electronic devices through web having a significant impact in the betterment of our traditional living style. The use of these web connected embedded devices, as Information and communication technologies for re-shaping modern cities, lead to the concept of smart cities. This chapter surveys the most important domains of smart cities and illustrates the recent research and development in them. After identifying critical areas, the chapter also highlights and discusses the issues and research gaps in recent technologies. Finally, it presents the opportunities and research directions in future advancements of intelligent cities.

Keywords Intelligent cities · Smart grids · Applications of internet of things (IoT) · Smart homes

A. Shahid (✉)
School of Computer Science and Informatics, University College Dublin, Belfield Dublin 4, Ireland
e-mail: arsalan.shahid@ucdconnect.ie

B. Khalid · S. Shaukat
Department of Electrical Engineering, HITEC University Taxila, Taxila, Pakistan

H. Ali
Department of Computer Science & Engineering, HITEC University Taxila, Taxila, Pakistan
e-mail: hashim.ali@hitecuni.edu.pk

M.Y. Qadri
School of Computer Science and Electronic Engineering, University of Essex, Colchester CO4 3SQ, UK
e-mail: yasirqadri@acm.org

N. Dey et al. (eds.), *Internet of Things and Big Data Analytics Toward Next-Generation Intelligence*, Studies in Big Data 30,
DOI 10.1007/978-3-319-60435-0_14

1 Introduction

1.1 Background of IoT

Nowadays, around two billion individuals across the globe utilize the Internet for browsing the web, sending and receiving messages, accessing multimedia content, playing games, using social networking applications and numerous other activities [1]. While more individuals will access such a worldwide data and correspondence foundation, another huge jump forward is coming, identified with the utilization of the internet as a worldwide stage for letting intelligent devices impart, dialog, process and organize. It is unsurprising that, within the following decade, the Internet will exist as a consistent framework of great systems and organized items [2]. Sensors and embedded electronic devices will be surrounding us, generally accessible, ready to make new applications, empowering better approaches for working, interfacing, amusement and living. In such a perspective, the traditional idea of the Internet as a framework system connecting with end-clients' terminals is fading, leaving space to a thought of interconnected "smart" objects forming pervasive computing environments [3]. The Internet foundation is not vanishing. Despite, it is holding its fundamental part as global backbone for overall data sharing and dispersion, interconnecting physical objects with registering/correspondence abilities over an extensive variety of administrations and innovations. This development is getting empowered by the embedding of electronics into everyday physical objects, making them "smart" and letting them flawlessly coordinate within the global cyber-physical infrastructure. This will offer ascent to open new doors for the information and communication technologies (ICT) segment, paving the way to new administrations and applications able to influence the interconnection of physical and virtual domains. Within such perspective, the term "Internet of things" (IoT) is comprehensively used to refer the subsequent worldwide system of interconnecting smart objects by means of developed internet enhancements and prearrangement of supporting technologies to acknowledge such dream [4, 5].

Figure 1 shows a typical representation of IoT in which different physical and virtual devices are connected to the web and ultimately sharing information to form a smart network of objects.

1.2 Applications of IoT

From the past few years, the research in IoT development has enabled over billions of devices to communicate and work efficiently via internet. Some of very well-known and recent applications of IoT includes smart home appliances, intelligent wearables such as smart watches, IoT in poultry, connected automotive, efficient retailing, smart farming, smart environment control including early earth

Fig. 1 A typical concept of internet of things

quake detection and snow level monitoring, smart agriculture enabling effective monitoring of green houses and quality of water and much more. The research towards being smarter has evolved to the extent that intelligent machines, i.e., are now getting integrating with qualities of having sensations like human [6]. A very recent emergence of robot pain (can a robot feel pain?) has attracted the attention of so many researchers. Ultimately, the aim is to have the world with smartest machines interconnected with each other via internet to bring human living standards to a far next level. In addition to this, IoT has also enabled eHealth that includes automatic patient surveillance, medical refrigerators to keep medicines/vaccines in a very controlled environment and fall detection systems in rooms that help elderly or disabled people who are living independently. In short, IoT development has been making human life easier and much more comfortable than ever.

1.3 Emergence of Smart Cities

With the ever increasing trend towards development of intelligent objects, the modern world is evolving with the concept of a broader term known as "smart cities". In simple, efficient utilization of resources, enhanced quality with improved performance are the key characteristics of devices; embedded together with the sensor networks, lead to IoTs and their use, as ICT results, in shaping smart cities [7].

This chapter presents a survey on most important components of smart cities by exploring advancement in smart technology; featuring a new role model for people

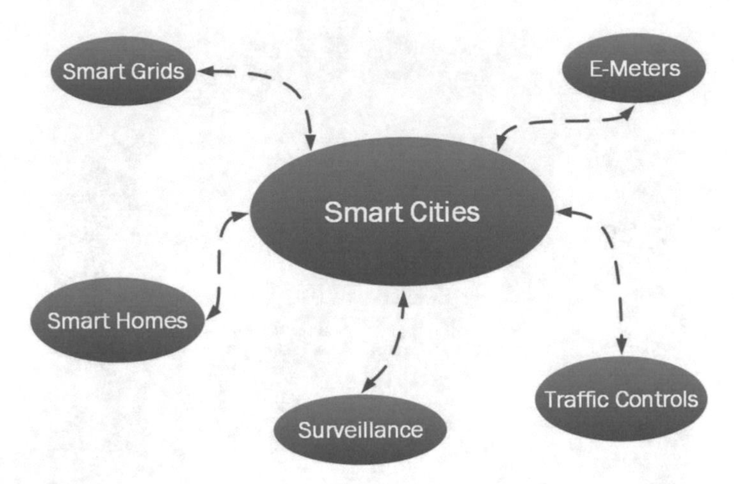

Fig. 2 A typical concept of internet of things

to make better use of the resources followed by research gaps and suggestions to further improve the living standards (see Sects. 2 and 3). Figure 2 shows the different aspects of smart cities covered in our survey. This chapter also highlights the challenges and critical shortcomings in the development in the area of smart city.

The rest of the chapter is organized as follows. Section 2 covers the major part of this chapter; starting with a detailed review of recent researches in the development of smart grids (see Sect. 2.1). Followed by, one of the major contribution of IoTs in the development of smart cities i.e. intelligent e-meters, in Sect. 2.2. Then, we illustrate smart homes that are featuring a new role model for people to make better use of the resources and improve living standards (see Sect. 2.3). In Sect. 2.4, surveillance cameras, an important product for ensuring security from thefts and serve in many other daily life applications, as well as for military purposes, have been covered. Lastly, advancements in smart traffic controls as an application of IoTs have been presented (see Sect. 2.5). In Sect. 3, we have discussed the challenges, opportunities and future directions in smart cities. Finally, in Sect. 4, conclusion is presented.

2 Smart Cities

The term "smart cities" refers to the aspect of availability and use of technology in a society or a region, which determine how much energy is being consumed and what ratio of output is being achieved by the use of such technology. All data, collected from embedded devices, sensors and other machines is gathered and analysed to extract meaningful information to output intelligent set of systems and to provide new services to the business, citizens and public administrations. The idea of smart city is not only restricted to the industrialization and administrative domains but is effectively related to the environment and atmosphere as well. Monitoring and control over the contaminants, caused by the industrial or instrumental components, is a major consideration which smart cities developers are taking into account. The ever increasing population and geographical boundaries of cities and towns also plays an important role in the communication of humans and thus effecting their daily life actions [8, 9]. In this section, we present the most trending R&D in the development of smart cities and its main features.

2.1 Smart Grids

Technological advancements are not restricted to any particular field or industry. Modern cities demand an infrastructure that is completely integrated and can balance efficiently the potential to meet the innovative challenges and competitions [10]. These challenges include the way from how we drive cars to how we make purchases and even how we get energy for our homes, from sharing of large data to the utilization of information into work, from management and administration and implementation of plans, from transportation to the medical services and many more. Internet infrastructural advancement along with the technological limits and its modification is the interest of scientists and researchers in modern world. In this section, we discuss a recently emerged form of power grid, i.e., smart gird (SG).

Generally, the term grid or power grid is used for electrical power infrastructure that may control all or a portion of the accompanying four electricity operations, i.e., (1) power generation, (2) power transmission, (3) power distribution, and (4) power control. A smart grid, is also referred to as intelligent grid, which is an upgrade of the classical twentieth century power grid [11]. The conventional grids are by and large used to transfer power from a couple of focal generators to clients. Interestingly, the SG utilizes two route path of power as well as data to make a robotized and efficient network. Apart from two-way communication, some other important features of SG include its digital design nature, complete sensors integration, distributed generation, self-monitoring and provision of many customer choices [12]. Although a complete and accurate definition of smart grids has not been presented yet; but, National Institute of Standards and Technology (NIST) has presented a conceptual model of intelligent power grids, shown in Fig. 3. This applied model partitions the SG into several functional areas. Each area incorporates one or more SG on-screen characters, including gadgets, frameworks, or projects that settle on choices and trade data that is vital for performing applications. The customers are the end users of all the smart power transmission facilities whereas the service providers are the companies providing electrical facilities. Markets are the operators of these smart facilities who regulate the transmission and supply them to individuals through distribution section.

Unlike traditional power grids, smart grid transmission enable two-way electricity and information flow. An important power generation mechanism adapted in smart grids is distributed generation (DG). The key advantage of DG is the

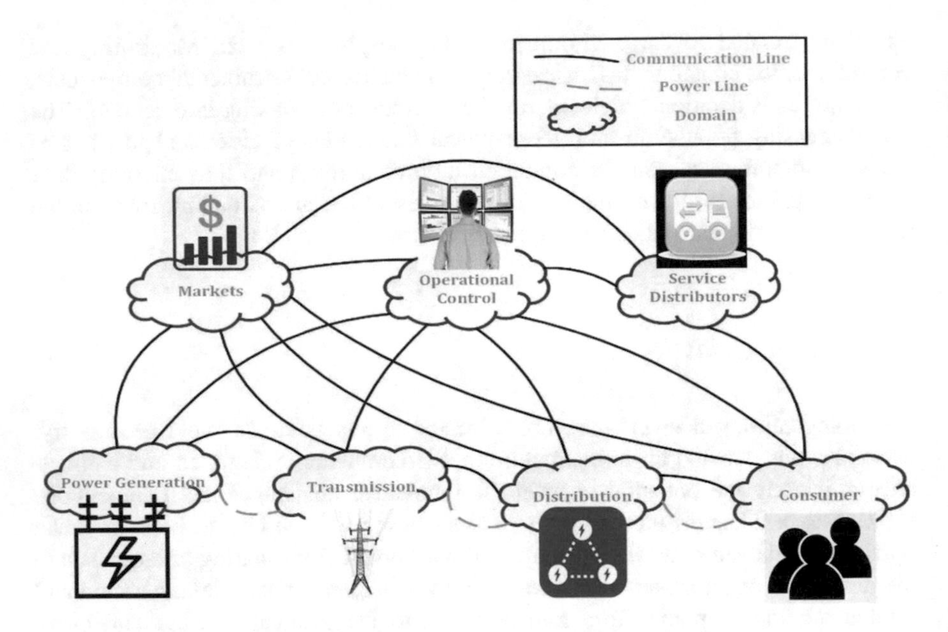

Fig. 3 Conceptual design of smart grids by NIST (adapted from [13])

distributed energy resources (DER) in the form of micro grids, i.e., small solar panels and wind turbines (ranging from 4 kW up to 10,000 kW) to improve quality, reliability and efficiency. This prevents the power distribution problems among users even if some small micro-grid fails operation or is removed from the system. Just like the advancements of power distribution systems, the transmission section of these modern grids are also improved driven by the modern power electronic technology and infrastructure challenges such as increasing load demands. As described in Byun et al. [10], on the basis of functionality, transmission part of SG can be further divided into three subsections, i.e., smart control centers, smart substations and intelligent transmission networks. The modern control centers are smart in a sense that they provide additional facilities of analysis and visualization with power flow monitoring. The smart substations are yet to be evolved with an expectation to have features like digitalization, self-healing and automatization. Finally, the new transmission networks have been enhanced by the improved electronics, sensing and communication technology helping to obtain improved quality of power and system security.

With the introduction of telecommunication phenomena in power grids, there comes a need of dealing with possible cyber security threats that may arise in smart grids [14]. Security analysts are required to pay attention on mechanism of power grids to meet the arising security issues. Therefore, cyber security is one of the main concerns in smart grids. In addition to this, the grid developers and operators also need to have a strong understanding of cyber security problems of SG era. Some of the very well-known attacks includes *stuxnet* and *night dragon.* Stuxnet causes harm to control system by reprogramming the system and hiding the specific changes. Whereas, night dragon is a mixture of techniques like social engineering, phishing, and system exploitation to harm sensible SG information.

Genge et al. [15] presented a middleware technique (SCYAMIX) having common features of HERMIX [16] and AMICI [17] platform for bringing in the real-time simulation capabilities. SCYAMIX was aimed to specifically ensure the real-time cyber security of smart grid architecture by considering its advanced features of energy management, distribution and metering. Authors claim SCYAMIX to be the first presented system for bringing in a Sensei/IoT compliant implementation. The system is to view communication protocols in detail with real-time software simulation. Their work also presented the efficiency of middleware through detailed demonstrations; and, a case study to overview the integration of other platforms such as ADHOC networks and HERMIX modules interaction with extensible messaging and presence protocol (XMPP) servers. However, their work lack the proper explanation of performance capabilities and tests in relation to real-world scenario.

Furthermore, security issues and challenges for the IoT-based smart grid for creating a secure network in order to keep track the real-time performance of the energy consumption and energy generation are discussed by the Bekara in his survey [18]. His work proposed a series of security challenges, i.e., the identity spoofing of smart meters, data modification to gain unauthorized rights by damaging them physically or remotely, and malicious software injections in order to manipulate smart grids' software. However, there are other challenges that need to be addressed such as scalability, mobility and development of the networks at a large scale. Further challenges are the incorporation of the already deployed system and adaptation of the existing gateways and the accommodation of the constrained resources.

Another survey in the field of cyber security of the smart grids was presented by Wang et al. [19]. Their work briefs the effectiveness of network protocols and web standards such as World Wide Web (WWW), distributed network protocol (DNP3), International Electrotechnical Commission (IEC 61850), Internet protocol version 4 (IPv4) and Internet protocol version 6 (IPv6) and the challenges that are being faced in smart grid networking. Major components for the network security and integration, such as availability of network and its confidentiality, are also presented; preceding the methods of cyber-attacks. Among the cyber-attacks various layers of the attacks are explained targeting the layer layout of networks. Denial of service (DoS) attack is a method in which author explains the testing of the networks integrity by investigating the vulnerabilities in the smart grids at different layers, such as physical layer's channel jamming, to show how attackers can use it to connect to the networks and exploit it. Network and transport layer are one of the main component of network which follows the TCP/IP protocol model and can be affected by DoS attacks.

In [20], Marta presented the ways of handling big amount of data which is generated as a result of computations in intelligent grids. Marta conducted his research on a local smart grid to present realistic solutions. The work described the effective role of IoT in smart grid development for improving energy consumption of grid. The work also showed that by using IoT as an alternative for the reduction in energy consumption and increasing the output efficiency of smart grid technology can be improved as it involves insertion of sensors and smart networks. These smart networks include renewable energy sources (RES) and radio-frequency identification (RFID) for the monitoring of data flow and usage at consumer end. A multilayer schematic proposal was presented to incorporate the demand response and the management for the system at different levels such as industry, aggregator and household usage. An economic overview for the IoT in smart grids is presented by Cedric [21] for advanced technology in the market. The study showed how smart grid is promoting the use and integration of RES and advanced technologies for optimal usage of the new energy products. They also highlighted the trend of smart meters' implementation in the European states. Policies and consumer demand management presenting the investors' interest and the consumer response, towards the pricing variation and their impact on the market, are detailed by various percentage ratios of strategies and plans.

In [22], Momoh presented design and control of flexible power network systems and optimization and adaptability methods to achieve reliable and efficient power systems. Multiple techniques and methods such as adaptive dynamic programming (ADP), action network and critic network methods were provided. The work presents some of smart grid challenges including supply side and demand side management technologies. The author also discusses the cases for the development of the stochastic dynamic optimal power flow, which includes different applications such as as assessment of power system reliability, adaptive control and optimization of power systems. The concluding part explains some possible solutions and strategies and an ongoing research in the field.

Energy management and its optimization on producer, buffer and consumer level can play a vital role in having a global control over the reliability of sources and its supply. Molderink et al. [23] in their work propose an algorithm for controlling and storing of the electricity and heat demand by dividing the horizon into intervals. The algorithms set an optimal combination of sources to the cost functions by selecting the number of devices and comparing it to the transients of the devices. It uses integer linear problem (ILP) to find the best solution modelled for the heat demand. However, their research gap lies in explaining the implementation of smart application devices with the system. Also, all the resources were switched-on; considering full load on the system and assuming that there was no short fall or surplus storage capacity, which could lead to a false identification of the current consumption on consumer end. The overall work presents a good model in making a choice for selecting the cost effective and optimized solution for the controlling the electricity consumption and consequently providing a margin on pricing to the consumers.

2.2 Residential E-Meters

Electricity consumption is increasing day by day at residential sector of modern cities as reported in many of reports and researches [24–26, 66]. The increase in electricity demand has urged researchers and power engineers to work and plan for overcoming the needs. Therefore, many energy efficient programs have been launched. But, to effectively plan for these programs, a strong understanding of household equipment participating in electricity consumption is needed and there should be a reliable mechanism to record those readings. In such a scenario, special energy monitoring machines is a requirement. Therefore, smart meters have been introduced in market.

The trend of using mechanical or analog energy meters that uses a rotating metallic disk, nonmagnetic in nature, to measure power passing through it, is now being replaced by electronic energy meters that shows the reading on a screen or a LCD display [27–29]. With the introduction of these meters in the industry, domestic sector has brought a smart change since these meters are capable of recording the maximum usage at peak and off-peak hours. Meter voltage and current levels are processed digitally to determine different parameters such as power factor and reactive factor. One feature that makes this device very much scalable is the capability of measuring reading at every instant of time which makes it possible to have electricity usage data record at peak and off-peak timings, thus providing a benefit to the consumers to have defined pricing for the time of the day. Some of recent advancements in the field of smart energy metering, their short-comings and research trends have been described in this section.

One of the major issues in smart e-meters is the ever existing problem of security [30]. Li et al. [31] articulated the homomorphic encryption methods to perform data aggregation on smart meters. A distributed aggregation method was presented to collect data at instantaneous points at multiple levels in order to allocate and balance the load along with resources. It also provides with the essential information in controlling and monitoring the power consumption. The homomorphic encryption used was Paillier cryptosystem [59], which uses additive homomorphic encryption functions. An aggregation tree based network topology ensures the transmission of data from destination to the nodes where destination is the in-network operations of smart meters and grids which passes through some keys which serve as the encryption for the data moving towards the destination node. This method ensures the privacy of electricity data usage along with the correct implementation of data aggregation.

Flavio et al. [32] discusses the security issues of smart e-meters with a special focus on secure communication and fraud detection. They propose a technique to obtain friendly and secret meter readings without interference in the privacy of users. The proposed idea is to measure the energy consumption of neighborhood level rather than at individual level, which would also help in identifying the behavior of consumer's energy use. Particularly it checks the time, in a day, when electricity usage is maximum, and further, it could also predict the type of devices used at a particular region. Paillier's additive encryption technique [59] has been incorporated with additive sharing methodology to perform the data measuring task at the same time to keep the privacy of the consumer. The scheme also takes care for the current leakage by reporting back reading after some instants, which can be compared to the pre-allocated factor assigned to an area by the supplier. One issue related to this scheme is the assumption that two out of N consumers are uncorrupted free consumers since the system is dealing with an order of hundreds of consumers at a time which could switch at any time. So, if the number of users fall below the above figure the privacy could be compromised. Since, the behavior of the consumer could be identified by subtracting the usage from the rest.

Another survey was presented by Le Blond et al. on a project in UK on 3eHouses in [33], regarding the smart metering implementation in domestic sector in order to demonstrate the decrease in the energy consumption by the use of information and communication technology. The paper discusses the ZigBee [34] technology, used for the meters, their developed system use IEEE 802.15.4 standard for connecting to the meters via an Ethernet link. One problem that was faced in this approach is the user's firewall protection, which prevents from sending interrupts in a public sector, however, it can be resolved by adding some additional rules to the firewall properties. Secure Shell (SSH) protocol is connected to R-tunnel which monitors the participant networks by setting up a TCP port on a ZigBee access point.

Travis et al. [35] presented Api-do, i.e., tools in their paper for exploring the wireless attack surface in smart meters to explain the methods and techniques along with a number of tools that are used for network exploitation, which could cause a security challenge for the smart meters. To start with, network identification is a first step leading towards the network hacking, different tools were used in order to identify the network; among which, OpenEar is a tool which monitors 16 channels simultaneously but at the same time assigns a unique number to each device. In this way, all connected devices become able to log data from SQL database. Another device which they present is zbWarDrive to capture the network traffic by injecting a beacon request to connected devices for monitoring purposes. For a more advanced analysis of network, ArcMap was used to interpolate the network range; even if a limited number of sample and strength points are available.

Energy metering policies is another factor that counts in when smart meters are discussed. Pupillo et al. [36], put forth the details in this area by highlighting energy metering policy approaches. Their work focuses on the details and results of a project called "E-cube project" funded by Italian Ministry and is a joint consortium of 12 major universities related to the energy fields. The main goal of the project was to create an equilibrium in rationalizing energy consumption by creating optimized and scalable infrastructure for smart meters. The project also aims to extend their domain at worldwide level and present an overview with various systems already working in leading world countries. Certain policies were assessed in order to tell the consumer response and the effect on the stakeholders to maintain a balance in the prices of metering, which in turn would make it feasible for the system implementation at a wider range. Policies regarding the system identification and consumer identity protection along with determining the personal behavior pattern, performing real time surveillance, providing accidental invasions and determining the use of specific devices are some key points to be considered by the policy makers.

Another system for smart metering is RF mesh system presented by Lichtensteiger et al. [37] for describing the smart metering systems' performance and architecture which works on frequency hopping spread spectrum (FHSS) and provides an advantage over direct sequence spread spectrum (DSSS) by having increased sensitivity and improved link budget. Figure 4 shows their typical representation of RF mesh network architecture.

The presented work also describes the simulation behavior of the performance of Landis+Gyr's GridstreamTM RF mesh based metering systems, showing the minimum time intervals of meter reading and the maximum amount of data transferred. Ad hoc networking communication is one key feature of RF mesh systems making it a valuable asset for dynamically linking with the neighboring nodes. One main thing that is considered in design is having established nodes in a neighborhood, when a node transmits data it waits to get acknowledgement and if fails to retrieve the information, it needs to resend the data or choose an alternative node to transmit data; which is a more efficient choice, practically. Hence, the down link could be altered without wasting time on resending data to a particular node again and again.

In the previous decade a lot of research has been done by companies and individual researchers to develop the suitable intelligent components that could take human living standards to a far next level. Sensors starting from wash basins to Bluetooth aware smart sofas and comfortable beds have turned out to be the most captivating innovations for everyone. The following section illustrates one of the most interesting components of smart cities, i.e., smart homes.

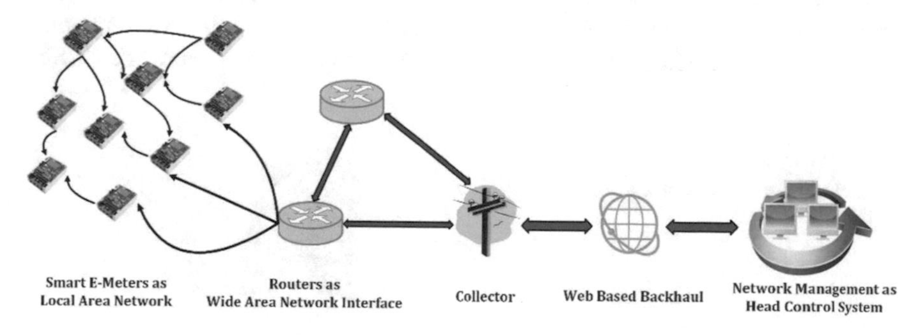

Fig. 4 RF mesh network representation

2.3 Smart Homes

Smart homes constitute an important branch of computing that includes bringing smartness into home appliances, medicinal services, wellbeing, and security [38]. Remote monitoring frameworks are regular segments of intelligent homes, which use telecommunication and web innovations to control home remotely and help the patients who are away from their caregivers. Home automation network provides a communication protocol for electronic devices that are connected to sensors and switches via a central hub or a gateway. Earlier concept of smart homes devices can be seen as automatic garage doors, coffee makers, timers and many other remote control devices [39–41]. But, technology advancements of present era have brought sensor networks integrated with many or almost all electronic devices being controlled through some application on a smartphone or through web. Which ultimately provides a seamless connection throughout a home system rather than devices working individually. In short, smart homes offer a superior personal satisfaction by presenting computerized apparatus control and assistive administrations [42].

A home needs three most important things to make it intelligent, i.e., (1) an internet network-either wired or wireless, (2) an intelligent gateway to manage sensors and system and (3) smart sensors or home automation in general with a link to devices inside home and external services. Figure 5 shows some of major concerned areas and focuses that motivates to contribute in smart homes. Three of the services are comfort, healthcare and security. One of the main objectives of smart homes, i.e., comfort can be achieved by mainly two things. First is human or gadgets activity identification and event automation and the second one is remote access and control of different areas of home. Second major objective i.e. healthcare can be obtained by intelligent indoor surveillance systems and local or remote monitoring. Finally, security services are always considered to be desired objectives and can be active by different identification methods.

Optimizing sensor technology is one of the key concern in research on smart homes. Byun et al. [44] propose an intelligent smart self-adjusting sensor for smart home services based on ZigBee communication using ZiSAS for the implementation of real-time situational based intelligent wireless sensor network (WSN).

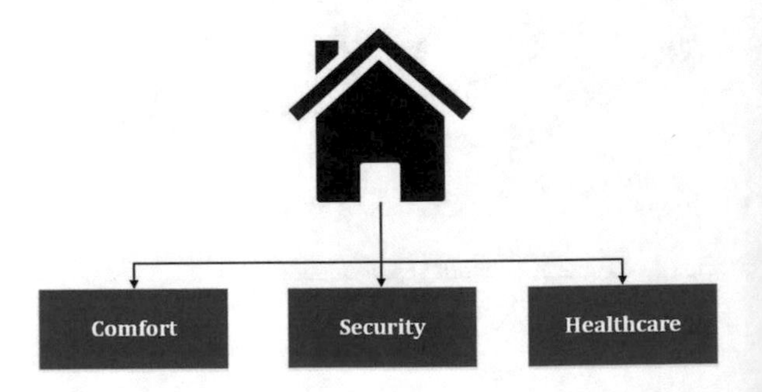

Fig. 5 Components/services of smart homes (adapted from [43])

However, some challenges considering the wireless sensor networks are the power consumption and battery lifetime along with some hardware limitations. Also the bandwidth of the network plays important role in defining the reliability and working. Some key features which ZigBee technology provides to the solution of above issue are as:

- Flexible architecture to reduce the hardware limitations.
- A situational event based control system behaviour to gather data and sense node density which was not the case in previous versions of WSN.
- A context awareness system to provide adaptable services by analysing the surroundings.
- Situational based self-adjustment so that the network can judge various parameters such as network topology, sensor density and sensitivity to the environment and can adapt to it.

ZigBee based ZiSAS architecture comprises of two parts one is self-adjusting sensor (SAS) which plays its role in sensing sensor rate, modifying topology and also gathering data and transmission according to the situation and the other is sensor management agent (SMA) which plays its role in sensor management, pattern learning and reasoning. The design of system is classified into three layers as management, network and interface. The bed test of the device shows that overall consumption of energy due to use of SAS and SMA is 3–12 and 8–34% depending on the number of devise used.

In [45], Darianian et al. presented an RFID based IoT system and services for smart homes. Their work introduces a system known as RF Energy Generator which limits the power consumption at room or appliances level. RFID master slave architecture allows the system to carry out reader services provided by the smart home, to initialize the wake up or a read process and then manipulate the ID information tags. Then, master performs the former task and latter one is done by the slave. However, many master readers can be interconnected in a system to collect item information and transmit them to further analyze and process. Another aspect of this system is mobile RFID (MRFID) reader which decreases the power consumption of multiple tag collision processes by using a master slave proxy network to answer the wake ups for the right tag by navigating to the location of desired appliance or a device. Authors [39] also gave examples to show how these devices could be connected to machines and devices to facilitate household services such as washing, cooking and even healthcare.

2.3.1 Human Activity Detection

The craving to enhance the personal satisfaction for handicapped and the fast growth of elderly individuals has provoked a tremendous effort from both scholarly world and industry to create smart home innovation [46]. For such individuals one of the most important development in smart home technology is the recognition of daily routine activities, i.e., sleeping, bathing, eating, drinking, exercising [47, 50, 51, 67] etc. It has been demonstrated that the ability to accurately distinguish the everyday exercises of people may have critical applications in healthcare [48]. For instance it might assist independent living of oldies at low medicinal services costs. Furthermore, detection of daily life activities may also contribute in indication of arising medical problems.

The development of system to detect the human activity and adapt itself accordingly to provide the most optimal services can be achieved by using adaptive neural networks (ANN). A challenging task is to track the human behavior since humans have a periodic variation in their daily life activities; and they also perform simultaneous tasks at the same time which is difficult to track. A very powerful method to tackle such problem is the data mining techniques for which different sensor networks are used to study the behavior through collected data. Zheng et al. [49] presented a growing self-organizing maps (GSOM) based on self-adaptive neural networks (SANN), which provides a brief study on data mining by using computational approach to monitor behavior of human activity within a closed environment of smart home. GSOM was proved to be dynamic since it can branch out various nodes after recognizing and visualizing patterns and thus has a self-evolving nature, which provides a good measure to handle large sets of data. The experiment conducted showed the detailed behavior of human activities at different time in various parts of home. One good feature of this system is having a record of activities and its ability to tell if there is any abnormal activity performed, that helps in maintaining and securing homes.

2.4 *Surveillance Cameras*

The growth of the IoTs in the development of smart cities has seen a massive advancement in wearable, non-computing devices, business to business (B2B) IoT technology and consequential momentum in the last few years. One subset of the IoT that has brought a significant growth is the internet-enabled surveillance cameras [40, 41]. A video camera used to monitor an area or building by transmitting recorded images to a central control room is known as surveillance camera. In this section we present some of the developments and researches in the advancements of these security cameras.

Perpetual video camera for IoT is proposed by Chen et al. [52]. In their tutorial, they highlighted the importance of perpetual video camera whose net energy consumption is approximately zero. They also presented the design challenges of such cameras, which include:

- High power consumption
- High data rate
- Heavy loading and maintenance cost of distributed cameras

They also provided the possible solutions of the above mentioned issues to design a camera with the properties of energy harvesting module, low power consumption, distributed video analysis engine and distributed video coding.

Chien et al. [53] introduce the concept of distributed computing and SoC design for smart cameras to reduce the transmission bandwidth requirement and to off-load the computation from the cloud servers. Figure 6 shows a simplified smart camera based distributed architectures in a smart city.

The end gadgets of this system are smart cameras. Advances in semiconductor technology has enabled to integrate more complex computations in intelligent cameras. Figure 7 shows that the cameras are associated locally to an aggregator or a passage. Next, they are further attached to a cloud networks. It is basically a demonstration of distributed resources (cameras) that are attached to each other in a city. By the help of different case studies, they observed that their proposed design can not only achieve greater coverage area but also power efficiency. They use the methodology to embed more computations into sensors and aggregators so that on every node computation can be distributed and we get rid of to employ centralized solutions on cloud servers.

Satyanarayanan et al. [54] analyze the technical issues that are implicated in creating an Internet-scale searchable repository for crowd sourced video content in edge analytics in a machine-to-machine/IoT network. The authors consider a hybrid cloud architecture to overcome the high cumulative data rate of incoming videos from different cameras and automated modification of video streams to sustain privacy. Methodology used was of sampling video frames, metadata-based filters with low computational complexity.

In [55], Liu et al. introduces the technology that represents video from different perspectives and describe the principle and coding structure of multi-view video. They pay a great attention on the application of intelligent security system and the

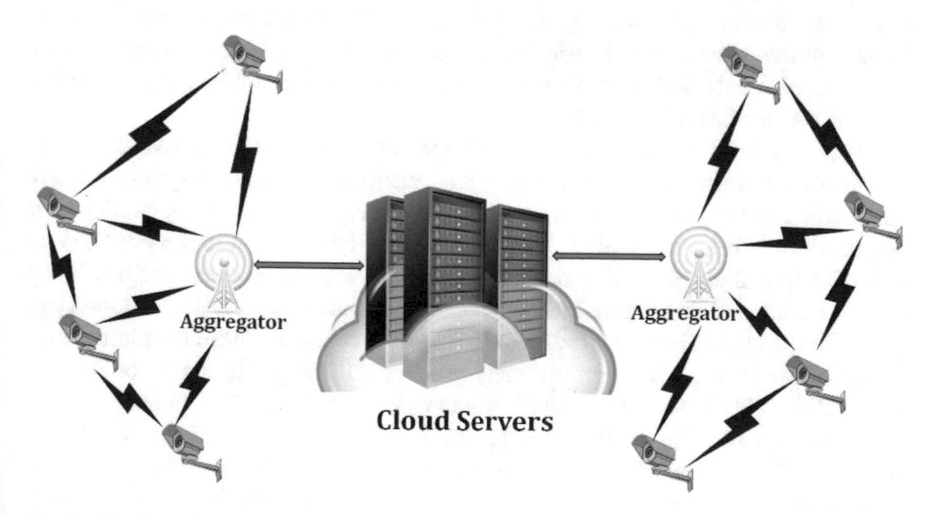

Fig. 6 Distributed smart cameras in a network

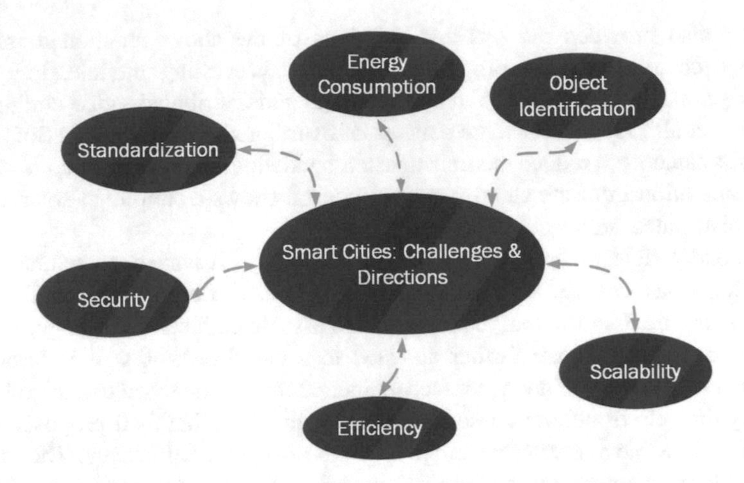

Fig. 7 Major challenges, opportunities and directions of IoTs in smart cities

challenges faces in this field. They observe the results by comparing data of single view video with multi-view video and proposed the concept of multi-view video coding. The research gap is large capacity data compression, information data extraction, and information data security management.

2.5 Traffic Controls

Traffic congestion has become one of the major problems in downtown areas. Due to rash traffic, horrible incident takes place and urban areas are facing several challenges because of alarming trend of growing populations. Therefore, smart traffic controls systems are considered to be a promising approach to avoid many road side accidents. IoT serves as the major component in making traffic control mechanism smart and intelligent.

Many solutions for improving traffic light systems have been introduced by researches and many of these techniques have been employed in the system. An IoT based solution for traffic light, where the traffic behavior changes dynamically is presented by Misbahuddin et al. [56]. This paper proposes that traffic flow can be dynamically controlled by on site traffic officers through their smart gadgets. They used an intelligently configured web server on hardware machine, i.e., Raspberry Pi (RPi), for the traffic signal controlling algorithm. In order to control the traffic lights in one of the patterns or to enforce any new pattern, the onsite traffic officer can access RPi's GPIO interface through a smart phone, utilizing the WebIoPi framework installed on RPi. In their work, one of the shortcoming is that the traffic situational information does not automatically pass to the RPi unit controlling the lights at an intersection. So, the authorities cannot take a quick decision. It is likely that prioritizing certain roads or certain directions for longer durations may cause traffic problems on other roads.

Nowadays, the navigation aids that do not work properly may cause fatal maritime accidents due to the rapidly increasing maritime traffic volume. Lighthouse and buoy tender method cannot check the real-time condition of maintenance aids. In [57], Cho et al. highlight the issues and complexity of maritime transportation and also suggests an efficient method to manage navigation aids based on IoT technologies in their work. The IoT-based navigation aids management system provides relevant data to comprehensive navigation aids management center and maritime traffic control center. Particularly, it sends a relevant data such as the identification service, the location identification and tracking service, as well as the operation status notification service to major relevant entities with the help of real-time information delivery service. By the use of IoT technology, time and expenses which can be correlated with extensive amount of maritime accidents are expected to be reduced. Moreover, the reliability and stability of the navigation aids will rise because no loss of information will occur through this method. In addition, on real-time basis a research on processing, analysis, and utilization of the various data collected using an IoT equipment will be carried out in the future.

The advancements in cloud computing and IoT have provided a promising opportunity to resolve the challenges caused by increasing the transportation issues. He et al. [58], supports the advances in IoT to resolve the transportation issues by using cloud computing and IoT technologies. Two innovative vehicular data cloud services describe a number of benefits such as car maintenance, to overcome road congestion, traffic management and enhancing road safety. As a consequence, remote security for disabling engine and remote diagnosis, have been developed to enhance driver's safety, convenience, and enjoyment. The proposed IoT-based layered architecture supports three new cloud services as follows:

- Vehicle provide their networking and data processing capabilities to other vehicles through the cloud.
- Some vehicles may need some specific application that require large amount of storage space.
- Thus vehicles that have unused storage space can share their storage space as a cloud-based service.

As a community, vehicular data clouds offer a variety of cooperative information, hazard location warning, lane change warning and parking availability. By using the two modified data mining models, the authors [49] demonstrate how data mining cloud service could be used to identify hidden potential issues to avoid accidents. But a research gap in their work is not to address the solution of complexity involved to enable the vehicular clouds. Due to this a number of challenges such as scalability, security, reliability, privacy, quality of service, and lack of global standards still exist.

In the field of IoT the metropolitan traffic management act as a vital role for highly integrated organizational processes. Foschini et al. [59] present and discuss the design and implementation of an M2M application in the field of road traffic management that assimilate, for the sake of efficiency. Their work demonstrates both the technical and organizational processes viewpoints which is already available service in IoT technology. In restricted city areas based on standard infrastructure and software components, the design of an integrated IoT retractable bollard management is present. Precisely, to simplify system maintenance and management a standard solution of integrating IoT-based services is presented.

Chen et al. [60] introduced a cyber-transportation systems for convalescent of road safety and efficiency. They also proposed the machine to machine (M2M) system consolidating intelligent road with unmanned automobile. In this article, the authors propose a solution which is applicable on an unmanned vehicle in the form of cyber physical system (CPS). They mentioned the cellular and capillary M2 M challenges, which comprise of the complexity, power, data rates, delays, security and peer-to-peer traffic. For the future work, the speed of the unmanned vehicle is correlated with performance of the system. By increasing speed many factors affect its response. To counter this, a new technique for the safety of system should be established.

3 Smart Cities: Challenges and Opportunities

Communication and connectivity is a major concern in any professional, commercial or industrial field [2]. Modern trends in smart cities have brought a revolution in communication worldwide; further contributing to change and improve living perspectives of entire human race. In the greater part of the presented areas in this chapter, we have highlighted some of the most important researches that have contributed to develop intelligent urban communities (smart cities). In this section, we discuss and summarize the most challenging areas of smart cities, the research opportunities and the future directions that would further take the concept of smart cities to a far next level. Figure 7 highlights the recent research opportunities, major challenges and directions in various applications of IoTs in smart cities.

3.1 Standardization Limitation

Internet of things is entirely related to internet, web protocols, and host identity protocols (HIP); but, due such large number of consumers, a lot of limitations occur while using Internet protocol version 4 (IPv4) [61]. One of such limitations involves the problem of identification and naming management to the users, as each consumer has to be assigned a unique identity [62] co-existence of the networks creates a problem for the identification of mobile and radio broadcasting, and other service spectrums at various nodes by use of techniques as dynamic spectrum access (DSA). Standards for bilateral communication techniques are being considered to interpolate the semantic web cross platforms such as XML, ASN.1 and many other web services. Research in the past decade has been conducted in this area on a very large scale and a lot of improvement is brought forward to replace IPv4 with IPv6. The essential capacity of IPv6 is to take into account more unique TCP/IP address identifiers, as we have come up short on the 4.3 billion addresses made with IPv4. Therefore, IPv6 is such a critical advancement for the Internet of things (IoT). Web associated items are turning out to be progressively prominent, keeping in mind IPv4 addresses couldn't take care of the demand for IoT items, IPv6 gives IoT items a stage to work on for quite a while. While IPv6 is a phenomenal and important overhaul from IPv4, it is still not an end to all research problems for the IoT. One major issue that organizations face is deciding how to get IPv4 and IPv6 gadgets to impart and dialogue appropriately and meaningfully.

3.2 *Object Identification*

Another limitation in this sector is object/human identification and detection [63]. Numerous electronic devices are embedded by RFID tags and contain huge amount of data, to be sensed and identified by other devices. And, the information is processed to detect whether it is about location, orientation, motion, temperature or any chemical change. IoT devices identify the objects and determines the objects on the basis of their identification capabilities, since all the objects would be containing the identification tags so it is a demanding challenge for the network to differentiate among such a huge number of objects.

3.3 *Security and Ethics Challenges*

With billions of new intelligent items being developed every day, security is an essential thought in the back of all IoT architects' minds [19, 64]. The earlier security networks were designed to only incorporate the privacy of humans; but, with the emerging concept of IoT, the security of the devices is also considered to be an important issue. As this technology is capable of human to things (H2T) communication and hence is not a suitable measure for this system in most of the cases [65], a proper encryption is needed to make sure that who can access the data and to prevent external sources to monitor or interfere.

3.4 *Transmission Reliance and Efficiency*

Wireless sensor networks (WSNs) and communication flows have become an integral part of IoT based communication in intelligent devices. Therefore, another considered research problem is the transmission of data within a specified spectrum. As data is to be transferred via many mediums like wireless networks, GPS, and Bluetooth etc., the communication should be efficient enough to transmit data without any loss of information. Efficiency of data transmission with a particular hardware is a crucial criterion for a wireless system and hence system efficiency and better/reliable communication flows is also under development.

3.5 *Energy Consumption*

The billions, and eventually trillions, of sensors and electronic devices are interacting with each other in modern IoT based cities. The efficiency, i.e., the consumption of resources among which the energy being the most important, is to be taken into account while designing an IoT based architecture of the system [57, 58]. The consumption of energy determines which type of technology is being used and how much it is contributing towards green environment.

3.6 *Scalability and Existing Networks*

As per a report put out by Gartner, 25 billion "smart devices" will be associated with the web by the year 2020. That is a really staggering estimation, considering the same report noticed that 4.9 billion intelligent gadgets will be associated in 2015. It can be estimated that about 400% expansion in development in just five years reveals some insight into the amount of exponential IoT development; we can hope to find in the following 10, 20, or even 50 years. Due to an already existence of a large number of smart products, another focused research domain is the incorporation of these devices and their scalability at such a large network. An efficient middleware is to be incorporated to effectively manage these devices at a common place; also a proficient semantic user interface is required to easily differentiate between the existing and new devices.

4 Conclusion

The current research is investigating the concept of smart city and related factors, which have enormous impact in our daily life, with an emphasis on improving its framework's ability to provide customized IoT functionalities. In this survey, we have presented some of the most advanced and trending researches in the fields of smart IoT based products that are ultimately shaping smart cities. The chapter covered five distinct areas in development of a smart city. We started with highlighting the improvements in smart grids' architecture and its research motivation;

followed by a very similar product of modern world, i.e., intelligent e-meters. We have discussed about how researchers are focusing to find out the solutions related to security threats in these smart e-meters. At that point, we examined the most intriguing advancements in smart homes and also illustrated research gaps in them. Moreover, we presented some critical features of surveillance cameras and traffic control systems. Finally, we discussed overall challenges, research directions and opportunities in developing smart cities.

References

1. Aazam, M., Khan, I., Alsaffar, A. A., & Huh, E.-N. (2014). Cloud of things: Integrating Internet of things and cloud computing and the issues involved. In *Proceedings of 2014 11th International Bhurban Conference on Applied Sciences & Technology (IBCAST) Islamabad, Pakistan, 14th–18th January*, 2014 (pp. 414–419). IEEE.
2. Alam, M. R., Reaz, M. B. I., & Ali, M. A. M. (2012). A review of smart homes—Past, present, and future. *IEEE Transactions on Systems, Man, and Cybernetics, Part C (Applications and Reviews), 42,* 1190–1203.
3. Aldrich, F. K. (2003). Smart homes: Past, present and future. In *Inside the smart home* (pp. 17–39). Berlin: Springer.
4. Amin, S. M., & Wollenberg, B. F. (2005). Toward a smart grid: Power delivery for the 21st century. *IEEE Power and Energy Magazine, 3,* 34–41.
5. Atzori, L., Iera, A., & Morabito, G. (2010). The internet of things: A survey. *Computer Networks, 54,* 2787–2805.
6. Augusto, J. C., & Nugent, C. D. (2006). Smart homes can be smarter. In *Designing smart homes* (pp. 1–15). Berlin: Springer.
7. Bekara, C. (2014). Security issues and challenges for the IoT-based smart grid. *Procedia Computer Science, 34,* 532–537.
8. Benzi, F., Anglani, N., Bassi, E., & Frosini, L. (2011). Electricity smart meters interfacing the households. *IEEE Transactions on Industrial Electronics, 58,* 4487–4494.
9. Berbers, Y., De Decker, B., Moons, H., & Verbaeten, P. (1987). The design of the HERMIX distributed system. In *Proceedings of the 34th ISMM International Symposium Mini and Microcomputers*.
10. Byun, J., Jeon, B., Noh, J., Kim, Y., & Park, S. (2012). An intelligent self-adjusting sensor for smart home services based on ZigBee communications. *IEEE Transactions on Consumer Electronics, 58,* 794–802.
11. Chan, M., Campo, E., Estève, D., & Fourniols, J.-Y. (2009). Smart homes—Current features and future perspectives. *Maturitas, 64,* 90–97.
12. Chan, M., Estève, D., Escriba, C., & Campo, E. (2008). A review of smart homes—Present state and future challenges. *Computer Methods and Programs in Biomedicine, 91,* 55–81.
13. Chen, S., Xu, H., Liu, D., Hu, B., & Wang, H. (2014). A vision of IoT: Applications, challenges, and opportunities with china perspective. *IEEE Internet of Things Journal, 1,* 349–359.
14. Chen, Y.-K., & Chien, S.-Y. (2012). Perpetual video camera for Internet-of-things. In *Visual communications and image processing (VCIP), 2012 IEEE* (pp. 1–7). IEEE.
15. Genge, B., Haller, P., Gligor, A., & Beres, A. (2014, May). An approach for cyber security experimentation supporting sensei/iot for smart grid. In *2nd International Symposium on Digital Forensics and Security*.

16. Chien, S.-Y., Chan, W.-K., Tseng, Y.-H., Lee, C.-H., Somayazulu, V. S., & Chen, Y.-K. (2015). Distributed computing in IoT: System-on-a-chip for smart cameras as an example. In *The 20th Asia and South Pacific Design Automation Conference* (pp. 130–135). IEEE.
17. Cho, M., Choi, H. R., & Kwak, C. (2015). A study on the navigation aids management based on IoT. *International Journal of Control and Automation, 8,* 193–204.
18. Chourabi, H., Nam, T., Walker, S., Gil-Garcia, J. R., Mellouli, S., Nahon, K., et al. (2012). Understanding smart cities: An integrative framework. In *System Science (HICSS), 2012 45th Hawaii International Conference on* (pp. 2289–2297). IEEE.
19. Cities, S. (2013). Trace analysis and mining for smart cities: Issues, methods, and applications. *IEEE Communications Magazine, 51,* 121.
20. Clastres, C. (2011). Smart grids: Another step towards competition, energy security and climate change objectives. *Energy Policy, 39,* 5399–5408.
21. Darianian, M., & Michael, M. P. (2008). Smart home mobile RFID-based Internet-of-things systems and services. In *2008 International Conference on Advanced Computer Theory and Engineering* (pp. 116–120). IEEE.
22. Fang, X., Misra, S., Xue, G., & Yang, D. (2012). Smart grid—The new and improved power grid: A survey. *IEEE Communications Surveys & Tutorials, 14,* 944–980.
23. Farhangi, H. (2010). The path of the smart grid. *IEEE Power and Energy Magazine, 8,* 18–28.
24. Fischer, C. (2008). Feedback on household electricity consumption: A tool for saving energy? *Energy Efficiency, 1,* 79–104.
25. Foschini, L., Taleb, T., Corradi, A., & Bottazzi, D. (2011). M2M-based metropolitan platform for IMS-enabled road traffic management in IoT. *IEEE Communications Magazine, 49,* 50–57.
26. Garcia, F. D., & Jacobs, B. (2010). Privacy-friendly energy-metering via homomorphic encryption. In *International Workshop on Security and Trust Management* (pp. 226–238). Berlin: Springer.
27. Genge, B., Haller, P., Gligor, A., & Beres, A. (2014). An approach for cyber security experimentation supporting sensei/IoT for smart grid. In *2nd International Symposium on Digital Forensics and Security.*
28. Genge, B., Siaterlis, C., & Hohenadel, M. (2012). Amici: An assessment platform for multi-domain security experimentation on critical infrastructures. In *International Workshop on Critical Information Infrastructures Security* (pp. 228–239). Berlin: Springer.
29. Goodspeed, T., Bratus, S., Melgares, R., Speers, R., & Smith, S. W. (2012). Api-do: Tools for exploring the wireless attack surface in smart meters. In *System Science (HICSS), 2012 45th Hawaii International Conference on* (pp. 2133–2140). IEEE.
30. Guesgen, H. W., & Marsland, S. (2016). Using contextual information for recognising human behaviour. *International Journal of Ambient Computing and Intelligence (IJACI), 7,* 27–44.
31. Hall, P. (1988). *Cities of tomorrow*. Oxford: Blackwell Publishers.
32. Harrison, C., & Donnelly, I. A. (2011). A theory of smart cities. In *Proceedings of the 55th Annual Meeting of the ISSS-2011, Hull, UK.*
33. He, W., Yan, G., & Da Xu, L. (2014). Developing vehicular data cloud services in the IoT environment. *IEEE Transactions on Industrial Informatics, 10,* 1587–1595.
34. Heer, T., Garcia-Morchon, O., Hummen, R., Keoh, S. L., Kumar, S. S., & Wehrle, K. (2011). Security challenges in the IP-based Internet of things. *Wireless Personal Communications, 61,* 527–542.
35. Internet of things and big data technologies for next|Chintan Bhatt| Springer, n.d.
36. Jabłońska, M. R. (2014). Internet of things in smart grid deployment. *Rynek Energii, 2,* 111.
37. Kimbahune, V. V., Deshpande, A. V., & Mahalle, P. N. (2017). Lightweight key management for adaptive addressing in next generation Internet. *International Journal of Ambient Computing and Intelligence (IJACI), 8,* 50–69. doi:10.4018/IJACI.2017010103
38. Kinney, P., et al. (2003). Zigbee technology: Wireless control that simply works. In *Communications Design Conference* (pp. 1–7).
39. Kinzer, S. (2004). Chicago moving to "smart" surveillance cameras. *New York Times,* A18.
40. Kopetz, H. (2011). Internet of things. In *Real-Time Systems* (pp. 307–323). Berlin: Springer.

41. Kortuem, G., Kawsar, F., Sundramoorthy, V., & Fitton, D. (2010). Smart objects as building blocks for the internet of things. *IEEE Internet Computing, 14,* 44–51.
42. Le Blond, S. P., Holt, A., & White, P. (2012). 3eHouses: A smart metering pilot in UK living labs. In *2012 3rd IEEE PES Innovative Smart Grid Technologies Europe (ISGT Europe)* (pp. 1–6). IEEE.
43. Li, F., Luo, B., & Liu, P. (2010). Secure information aggregation for smart grids using homomorphic encryption. In *Smart Grid Communications (SmartGridComm), 2010 First IEEE International Conference on* (pp. 327–332). IEEE.
44. Lichtensteiger, B., Bjelajac, B., Müller, C., & Wietfeld, C. (2010). RF mesh systems for smart metering: System architecture and performance. In *Smart Grid Communications (SmartGridComm), 2010 First IEEE International Conference on* (pp. 379–384). IEEE.
45. Liu, Z., & Yan, T. (2013). Study on multi-view video based on IOT and its application in intelligent security system. In *Mechatronic Sciences, Electric Engineering and Computer (MEC), Proceedings 2013 International Conference on* (pp. 1437–1440). IEEE.
46. Misbahuddin, S., Zubairi, J. A., Saggaf, A., Basuni, J., Sulaiman, A., Al-Sofi, A., et al. (2015). IoT based dynamic road traffic management for smart cities. In *2015 12th International Conference on High-Capacity Optical Networks and Enabling/Emerging Technologies (HONET)* (pp. 1–5). IEEE.
47. Molderink, A., Bakker, V., Bosman, M. G., Hurink, J. L., & Smit, G. J. (2009). Domestic energy management methodology for optimizing efficiency in smart grids. In *PowerTech, 2009 IEEE Bucharest* (pp. 1–7). IEEE.
48. Molina-Markham, A., Shenoy, P., Fu, K., Cecchet, E., & Irwin, D. (2010). Private memoirs of a smart meter. In *Proceedings of the 2nd ACM Workshop on Embedded Sensing Systems for Energy-Efficiency in Building* (pp. 61–66). ACM.
49. Momoh, J. A. (2009). Smart grid design for efficient and flexible power networks operation and control. In *Power Systems Conference and Exposition, 2009. PSCE'09. IEEE/PES* (pp. 1–8). IEEE.
50. Najjar, M., Courtemanche, F., Hamam, H., Dion, A., & Bauchet, J. (2009). Intelligent recognition of activities of daily living for assisting memory and/or cognitively impaired elders in smart homes. *International Journal of Ambient Computing and Intelligence (IJACI), 1,* 46–62. doi:10.4018/jaci.2009062204
51. Odella, F. (2016). Technology studies and the sociological debate on monitoring of social interactions. *International Journal of Ambient Computing and Intelligence (IJACI), 7,* 1–26. doi:10.4018/IJACI.2016010101
52. Poland, M. P., Nugent, C. D., Wang, H., & Chen, L. (2011). Smart home research: Projects and issues. In *Ubiquitous Developments in Ambient Computing and Intelligence: Human-Centered Applications* (pp. 259–272). IGI Global.
53. Pupillo, L., D'Amelia, B., Garino, P., & Turolla, M. (2010). Energy smart metering and policy approaches: The E-cube project.
54. Satyanarayanan, M., Simoens, P., Xiao, Y., Pillai, P., Chen, Z., Ha, K., et al. (2015). Edge analytics in the internet of things. *IEEE Pervasive Computing, 14,* 24–31.
55. Shapiro, J. M. (2006). Smart cities: Quality of life, productivity, and the growth effects of human capital. *The Review of Economics and Statistics, 88,* 324–335.
56. Sigg, S., Shi, S., & Ji, Y. (2014). Teach your WiFi-device: Recognise simultaneous activities and gestures from time-domain RF-features. *International Journal of Ambient Computing and Intelligence (IJACI), 6,* 20–34.
57. Solanki, V. K., Katiyar, S., BhashkarSemwal, V., Dewan, P., Venkatasen, M., & Dey, N. (2016). Advanced automated module for smart and secure city. *Procedia Computer Science, 78,* 367–374.
58. Tamane, S., Solanki, V. K., & Dey, N. (2001). *Privacy and security policies in big data.* IGI Global.
59. van Rysewyk, S. (2014). Robot pain.

60. Vlacheas, P., Giaffreda, R., Stavroulaki, V., Kelaidonis, D., Foteinos, V., Poulios, G., et al. (2013). Enabling smart cities through a cognitive management framework for the internet of things. *IEEE Communications Magazine, 51,* 102–111.
61. Wang, W., & Lu, Z. (2013). Cyber security in the smart grid: Survey and challenges. *Computer Networks, 57,* 1344–1371.
62. Weiss, M., Mattern, F., Graml, T., Staake, T., & Fleisch, E. (2009). Handy feedback: Connecting smart meters with mobile phones. In *Proceedings of the 8th International Conference on Mobile and Ubiquitous Multimedia* (p. 15). ACM.
63. Whitmore, A., Agarwal, A., & Da Xu, L. (2015). The Internet of things—A survey of topics and trends. *Information Systems Frontiers, 17,* 261–274.
64. Wolde-Rufael, Y. (2006). Electricity consumption and economic growth: A time series experience for 17 African countries. *Energy Policy, 34,* 1106–1114.
65. Wolf, W., Ozer, B., & Lv, T. (2002). Smart cameras as embedded systems. *Computer, 35,* 48–53.
66. Yoo, S.-H. (2005). Electricity consumption and economic growth: Evidence from Korea. *Energy Policy, 33,* 1627–1632.
67. Zheng, H., Wang, H., & Black, N. (2008). Human activity detection in smart home environment with self-adaptive neural networks. In *Networking, Sensing and Control, 2008. ICNSC 2008. IEEE International Conference on* (pp. 1505–1510). IEEE.

Big Data Analytics for Smart Cities

V. Bassoo, V. Ramnarain-Seetohul, V. Hurbungs, T.P. Fowdur and Y. Beeharry

Abstract The main objectives of smart cities are to improve the well being of its citizens and promote economic development while maintaining sustainability. Smart cities can enhance several services including healthcare, education, transportation and agriculture among others. Smart cities are based on the ICT framework including the Internet of Things (IoT) technology. These technologies create voluminous amount of heterogeneous data, which is commonly referred to as big data. However these data are meaningless on their own. New processes need to be developed to interpret the huge amount of data gathered and one solution is the application of big data analytics techniques. Big data can be mined and modelled through the analytics techniques to get better insight and to enhance smart cities functionalities. In this chapter, four state-of-the-art big data analytics techniques are presented. Applications of big data analytics to five sectors of smart cities are discussed and finally an overview of the security challenges for big data and analytics for smart cities is elaborated.

V. Bassoo (✉) · T.P. Fowdur
Department of Electrical and Electronic Engineering, Faculty of Engineering,
University of Mauritius, Réduit, Mauritius
e-mail: v.bassoo@uom.ac.mu

T.P. Fowdur
e-mail: p.fowdur@uom.ac.mu

V. Ramnarain-Seetohul
Department of Information Communication Technology, Faculty of Information,
Communication and Digital Technologies, University of Mauritius, Réduit, Mauritius
e-mail: v.seetohul@uom.ac.mu

V. Hurbungs
Department of Software and Information Systems, Faculty of Information,
Communication and Digital Technologies, University of Mauritius, Réduit, Mauritius
e-mail: v.hurbungs@uom.ac.mu

Y. Beeharry
Faculty of Information, Communication and Digital Technologies,
University of Mauritius, Réduit, Mauritius
e-mail: y.beeharry@uom.ac.mu

N. Dey et al. (eds.), *Internet of Things and Big Data Analytics Toward Next-Generation Intelligence*, Studies in Big Data 30,
DOI 10.1007/978-3-319-60435-0_15

Keywords Big data · Internet of things · Smart cities · Predictive analytics · Semantic analytics · Edge analytics · Security analytics

1 Introduction

Half of the world's population currently lives in cities [1]. The World Health Organisation (WHO) has predicted that 70% of the global population will live in urban areas by 2050 [2]. Moreover, it is anticipated that by 2025, there will be 37 megacities of over 10 million inhabitants and 22 of those megacities will be located in Asia [3]. It is evident that many cities will not be able to cope with the increase in population. Most of the operations in those developing and expanding cities are uncoordinated and inadequate. Additionally, there are practically no automatic data capture processes and the data captured are not analysed adequately [4].

The implementation of smart cities will help alleviate the burdens on those growing cities and significantly improve the quality of life of their citizens. Smart cities will employ a large number of Internet of Things (IoT) devices connected by high-speed networks and managed by intelligent systems. It is forecasted that by 2019, 35 billions objects will be connected to the Internet [5] and this will generate massive volumes of data. According to a recent study [6], it is believed that by the year 2020, the digital universe would comprise of 40 Zettabytes of data and that it would double every 2 years.

This massive surge of structured and unstructured data will require the ability to store, manage and analyse the data securely. The representation of the input data and generalization of the learnt patterns for use on future unknown data is the general focus of machine learning [7]. Data can be classified as structured, semi structured and unstructured [8]. Relational database can accommodate only structured data. However, the semi structured and unstructured data can also be analysed to provide new opportunities. In this information age, querying databases is no longer the right solution to obtain right and timely information. Big data analytics, which is the use of advanced techniques to examine vast amount of structured and unstructured data to get hidden patterns, correlations and other insights [9], is a solution and key to success nowadays. Big data analytics can help authorities get near real-time assessment of the situation in their respective cities and take relevant corrective measures. These technological innovations will lead to the birth of sustainable, scalable and smart cities.

In this chapter, we have presented four state of the art analytic techniques. We have also provided in depth reviews of the applications and benefits of big data analytics in the sectors of healthcare, education, transport, weather forecasting and agriculture. The last section deals with the security aspect of big data.

2 Big Data Analytics for Smart Cities

Big data can be very beneficial to smart cities if crucial information can be extracted and analysed from the volume of data. However, to extract value from the big data, data must be explored and processed at the right time. This can be achieved with the proper combination of people, process and technology. Big data analytics, which is concerned with analysing trends and finding pattern to model data, can be very helpful to smart cities departments. Using statistical analysis, predictive modelling and other techniques, business decisions can be enhanced. Also, analytics can help to improve the efficiency of essential processes, operations and roles. Analytics can upgrade business performance by transforming information into intelligence [10]. Some important analytics and techniques that can be used by smart cities are listed below and shown in Fig. 1:

- Edge Analytics
- Semantic Analytics
- Security Analytics
- Predictive Analytics

Fig. 1 Big data analytics for smart cities

2.1 Edge Analytics

Edge analytics also known as distributed analytics refers to the processing of raw data performed at the data collection point or close to it. The sheer volume and velocity of data collected by modern systems is difficult to manage by most infrastructures. The concept of edge analytics is appealing as it helps to reduce the load on network systems and reduce delays due to network congestion when dealing with time sensitive data, therefore providing faster insights [11]. Some IoT devices generate continuous stream of data that might not always be of value, therefore it is preferable that the data is analysed at the source and useless data is discarded freeing up expensive resources such as storage and bandwidth. Many IoT systems have implemented a three-tier distributed analytics architecture where simple analytics are performed by the smart device, complex multi-device analytics are performed at the servers or gateways and big data analytics performed on the cloud [12].

High-speed powerboat racing uses edge analytics to examine boat performance data, which needs to be fed back to the crew in near real-time. Analytics performed within the boat removes network delays and provides instant feedback giving the team a split-second advantage [13]. A potential application of edge analytics is in the management of localised pollution levels caused by traffic. If the air quality plummets below a threshold for a certain area, the traffic could be regulated to reduce pollution [13]. It is predicted that by 2019, edge analytics will be applied on 40% of IoT generated data [14].

2.2 Semantic Analytics

Semantic analytics refers to the ability of technology to understand the context of words and the contextual relationship between a group of words. Advances in natural language processing have helped to improve semantic analysis. The Internet and social media generate a huge amount of data and the challenge of semantic analytics is to extract meaningful and useful information out of the raw unstructured data [15]. Often, texts that are posted online are written in colloquial language and semantic analytics should be able to uncover the meaning behind the informal language. An extension of semantic analytics is sentiment analytics, which monitors public's feelings about a particular public personality, event, brand or product. Recent studies have applied sentiment analytics to posts made on the social media platform Twitter [16].

Semantic analysis is used in the healthcare industry to support diagnostics by searching medical documentations for relevant data to identify diseases. Semantic and sentiment analysis is used in market analysis and advertising sectors [15].

2.3 *Security Analytics*

It is becoming increasingly important for enterprises to review their security controls nowadays as cybercrime and other malicious activities are on the increase on the Internet. Preventive and detective controls can help to mitigate existing risks. However, more in depth analysis are required to protect important assets [17]. Fortunately, big data security analytics can be helpful in this process. Big data security analytics exploit the approaches for gathering, examining and controlling voluminous and high velocity data in real time. Hence, new incidents that are linked with those that happen in the past can be identified. Big data security analytics can decrease huge amount of security events to a more manageable number by removing statistical noise. In addition, all historical details are archived so that they can be analysed by forensic experts later. In cases of threat detection, big data security analytics can respond automatically through workflows and this can be a driving factor for organisations to adopt security analytics. Hence, big data security analytics can assist smart cities to enhance their cyber resilience [18].

2.4 *Predictive Analytics*

Predictive analysis is based on various approaches that forecast outcomes based on archived and actual data. Predictive analytics try to establish patterns and relationships in data and they are generally based on statistical techniques. However, traditional statistical techniques cannot be applied on big data, as firstly big data samples are very large. Secondly big data consists of heterogeneous data with uncorrelated variables and finally big data contains noise [19]. Hence an inductive approach, which is based on data discovery rather that models and association should be adopted for big data as inductive methods operate on set of rules to perform intricate computation on heterogeneous and voluminous, set of data. Hence predictive analytics use advanced quantitative methods such as neural networks, machine learning, robotics, artificial intelligence and computational mathematics to study the data and uncover patterns and associations [20]. Consequently, the application of these methods on real life cases in organization can bring better insight from predictive model.

3 Applications of Big Data Analytics in Smart Cities

In this section of the chapter, five important aspects of smart cities are examined and the role of big data analytics is highlighted. The five sectors that have been chosen are transport, education, healthcare, agriculture and weather forecasting as shown in Fig. 2.

Fig. 2 Sectors of smart cities

3.1 Big Data Analytics in the Transport Sector

With the increasing population in cities, congestion is a massive problem for authorities. A report by McKinsey Global Institute concluded that around 400 billion dollars could be saved worldwide annually if existing transport infrastructure could be used optimally [21]. Proper analysis of data could help to improve reliability and efficiency of transport infrastructure. Enormous amount of transport data is available and is gathered through various methods. Fixed sensors such as loop detectors are placed on roads to measure number of vehicles having passing through a point giving useful insights on traffic flow and congestion. A range of mobile sensors such as mobile phones, GPS navigation systems and on-board diagnostics systems can provide data on speed, location and direction of travel of vehicles, provided that users give access to their data. There are also a number of mobile transport applications such as Moovit and Google Waze, which allow authorities to gather crowd-sourced information [22]. In the rest of this section, we showcase how a few cities around the world are using data gathered by the abovementioned sources to improve reliability and efficiency of their transport infrastructure and enhance the quality of life of their city dwellers.

Researchers have developed a complete transportation decision-making system called the TransDec for the city of Los Angeles. The system acquires data from different sources in real time and the amount of data that arrives per minute is around 46 megabytes [23]. The system gathers traffic occupancy and volume data from more than 8900 traffic loop detectors. Data from buses and train are collected; the data is detailed and contains GPS location updated every two minutes and delay information calculated by taking into account pre-defined schedules. Information from ramp meters that are located at the entrance of highways is also used. The system also accepts text format information about traffic incidents. All the data is then cleaned, reducing the input rate from 46 megabytes per minute to 25 megabytes per minute. Analytical techniques are applied to produce precise traffic patterns. TransDec can also predict clearance times and resulting traffic jams after accidents [23].

London is a growing city with a population of around 8.6 million and is expected to hit 10 million by 2030. Transport for London is using big data to keep London moving and plan for challenges arising from a growing city. The system gathers data from the 'Oyster' cards, GPS location of around 9200 buses, 6000 traffic lights and 1400 cameras [24]. Transport for London uses big data tools to develop accurate travel patterns of customers across rail and bus networks. Passengers do not touch their Oyster card when the alight from buses, however, authorities can infer where passengers have alighted by using information from the card and the location of the bus. This application of big data analytics enables Transport for London to map the end-to-end journey of the passengers. The information can be used by authorities to plan closures and diversions. The system also enables authorities to send targeted emails to customers providing them alternative routes thus minimizing the impact of scheduled and unscheduled changes [25].

The Land Transport Authority (LTA) of Singapore is using big data to better serve their customers. LTA uses data generated from logins to their public WIFI system available in the MRT stations to produce a real time crowd heat map on each of their platforms. Additional trains are added to the network if needed [26]. The arrival times of a fleet of 5000 buses fitted with sensors are also monitored. The real time position of buses is sent to a central authority where predictive analysis is used to evaluate whether the supply matches the demand. Corrective measures are applied if necessary. Singapore is planning to further delve in big data analytics in the near future. The Director of Innovation and Smart Nation Office declared that they have plans to equip signal crossings, electronic car parks and thousands of lamp posts with sensors to generate even more data and therefore improve the transportation system [26].

Rio de Janeiro is the second most populous city in Brazil. The population of Rio de Janeiro and the neighbouring suburbs is estimated to be around 11.6 million and is expected to reach 13.6 million by the year 2030 [27]. The metropolis suffers from severe congestion [28]. Since 2013, the city has partnered with the GPS-based geographical navigation application program, Google Waze, to collect real time data from drivers. Data is also collected from pedestrians using the public transport network through the Moovit mobile application. The crowd-sourced information is superimposed with real time information from various sensors and cameras across the city to help authorities come up with plans to ease traffic congestion and reduce travel time for citizens. The city has plans to extend its crowd sourcing data gathering strategy by using information collected from Strava, which is a cycling and running mobile application [29].

3.2 Big Data Analytics in Weather Forecasting

Climate change is an undeniable fact. Each one of the first 6 months of 2016 set records as the hottest respective months since the year 1880 [30]. The temperature rise for the twenty first century is expected to be 2–4 °C [31]. One of the serious consequences of climate change is extreme weather condition such as intense cyclones, heavier rainfalls and longer periods of droughts. Therefore, the ability to predict the weather accurately and in a timely manner can help save thousand of human lives. Moreover, changes in weather have significant impacts on businesses and accurate forecasts can improve operations.

IBM has developed a sophisticated weather forecasting model called Deep Thunder, which merges hyper local short-term forecast model with the global forecast model of The Weather Company. Around 100 terabytes of meteorological data is pulled every day from established weather stations, radars, satellites, aircrafts, spacecrafts and around 200,000 individual weather stations. The Deep Thunder model uses the collected data along with historical ones to predict the weather 24–84 h ahead, throughout the world, with high degree of accuracy [32]. Deep Thunder can also generate on-demand and personalised forecast for businesses allowing them to understand the impacts of changes in weather patterns on their operations and giving them ample time to identify and implement corrective actions. For example, a utility company can use Deep Thunder to determine the amount of power that will be added to the grid by renewable energy sources such a wind farm or a solar plant a few days in advance. IBM claims that Deep Thunder can deliver 95% accuracy for solar power forecasting and 93% accuracy for wind power forecasting [33]. IBM's Deep Thunder can help to boost profits across many sectors.

In Rio de Janeiro, Brazil, the city's operation centre uses high-resolution weather forecasting powered by IBM's Deep Thunder and hydrological modelling algorithms to predict torrential rains 48 h in advance with a high degree of accuracy. The algorithms use data from sensors in the city's river basin, topographical studies, historical precipitation records and feeds from radar [2]. The model also simulates surface flow and water build-ups and can therefore predict disasters such as mudslides and flash floods. The model gives authorities sufficient lead times to deal with these severe weather conditions.

In 2012, hurricane Sandy made landfall on the eastern seaboard of the United States. The prediction of the trajectory of a hurricane is not always accurate. However, the use of historical data, current weather conditions and appropriate computer models, allowed meteorologists to predict the trajectory of the hurricane

Sandy with a greater degree of precision [34]. With the help of big data technologies, the US national hurricane centre was able to predict hurricane Sandy's landfall 5 days in advance and within 30 miles [35]. This level of accuracy was unimaginable a few years before. Non-profit organization, Direct Relief, used big data to plan and manage emergencies in near real time. The organization distributes prescription medicines to the affected population in crisis time. During Hurricane Sandy, the organization used big data to identify vulnerable areas and stock up on most requested items and necessary medications [36]. Korean Meteorological Administration also uses big data technologies powered by IBM's computing power to predict the location and intensity of weather events such as dust storms and typhoons.

3.3 Big Data Analytics in Agriculture

The agricultural sector is facing some difficult challenges. Almost a billion people suffer from malnutrition or famine and by 2050 there will be an additional 2 billion people in the world putting an additional strain on the food production sector. The food consumption habits have changed and a big proportion of the population of the world is getting richer, therefore consuming more food. Moreover, climate change is affecting farming significantly. The use of technologies in agriculture can help maximise food production [37].

Big data is used in agricultural soil analysis. Samples of soil are taken at different depths and different locations of a farm. Computed tomography (CT) images of the soil samples are analysed to generate data. For proper analysis, a large number of soil samples are needed which leads to numerous CT files which can lead to Petabytes of data. Information such as level of nutrients, soil density and level of compaction are gathered [38]. Farmers are given precise information about the state of their land enabling them to use relevant fertilisers in a targeted and rational approach. Moreover, they are able to better decide which crop to plant in which area of their farm for maximum crop yield. Solum is a start up firm, which provides farmers with big data analytics and cloud technology platforms to give them in-depth information about their soil. After performing soil analysis, Solum give farmers access to a web portal with all the information and maps of their fields [39]. Using big data analytics in agriculture, farmers are able to increase crop production, save on cost of fertilisers and reduce the application of chemicals, which may harm the environment.

Climate change is another factor that has greatly affected crop yield over the past decades. Unpredictable and changing weather conditions are confusing farmers who can no longer adhere to their established farming practices. Crop yield data, weather forecast data and historical weather data are processed using big data analytics to generate information to help farmers improve their harvest. In Colombia, researchers used agricultural records on the production of rice such as harvest monitoring data on which algorithms were applied to find relevant patterns to discover relationships between weather conditions and yield. Using in-depth seasonal forecasts, the researchers were able to predict months ahead, which varieties of rice would give a better harvest for a particular location. They were also able to advice farmers when to sow seeds. It is believed that this research could boost production by 1–3 tons/ha [40].

3.4 Big Data Analytics in Healthcare

Every human being needs to enjoy a sound health. A healthy population can make huge contribution to economic progress by being more productive [41]. However in this digital era, there still exist problems in the healthcare and people are dying due to various reasons. For instance, the mortality rate of Americans from preventable medical error in hospital, yearly, varies between 44,000 and 98,000 [42]. The number of errors in medical prescriptions can be can be totalised up to 1.5 million in the US and studies revealed that about 5% of patients can be wrongly diagnosed [43]. In addition, the death rate associated with diseases is very high worldwide. In 2015, 56.4 million deaths were reported worldwide. Ischaemic heart disease and stroke combined accounted for 15 million deaths. These diseases are the main causal agent resulting in death worldwide for the last 15 years. Chronic obstructive pulmonary disease, lung cancer, diabetes, dementias, lower respiratory diseases, diarrhoea, and tuberculosis are among the top diseases that cause many fatalities around the world [44].

Furthermore, the breakthrough of various epidemics has become a major concern for our healthcare systems nowadays. Diseases such as Dengue, Chikungunya, Zika, Middle East Respiratory Syndrome (MERS) and Ebola have all been reported in 2015 [45]. These epidemics have spread across the world at an alarming rate, affecting people in different countries. Presently, there is no cure for Chikungunya, Zika and Dengue, MERS and Ebola. No vaccine or specific medicine has been commercialised yet and the treatment of these diseases is focused only on relieving the symptoms [45, 46]. Advances in research have contributed a lot towards the development of new drugs and vaccines, which have helped to save many lives. However, emerging diseases are mutating and becoming more resistant to existing antibiotics. Consequently, many people are dying because of lack of treatment. Moreover, every country is at risk in the present era where epidemics outbreaks have become a common problem. The death toll is on the increase in the European Union and the United States and much higher in the low and middle-income countries [47].

The aging population will have a significant impact on the healthcare systems. The United Nations has predicted that there will be increasing number of elderly citizens in most societies, particularly in western countries. Consequently, there will

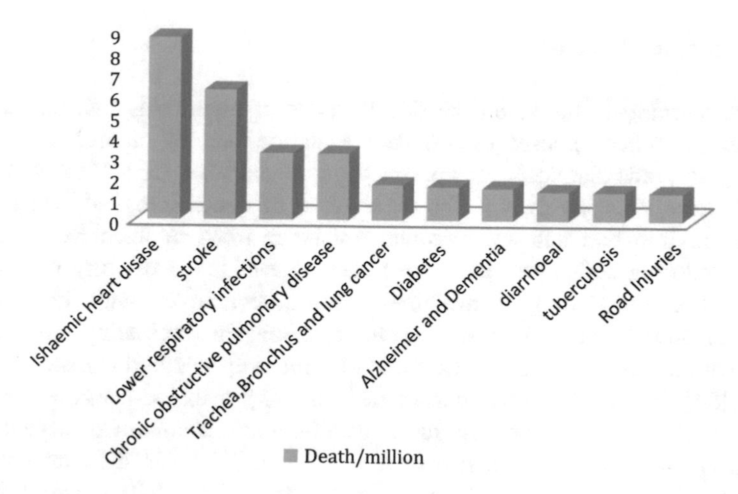

Fig. 3 Top ten causes of death in 2015

be an increase in diseases like dementia and other memory disorders. Those patients will require monitoring and special care [48–50].

Currently, most medical practitioners rely a lot on structured data. These data are easily collected, stored, and queried using databases [51]. However, public and private health care organisations generate massive amount of semi-structured and unstructured data pertaining to patients every day. These data are mainly from sensors, instrument readings, magnetic resonance imaging (MRI), computed tomography (CT) and radiography [43], social media research and other sources. Useful and crucial information can be extracted from these data if harnessed properly. Unfortunately, these data are not fully exploited and analysed (Fig. 3).

Nowadays, big data offers a different approach of collecting and storing structured, semi structured and unstructured data from divergent origin. However, just compiling all these data is not helpful; analytics should be applied to gain more insight. Hence, Big data analytics can provide a solution to enhance services and alleviate problems in healthcare. Big data analytics has recently begun to play a critical role in the progress of healthcare. Hence, it is helping in improving care delivery and in the study of diseases [51]. Big data analytics offer various potential in analysing trends and finding pattern to model data. Data concerning patients are crucial. It is important to find the best approach to propagate and access data for real time tracking and detailed analysis [47]. If the right information is available at the right time, better and timely decision can be taken. In the case of health care, real time sharing of data in a constructive, harmonized and collaborative way can be beneficial [47]. By finding association and modelling data, big data analytics can enhance healthcare, rescue millions of life and reduce cost associated with treatment [52]. Various data analytics exist and can be useful in healthcare.

3.4.1 Machine Learning

Machine learning is based on the development of algorithms that can improve automatically when exposed to new data. Machine-learning algorithms are being utilized to upgrade diagnostic, and prescription in healthcare [53]. Machine learning methods have been involved in the observation of microbiological data to detect variation in infection and antimicrobial resistance trend in intensive care. Also, machine-learning techniques have been used to evaluate morbidity after cardiac surgery. The integration of machine learning techniques with local logistic regression models has proved to give better results in forecasting the causes of death in the intensive care units compared to using only standard logistic regression models [54]. According to research carried out [55], machine-learning techniques can explore patient's charts in more details than doctors. Consequently, by amplifying more attributes, there was a notable expansion in the capacity of the model to differentiate between people with congestive heart failure with those who have not.

3.4.2 Image Analytics

Image analytics are concerned with algorithms, which can extract and inspect information found in image using digital image processing methods. Diagnosis and medical care can be enhanced using image analytics. In this digital era, advances in medical science are increasing rapidly and clinicians have difficulty in keeping the same pace with treatment and new approaches. Image analytics can help in minimising biased opinions and error, and at the same time optimising treatment in medical care. Image analytics system has the potential to handle and prescribe actions more exactly and hence can be beneficial in providing real time information to monitor patients [56].

3.4.3 Audio Analytics

Audio analytics refer to techniques that are applied on unstructured audio data in order to study and extract information from them [19]. Audio analytics help to diagnose and treat patients that have some medical problems, which can impact on their communication. Furthermore, audio analytics can assist in studying babies crying pattern, which can in turn determine the baby's health and psychological condition. Huge quantity of data is recorded through speech-driven clinical documentation systems, hence, the reason to adopt audio analytics in medical care [19].

Data analytics can bring a lot of improvement in the healthcare; however it works on very large volume and complex data, which is difficult to gather. Hence, its application is being hindered and many healthcare providers are not being able to harness the potential of data analytics [53]. In addition to that, healthcare data are very sensitive and several approaches have to be considered before the application of big data analytics on these data to prevent data violation.

3.5 Big Data Analytics in Education

The traditional education system is still common in this digital era. Teachers and students have face-to-face contact in classroom whereby students have to listen to teacher's explanation, followed by some group discussions. Students' performances are monitored by tests, assignment and examinations [57]. The educational organisations collect various source of information about students and teachers in this system. Face to face traditional system can be complemented by using computer based educational system. However, with the global use of Internet nowadays, new web based educational systems are available. These include among others online instruction system, e-learning platforms [57]. A lot of information is generated and stored by these systems; however, these data are scarcely utilized. These datasets offer various possibilities to assess learning, generate feedback and hence improve the learning systems [58]. Consequently, new approaches to exploit these data have to be developed and one solution is the use of big data and big data analytics.

Big data can have a big impact on education as it can assist in enhancing the performance of students by showing facets of learning that were difficult to discern previously [59]. Big data can help teachers to evaluate students' know- how and at the same time study new pedagogical approaches [60] which can in turn help particular students with special needs. Furthermore, big data analytics can give quick review to both teachers and students about their educational performance, by analysing data and filtering out useful information from them [60]. This can be done in real time to show relevant information according to students' need at a specific time [61]. This technique is known as adaptive learning and it is generating customized instructions for students. Lectures can be adjusted to the requirement of students, promoting their understanding and upgrading their results [59]. In addition to that, big data has the potential to promote a variety of functions pertaining to higher education and these include accreditation, quality assurance and developing pedagogic standards [62].

However, despite the benefits that big data and big data analytics offer in education, there are still numerous challenges that should be addressed before its proper implementation [62]. Educational organisations should review privacy, policy and legal matters when gathering, keeping, examining and divulging personal information pertaining to students to researchers. There should be a balance, so that students' privacy is not violated and at the same time data is accessible for research purposes. Confidential information relating to students records should be preserved, however researchers should be given the possibility to analyse data for identifying trends and patterns that can enhance learning [63]. There exists a federal law in the United States about the protection of the privacy of students' data, which is known as The Family Educational Rights, and Privacy Act (FERPA). However, this law can disclose information in case of authorised education interest [61]. Thus, it is of utmost importance to develop a framework, which promotes accessibility to more relevant and valuable data while protecting the privacy and intellectual property rights of students [62].

3.5.1 Data Mining

Educational data mining applies to approaches from machine learning and statistics to examine data gathered in classes. It evaluates learning concepts and enhances educational operations [61]. Big data helps in mining information to promote students' potential and learning approaches. Instead of depending upon students' performance in test, teachers can assess the know-how of students using big data analytics. In addition to that, online tools help to assess students' activities, which can provide useful information about students learning patterns. With the help of data mining techniques, students and instructors can get prompt feedback about their educational performance. Moreover, it can identify pedagogical techniques that are more suitable for particular groups of students [63].

3.5.2 Learning Analytics

Learning analytics are related to the study of variety of data pertaining to students that are generated and collected, in order to evaluate their learning patterns, forecast their progress and identify any possible problem. Educational data is collected from various sources including assignment, exams, educational games and simulations [64], online social interaction, blogs, forums or other sources, which are not directly related to the students' pedagogical activities. This information can be mined to uncover the need of the students and hence customise educational techniques that best suit them. This can in turn contribute in increasing their probability to succeed. Learning analytics do not only focus on improving students' performance but can also be helpful in evaluating syllabus, programs and academy. Furthermore, it can be used by students to improve both their official and unofficial learning activities. Universities are more interested in using learning analytics as this can assist students to enhance their learning, which may contribute to a decrease in the rate of failure [61].

4 Security Challenges with Big Data and Analytics for Smart Cities

Smart cities are known to enhance the quality life of citizens [65]. This would be possible only if the big data, generated by the ICT framework in smart cities, is available to be interpreted in a timely manner. Data is a critical asset in any organization [66] and therefore the confidentiality, privacy and integrity of data should be preserved [67]. Structured data can be stored in databases, but big data, which consist of massive dataset of unstructured data, semi-structured and structured data cannot be stored in traditional Relational Database Management Systems. Furthermore, big data cannot be analysed with existing standard techniques as it is incomplete, inconsistent and it has some privacy and security concerns [68]. Data from different sources, with different format and streaming of data can engender security weaknesses [69]. Hence, existing security mechanism is not sufficient to cater for big data [70]. Organisations using big data and analytics have to find a holistic and proactive approach to protect them from emerging threats and prevent data violation [71, 72].

4.1 Privacy

Data privacy is concerned with the sharing of data and at the same time preserving personally identifiable information [73, 74]. Smart cities rely a lot on sharing of data among different city departments and applications [75]. Consequently, privacy has become a significant concern [76] in smart cities. Moreover, the development of big data analytics has helped to derive and associate data, making privacy breaching even simpler [77, 78]. However, there exists some Acts that cater for privacy concern [77]. One example among others include the Health Insurance Portability and Accountability (HIPAA) Act which is concerned with strategies and recommendation to preserve the privacy and security of healthcare data [79]. However, regulations should be reviewed to ensure that the use and application of big data and analytics are in line with them. [68]. For instance, HIPAA is more concentrated on assuring the security approaches rather than executing them [80].

One probable solution to address the breaching of privacy with big data is to enforce laws and regulations in cases of illegal access, data sharing, fraud, and copyright issue [69]. Another way to address privacy concerns is to perform data anonymisation. However, this technique should be executed without impacting on the performance of the system or data quality. Existing anonymisation techniques involve a lot of computations, which take much time and involve expensive calculations [81]. However studies concluded that anonymisation can actually protect privacy [82].

Also with the wide use of big data nowadays, the existing infrastructure governing privacy of personal data is becoming a problem. Hence each department of the smart cities need to audit the execution of their privacy strategies to guarantee that privacy is adequately preserved [10].

4.2 Integrity

Integrity is about preserving and certifying that data is accurate and consistent during the whole life cycle [72]. To ensure the confidentiality of data, the data must first be reliable, i.e. have integrity. However, one main problem associated with big data is from where the data originates. Big data contains heterogeneous data from divergent sources and it is difficult to ensure that all data sources are reliable. Big data sources can contain inaccurate, ambiguous or incomplete data and noise [81]. To ensure that data has not been tampered, techniques like edit checks and similitude with other sources can be applied. Unfortunately, this is not effective for big data [83]. Alternately, data from the different sources must be filtered and prepared before modelling. This can help to get more reliable predictions and forecast [81].

4.3 Confidentiality

Confidentiality is about ensuring that only authorized people can access information and these people can access the information only to the extent granted [84]. If confidentiality is not well protected, it may engender damage to data subjects and the data provider [85]. In this digital era, smart cities are using technology to retain and examine voluminous amount of data. Consequently, a secure framework is needed to store, administer and explore the big data [76]. Since big data contains variety of data, this makes them more vulnerable to malicious activity [85]. Moreover, if security of big data is violated, this can cause major legal repercussions and can be detrimental to the entities reputation [76].

Application of data encryption to big data can be challenging due to its volume and variety. Existing encryption techniques as applied to *normal* data cannot meet the requirement of big data. Consequently, the categorization of information is very important and more robust and effective cryptography techniques shall be developed for big data [86]. In addition to that, effective strategy must be analysed for managing safety, access control and communications for big data [87]. Hence a secure infrastructure should be put in place to prevent any information disclosure while dealing with big data and analytics [88] [70] in smart cities.

On the other hand, big data analytics can be useful in detecting and preventing threats from malicious attacks. By examining different data sources and drawing patterns, big data analytics can assist in identifying risks and mitigate them at an early stage [76]. Smart cities can also rely on advanced dynamic security analysis to improve big data security. This analytics is based on inferring and examining real time or almost real time security events and the different users involved, promoting security online and hence deterring fraudulent attacks. The analytics on the generated big data can in turn help to uncover incidents at the right time, to observe clients' odd behaviour's, to mitigate risks, to identify emerging threats among others [81].

Data breaches are on the rise in this digital era. It is becoming more challenging to protect big data using existing standards. More advanced techniques are required to protect privacy, integrity and confidentiality of big data in smart cities. Data protection does not deal only with the encryption of the data but also making sure that the data is shared in an appropriate way. Big data analytics has in some way helped to reduce security frauds. However, a secure framework should be put in place to prevent any information disclosure while dealing with big data analytics [71, 89].

5 Conclusion

Rapid urbanisation has made smart cities a future reality for many urban agglomerations. Consequently, many more IoT devices will be installed and it is expected that these devices will generate massive amounts of data. Big data analytics are essential for harnessing the value that is present in the growing volume of data. Predictive analytics, edge analytics, semantic analytics and security analytics are four types of processes that can be applied to many sectors to examine the data being generated. This chapter has presented the application of big data analytics to five prominent sectors of smart cities namely transport, agriculture, weather forecasting, healthcare and education. Moreover, security is a major concern when dealing with large volumes of data and this chapter has presented an overview of the security aspects for big data in smart cities.

References

1. Dohler, M., Vilajosana, I., Vilajosana, X., & LLosa, J. (2011). Smart cities: An action plan. In *Proceedings of Barcelona smart cities congress 2011*. Barcelona.
2. Pretz, Kathy. (2014). An urban reality: Smart cities. *The Institute, 38*(2), 10.
3. Enbysk, L. (2013). SmartCities Council. http://smartcitiescouncil.com/article/smart-cities-technology-market-top-20-billion-2020
4. Deloitte. (2015). *Smart cities big data*. Deloitte.
5. Datameer. (2016). *Big data analytics and the internet of things*.
6. Gantz, J., & Reinsel, D. (2012). *The digital universe in 2020: Big data, bigger digital shadows, and biggest growth in the far east*. Framingham, MA: EMC Corporation.
7. Najafabadi, M. M., et al. (2015). Deep learning applications and challenges in big data analytics. *Journal of Big Data, 2*(1), 1.
8. Datameer Inc. (2013). The guide to big data analytics. In *Datameer*. New York: Datameer.
9. Sas. (2017). Sas. http://www.sas.com/en_us/insights/analytics/big-data-analytics.html
10. Ernst & Young. (2014). *Big data changing the way business compete and operate*. Ernst & Young.
11. Marr, B. (2016). Forbes. https://www.forbes.com/sites/bernardmarr/2016/08/23/will-analytics-on-the-edge-be-the-future-of-big-data/#3791f7d63644
12. Rachel, A. (2016). Travancore analytics. http://www.travancoreanalytics.com/can-edge-analytics-future-big-data/

13. Marr, B. (2016). IBM. https://www.ibm.com/think/marketing/will-analytics-on-the-edge-be-the-future-of-big-data/
14. MacGillivray, C., Turner, Vernon, Lamy, Lionel, Prouty, Kevin, & Segal, Rebecca. (2016). *IDC future scape: Worldwide internet of things 2017 predictions*. Framingham, MA: International Data Corporation.
15. Stracke, N., Grella, M., Chiari, B., Bechle, K., & Schmid, S. (2015). *Semantic technology: Intelligent solutions for big data challenges*. Germany: Advanced Analytics & EXOP GmbH Constance.
16. Saif, H., He, Y., & Alani, H. (2012). Semantic sentiment analysis of twitter. In *International semantic web conference* (pp. 508–524). Berlin.
17. Sullivan, D. (2015). Five factors for evaluating big data security analysis platforms. http://searchsecurity.techtarget.com/feature/Five-factors-for-evaluating-big-data-security-analytics-platforms
18. Business Application Research Center. (2016). BI-Survey.com. http://bi-survey.com/big-data-security-analytics
19. Gandomi, A., & Haider, M. (2014). Beyond the hype: Big data concepts, methods and analytics. *International Journal of Information Management, 35,* 137–144.
20. Intel IT Centre. (2013). Predictive analytics 101: Next-generation big data intelligence.
21. Neumann, C. S. (2015). McKinsey & Company, Capital Projects & Infrastructure. http://www.mckinsey.com/industries/capital-projects-and-infrastructure/our-insights/big-data-versus-big-congestion-using-information-to-improve-transport
22. Parliamentary Office Science and Technology. (2014). Big and open data in transport. In *Parliamentary office science and technology*. London.
23. Jagadish, H. V., et al. (2014). Big data and its technical challenges. *Communications of the ACM, 57*(7), 86–94.
24. Sager Weinstein, L. (2015). *Innovations in London's transport: Big data for a better customer experience*. London: Transport for London.
25. Sager Weinstein, L. (2016). How TfL uses 'big data' to plan transport services. *Euro Transport, 3*.
26. Infocomm Media Development Authority. (2016). Infocomm media development authority. https://www.imda.gov.sg/infocomm-and-media-news/buzz-central/2016/6/smart-nation-big-on-big-data
27. World Population Review. (2017). World population review. http://worldpopulationreview.com/world-cities/rio-de-janeiro-population/
28. BBC. (2017). Management of urban change. http://www.bbc.co.uk/education/guides/zqdkkqt/revision/5
29. International Transport Forum. (2015). *Big data and transport understanding and assessing options*. OECD.
30. NASA. (2016). Climate. https://www.nasa.gov/feature/goddard/2016/climate-trends-continue-to-break-records
31. IFRC. (2010). Climate change conference. http://www.climatecentre.org/downloads/files/conferences/COP-16/Fact%20and%20Figures.pdf
32. Treinish, L. (2016). IBM research. https://www.ibm.com/blogs/research/2016/06/deep-thunder-now-hyper-local-global/
33. Davis, J. (2016). Information week. http://www.informationweek.com/big-data/big-data-analytics/ibms-deep-thunder-shows-how-weather-is-big-business/d/d-id/1325946
34. Jackson, W. (2012). How models got a complex storm like Sandy mostly right. https://gcn.com/articles/2012/10/30/how-models-got-a-complex-storm-sandy-mostly-right.aspx
35. Anderson, A., & Semmelroth, D. (2014). Big data and weather forecasting. http://www.dummies.com/programming/big-data/data-science/big-data-and-weather-forecasting/
36. Rael, H. (2012). Direct relief. https://www.directrelief.org/2012/10/big-data-vs-big-storm-new-technology-informs-hurricane-sandy-preparedness-response/

37. Campbell, B. M., Thornton, P., Zougmoré, R., van Asten, P., & Lipper, L. (2014). Sustainable intensification: What is its role in climate smart agriculture? *Current Opinion in Environmental Sustainability, 8,* 39–43.
38. Alves, G. M., & Cruvinel, P. E. (2016). Big data environment for agricultural soil analysis from CT digital images. In *Tenth international conference on semantic computing (ICSC)* (pp. 429–431). Laguna Hills, CA.
39. Geron, T. (2012). Forbes tech. https://www.forbes.com/sites/tomiogeron/2012/06/27/solum-lands-17-million-for-big-data-analysis-of-farm-soil/#3436c9823436
40. CGIAR Research Program on Climate Change, Agriculture and Food Security. (2015). *Research program on climate change, agriculture and food security.* https://ccafs.cgiar.org/bigdata#.WNU-YBJ96Rs
41. WHO. (2017). World Health Organization. http://www.who.int/hdp/en/
42. Hippocrates. (2014). *Orthopedic service line optimization. How to use big data for your value-based purchasing.*
43. Sternberg, S. (2015). US news and world report. http://www.usnews.com/news/articles/2015/09/22/iom-study-shows-errors-in-diagnosis-harm-countless-patients-each-year
44. World Health Organization. (2017). The top 10 causes of death factsheet. http://www.who.int/mediacentre/factsheets/fs310/en/
45. Frontières, M.S. (2016). Médecins Sans Frontières. https://www.msf.ie/article/msf-alert-5-epidemics-watch-2016
46. Centres for Disease Control and Prevention. (2016). Centres for disease control and Protection CDC24/7. https://www.cdc.gov/dengue/
47. World Economic Forum. (2016). World Economic Forum. http://reports.weforum.org/global-risks-2016/global-disease-outbreaks/
48. Poland, Michael, Nugent, Chris, Wang, Hui, & Chen, Liming. (2009). Smart home research: Projects and issues. *International Journal of Ambient Computing and Intelligence (IJACI), 14*(1), 14.
49. Najjar, M., Courtemanche, F., Hamam, H., Dion, A., & Bauchet, J. (2009). Intelligent recognition of activities of daily living for assisting memory and/or cognitively impaired elders in smart homes. *International Journal of Ambient Computing and Intelligence (IJACI), 1*(4), 17.
50. van Rysewyk, S. (2013). Robot pain. *International Journal of Synthetic Emotions (IJSE), 4*(2), 12.
51. Belle, A., et al. (2015). Big data analytics in healthcare. *BioMed Research International.*
52. Raghupathi, W., & Raghupathi, V. (2014). Big data analytics in healthcare: Promise and potential. *Health Information Science and Systems, 2,* 3.
53. Fang, R., Pouyanfar, S., Yang, Y., Chen, S. C., & Iyengar, S. S. (2016). Computational health informatics in the big data age: A survey. *ACM Computing Surveys, 49*(1), 12.
54. Geert, M., Fabian, G.G., Jan, R., & Maurice, B. (2009). *Machine learning techniques to examine large patient databases.*
55. McDonald, C. (2016). How big data is reducing costs and improving outcomes in health care. https://www.mapr.com/blog/reduce-costs-and-improve-health-care-with-big-data
56. Venter, F., & Stein, A. (2012). Analytics driving better business solutions. http://analytics-magazine.org/images-a-videos-really-big-data/
57. Cristobal, R., & Sebastian, V. (2013). *Data mining in education* (Vol. 3). New York: Wiley.
58. Greller, W., & Drachsler, H. (2012). Translating learning into numbers: A generic framework for learning analytics. *Educational Technology & Society, 15*(3), 42–57.
59. Mayer-Schonberger, V., & Cukier, K. (2014). *Learning with big data the future*: *An eamon dolan book/houghton mifflin harcourt.*
60. Drigas Athanasios, S., & Panagiotis, L. (2014). The use of big data in education. *International Journal of Computer Science, 11*(5).
61. Office of Educational Technology U.S. (2014). *Department of education, enhancing teaching and learning through educational data mining and learning analytics: An issue brief.* Washington D.C.

62. Arwa, I. A. (2016). Big data for accreditation: A case study of Saudi Universities. *Journal of Theoretical and Applied Information Technology, 91*(1).
63. West, D. M. (2012). *Big data for education: Data mining, data analytics and web dashboards*. Governance Service at Brookings.
64. Markus, I., Deirdre, K., & Hamid, M. (2012). Big data in education assessment of the new educational standards.
65. Batt, M., et al. (2012). Smart cities of the future. *The European Physical Journal Special Topics*, 481–518.
66. Stephen, K., Armour, F., Alberto, E.J., & William, M. (2013). Big data: Issues and challenges moving forward. In *46th Hawaii international conference on system sciences, HICSS'13*, pp. 995–1004.
67. Odella, F. (2016). Technology studies and the sociological debate on monitoring of social interactions. *International Journal of Ambient Computing and Intelligence (IJACI), 7*(1), 26.
68. Philip Chen, C. L., & Chun-Yang, Z. (2014). Data-intensive applications, challenges, techniques and technologies: A survey on big data. *Information Sciences* (in press).
69. Duygu, S.T., Terzi, R., & Sagiroglu, S. (2015). A survey on security and privacy issues in big. In *The 10th international conference for internet technology and secured transactions*.
70. Cloud Security Alliance. (2012). *Top ten big data security and privacy challenges*.
71. Wiggins, L. (2015). Security intelligence analysis and insights for security professionals. https://securityintelligence.com/big-data-opportunities-need-big-data-security/
72. Ernst & Young. (2014). *Cyber insurance, security and data integrity, part1: Insights into cyber security and risk*.
73. Rajkumar, N., Vimal, Karthick R., Nathiya, M., & Silambarasan, K. (2014). Mining association rules in big data for e-healthcare information system. *Research Journal of Applied Sciences, Engineering and Technology, 8*(8), 1002–1008.
74. Fagan, D., Caulfield, B., & Meier, R. (2013). Analyzing the behavior of smartphone service users. *International Journal of Ambient Computing and Intelligence (IJACI), 5*(2), 16.
75. Al Nuaimi, E., Al Neyadi, H., Mohamed, N., & Al-Jaroodi, J. (2015). Applications of big data to smart cities. *Journal of Internet Services and Applications, 6*, 25.
76. Venkata, N. I., Sailaja, A., & Srinivasa, R. R. (2014). Security issues associated with big data in cloud computing. *International Journal of Network Security& its Application (IJNSA), 6* (3), 45.
77. Cárdenas, A. A., Manadhata, P. K., & Rajan, S. P. (2014). InfoQ. https://www.infoq.com/articles/bigdata-analytics-for-security
78. Hashem Ibrahim, A. T., et al. (2014). The rise of "big data" on cloud computing: Review and open research issues. *Information Systems*, 98–115.
79. For Health Information Technology The office of the National Coordinator. (2015). *Guide to privacy and security of electronic health information*. USA: Department of Health and Human Services.
80. Patil, H.K., Seshadri, R. (2014). Big data security and privacy issues in healthcare. In *IEEE international congress on big data*.
81. Fatima-Zahra, B., Lahcen, A. A. (2014). Big data security: Challenges, recommendations and solutions. In *Handbook of research on security considerations in cloud computing*: *IGI global*.
82. Sánchez, D., Martínez, S., & Domingo-Ferrer, J. (2016). How to avoid re identification with proper anonymization comment on "Unique in the shopping mall: On the reidentifiability of credit card metadata". *Science, 351*(6279), 1274.
83. Juan, Z., Yang, X., & Appelbaum, D. (2015). Toward effective big data analysis in continuous auditing. *Accounting Horizons, 29*(2), 469–476.
84. Schneider, G., & Hammer, J. H. (2007). *On the definition and policies of confidentiality*, pp. 337–342.
85. Kompelli, Swetha, & Avani, Alla. (2015). Knowledge hollowing in enormous information. *International Journal of computer science and Electronics Engineering, 5*(4), 88–92.

86. Kimbahune, V., Deshpande, A., & Mahalle, P. (2017). Lightweight key management for adaptive addressing in next generation internet. *International Journal of Ambient Computing and Intelligence (IJACI), 8*(1), 20.
87. Min, Chen, Shiwen, Mao, & Yunhao, Liu. (2014). Big data: A survey. *Mobile Network Application, 19,* 171–209.
88. Hu, H., Wen, Y., Chua, T. S., & Li, X. (2014). Toward scalable systems for big data analytics: A technology tutorial. *IEEE Access, 2,* 652–687.
89. Lagoze, C. (2014). Big data, data integrity, and the fracturing of the control zone. *Journal of the American Society for Information Science, 63*(6), 1–40.

Bigdata Analytics in Industrial IoT

Bhumi Chauhan and Chintan Bhatt

Abstract In IoT, technology is on an gradually developed both in terms of software and hardware. The high speed with which humans interact with the internet, use social media and interconnect their devices with another device is growing quickly. Due to the interaction between machine to human and machine to machine communication the massive amount of data will be generated and to store this generated data in the database becomes more difficult to store, manage, process and analyses. The data management is a biggest problem in IoT due to the connectivity of billions of devices, objects which are generating big data. With the help of big data technology we can handle that data. However due to the nature of Bigdata it has become important challenge to achieve the real-time capability using the traditional technologies. Big data is some technologies to capture, manage, store, distributed and analyses petabyte or larger sized datasets with highest velocity and different structure. Hadoop is a best platform for structuring Bigdata. It is a best tool for data analysis as it works for distributed big data, Time stamped data, structured, unstructured and semi-structured data, streaming data, text data etc. This paper represents the layer architecture of big data system. In addition, how to use FLUME and HIVE tool for data analysis. For NoSQLdatabse we use Hive which is SQL like query language is used for some analysis and extraction. Flume is used to extract real time data into HDFS.

Keywords Big data · IoT · Industrial IoT · Hadoop · MapReduce · Flume · Hive

B. Chauhan (✉) · C. Bhatt
Department of Computer Engineering, Charotar University of Science and Technology, Changa, Anand 388421, Gujarat, India
e-mail: 15pgce007@charusat.edu.in

C. Bhatt
e-mail: chintanbhatt.ce@charusat.ac.in

N. Dey et al. (eds.), *Internet of Things and Big Data Analytics Toward Next-Generation Intelligence*, Studies in Big Data 30,
DOI 10.1007/978-3-319-60435-0_16

1 Introduction IoT

> When we talk about an Internet of Things [20, 21], its's not just putting RFID tags on some speechless thing so we smart people know where that speechless thing is. It's about embedding intelligence so things become smatter and do more than they were proposed to do.

The basic concepts behind the Internet of Things is that when we discuss about 'things', we refer to objects, devices like wristwatches and medical sensors, actuators etc. These devices are such that they able to connect with users by yielding information from respective environment sensors along with actuators such that users can operate over a span of various interfaces. (see Fig. 1).

When we referring IoT and its technology the concept behind IoT is that the elements are connected with the realistic environment of the internet by reference of sense, monitoring and eventual tracking the objects and their environment. To enable themselves build web application, the developers in hand with the users assemble few components, allotting them sense and networking capabilities for communication and program them into execution by denoting a particular task.

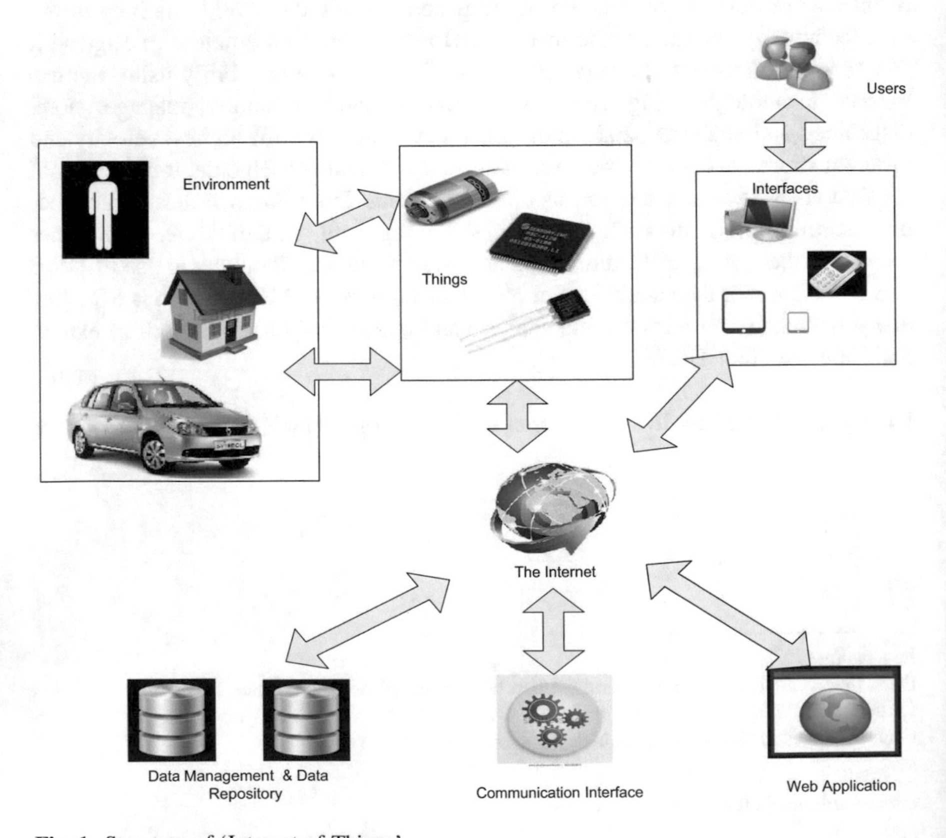

Fig. 1 Structure of 'Internet of Things'

The following are the characteristics of a devices that doing as a component of an IoT networks.

- *Collect and Transmission of data* In environment, the device can sense the things and collecting some information related to it and then transferring to a various device or to Internet also.
- *To Operating the devices based on triggers* It can be operating to abet other devices (e.g., turn on the lights or turn off the heating) base on condition set by you. For example, you can program the device when you enter in room the lights are turn on automatically.
- *To Collecting the information* This is most implement feature for IoT device that they collect the information from the network to the other devices or through the Internet.
- *Communication support* In IoT device there are members of a device network can also nodes of the same network.

The above diagram shows the structure of "Internet of Things". Things are referred as various sensors, actuators and even microcontrollers. These things interact with environment and Internet, thus allowing users to manage their data over various interfaces.

1.1 Industrial IoT

The Industrial Internet of Things is a network of Physical objects, Systems, platforms and application that contain embedded technology to communicate and share intelligence with each other, the external environment and with people.

The main objectives of IIoT is to improved affordability and Availability of processors, sensors microcontrollers and other technologies are help to facilitate capture and access to real-time information.

GE introduce Industrial Internet (IoT) as a term which means "integrating complex physical machinery together with networked sensors and software. Industrial internet joins fields such as the IoT, Bigdata, machine learning and M2M (Machine to Machine) communication, to collect and analyses data from machines and use if for adjusting operations." [1] Industrial Internet contains three key elements which are represent the following figure (Fig. 2).

The first key element is Intelligent Machines that reparents connected the machines, networks and facilities with advanced controls, sensors and software applications. The second key element is advanced analytics that represent the combination of domain expertise, physical-based analytics, automated and precative algorithms for understanding the operation of systems and machines. The third key element is People at work represent, connecting people at any time and anywhere for supporting intelligent operations, maintains and high service quality and safety. [2, 3]

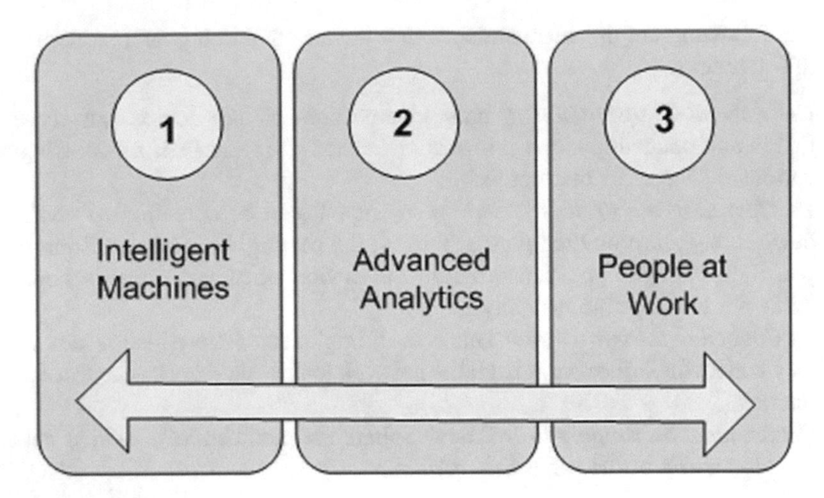

Fig. 2 Key elements of industrial IoT

1.2 IoT Application and Techniques

- **Smart City**

Smart City is one of the strongest application of IoT in subject to World Population's. Smart surveillance, smarter energy management systems, automated transportation, urban security and environmental monitoring and water distribution are some of the real-time examples of IoT for making.

IoT take care of huge issues confronted by mankind living in urban communities like traffic blockage and deficiency of vitality supplies, pollution etc. For example, a product named Cellular Communication if get enabled by smart belly trash then it will send alerts to municipal administrations when a container should be exhausted. By installing sensors and web applications, individuals can undoubtedly see accessible stopping spaces over the city. In additional sensor, can likewise distinguish meter altering issues, general malfunctions and establishment issues in any power framework.

- **IoT in Agriculture**

Nowadays, the demand of Food supply has been extremely raising due to the continued rise in the world's population. In this way, by propelling propelled method and researcher can take greatest yields. Keen cultivating is one of the quickest developing field in IoT. For better quantifiable profit farmers, have begun utilizing important insights from the information to yield. Detecting of soil dampness and supplements, controlling water usage for plant development and deciding customer fertilize are some examples of smart farming of IoT.

- **Smart Retail**

The prospects of IoT in the retail sector is simply amazing. To raise the in-store experiences. IoT gives a great chance to retailers to interface with the customers. Advance mobile phones will make too simple for retailers to get associated with their purchasers at whatever time level out of store to serve them better and also in more efficient way. By this they can undoubtedly track shopper's way through a store to enhance store design and place items appropriately.

- **Energy Engagement**

Power grids won't just be sufficiently keen additionally exceptionally solid for up and coming era. By keeping in that center Smart grid idea pulls in us and getting to be distinctly well known all over world. The principle objective of the Smart grid is to gathering information in a mechanized manner and examining the conduct or power consumers and suppliers for enhancing productivity also financial aspects of power utilize. It also detects sources of power outages more rapidly and at particular household levels like nearby solar panel, making possible distributed energy system.

- **IoT in Healthcare**

Connected healthcare yet remains the resting giant of IoT applications. The concept behind the healthcare system and smart medical devices brings huge prospects together not just in corporate but also in general for well-being of people. Research exhibit IoT in healthcare will set sound standard in coming years. The principle reason for IoT in healthcare is to engage individuals to live more beneficial life by wearing connected gadgets. The gathered information in customized examination will help to analysis individual well-being and give tailor made systems to battle disease.

- **Industrial Internet**

Industrial internet is another powerful application which has already created new buzz in the industrial sector which additionally named as Industrial Internet of Things (IIoT). It is empowering industrial engineering with sensors, software and big data analytics to make extraordinary machines. According to Jeff Immelt, CEO, GE Electric, IIoT is an "Excellent, desirable and investable asset of every organization. Compare to humans, Smart machines are more accurate and precise while communicating through data. In additional this data can help companies to recognize any fault and solving it in more efficient way.

When it comes to control quality and sustainability IIoT holds very strong and remarkable position. As indicated by GE, change industry profitability will create $15 trillion to $20 trillion in GDP worldwide over next 15 years. Application of keeping tracking over goods, continuous data trade about stock among provides and retailers and computerized conveyance will expand the store network effectiveness.

- **Connected Cars**

The internal functions of the vehicles have been optimized by the continual focus the automotive digital technology. Since then for now the attention is biased towards the enhancement of the in-car experience. A connected car is a vehicle which is proficient to upgrade its own operation, upkeep additionally solace of travelers utilizing installed sensors and web availability. Substantial car creators and additionally some valiant new businesses are being chipping away at associated car arrangements. As indicated by the most recent news, the brands like BMW, Apple, Testa and Google are being attempting to acquire new transformation car division.

- **Wearables**

Wearable have an explosive demand in global market companies like Google, Samsung have heavily invested in producing such devices. Wearable gadgets are introduced with sensors and software's which collect information and data about the clients. At that point this all information will be pre-prepared to concentrate fundamental data about client. These devices for the most part cover wellness, well-being and entertainment necessities. These devices mainly cover fitness, health and entertainment necessities. The pre-imperative from IoT innovation for wearable applications is to be exceptionally vitality proficient or ultra-low power and little estimated.

- **Smart Home**

Smart home is the most sought word on Google which is associated with IoT. Let's see how IoT will surprised you by making your routine life simpler and convenient life at your own way by having the real-time example of smart home. Wouldn't you cherish in the event that you could switch on air conditioning before reaching home or turn off lights subsequent to leaving home? Or, on the other hand open the entryways. If someone arrived at your home and want to access even when you are not at home. It will not surprise us if one day smart homes will become a common use in human life as to smart phones. If we can see nowadays, the cost of owning a house is the greatest and unavoidable cost in a property holder's life. So, to make their life smoother, smart home organizations like Nest, Ring, Eco bee and couple of more are wanting to convey a never observed or deal and to set their family rand in Real Estate Sector.

- **Techniques**

To making the IoT possible several technologies are to be used. This section in represent different techniques which are used in IIoT.

The first technologies are identifications. In the IoT system there will be trillions of devices which are connect to internet, to identify these all devices, each one required a unique identification and this is possible by using Ipv6 enabled. The Second technology is, an IoT device needs to sense. It is possible by putting sensors on the system, here sensor able to measured different aspects of an object. These objects will be able to communicate with another object which are outside the world or other similar object around it. For device identification RFID and QR code are mostly used.

For communicating to each other there are some protocols to be used. By helping this protocol device could be able to collect the data and then sent to another device. If required is sent back to devices with other information also. For this purpose, IIOT use different protocols such as MQTT (Massage Queue Telemetry Transport), XMPP (Extensible Messaging and Presence Protocol), AMQP (Advanced Message Queuing Protocol), CoAP (Constrained Application Protocol) and DDS (Data Distribution Service)

1.3 M2M Communication

M2M Communication is a one type of data communication. It includes one or more entities where the process of communication is done without human interaction. M2M is known as Machine type communication (MTC) in 3GPP (3rd generation partnership project). M2M Communication is varied from the existing communication model It involved

- Less efforts and costs.
- Different market scenarios.
- Communicating terminal in very large number.
- Minor traffic per terminal.

M2M Communication is depending on mobile networks (e.g. CDMA EVDO networks, GSM–GPRS networks). The main responsibility of mobile network is to serve as a transport network in M2M Communication. M2M Communication makes enabling machines such as mobile devices, computers, embedded processors, actuators and smart sensors to communicate with each other, and make some decisions without human interaction [4]. Devices can be communication via wireless or wired network. M2M is a view of the IoT where objects are connected to the whole environment and managed with devices, networks and cloud based servers [5].

The main goal behind to develop M2M for cost reducing, increasing safety and security and achieving productivity gains (Fig. 3).

There are three stages of M2M communication: collection, transmission, and data processing. The data collection stage is the process for collecting the physical data. The transmission stage is the process for transmit the data to the external server. The data processing stage represent dealing and analyzing with data and also provide feedback or controlling the application.

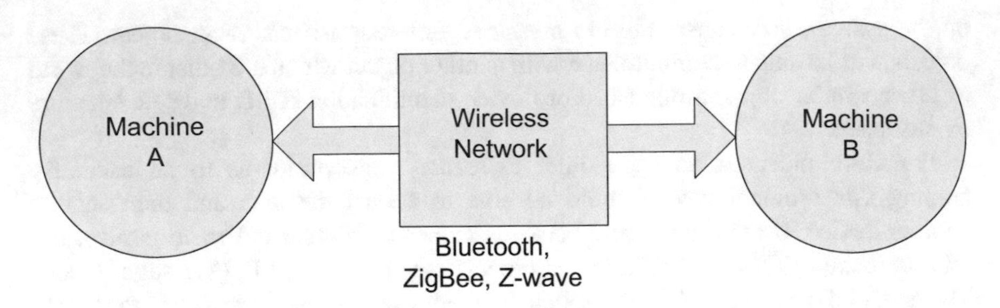

Fig. 3 M2M Communication

Table 1 Area of application of M2M Communication

Area	Application
Health	Remote diagnostics, to monitoring vital signs, supporting handicapped or aged
Tracking & tracing	Traffic information, traffic optimization, pay as you drive, road tolling, order management, navigation
Security	Access control, car security, alarm systems
Metering	Industrial metering, grid control, power, water, gas, heating
Facility management	Building or home or campus automation
Payment	Vending machines, point of sales

1.4 Application of M2M Communication

There are many areas where M2M application used. The following table shows the area in which M2M is used recently (Table 1).

The general requirements of M2M is that, using multiple communication Such that, SMS, IP Access and GPRS. The M2M system will be capable to provide communication between the M2M devices or M2M gateway or network and application domain.

- To deliver the message for sleeping devices. The M2M system will be able to manage the communication between sleeping device.
- M2M system should be support the different types of delivery nodes like any cast, unicast, multicast and broadcast communication modes.
- M2M system should be capable to manage the schedule of messaging and network access also.
- M2M system should be capable to decide communication path based on cost of network, delays or transmission failures when another communication paths exist.

- When number of devices are connected with object then M2M system should be scalable.
- M2M system should be capable to give any notification when it notified of any failure to deliver the message.

1.5 Bigdata and IoT

The use of big data analytics has grown extremely in just the past petty years. At same time, the IoT has introduce the public awareness, people's imagination on what a fully interconnected world can offer. For this reason, where big data and IoT are almost made for each other.

Big data and IoT basically go together. One study predicts that by the year 2020, IoT will be generated 4.4 trillion GB of data in the world. In year 2020, tens of billions of devices and sensors will have some specific type of connection to the internet. All of these devices will be producing, analysing, sharing and transferring data all in real time. Without this data, IoT devices would not have the functions and capable that have caused them to gain so much worldwide observation.

The IoT and big data are so nearly link. For example, wearable technology. Other is fitness band are part of IoT, it taking personal health data, giving the user calculations on recent fitness levels and tracking and noticed them about progress.

2 Bigdata

Newly companies, Government and academia become devoted in the big data. Using Traditional data management techniques are very difficult to process or manage bulky, multifarious and unstructured information. [6] Still, a majority issue for IT analyst and researchers is that this data expansion rate is fast excess their ability to both: (1) analyze is to extract suitable meaning for decision making, to lead and analysis of such datasets and (2) Design proper system to handle the data effectively. Bigdata can be describe using 5V model [7] represent in Fig. (4). This model is an expansion of 3V's model [8] and that includes:

1. *Volume (Size of data)* To generating and collecting huge amount of information (data), data rate take place big. Now-a-days, the data will be producing in the order of zettabytes, and it's expanding almost 50% every year. [9]
2. *Velocity (Streaming Data)* This represent the timeless of Bigdata. Data collection and analyzation both are immediately and timely managed. So, to magnified the use of the financial value of big data.
3. *Varity (Unstructured Data)* This model indicating the version type of data which contain data such as video, audio, webpages, images and text.

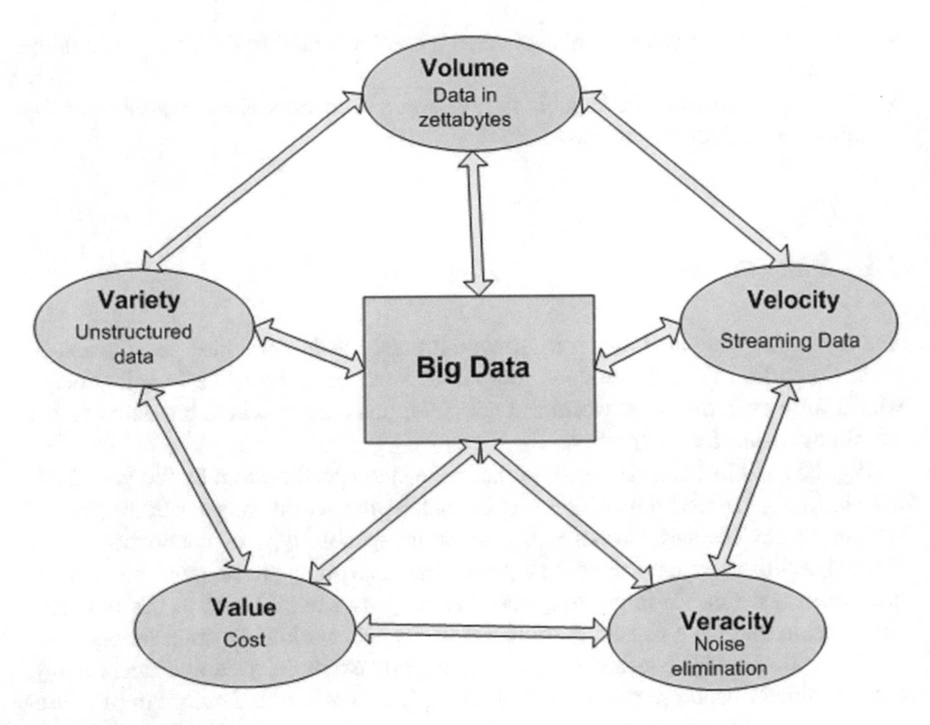

Fig. 4 5V's of big data

4. *Value (Cost of Data)* Data will be producing, collecting, and analyzing from the heterogeneous sources. Now a day the data having some cost. To make budget decision for estimation of data storage cost is very important.
5. *Veracity (Messiness of the data)* It is referring to the uncertainty or distrust around data, which is due to data incompleteness and inconsistency with many types of big data, quality and accuracy are less controllable but big data analytics technology permit us to work with these types of data.

The data will be generating as big impacts on our daily lives as internet has done. Data comes from social network Facebook, twitter, finance for example stock transaction, E-commerce system and specific research and sensor network etc. The given Fig. (5) shows the Layered Architecture of big data system. It is differentiating in three layers from top to bottom that consist infrastructure layer, computing and Application Layer.

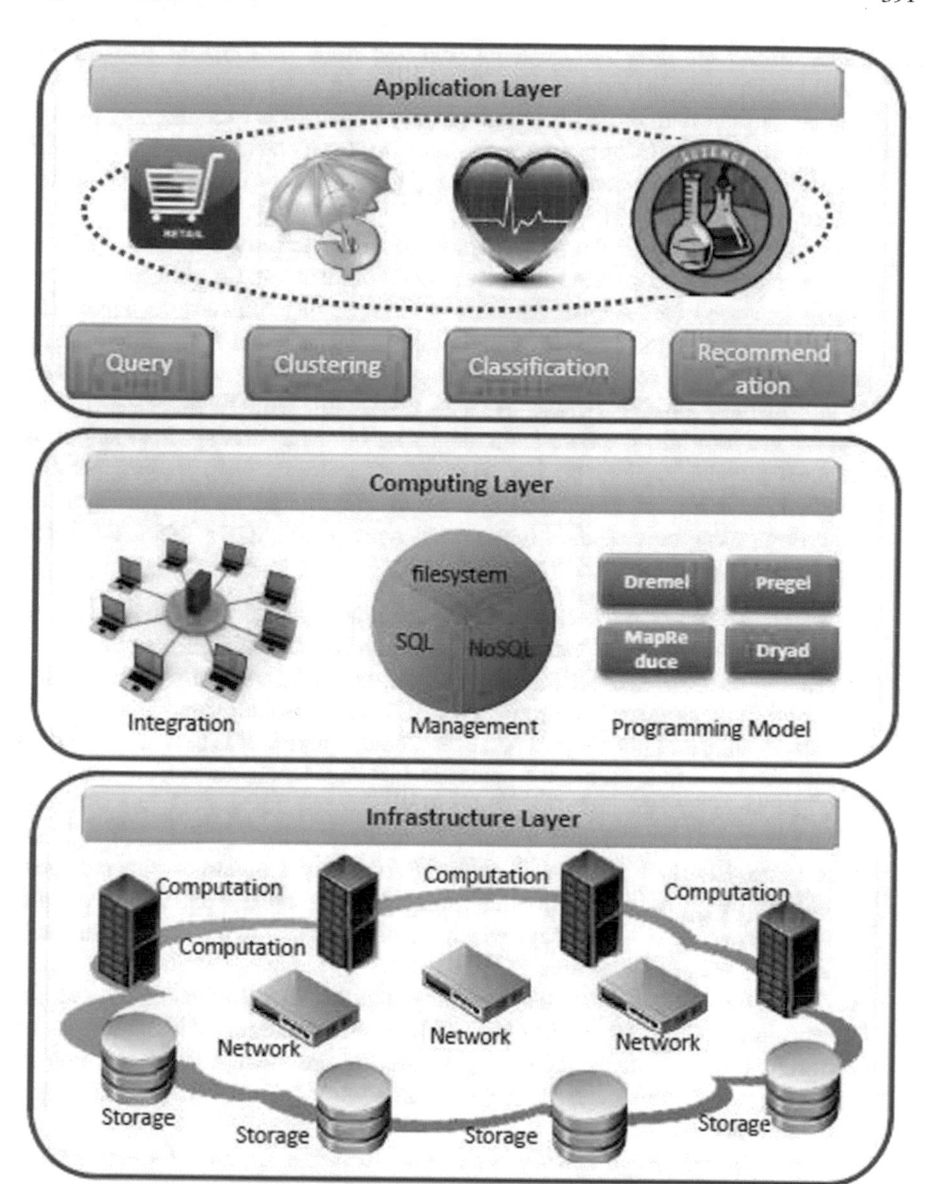

Fig. 5 Layered architecture of big data

2.1 Bigdata Challenges

There are a lot of challenges when manage Bigdata problem, during capturing the data, storing the data, sharing the data and searching the data and analysis and visualization the data. [10, 11]

1. *To Capturing the data* Data can be collected from the different sources like social media, sensors, experiments, meta data and transaction. Because of the different of data sources and the total quantity of data, it is exceptionally hard to gather and integrate information with versatility from conveyed areas. Different issues like this test are data pre-processing, transmission of data and automated created of metadata. [12]
2. *Data Storage* To Store the large data it not requires only a large amount of storage place but also require management of data on large heterogeneous system because of traditional database systems having difficulty to handle big data. As result of the great properties like being schema free, processing a simple API, simple replication, consistency and supporting a colossal measure of information. Apache HBase, Apache Cassandra, NoSQL database this all projects are turn into the core technologies of big data. The apache Hadoop is the most popular implementation open source software. Map Reduce provide automatic scalable and parallelization data distribution across many computers. Data Storage can be classified into several types as: NoSQL DBMS, RDBMS, DBMS based on HDFS and Graph DBMS.

   - *NoSQL DBMS* NoSQL DB is free in schema structure. It can be providing some features, such as distributed storage, horizontal scalability, dynamically schema etc. NoSQL DBMS storage the unstructured data in a key-value model. It is not good for Atomicity, Consistency, Isolation and Durability of data. Also, it can not support some distributed queries. It is new style for data management, that utilize the semantic information represented in ideology when searching the IoT data stores. Thus, some related work has tested to integrated the feature of distributed file system to the IoT data stores. [13]
   - *RDBMS* RDBMS is a basic platform for many structured data storages. RDBMS uses SPARQL queries to get the data on the existing relational stored views and ultra-wrap, which encodes logical representation of each RDBMS as an RDF graph. [14]
   - *DBMS based on HDFS* Here we integrated the HDFS with distributed file repository, which processing huge amount of unstructured file efficiently. In the area of IoT, many data are generated in the XML format, so how to handle these small sized, large-volume XML files becomes a major challenge.
   - *Graph DBMS* Graph DBMS is database. To represent and store the data it uses nodes and edges. Sensor data can be managed efficiently using Graph DBMS. [15] Graph DBMS provides a large graph that consists of large scale nodes and edges and high performance Graph DBMS management system (Table 2).

3. *Data Search* Today data must be available in exact, timely manner and completed and it becomes necessary. [16] At last data are to be use to aside a few minutes. Query optimization have been essential for responding effectively of complicated SQL queries. MapReduce is an exceedingly strong platform. It also provides high level query language such as PigLatin, HiveQL and SCOPE. This

Table 2 Comparison between data storage types

Features	NoSQL DBMS	RDBMS	DBMS based on HDFS	Graph DBMS
Support for Scalability	Yes	Not well	Yes	Yes
Support for Structured data	Not well	Yes	Yes	Use graph structure with nodes, edges
Support for semi and unstructured data	Yes	Not well	Yes	Use graph structure with nodes, edges
Support for ACID	Yes	Not well	Yes	Common
Support for massive & distributed processing	Yes, but not well	Not well	Yes	Yes

profoundly parallel platform, the cost of drag data across modes is capable and there for query optimization and physical structure have a tendency to be analyse components of the architecture.

4. *Sharing Data* Today, data sharing is become more important. In market thousands of data will be generated and absorb important data that is specific to their own business requirements but now it is produced in a way that could be interacted and shared to others requirements of an organization. Researchers facing issues regrading data sharing and presentation in several specific field like medicine, ecology and biology. [17]
5. *Data Visualization* Data visualization is a very challenging task for big data. As indicated to Fisher et al. [19] visual interfaces are reasonable for (1) to recognize patterns in data streams, (2) giving to deal with context by indicating data as a subset of a tremendous bit of data, and represents related variables, (3) data with statistical analysis. Visualization is an extraordinary field in which huge number of datasets are represented to users in visually powerfully irresistible ways with the faith that users would be capable to developed manageable relationships. Visual analytics required to developed many visualization across many datasets.
6. *Data Analysis* The magnetism of system expandability and hardware replication are represent by cloud computing along with MapReduce and Message Passing Interface (MPI) parallel program system suggest one resolution of this challenge by using a distributed approach. [18] Belonging to this, few issues are performance optimization of the software framework of big data analytics and dispersed and scalable data management ecosystem for profound analytics, i.e. the implementation of machine learning, different mining algorithms and statistical algorithms for analysis.

According to some kind of analysis big data tools are classified into (1) Batch analysis, (2) Stream analysis and (3) Interactive analysis. [20] In Batch analysis, data are primarily stored and after that analysed. In stream analysis, to extract the important information and to discover the knowledge from the generated data. In Interactive analysis, allow to users to lunch their own analysis of information.

2.2 Framework, Tools, Techniques, and Programming Language Supported by Big data

The tools of big Data develop for manage and analyse big data. This tools are efficiently process huge amounts of data inside sufficient expired times. The big data uses mostly open source software framework. Big organization give to open sources project and give some profit such as Yahoo!, Twitter, Facebook and LinkedIn. We represent how to deal with big data and their tools, techniques, framework and programming language in this section.

2.3 Big Data Techniques

In Bigdata many techniques are used to analyse the large amount of data. Here we describe some exclusive techniques which are generally use in Bigdata and their projects.

1. *Data Mining* Data mining is multidisciplinary area. It analyses and explore to huge amount of data to discover meaningful patterns. Data mining field use three rules which are Machine Learning, Pattern Recognition rule and Statically rule. Data mining is applicable in security analysis and intelligence, genetics, business and the social and natural sciences. Some techniques consist cluster analysis, association rule mining, classification and regression.
2. *Statistics* This technique are generally used for judgments about relationship between variables could and relationship variables. Variables occurred by chance and variables likely outcome from some kind of understanding causal relationships.
3. *Machine Learning* Machine Learning is the basis method of data analysis. It was well known from pattern recognition. The theory behind the machine learning is that computers can learn to perform specific tasks without being programmed. It has the ability to apply complex mathematical calculation to big data. It's goal to build analytical model automatically and it allows computers to find unread data into model. Right now, machine learning is expanded widely.
4. *Signal Processing* In this research, big data challenges offer extensive opportunities. Where the data are analyses in real-time dictionary learning, principle component analysis, compressive sampling, have already arranged on time/date adaptivity, robustness and dimensionality reduction. The main task of computer science in big data is disputed.
5. *Visualization Techniques* This visualization technique is analysis of high dimensional it is mostly used as sufficient data abstraction tools in searching knowledge, information awareness and making decision process.

2.4 Big Data Tools

According to the kind of analysis big data tools are classified into three types (1) batch analysis, (2) stream processing, (3) interactive analysis. In batch analysis, data are stored after that it is analysed. In stream processing, as early as possible which is analysis and it give results. In interactive analysis allow to users to perform their own analysis of data. Following table represent the taxonomy of the tools, frameworks for applications of big data (Table 3).

- *Based on batch analysis* there are four tools are used such as Google MapReduce, Apache Hadoop, Apache Mahout and Apache Microsoft Dryad.

(1) **Google MapReduce**

It is parallel computing batch oriented java based model. As Master, which responsible for assigning the work to the workers. Master assign to map workers from input data divide into splits. Each worker processed the individual input split and it is also generating key value pairs and write then on disk or in memory (intermediate files). The master gives the location of that files to the reducer workers. Reduce read data, process it and ultimately write data to output files. Map Reduce have three primary features scalability, simplicity and fault tolerance.

Table 3 Tools and framework for big data application

Name	Specific use	Advantages
Batch analysis		
Google MapReduce	Processing of data on large cluster	Fault tolerant, simple, scalable
Apache Hadoop	Platform and Infrastructure	Completeness, scalable, reliabilities, extensibility
Apache Mahout	Machine Learning Algorithm	Scalable
Microsoft Dryad	Platform and Infrastructure	Fault tolerant, good programmability
Stream analysis		
Apache Spark	Use for data processing	Easy to use, fast
Apache Strom	Real-time computing system	Simple, scalable, efficient, fault tolerant, easy to use and operating easily.
Apache S3	Platform for Stream computing	Scalable, extensible, fault tolerant
MOA	Use for Data stream mining	Scalable, extensible
Interactive analysis		
Apache Drill	SQL query engine for Hadoop and NOSQL	Familiar and quick, flexible
Apache BI	Business Intelligence	Big data streaming on Real-time BI
D3	Interactive	Scalable

In a cloud environment MapReduce is a good example for Bigdata processing. It allows develop parallel programs to a newer programmer. MapReduce processing a large amount of data in a cloud. So, that it is preferred for computation model of cloud provides.

MapReduce also have some limitations, MapReduce is suitable only for batch processing job, it is impossible for iterative jobs and models. It became very expensive while implementing iterative map reduce jobs for huge space consumption by each job.

(2) **Apache Hadoop**

Apache Hadoop is used for distributed processing of very huge amount of data sets and it is an open sources framework. Apache Hadoop inspired by Google File System, MapReduce and Google Big Table. It consists processing part called Map Reduce and Hadoop Distributed File System(HDFS) known as storage part that can usually replace SAN devices. The following Fig. (6) shows architecture of Hadoop stack.

Hadoop procures users to distributed queries across multiple datasets on large clusters. HDFS stores the large files across the multiple machines and that files running parallel MapReduce computing on the data. Hadoop support some kind of

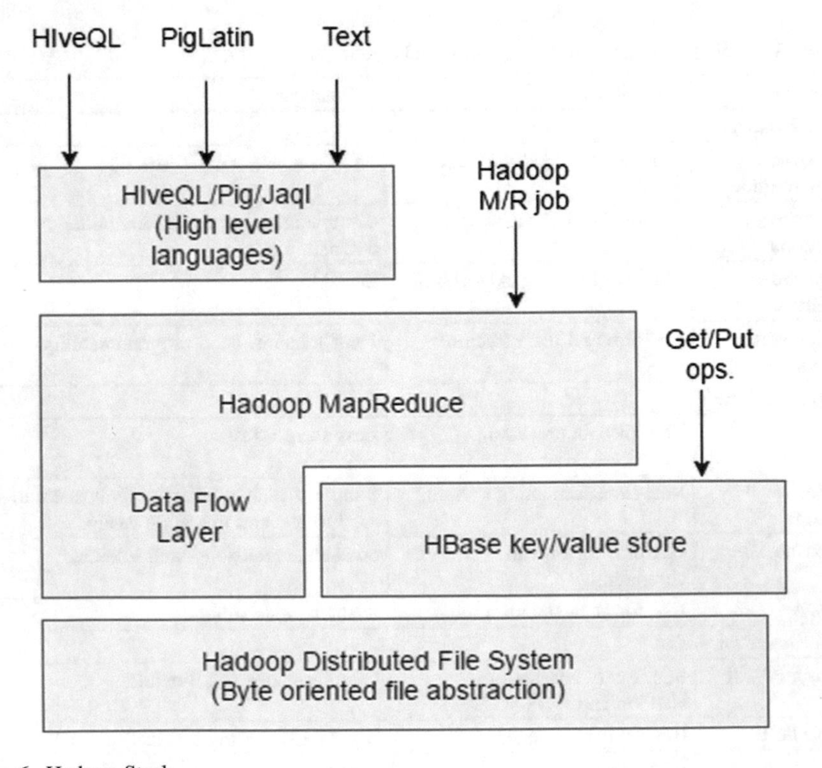

Fig. 6 Hadoop Stack

languages such as for domain specification language like Java and Python, for Query language like SQL, HiveQL and Pig Latin, this all languages Hadoop is suitable for deep analytics. In additional, Hadoop sub-project, HBase and Hive offer data management solution for unstructured and semi-structured data.

Today, Hadoop has risen to be "gold standard" in the industry as a platform that is feasible with scaled data intensive Map Reduction platform, it is highly used for large scale information extraction, Web indexing, Machine Learning, Clickstream and log analysis. Facebook, Twitter, Amazon and LinkedIn this companies used Hadoop technology.

(3) **Apache Mahout**

Apache Mahout is a Machine Learning framework. Apache Mahout supports three utilize cases: Clustering, Classification and Recommendation. It has aims to make intelligent application faster and easier.

(4) **Microsoft Dryad**

Dryad is a framework which allow a software engineer to used the resources of data centre and computer cluster for running data parallel programs. A Dryad programmer can utilize a huge of number of machines, each machine with multiplcs processors without knowing anything about simultaneous programming. It contains a runtime parallel system called Dryad and two high level programming models.

The process of Dryad is that the dryad programmer write programs sequentially and linked them using one way channels. As a directed graph the computing structured is that the program is known as graph vertices and channels are known as graph edges. It is responsible for generating graph which can synthesize all directed acyclic graph.

Dryad is similar to other frameworks like Google MapReduce or relational algebra. In additional, it handles job mentioning and visualization, fault-tolerance, resource management, scheduling, re execution and job creation and management.

- *Based on Stream Analysis* here we describe four tools they are Apache spark, Apache Strom, Apache S4 and MOA (Massive Online Analysis). This all tools are used for real time streaming in big data. This tools have to ability to handle large amount of data, using compute clusters to balance the workload.

(1) **Apache Spark**

It is large scale data processing engine. The speed of spark is faster than Hadoop Map Reduce in memory. Spark include the stack of libraries such as SQL and Data frames, MLib for Machine Learning, GraphX and Spark streaming. You can join this library consistently in the similar application. Spark can run all around. Using spark standalone cluster mode, you can run it on EC2, on Hadoop Yarn or on Apache Mesos. Spark can access data in HBase, Hive, Cassandra. Spark gives APIs in Python, Scala and Java.

(2) **Apache Strom**

Apache Strom is open sources free distributed real time computation system. Strom doing real time processing what Hadoop did for batch processing. It makes easy to reliable process of data. Strom is very easy and any programming language is used with storm. It is very fast tool for stream analysis. It can be processed millions of tuples per second each node.

(3) **Apache S4**

Apache S4 is a Distributed stream computing platform. It allows programmer to easily develop applications for processing continuous unbounded stream of data. It was initially released by Yahoo! It is fault tolerant stand by server is automatically activated to take over the task. It is extensible on which applications will be easily written and deployed via simple API.

(4) **MOA (Massive Online Analysis)**

MOA is an open source framework for data streaming. It is related to Weka project that allows to build and run experiments of data mining and machine learning algorithms. It includes Classification, Clustering and Frequent item set mining. By combing this two tools S4 and MOA software we can also use for distributed stream mining.

- *Based on interactive analysis* here we describe three tools they are SpagoBI, Apache Drill and D3.

(1) **SpagoBI**

SpagoBI is an open source business intelligence on big data. It offering large range of analytical function, it is an advance data Visualization for geospatial analytics.

(2) **Apache Drill**

Apache Drill is use for analysis of read only data. It is scalable, interactive ad hoc query system. Apache Drill is to able to manage more than 12,000 servers and processed petabyte of data in seconds. It is a framework for data intensive distributed application for interactive analysis of large datasets.

(3) **D3 (Data-Driven Documents)**

D3 stand for Data-Driven Documents. It is a java script library for developing dynamic and interactive data visualization in web browsers. D3 is generally implemented in HTML5, SVG and CSS standard.

3 To Collect, Organize and Analysis the Data

The Internet of Things gather outstanding consideration over the past certain years with implementing latest technology like sensor hardware technology and micro-controller materials that are not expensive. Sensing devices connected to all the elements surrounded such that they can be able to interact with each other without human interaction. To undertake the concept of sensor data is something that is a biggest challenge that will come across IoT.

IoT is creating tremendous volume of data, also rising analytics tools and mechanisms, that are affording that to allow us to operate this all machines absolutely new ways, and efficiently way. The data which are generated from all this resources which are linked to the Internet will apparently of this world to fetch advantages. Here the big data analysis needed to take advantages of it's very high-level engineered knowledge.

This section is describing the research work that how to collect data using Flume, how to organize data in Hadoop and how to analyses data using Hive. Here we use some sensor data like temperature sensor data. We collect the temperature datasets which generate the maximum temperature dataset by the year 1900–2016.

3.1 *To Acquire Data*

Data has to be collect from various sources. They are listed below.

(1) *Sensor or Machine Generated Data* it is including Smart meters, Call Detail Records, Sensors, Weblogs and trading system data.
(2) *Social Data* Twitter and Facebook. It including customer feedback stream and blogging sites.
(3) *Traditional Organization Data* It including ERP data, Web store truncations and general data. It also includes customer information from CRM systems.

- **Flume**

Flume is distributed open source software. From the many various sources this system collecting, assembly and moving huge number of logs data and store to be centralized data. Here using Flume, we moving and aggregating very huge amount of data around a Hadoop Cluster. In Hadoop cluster Flume is collect log file from all machines and then continues in a HDFS which store centralized. By developing chains of logical nodes, we have to creating data flows and then connect them to source and sinks.

Mostly Flume have three tire architecture: (1) Agent tire, (2) Collector tire, (3) Storage tire. The Agent tier, it has flume agent that collect the data from various sources which is to be moved. Collector tire, includes multiple collectors which collect the data comes from multiple agents and forward it on to storage tier which have file system like HDFS or GFS.

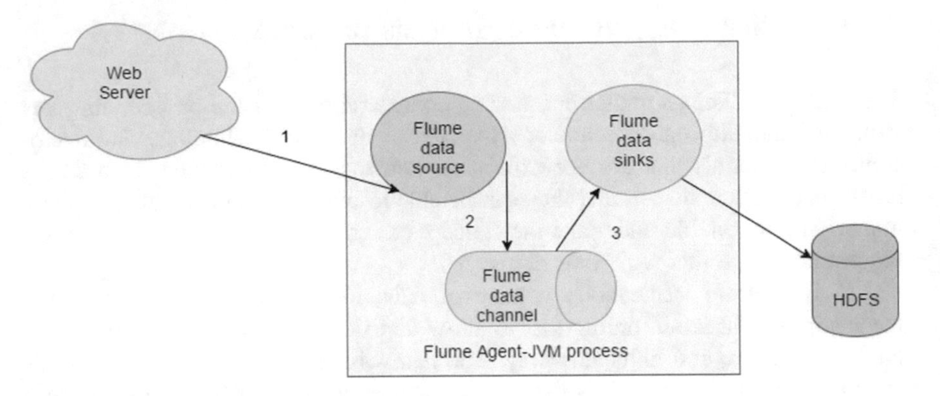

Fig. 7 Working of flume

Flume has three components, *Flume source, Flume channel* and *Flume sink.* Flume agent is a JVM process (Fig. 7).

(1) The above Fig. (3) the working of Flume. The events are generated from the external sources (Web server). The external source sends events to the flume source.
(2) Flume source receives that event which are send by external sources and store it into different channels. The channel performs as a storage system which place the event till it absorbs by the sink.
(3) All events are remove from channel in the sink and store it at HDFS or external database. There will be various flume agents in which case flume sink transfer the events to the flume source of next flume agent in the flow.

3.2 To Organize Data

After collecting data, it has to be organized using a distributed file system it has to be organized by collecting data. Here we have to split this data into fixed size blocks and store it into HDFS file systems. Figure (8) illustrate the architecture of HDFS.

Hadoop distributed file system is a scalable, portable and distributed file system which is written in java. In Hadoop cluster, every cluster has a single name node and many data nodes. Data node has many blocks by default each block have 64 MB size. For communicate with each other clients use RPC call. There are three copies of replication of data are present in every data node. Data nodes are communicating to each other to copy data or to rebalance data or to keep high replication of data. In case, if node goes down the HDFS has able to allowing name node to be failed over to backup.

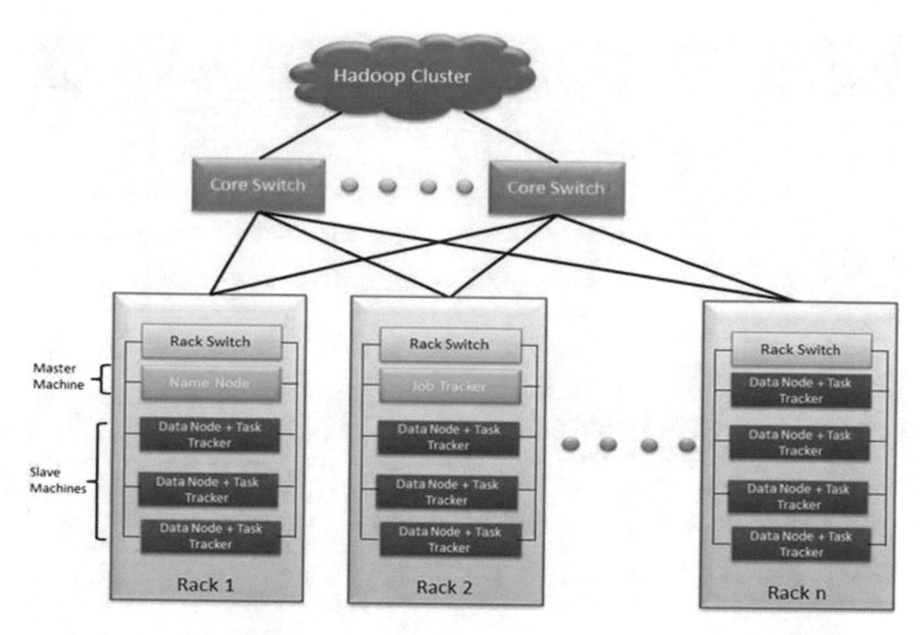

Fig. 8 Architecture of HDFS

Today, automatic failover is also establishing. For this we use a secondary name node which takes the snapshots continuously of primary name node so that is can be active when failure of node occurs.

3.3 To Analyze Data

To analyze data here, we collected the temperature datasets and perform some analysis using Hive. For this purpose, we install Hadoop, Hive on cloud era. The following Fig. (9) represents temperature datasets which are present in HDFS.

- **Hive**

Hive know as a data warehouse to process structure data in Hadoop. It builds on top of the Hadoop and providing query and analyze easily. It was developed by Facebook and it is used by various components for example Amazon Elastic Map Reduce which is used by Amazon. The Architecture of Hive is representing following Fig. (10) it contains many components which are describe in following table (Table 4).

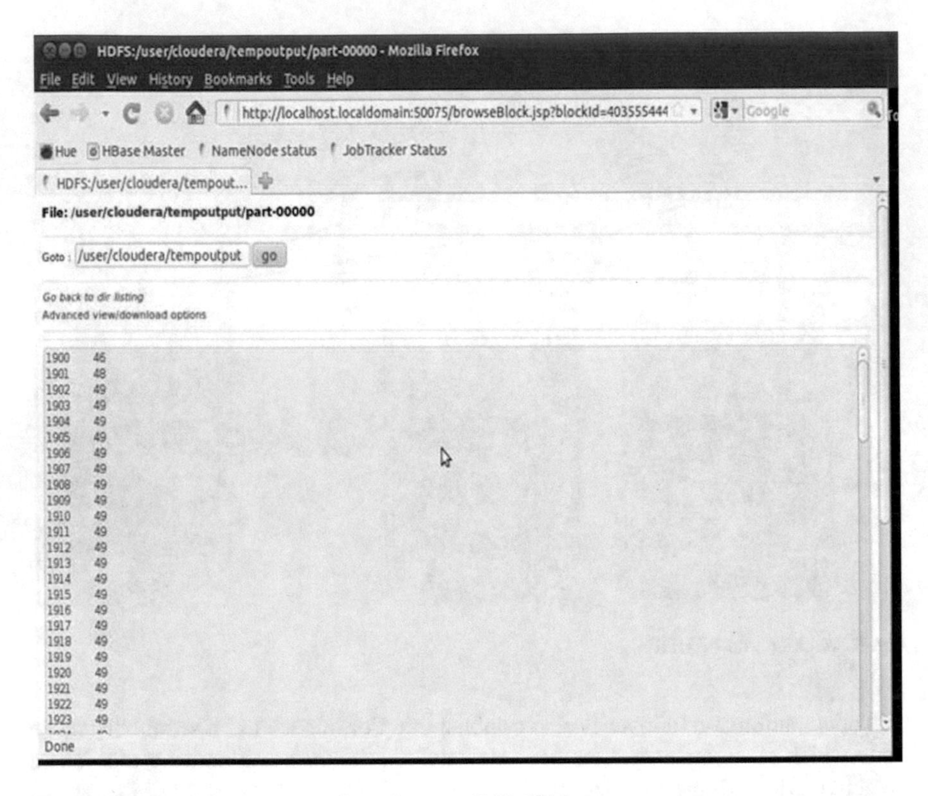

Fig. 9 Showing the temperature from the year 1900–2016

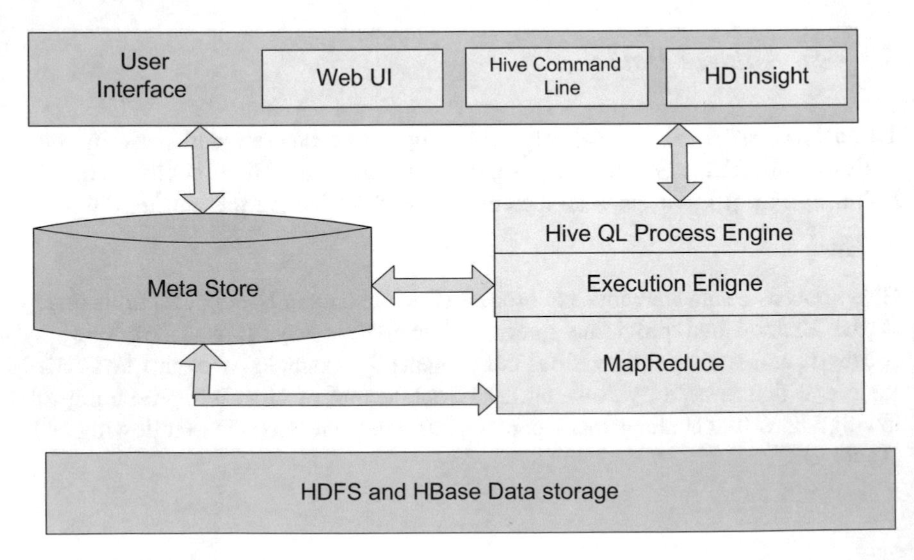

Fig. 10 Architecture of hive

Table 4 Description of Hive components

Component name	Operation
User interface	This component interacts with user and HDFS. The user Interfaces that supports are Hive HD Insight, Hive Command line and Hive web UI
HiveQL process engine	It is use for querying on schema information on the types and it is similar to SQL
Meta store	Hive select respective database server to store the meta data of tables or schema, databases, HDFS mapping and columns in a table and their data types
Execution engine	This component processes the query and produce the results as same as MapReduce results
HDFS or HBase	It is a technique to store data into file system

Fig. 11 Hive screen

There are three different ways to configured Hive

1. By using set command.
2. By hiveconf option in Hive command shell
3. By editing a file hive-site.xml

Here more than ten lacs database record which is analyzed by using Hive to get a temperature. We perform the analysis by using following Hive query language commands (Fig. 11).

- Create Database,

  Create database temp_records;
  Use temp_records;

- To store the records, we create table.
- *Create table temp_record (year INT, year INT);*
- Loading the data into table
- Describe metadata or schema of the table,

Describe temp_record;

- Select data
  *Select*from temp_record;*

4 Open Issues and Challenges in Big Data Analytics

To solve many issues regarding to data cleaning, security, privacy and Communication between systems big data become latest topic in the research field. This all issues are describing briefly in this section.

- **Data Cleaning**

Data Cleaning playing the main role in the data handle, management and analytics and still it has quick development. Furthermore, in the age of Bigdata Data Cleaning consider a major challenge. Because of the growth of volume, variety and velocity of data in so many applications. Here we introduce some issues to be repaired to needs are: (1) Usability, this feature is very important of data management, to interact with machines for human is very challenging task, (2) Tool Selection, the question arise here is that which tool to be use for given database to perform specific task for data cleaning, (3) Rule discovery, another challenge for cleaning data which is very difficult process.

This all issues are extremely hard to deal with when goes to the specific topic of data cleaning. All things considered, given detected errors, there is no right and which aren't right, even for human also.

- **Data Security**

In big data analytics, what way to handle generated data without a security it also biggest challenge of big data analytics. Related to our examination, big data analytics in security issues can ordered into four parts: input, data analysis, output and communication with different framework. In the input part, security issues of big data analytics are to confirm that the sensors won't influenced by the assaults. In the output part and analysis, it could be connected as the security worry of a framework. For communication with rest of the systems, the security concern has been between various external system and big data analytics. Because of issues all issues, security becomes most important issues in big data analytics.

- **Communication Between System**

For the most part, big data analytics systems has been constituted for computing parallels and would be actualized on different systems like cloud computing or else web index or else learning focused, the communication between the other systems and big data analytics would be heavily affect the execution of the entire procedure of KDD. The cost of communication between the system is first challenge of data analytics. The first responsibility of data scientists is that how can there be reduction in cost of communication. Another open issue on the communication between system is that the data consistency between various systems, operators and modules is also and very important issue, and second open issues of big data analytics is which to make the communication between these frameworks as dependable as could be allowed.

- **Data Privacy**

In big data applications, the data privacy will be big issues specially if an organization does not give the conformation that their personal information's are secure with them the analysis process needs the data so privacy is most concern issues.

5 Conclusions

In IoT, it is very difficult to manage big data which are generated from different sources using traditional database management systems. We also provide a varied diversification of big data challenges which are Capturing, storing, Searching, Sharing, Analyzing and Visualizing. Here we conclude that Analysis is very critical challenge for research relative to big data, since its application is in the areas of earning potential knowledge for big data. We also represent the tools and frameworks which are used mostly and programming language for application of big data.

In addition, to handle big database we use some intensely parallel software. First of all, we collect data from heterogeneous sources which can be done easily by using flume. Then we can be organized data using distributed file system like HDFS. After this we analyzed that data using Hive. Hive can be analyzing a database of ten lacs records in just 35 s. So, using these all components it makes possible to manage and to use big database in an easy and efficient manner.

Acknowledgements Thank you for your cooperation and I would also thank to my guide for his imitable guiding constant monitoring and encouraging support throughout the task. His blessing continual guidance shall help me achieve my goals in journey of my life over which I am.

References

1. General Electric Company. (2014). GE intelligent platforms, Industrial Internet. Retrieved from http://www.ge.com/industrial-internet
2. Industrial Internet Consortium. (2014). Engineering: The first steps. Retrieved from http://www.iiconsortium.org/pdf/IIC-frirst-steps_2014.pdf
3. Annunziata, M., & Evans, P. C. (2012). *Industrial internet: Pushing the boundaries of minds and machines*. Boston: General Electric Co.
4. Watson, D. S., Piette, M. A., Sezgan, O., & Motegi, N. (2014). Machine to Machine (M2M) technology in demand responsive commercial buildings Washington: American Council for an Energy Efficient Economy.

5. Chen, C. L. P., & Zhang, C.-Y. (2014). Data-intensive applications, challenges, techniques and technologies: A survey on big data. *Information Sciences, 275,* 314–347.
6. Taila, D. (2013). Clouds for scalable big data analytics. *Computer, 48*(5), 98–101.
7. Lomotey, R. K., & Deters, R. (2014). Towards knowledge discovery in big data. In *Proceeding of the 8th International Symposium on Service Oriented System Engineering IEEE Computer Society.*
8. Lancy, D. (2011). 3-D management: Controlling data volume, velocity and veracity application. *Delivery Strategies.*
9. Fan, W., & Bifet, A. (2012). Mining big data: Current status, and forecast to the future. *SIGKDD Explore.*
10. Chen, P. C. L., & Zhang, C. Y. (2014). Data intensive applications, challenges, techniques and technologies: A survey on big data. *Information science, 275,* 314–347.
11. Alshammari, H., Lee, J., & Bajwa, H. (2015). Improving current Hadoop mapreduce workflow and performance. *International Journal of Computer Applications, 116*(15), 38–42.
12. Hu, H., Wen, Y., Chua, T. S., & Li, X. (2014). Toward scalable systems for big data analytics: A technology tutorial. *IEEE Access, 2,* 652–687.
13. Cure, O., Kerdjoudj, F., Faye, D., Le Due, C., & Lamolle, M. (2013). On the potential integration of an ontology based data access approach in NoSQL stores. *International Journal of Distributed System & Technologies (IJDST), 4*(3), 17–30.
14. Yaish, H., Goyal, M., & Feurelicht, G. (2014). Multi-tenant elastic extension tables data management, proedria computer. *Science, 29,* 2168–2181.
15. Martinez-Bazen, N., Gomez-Villamor, S., & Escale-Claveras, F. (2011) Dax: A high-performance graph database management system. In *Data Engineering workshop (ICDEW), 2011 IEEE 27th International Conference on IEEE 2011* (pp. 124–127).
16. Hameurlain, A., & Morvan, F. (2015). Big data management in the cloud: Evolution or crossroad? In *International Conference: Beyond Databases, Architectures and Structures.* Springer, Cham.
17. Naseer, A., Laera, L., & Matsutsuka, T. (2013). Enterprise big graph. In *46th Hawaii International Conference on System Sciences. IEEE Computer Society* (pp. 1001–1014).
18. Gropp, W., et al. (1996). A high-performance, portable implementation of the MPI message passing interface standard. *Parallel computing, 22*(6), 789–828.
19. Rossi, R., & Hirama, K. (2015). Characterizing big data management. *Issues in Informing Science and Information Technology, 12,* 165–180.
20. Bhayani, M., Patel, M., & Bhatt, C. (2016). Internet of Things (IoT): In a way of smart world. In *Proceedings of the International Congress on Information and Communication Technology* (pp. 343–350).
21. Bhatt, Y,. & Bhatt, C. (2017). Internet of Things in healthCare—Internet of Things and big data technologies for next generation healthcare (pp. 13–33).

Part IV
Internet of Things Security and Selected Topics

High Capacity and Secure Electronic Patient Record (EPR) Embedding in Color Images for IoT Driven Healthcare Systems

Shabir A. Parah, Javaid A. Sheikh, Farhana Ahad and G.M. Bhat

Abstract In this chapter a very high capacity and secure medical image information embedding scheme capable of hiding Electronic Patient Record (EPR imperceptibly in color medical images for an IoT driven healthcare setup has been proposed. A fragile watermark has been embedded besides the EPR to make the proposed scheme capable of detecting any tampering of EPR during transit. At the receiving end, firstly we extract the fragile watermark, if it is same as one embedded, signalling that EPR has not been attacked during transit, EPR is extracted otherwise system uses a retransmission request instead of wasting time to extract the compromised EPR. For ensuring security of the EPR we hide it pseudo-randomly in cover medium. Two address vectors have been used to embed the EPR pseudo-randomly in host image planes and hence secure the information. Further hiding is carried out in spatial domain and hence is computationally efficient. Given the merits of our technique with regard to payload, imperceptibility, security, computational efficiency and ability to authenticate EPR at the receiver it can be used for real time medical information exchange like in case of Internet of things (IoT) driven healthcare and health monitoring systems for effective transfer of medical data during transit. A detailed experimentation carried out and the results obtained prove that our technique is proficient in hiding very high Payload EPR imperceptibly. We have further tested our scheme to several attacks like noise addition, filtering and rotation etc. Experimental investigations show that our scheme effectively detects the tamper caused by any of the mentioned attacks.

Keywords Data hiding · Electronic patient record · Internet of things · Embedding

---

S.A. Parah (✉) · J.A. Sheikh · F. Ahad
Post Graduate Department of Electronics and Institute Technology,
Kashmir University, Srinagar 190006, Jammu and Kashmir, India
e-mail: shabireltr@gmail.com

G.M. Bhat
Department of Electronics and Electrical Engineering,
Institute of Engineering, Zakoora 190006, Jammu and Kashmir, India

N. Dey et al. (eds.), *Internet of Things and Big Data Analytics Toward Next-Generation Intelligence*, Studies in Big Data 30,
DOI 10.1007/978-3-319-60435-0_17

1 Introduction

In present world, conventional healthcare is being replaced by the electronic healthcare. The electronic healthcare makes distance between the patient and a doctor irrelevant. The concept of Internet of things (IoT) driven healthcare and health monitoring is further enhancing the growth of electronic healthcare sector. In an IoT driven healthcare setup, the data collected originates from various things deployed for facilitating the proper diagnosis [1–4]. Implementation of IoT in healthcare sector besides solving many issues is resulting in analysis or even predictions of patients' health status. As the IoT ensures linkage between virtual and real world entities and thus enables anytime connectivity of anything [5]. Hence, the IoT approach in healthcare is proving valuable in assuring the compliance with some of the right principles in healthcare: right place and right time. IoT results in improved communication between medical staff and patients to solve problems more effectively. This is because in IoT approach the ubiquitous sensors and connected systems, can provide valuable and right information at right time for better healthcare delivery. Thus, with more complete patient information healthcare providers could focus on preventive rather than reactive medicine. Hence, Internet of Things paves way for new opportunities for right care, another important paradigm for proper healthcare. Usage of IoT in healthcare is bound to improve the health sector tremendously, but, this evolution poses a few serious problems that need to be addressed to reap the benefits. Some of the important issues include: (a) Security and privacy of the data transferred from source to destination. (b) Development of proper communication standards, as there exist many standards as of now. (c) Addressing Scalability concerns. This chapter focuses on the first challenge- security and privacy of the data during its transit from source to destination [6, 7].

Security and privacy of Electronic Patient Record (EPR) is one of the most important issues for successful implementation of IoT driven healthcare systems. This is due to the fact that in a typical IoT based healthcare/health monitoring setup, sensitive data has to be transferred between various devices/patient monitoring system and command and control unit connected to Cloud. The successful reception of medical data (Electronic Patient Record) is directly linked with quality of diagnosis, thus, it is imperative that critical medical data may be secured throughout its transit from source to destination. Thus security of the EPR data during transit is of prime importance [8]. Though internet security protocol (IPSec) is being used for security of data however it has been reported that use of IPSec reduces the throughput. Thus there is a scope to look out for alternative ways for securing medical information in a typical IoT based healthcare setup.

Data hiding in various multimedia objects like text, images videos, of late has flourished as a potent tool for securing information [9–13]. Recent research reports show that images are the main multimedia data being transferred over internet. This has triggered a massive use of images as cover media for hiding data in them and hence securing the data. As medical images form an imperative part of patient health record for diagnostic purpose, hiding EPR in medical images is receiving huge attention from research community. Some of the fundamental requirements of a data hiding algorithm are: (1). It should be robust to attacks (2). The hiding algorithm should be computationally efficient, (3) The algorithm should have high payload.

The payload of an information hiding scheme is directly proportional to size of the cover image. A Color images can be viewed as an equivalent to three grey scale images due to their RGB plane structure. Thus color images serve as best containers when payload is a priority in an information hiding scheme. Some of the main applications of color image based data embedding, are in the fields of secure data and EPR communication [14–33, 59, 60]. A high capacity and secure information hiding technique capable of hiding basic diagnostic information and case history imperceptibly has been presented in this chapter. The security of EPR has been ensured by embedding it at locations pointed to by PAV and CAV. The implementation is done in spatial domain and hence is computationally efficient. Further the scheme is capable of detecting any tamper to embedded data, thus informing the receiver in case the data has been fiddled with, during its transit from source to destination. Given the merits of the scheme it can be used for real time medical information exchange and as such it can be used in Internet of things (IoT) driven healthcare and health monitoring systems for ensuring security for medical data during transit.

2 Related Work

The security of information has been one of the most important factors while communicating over insecure channels [34, 61]. Data scrambling (encryption) is one of the important and cost effective methods for ensuring security to a data in transit. Steganography has also been given a due attention by research community in last decade as a viable and reliable option for communicating securely [35, 36]. Steganography is often used in conjunction with cryptography to improve the security provided to the embedded data. Strength of a steganographic algorithm is enhanced to a greater extent by combining it with cryptography [37, 38].

Color has always been a fascinating medium for mankind. The exponential growth of multimedia technology and its presence in everywhere has made world more colorful. Images are the most common and convenient means of conveying or transmitting information. Color images, because of their, three plane structure, form a very good cover medium for data hiding. A color image for all data hiding purposes can be thought of three equivalent grayscale images. An early technique for embedding data in a cover medium is least significant bit (LSB) insertion [39–42, 62]. In this data hiding technique LSB of few or all of the pixels of host image is altered by information to be concealed. When a color image whose bit depth is 24-bit is used as cover, each of the constituent planes holds the secret data. In a well-designed embedding algorithm, one besides embedding data in LSBs can embed the secret information in intermediate significant bit planes in a controlled manner and as such tremendously increase the hiding capacity of the data hiding system.

Embedding of EPR in medical images to be used in a typical IoT based healthcare system, deteriorates the cover medical image and affects the perceptual quality of cover image that could lead to wrong diagnosis. Data hiding in medical images thus needs a different approach from conventional hiding approaches. For proper diagnosis prior to the cover image is divided into Region of Non Interest (RONI) and Region of Interest (ROI). One more approach used in medical images is reversible data hiding [43–45].

Numerous medical image based data hiding schemes utilizing concepts of RONI and ROI could be found in literature. A robust information hiding technique utilizing ROI and RONI concept has been presented in [46]. Though the technique does not deteriorate the diagnostic information of the cover medical image but it has very small embedding capacity. The scheme makes use of Contourlet transformation prior to embedding in ROI portion of images. A blind and fragile steganographic system for various images of different modalities has been reported in [47]. SHA-256 algorithm has been used for encrypting data prior to embedding for enhancing security. The imperceptivity of stego-images is very good as the PSNR has been shown in excess of 56 dB. A robust medical image data hiding system has been reported in [48]. The system makes use error correction codes to minimise the number of erroneous bit due to noisy environment. This scheme however suffers from a low payload issue.

A region based information hiding scheme has been reported in [49]. Encryption using RSA algorithm been done to increase the EPR security. A multiple hybrid transformation technique has been reported in [50]. The cover medical images have been divided in ROI and RONI. Two level DWT has been applied to the ROI. The low frequency coefficients have been quantized and watermark embedded into ROI. The scheme has been reported to be robust to common attacks and has the ability to detect the tampered region. This scheme however suffers from low payload drawback. A fragile and ROI lossless EPR hiding technique has been presented in [51]. It combines scrambling and compression techniques to embed EPR, image hash to provide solution to IPR and content authentication issues.

The chief limitation when using segmentation based on ROI, RONI is low payload. Any attempt to increase the payload by embedding data in ROI leads to wrong diagnosis as ROI is deteriorated. Reversible schemes are a better choice in situations no loss of critical information is a must. A reversible algorithm is one where in we can retrieve the cover image in its true form after data extraction. A reversible steganographic algorithm for images is presented in [52]. The work uses the principle of interpolation of the neighbouring pixels, using maximum difference values. An additive interpolation-error expansion technique which is fully reversible can be found in [53]. Average payload of 38,545 bits has been embedded and PSNR of 49 dB has been reported.

Some of the key requirements of EPR hiding system for successful and secure transfer of medical data to make real time decisions involve less computational complexity, ability to transfer large payload in one go, high imperceptibility, ability to detect tamper if any caused to EPR during transit and security of embedded data. In the above cited literature it could be concluded that though ROI/RONI separation of cover images is an efficient way-out while dealing with medical images but it reduces the embedding capacity. Further reversible data hiding schemes also suffer from the same issue. Towards this end this chapter proposes a very high payload steganographic system for color images. RGB planes have been used to hide EPR information. The address vector table and complementary address vector table determine the locations where secret data is embedded. The data in the three different planes is embedded using different embedding strategies, to thwart the adversary. Besides data in one of the planes has been embedded using crypto-domain embedding [6, 54, 55], which further enhances the security of the system. The proposed system utilizes spatial domain embedding thus is computationally efficient and further it is capable of detecting any tamper caused to the EPR during transit to receiver. Rest of the chapter has been organized as follows. The proposed system has been discussed in detail in Sect. 3. Experimental investigations have been discussed in Sect. 4. The chapter concludes in Sect. 5.

3 Proposed Technique

The block diagram of proposed data hiding system using color images as cover medium for covert communication is shown in Fig. 1. The RGB color space cover image in which data is to be hidden is broken down into its three constituent color

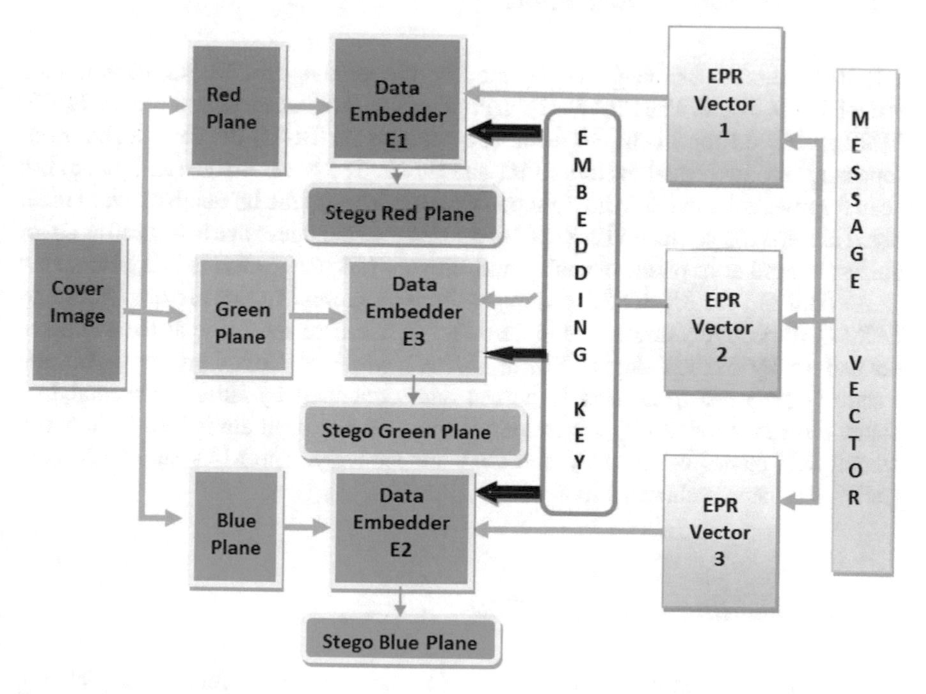

Fig. 1 Hiding in RGB planes of color image

planes: Red plane, Blue plane and Green plane. The message vector containing the EPR to be hidden into cover image is broken down into three equal length message vectors, EPR1, EPR2 and EPR3. The data embedder E1 embeds the EPR1 in the plane of the cover image. Similarly embedder E2 embeds EPR2 in the green plane and embedder E3, embeds EPR3 in the blue plane of the cover image. Although same Key is used to embed data in all the cover images but different embedding strategies have been used to hide data in the various constituent planes of the color cover image.

The proposed high capacity system for data hiding uses (512 $\times$ 512 $\times$ 3) size color test images as cover medium and hides '262,144 $\times$ 3' bits of EPR data in every constituent color plane of the RGB cover image. The implemented technique uses three bit planes (First three least significant bit planes) of every constituent color plane for hiding EPR data and as such is referred to as 3R-3G-3B technique. The proposed technique has been also evaluated for '262,144 $\times$ 2' bits of EPR data in every constituent color plane of the RGB cover image. The implemented technique uses two bit planes (First two Intermediate significant bit planes) of every constituent color plane for hiding EPR data and as such is referred to as 2R-2G-2B technique. The embedding techniques pertaining to various color planes are described below.

3.1 Embedding in Red Plane

The data embedding strategy in red plane of the cover image for 3R-3G-3B technique is depicted in Fig. 2. The EPR has been divided into three equal length vectors EPR1, EPR2 and EPR3 for 3R-3G-3B technique while as for 2R-2G-2B the EPR has been separated into two equal vectors EPR1 and EPR2. The cover image (Red plane) has been fragmented down into its eight constituent planes as the bit depth of every pixel of red plane is eight. Since 512 $\times$ 512 sized, color test images have been used as cover medium, number of pixels of each plane of every test image used is 262,144.

As depicted for 3R-3G-3B technique the data present in the vectors EPR1 and EPR2 is respectively embedded in LSB and first ISB of red plane at the addresses pointed to by Main Address Vector (MAV) while as Complementary Address Vector (CAV) has been used to embed data contained by EPR3 in second ISB plane. However for 2R-2G-2B technique the data has been embedded in first and second ISB planes using MAV and CAV respectively. The MAV and CAV generation has been elaborated in Sect. 3.4.

3.2 Embedding in Green Plane

The EPR data embedding strategy in green plane of the cover image is depicted in Fig. 3. Like red plane this plane also hides 262,144 $\times$ 3 = 786,432 bits of secure

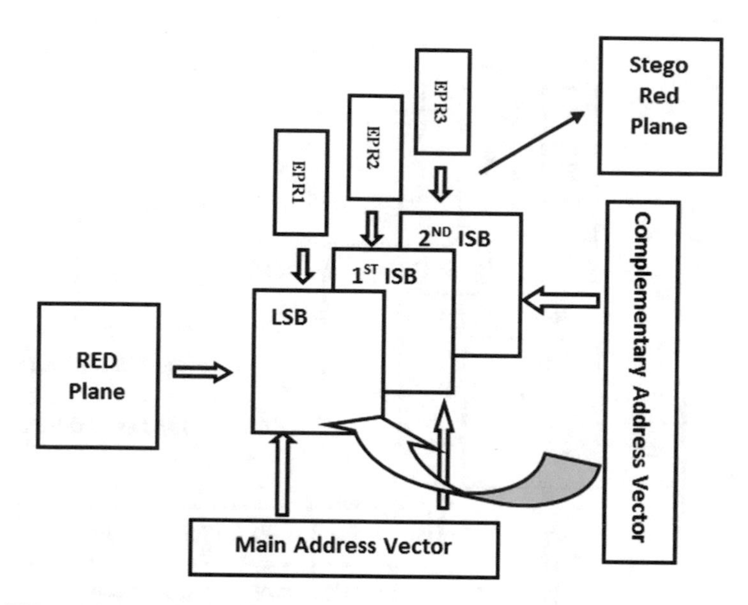

Fig. 2 EPR embedding in red plane

data when 3R-3G-3B technique is used while as 262,144 × 2 = 524,288 bits of secure data are embedded when 2R-2G-2B embedding technique is used. The embedding strategy however is quite different. The data to be embedded using 3R-3G-3B technique in this plane is divided into three EPR vectors (EPR1, EPR2, EPR3). The lengths of all data vectors are equal, and are equal to 262,144. However, to thwart an adversary and keep it guessing the embedding scheme in green plane is quite different from that of the red plane. In green plane alternate top down and down top embedding has been used to embed data: data is embedded in the first bit plane of green plane at locations determined by addresses contained by the MAV by traversing it from First Address (FA) location to Last Address (LA) location. This embedding is referred to as top down embedding. Data embedding in the second bit plane of green plane has been carried out using CAV in reverse order that is from Last Address (LA) location to First Address (FA) location (bottom top embedding). Embedding in third bit plane also uses top down embedding. The process has been depicted in Fig. 3.

3.3 *Embedding in Blue Plane*

The EPR data hiding strategy for blue plane is different from red and green plane. The cover medium (blue plane) is first encrypted by scrambling it as per address locations pointed to by the main address vector. The EPR embedding is carried out in encrypted blue plane to thwart adversary. The data embedding has been carried

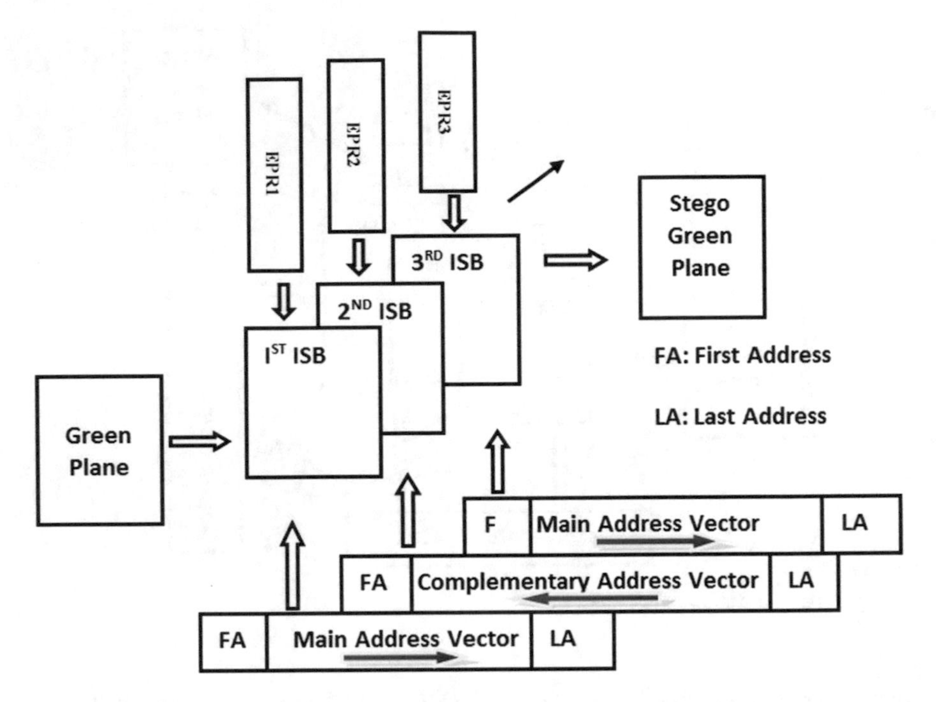

Fig. 3 EPR hiding in green plane

out using same strategy as has been used in case of red plane. Once the embedding is complete in all the specified planes of blue plane, the encrypted cover medium (blue plane) containing hidden data is converted back to its original form by decrypting it using same Key as has been used during encryption. The whole process has been depicted in the Fig. 4.

Once the embedding in all the three constituent planes of a cover image is complete, the constituent stego planes are concatenated to obtain the final stego-image. The process has been depicted in Fig. 4b.

3.4 Address Vector Generation

To address all the pixel locations a Key has been used to generate a pseudorandom address table containing as many as 262,144 addresses. A typical example showing how the pseudorandom addresses have been generated is presented in Fig. 5.

The shown scheme represents a three bit Pseudo Noise code generator [56–58] which is capable of generating a pseudorandom address vector containing seven addresses. Assume initial value of Q1, Q2 and Q3 is 111(7), the successive clock cycles will make the generator to run through 011(3), 101(5), 010(2), 001(1), 100 (4), 110(6) and back to 111(7). The set of all the pseudorandom addresses generated

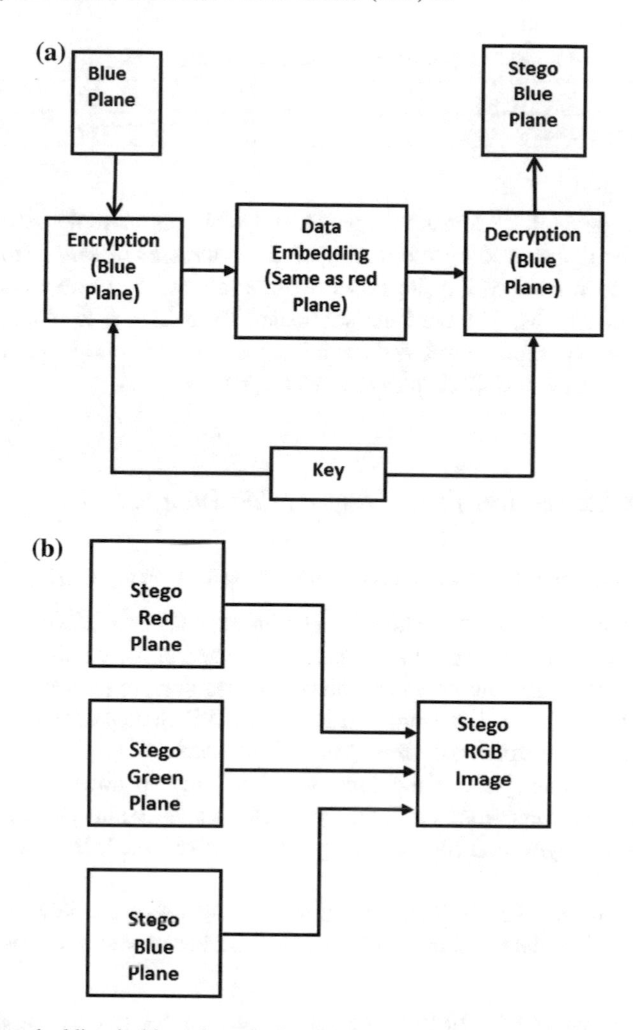

Fig. 4 **a** EPR embedding in blue plane. **b** Obtaining color image from RGB planes

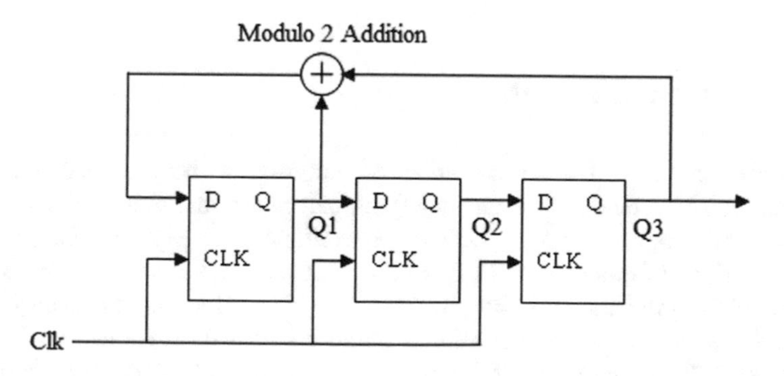

Fig. 5 PN code generator used for Main and Complementary Address Vector Generation

Table 1 Pseudorandom address vector and its complementary vector

Pseudo random address vector	7	3	5	2	1	4	6
Complementary address vector	1	5	3	6	7	4	2

is referred to as Main Address vector (MAV) while as Complementary Address Vector has been generated by subtracting all the addresses of MAV from a number 2n; n being the number of flip-flops used to generate MAV. In the current example all the addresses of MAV have been subtracted from 23 = 8 to obtain CAV.

A PAV and corresponding CAV have been depicted in Table 1. For a detailed explanation of PAV and CAV please refer to [10, 54].

3.5 *Data Extraction from Stego-Color Images*

The EPR is extracted from the stego image using following steps:

1. The stego-color image is divided in RGB planes at the receiver.
2. Same seed word as that used at transmitter is used at the receiver as key for the generation of Main and complementary address vectors (MAV and CAV).
3. Use MAV to extract data from first and third ISB planes of red plane while as CAV is used for extraction from second ISB plane.
4. The extraction in green plane uses MAV in the top down configuration for extracting data bits from first and third ISB planes while as CAV is used in down-top configuration for extracting data from second ISB plane of the green plane.
5. The stego-blue plane is firstly scrambled using the same key as used at the transmitter. The data is extracted later in the similar way as is extracted from red plane.

It is to be noted that for 2R-2G-2B scheme data is extracted from first and second ISB planes only.

4 Experimental Results

Data hiding for covert communication is supposed to be accompanied by the attributes like high data hiding capacity (payload) perceptual transparency and high security. The proposed system, besides providing a very high payload, uses different strategies to embed data in three different planes and as such improves the security of the information hiding system. The subjective quality analysis of the proposed system shows good results as could be seen from subjective quality of stego-images. The objectivity of the proposed system has been checked in terms of

Peak Signal to Noise Ratio (PSNR) and Structural Similarity Measure Index (SSIM). The scheme has also been tested for content authentication ability. For this a fragile watermark has been embedded besides EPR in the images. The watermark is retrieved from the attacked images and its quality has been evaluated from the computed values of Bit Error Rate (BER %) and Normalized Cross-Correlation (NCC). BER, NCC, PSNR and SSIM have been defined as in [3].

(a) **Imperceptibility Analysis for medical images**

We have carried out subjective and objective analysis for various test medical images which are generally transferred in an IoT driven healthcare system. We have compared 3R-3G-3B and 2R-2G-2B embedding scheme which embed 3 bits per pixel per plane (9 bits per color image pixel) and 2 bits per pixel per plane (6 bits per color image pixel). The subjective and objective analysis has been carried out and the results are provided below. The objective quality metrics (PSNR and SSIM) for various medical images are used have been reported in Table 2.

As could be seen that PSNR (average) for 3R-3G-3B technique is 37.2464 dB and average SSIM is 0.76144 and for 2R-2G-2B technique it is 37.6832 dB while the average SSIM is 0.7803. Figure 6 depicts the original and corresponding watermarked images of both techniques. Clearly subjective quality of watermarked images is good.

We have compared our scheme with Das and Kundu [51]. The comparison has been carried out has been presented in Table 3.

The optimum payload of our technique is 1, 572,864 bits (2 bpp, per color image plane) which is much higher than the technique under comparison. Experimental investigations show that the proposed scheme provides stego-images with good degree of imperceptivity for a very high payload.

Table 2 Perceptual quality metrics

Image	3R-3G-3B technique		2R-2G-2B technique	
	SSIM	PSNR	SSIM	PSNR (dB)
Image 1	0.5372	36.2413	0.5796	36.9833
Image 2	0.7987	37.3811	0.8131	37.7672
Image 3	0.8745	38.0005	0.8781	38.1985
Image 4	0.9157	38.4202	0.9145	38.5483
Image 5	0.9055	37.9457	0.9097	38.1617
Image 6	0.6478	36.5931	0.6805	37.2487
Image 7	0.5668	36.1839	0.6073	36.9369
Image 8	0.8243	37.2554	0.8384	37.6705
Image 9	0.9504	37.9284	0.9525	38.1393
Image 10	0.5935	36.5147	0.6297	37.1779
Average	0.76144	37.2464	0.7803	37.6832

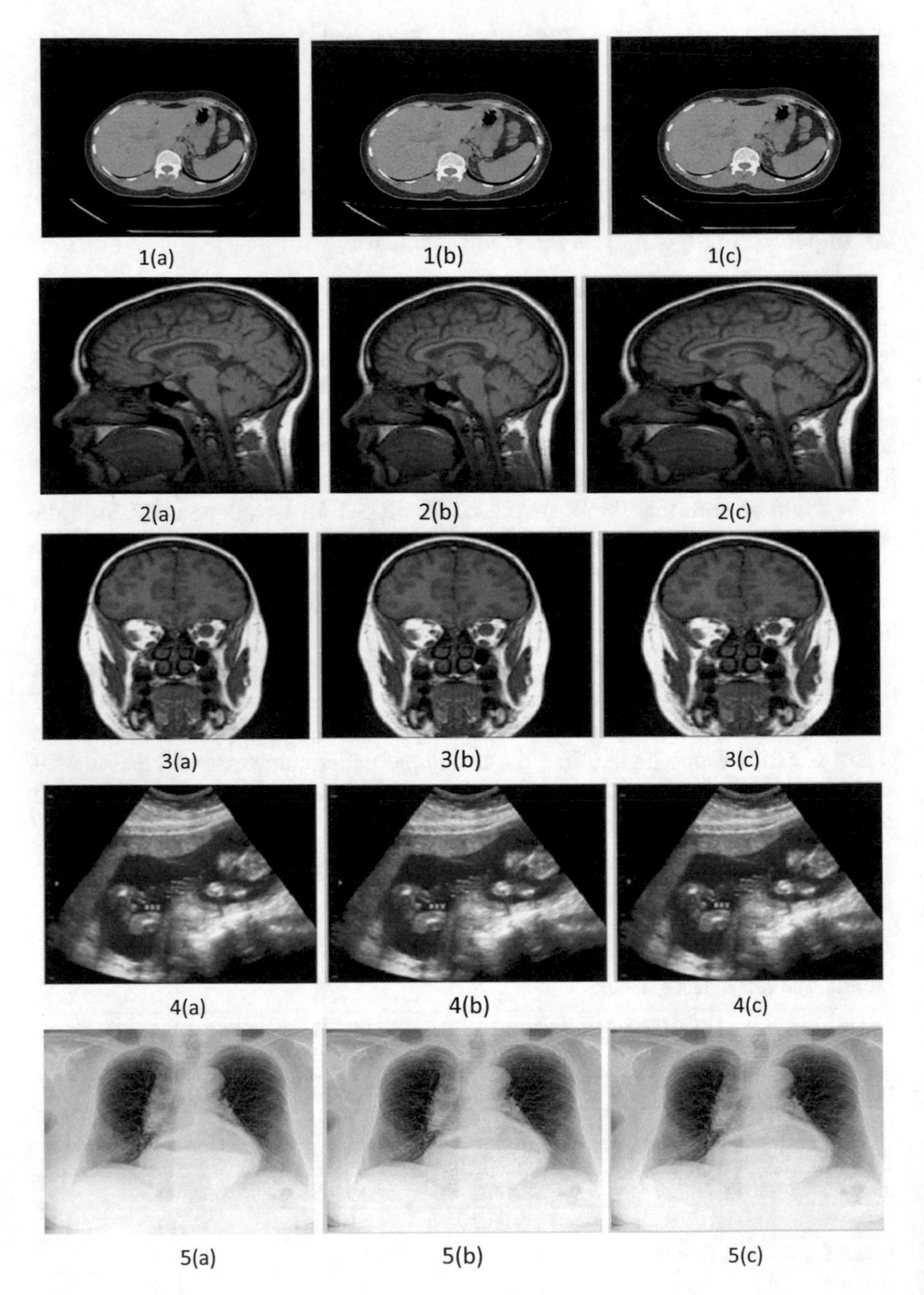

Fig. 6 **1a–10a** Original images, **1b–10b** stego-images for 3R-3G-3B technique, **1c–10c** stego-images for 2R-2G-2B technique

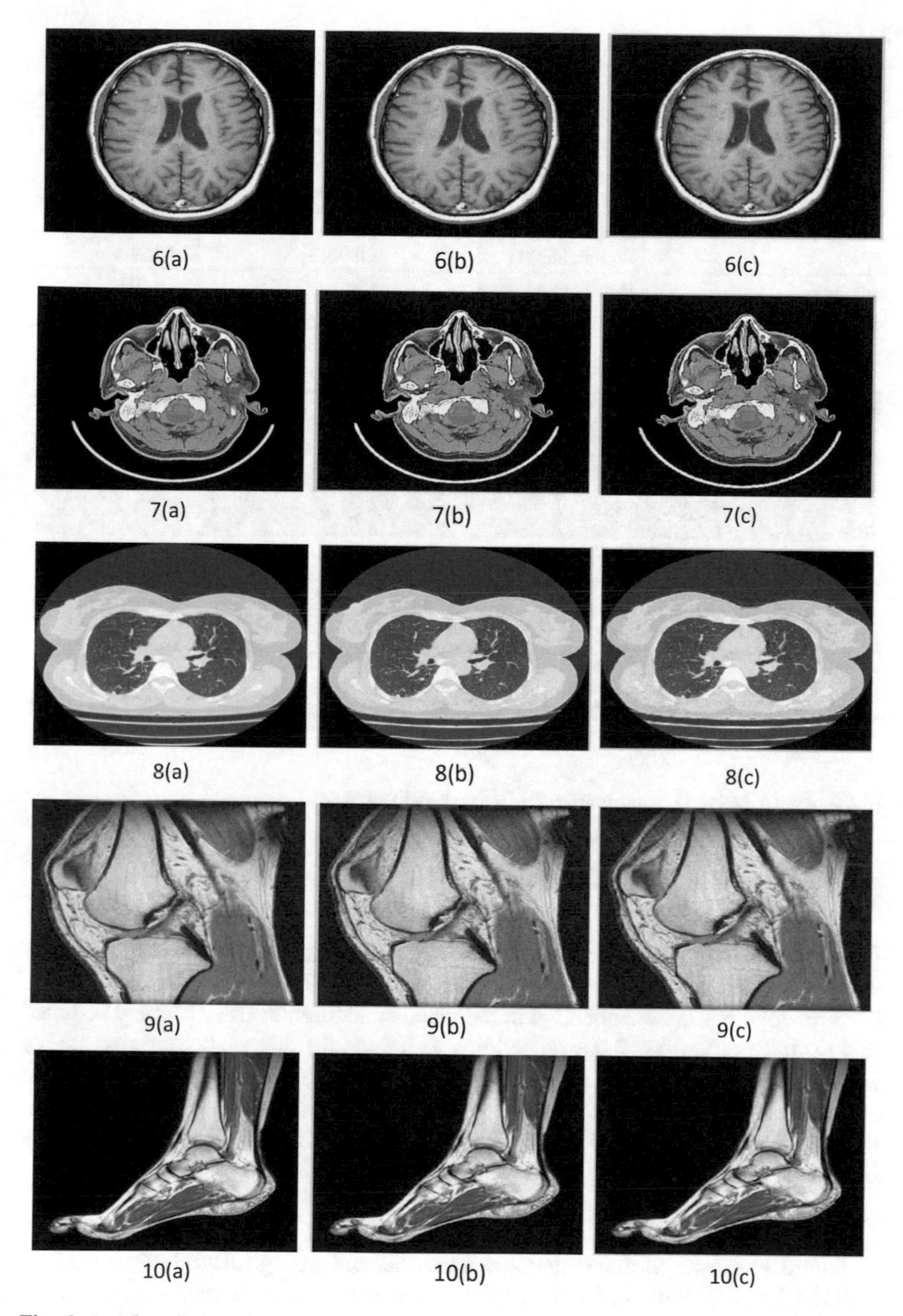

Fig. 6 (continued)

Table 3 Perceptual transparency comparison

Medical image	Schemes	BPP	PSNR
Image 1	Proposed scheme Das & Kundu	2 0.0052	36.9833 35.952
Image 2	Proposed scheme Das & Kundu	2 0.0052	37.7672 35.982
Image 3	Proposed scheme Das & Kundu	2 0.0052	38.1985 37.159
Image 4	Proposed scheme Das & Kundu	2 0.0052	38.5483 36.412

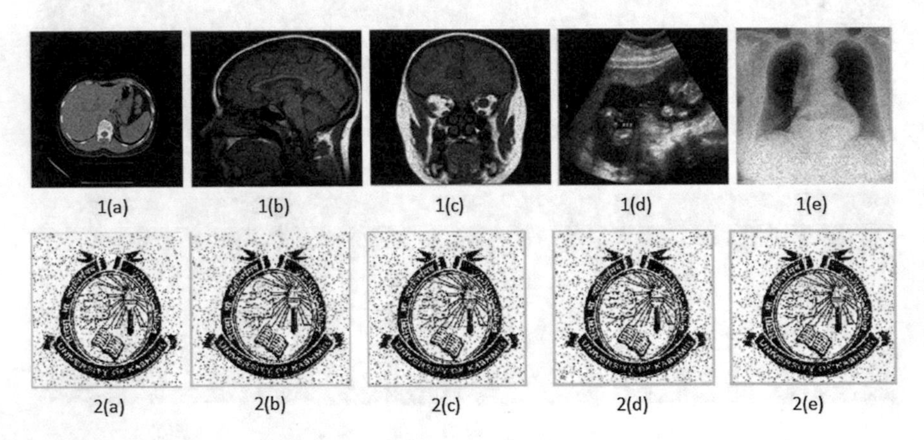

Fig. 7 **1a–1e** Attacked stego-images **2a–2e** extracted watermarks

4.1 Content Authentication Analysis

In order to analyse our scheme for its ability to detect tinker by a possible attack we have embedded a fragile water mark in addition to EPR. It is to be noted that a 64×64 logo has been embedded in red plane in addition to EPR1. It may be noted that we have used 2R-2G-2B embedding technique for this analysis. Same sort of results were also observed for 3R-3G-3B technique during experimentation. The stego-images obtained were attacked by attacks like Gaussian noise addition, various filtering process, JPEG compression etc. The results obtained various attacks have been presented below:

(b) **Salt and Pepper Noise (SPN)**

Figure 7 shows various stego-images obtained using 2R-2G-2B technique attacked by SPN with noise density 0.1. The extracted watermarks from all the attacked stego images are shown in Fig. 7. The quality metrics obtained have been presented in Table 4.

Table 4 Salt and pepper attack (density = 0.1)

Image	Watermark		Data (EPR)	
	NCC	BER (%)	NCC	BER (%)
Image 1	0.9507	5.04	0.9498	5.02
Image 2	0.9510	4.84	0.9502	5.00
Image 3	0.9492	5.10	0.9498	5.02
Image 4	0.9507	4.88	0.9500	4.96
Image 5	0.9498	5.00	0.9493	5.01
Average	0.95028	4.972	0.9498	5.00

Table 5 Additive White Gaussian Noise

Image	Watermark		Data (EPR)	
	NCC	BER (%)	NCC	BER (%)
Image 1	0.2728	61.73	0.3382	49.35
Image 2	0.3406	58.75	0.4055	49.50
Image 3	0.3512	57.76	0.4138	49.50
Image 4	0.3333	58.89	0.4091	49.58
Image 5	0.5031	49.87	0.5571	49.59
Average	0.3602	57.4	0.4247	49.50

The average NCC is 0.95028 for extracted watermark and 0.9498 for Electronic Patient Record data. The average BER (%) of the watermark and data is 4.972 and 5.00 respectively. The results show that watermark is not extracted exactly same as one embedded and hence one can simply detect that the content has been compromised during transit.

(c) **Additive White Gaussian Noise (AWGN)**

The stego-images have been subjected to AWGN having zero mean and variance of 0.02. Table 5 shows the NCC values and BER for different attacked stego-images.

Table 5 shows the average NCC value is 0.3602 for extracted watermark and 0.4247 for EPR data. The average BER (%) of the watermark and data is 57.4 and 49.50 respectively. Figure 8 depicts the fragility analysis of various images when attacked by AWGN. The high values of BER and low values of NCC besides subjective quality of various watermarks depict superiority of our scheme.

(d) **JPEG Compression**

The stego-images have been subjected to JPEG compression of quality factor 90. Table 6 depicts the quality parameters for JPEG compression for various images. The watermarks extracted from various watermarked and then attacked images have been shown in Fig. 9.

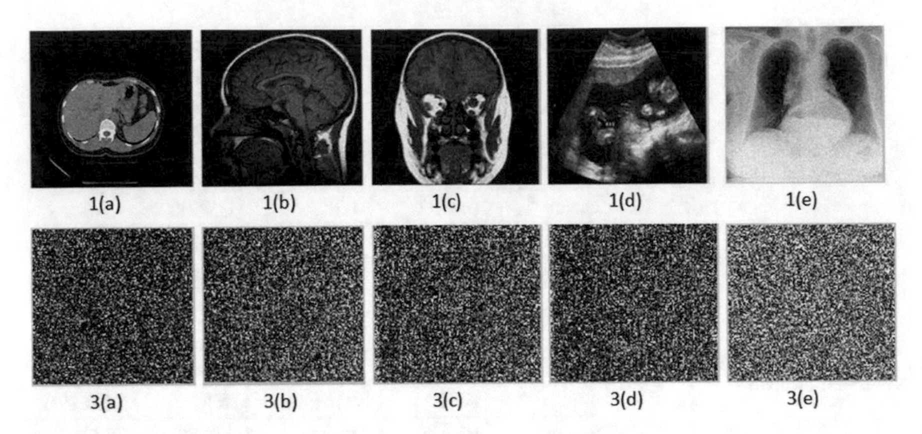

Fig. 8 **1a–1e** Attacked stego-images, **2a–2e** extracted watermarks

Table 6 JPEG compression (QF = 90)

Image	Watermark		Data (EPR)	
	NCC	BER (%)	NCC	BER (%)
Image 1	0.5676	47.92	0.4985	47.55
Image 2	0.5232	49.30	0.5172	47.69
Image 3	0.4792	51.83	0.5266	47.49
Image 4	0.4245	55.15	0.5313	47.66
Image 5	0.4911	50.95	0.5247	47.69
Average	0.4971	51.03	0.5197	47.62

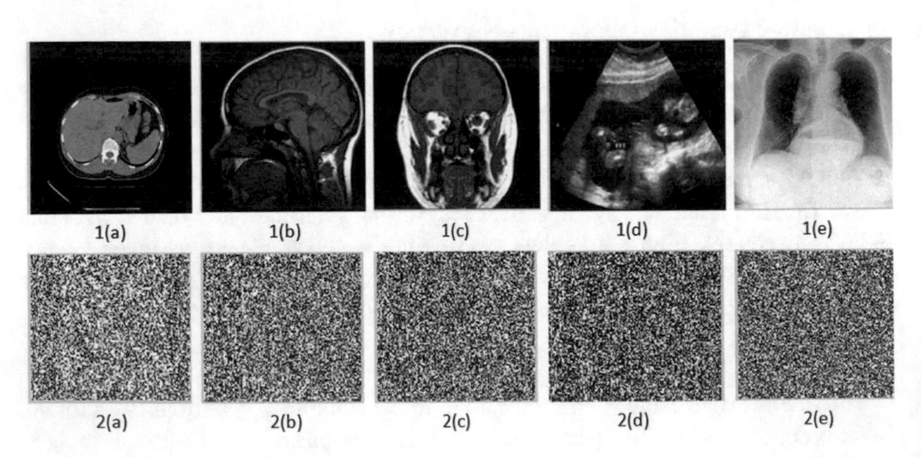

Fig. 9 **1a–1e** Attacked stego-images **2a–2e** extracted watermarks

The objective and subjective quality results shown in Table 6 and Fig. 9 clearly show the delicacy of our technique to JPEG compression. The average BER (%) of the watermark and Electronic Patient Data is 51.03 and 47.62 respectively which clearly indicate that this attack can also be detected at the receiver.

(e) **Filtering attacks**

A detailed account of results obtained after carrying out various filtering attacks is presented below

I. ***Median Filtering(MF)***

Watermarked images were subjected to Median filtering attack with kernel size [3 × 3]. Table 7 shows the quality performance parameters and Fig. 10 shows the various results after said attack. The average NCC value is 0.8412 for extracted watermark and 0.6103 for data. The average BER (%) of the watermark and data is 23.38 and 39.05 respectively.

Table 7 Median filtering

Image	Watermark		Data (EPR)	
	NCC	BER (%)	NCC	BER (%)
Image 1	0.8146	26.32	0.5982	40.16
Image 2	0.8509	21.78	0.6141	38.57
Image 3	0.8460	22.86	0.6280	37.43
Image 4	0.8392	23.69	0.6157	38.77
Image 5	0.8554	22.27	0.5954	40.33
Average	0.8412	23.38	0.6103	39.05

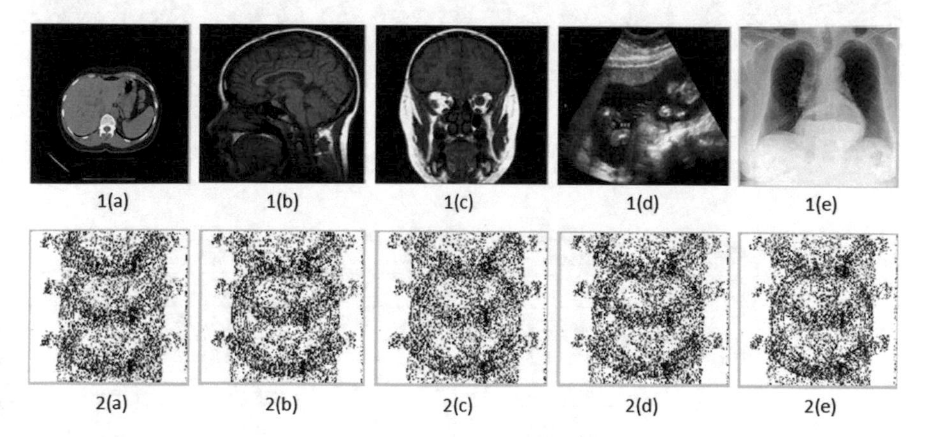

Fig. 10 **1a–1e** Attacked stego-images **2a–2e** extracted watermarks

II. ***Low Pass Filtering (LPF)***

A 3 × 3 LPF kernel has been used to attack the system. The filtered images and the respective watermarks obtained from them are depicted in Fig. 11. The objective indices have been reported in Table 8. The average NCC value is 0.3931 for extracted watermark and 0.5173 for EPR. The average BER (%) of the EPR and watermark is 48.47 and 58.49 respectively.

III. ***Weiner Filtering (WF)***

The image quality metrics obtained after carrying out this attack are shown in Table 9 while as related images have been depicted in Fig. 12. The average NCC value is 0.4096 for extracted watermark and 0.5312 for EPR data. The average BER (%) of the watermark and data is 57.11 and 47.11 respectively. The results reveal the fragility of the proposed technique.

IV. ***Sharpening***

The quality metrics are reported in Table 10 for this attack while as the subjective quality of various images obtained after this attack are presented in Fig. 13.

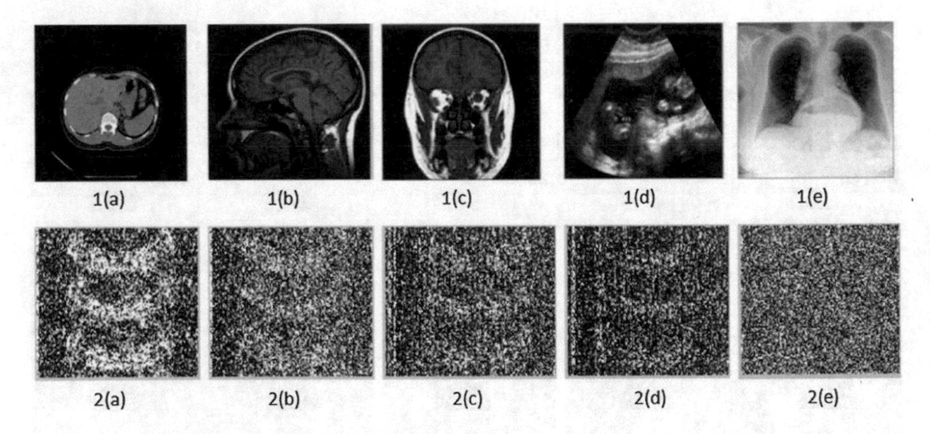

Fig. 11 **1a–1e** Attacked stego-images **2a–2e** extracted watermarks

Table 8 Low Pass Filtering

Image	Watermark		Data (EPR)	
	NCC	BER (%)	NCC	BER (%)
Image 1	0.4049	60.84	0.5170	47.81
Image 2	0.4211	57.40	0.5152	48.64
Image 3	0.3590	59.71	0.5188	48.53
Image 4	0.3030	62.29	0.5222	48.66
Image 5	0.4675	52.19	0.5135	48.73
Average	0.3931	58.49	0.5173	48.47

Table 9 Weiner filtering

Image	Watermark		Data (EPR)	
	NCC	BER (%)	NCC	BER (%)
Image 1	0.4648	53.42	0.5395	46.15
Image 2	0.3973	57.15	0.5426	46.22
Image 3	0.3159	61.51	0.5367	47.30
Image 4	0.4657	52.56	0.5158	48.50
Average	0.4096	57.11	0.5312	47.11

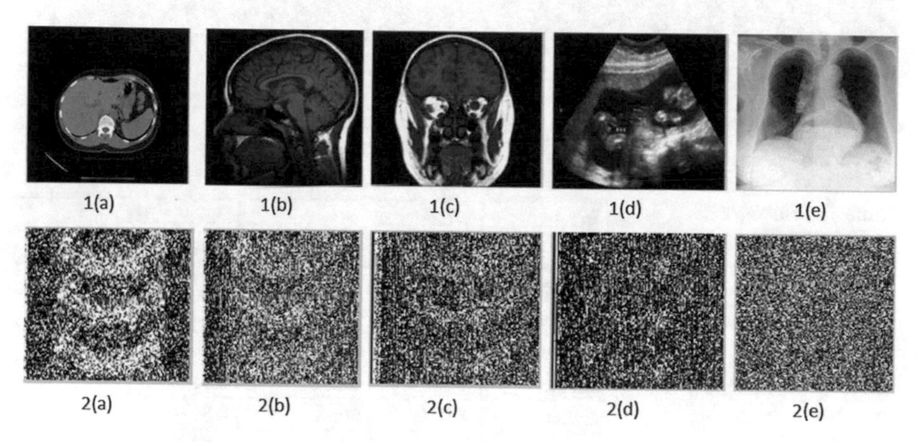

Fig. 12 **1a–1e** Attacked stego images **2a–2e** extracted watermarks

Table 10 Sharpening

Image	Watermark		Data (EPR)	
	NCC	BER (%)	NCC	BER (%)
Image 1	0.2671	62.28	0.4374	40.33
Image 2	0.3609	57.07	0.4757	45.38
Image 3	0.3651	56.67	0.4838	43.84
Image 4	0.3441	57.79	0.4701	45.10
Image 5	0.5025	49.70	0.5402	47.97
Average	0.3679	56.70	0.4814	44.52

The average NCC value is 0.3679 for extracted watermark and 0.4814 for EPR. The average BER (%) of the watermark and data is 56.70 and 44.52 respectively. It is clear from above results that our algorithm is fragile to Sharpening.

It is evident that the various results presented from Sects. 4.1 to 4.4 that our scheme that the proposed scheme is fragile to various filtering attacks. As such any filtering attack on the stego-images intentionally/un-intentionally can be easily detected at the receiver and hence content authenticated.

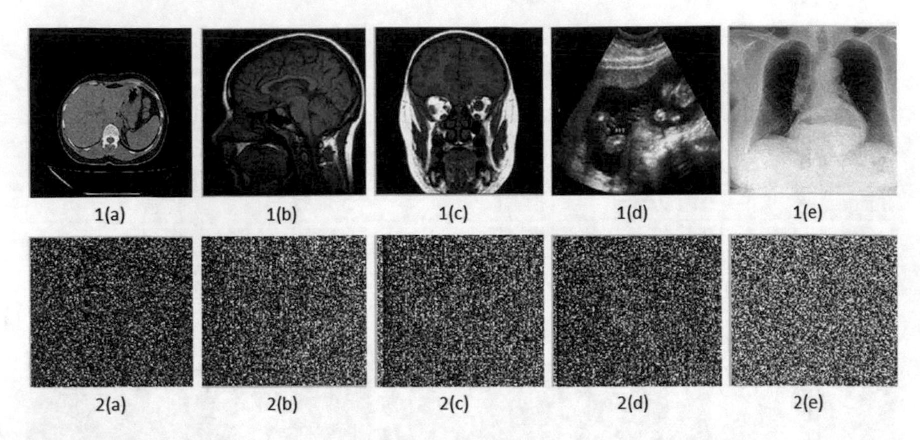

Fig. 13 **1a–1e** Attacked stego-images **2a–2e** extracted watermarks

Table 11 Histogram equalization

Image	Watermark		Data (EPR)	
	NCC	BER (%)	NCC	BER (%)
Image 1	0.4955	38.93	0.5751	44.85
Image 2	0.2602	62.45	0.5065	48.80
Image 3	0.2729	62.20	0.5396	46.02
Image 4	0.2573	62.90	0.5193	49.52
Image 5	0.1271	69.80	0.5119	48.61
Average	0.2826	59.26	0.5305	47.56

(f) **Histogram Equalization (HE)**

Various objective parameters for histogram equalization have been shown in Table 11 while as necessary images are shown in 14. Average NCC value is 0.2826 for extracted watermark and 0.5305 for EPR. The average BER (%) of the watermark and data is 59.26 and 47.56 respectively. The results show that our scheme is fragile to Histogram equalisation (Fig. 14).

(g) **Rotation**

We have rotated the watermarked by 5°. Image quality metrics and necessary images obtained are presented in Table 12 and Fig. 15 respectively. It is obvious from the objective indices presented that our technique is fragile to rotation. Given the quality of extracted watermarks available at the receiver it can be concluded with ease that the data has been compromised during transition.

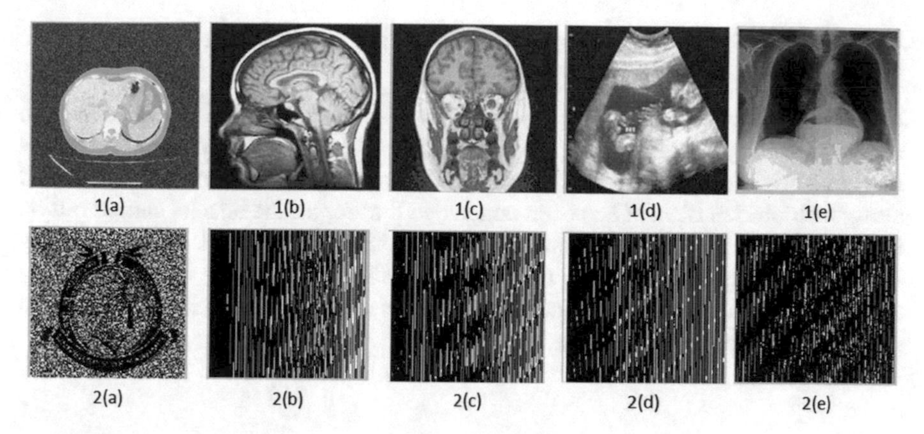

Fig. 14 **1a–1e** Attacked stego-images **2a–2e** extracted watermarks

Table 12 Rotation (5°)

Image	Watermark		Data (EPR)	
	NCC	BER (%)	NCC	BER (%)
Image 1	0.4718	52.29	0.4438	49.93
Image 2	0.4626	52.88	0.4733	49.98
Image 3	0.4650	52.59	0.4832	49.97
Image 4	0.4774	52.08	0.4933	50.01
Image 5	0.4578	53.16	0.4821	49.99
Average	0.4669	52.6	0.4751	49.98

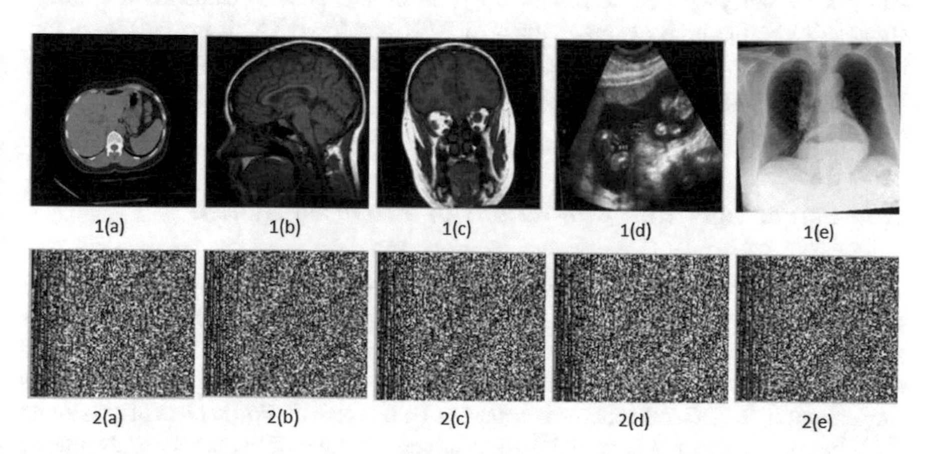

Fig. 15 **1a–1e** Attacked stego-images **2a–2e** extracted watermarks

4.2 Brief Discussion About Content Authentication Results

In the above section we subjected various stego-images obtained in our scheme to various image processing and geometric attacks. We embed a watermark besides EPR in a cover image at the transmitter. At the receiver the watermark is extracted along with hidden EPR. It has been concluded that for all the attacks carried out on the stego-images we have not been able to extract the watermark at the receiving end. This clearly indicates that the proposed scheme is capable of detecting any tamper caused during transit. Once tamper is detected an automatic retransmission request could be used to receive the data again.

4.3 Imperceptibility Analysis for General Standard Images

We have used a set of five generally used test images to further evaluate perceptual transparency of our scheme. Figure 16 shows the original standard images and corresponding stego-images. Figure 16 shows the perceptual quality of the proposed scheme with 3R-3G-3B and 2R-2G-2B embedding techniques.

4.4 Authentication Analysis for General Images

The proposed technique has been analysed for a set of generally used test images for content authentication ability. Table 13 demonstrates the objective quality analysis of the proposed technique for one of the general standard test image (Lena). Given the high average values of BER and low NCC it is evident enough that values that our scheme is fragile to different attacks. The subjective quality of the attacked images as shown in Fig. 17 justifies the conclusion (Table 14).

4.5 Advantages and Limitations of Proposed System

The system proposed in this chapter deals with secure and very high capacity EPR embedding in medical images for a typical IoT based healthcare system. The various advantages offered by the system include high payload, better imperceptivity and capability to detect tamper if any, at the receiver. Out of the two techniques implemented, 2R-2G-2B scheme is robust to LSB removal. Further use of MAV and CAV helps securing the data to thwart an adversary. One of the main disadvantages of proposed scheme is that the embedded data is fragile to various signal processing and geometric attacks. Our future works aims at improving the system robustness to

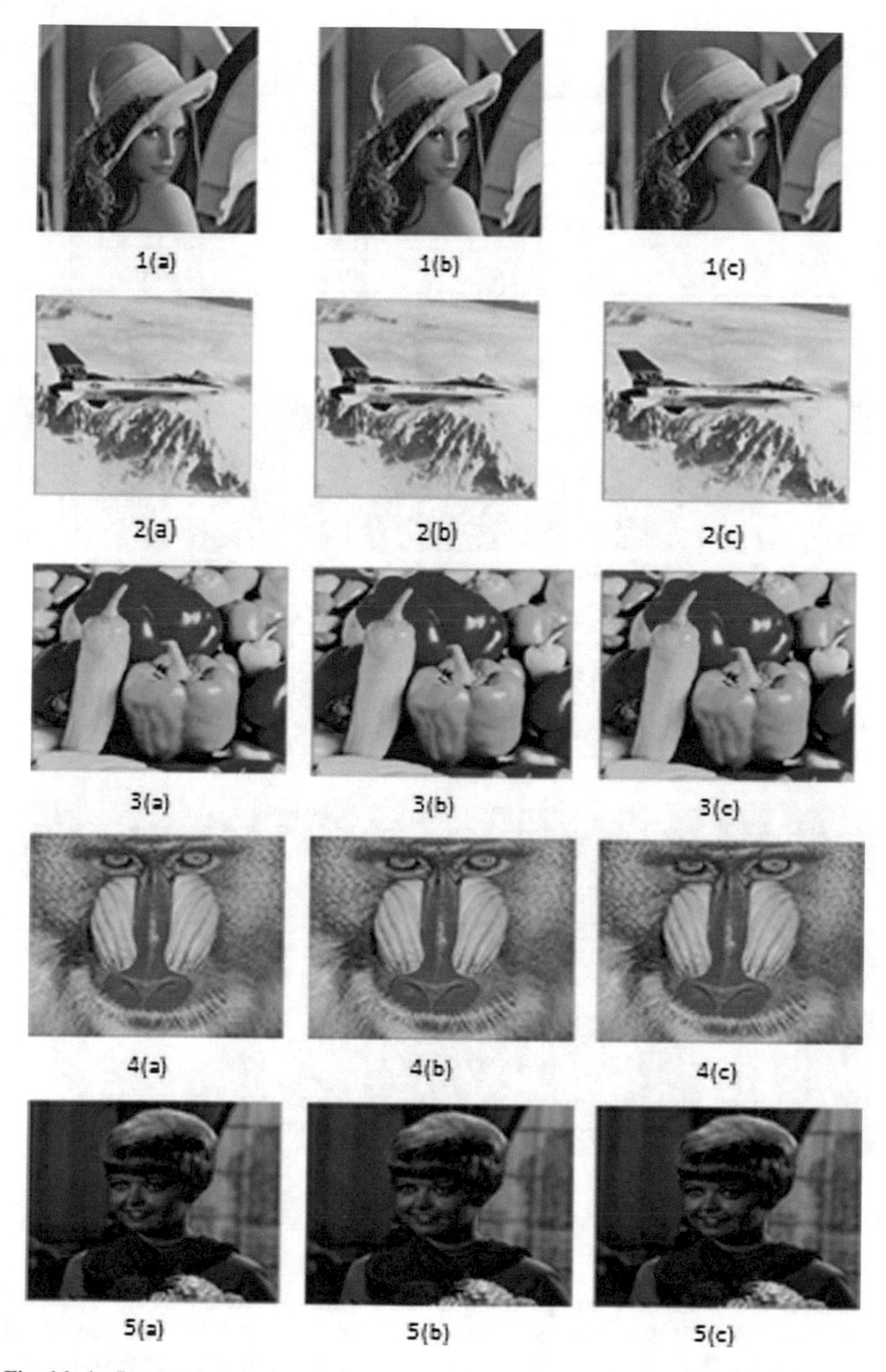

Fig. 16 **1a-5a** Original images **1b–5b** stego-images using 3R-3G-3B technique **1c–5c** stego-images using 2R-2G-2B technique

Table 13 Various color test image indices

Image	3R-3G-3B technique		2R-2G-2B technique	
	SSIM	PSNR	SSIM	PSNR (dB)
Lena	0.9956	37.9049	0.9958	38.1214
Plane	0.9344	37.8958	0.9373	38.1060
Peppers	0.9949	37.7927	0.9953	38.0493
Baboon	0.9934	37.9122	0.9937	38.1184
Girl	0.9531	37.9624	0.9551	38.1612
Average	0.76144	37.2464	0.7803	38.1109

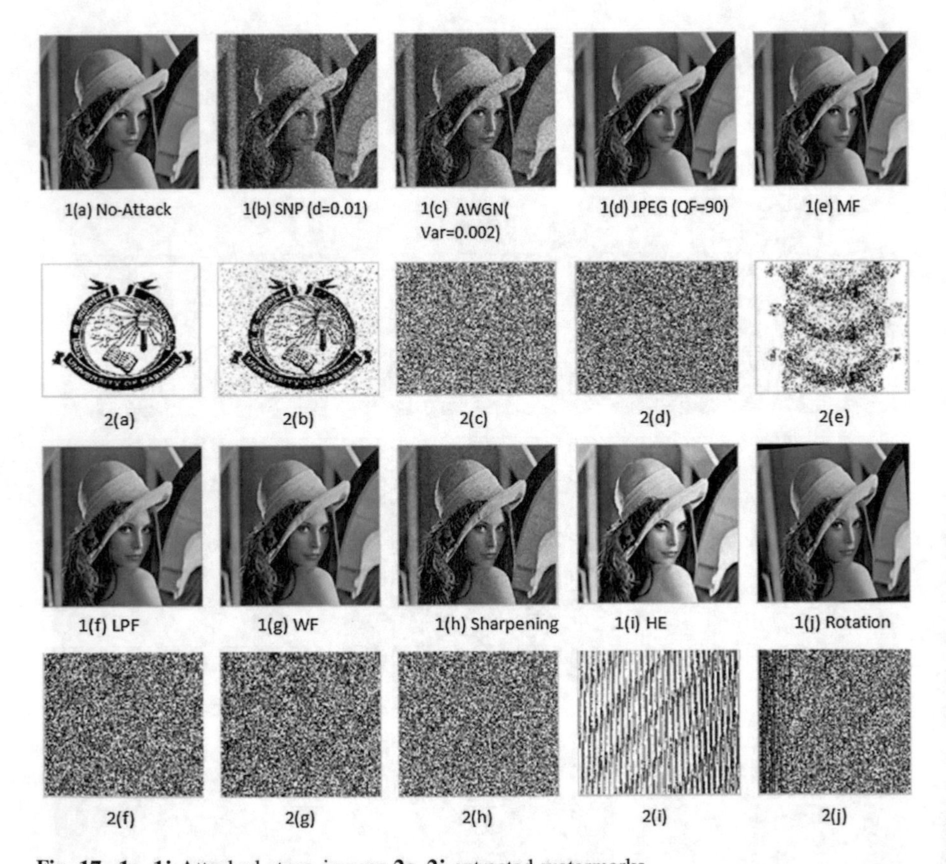

Fig. 17 **1a–1j** Attacked stego-images **2a–2j** extracted watermarks

various image processing and geometrical attacks while maintaining high payload capability. We further aim to implement the proposed scheme on Field Programmable Gate Array (FPGA) platform so that it can be used for developing a state-of-art ubiquitous real time system IoT based healthcare system.

Table 14 Image quality metrics for Lena

Noise added	Watermark		Data (EPR)	
	NCC	BER (%)	NCC	BER (%)
No attack	1.0000	0	1.0000	0
Salt and pepper	0.9498	5.07	0.9497	5.01
AWGN	0.5509	47.41	0.50276	49.78
JPEG	0.5009	49.98	0.5023	49.83
Median filtering	0.8370	24.82	0.5902	41.18
Low pass filtering	0.4705	52.33	0.5032	49.77
Weiner filtering	0.4751	51.79	0.5046	49.61
Sharpening	0.5140	49.26	0.5031	49.35
Histogram equalization	0.6671	40.86	0.4629	52.70
Rotation	0.4596	52.92	0.4843	49.93

5 Conclusion

A secure and high payload EPR hiding scheme using color images as cover medium for an IoT based healthcare setup has been presented in this chapter. The technique uses two/three bit planes of all fundamental planes (RGB) for hiding EPR. A fragile watermark has been embedded besides the EPR to make the proposed scheme capable of detecting any tamper of EPR during transit. At the receiver the embedded watermark is first extracted, if it is same as one embedded, signalling that EPR has not been attacked during transit; EPR is extracted otherwise system uses a retransmission request instead of wasting time to extract the compromised EPR. In order to secure the EPR from adversary, It has been embedded at locations in accordance with PAV. Further hiding is carried out in spatial domain and hence is computationally efficient. Given the merits of the proposed algorithm viz-a-viz payload, imperceptibility, security and computational efficiency and ability to authenticate EPR at the receiver it can be used for real time medical information exchange like in case of Internet of things (IoT) driven healthcare and health monitoring systems for effective transfer of medical data during transit. Besides testing the algorithm for various medical images, we have also tested it for some commonly used standard test images. Experimentation shows that our algorithm is able to hide a very high payload EPR imperceptibly. It has been concluded that from imperceptibility point-of view the scheme 2R-2G-2B performs better than that of 3R-3G-3B. Further, the former technique is also immune to commonly used LSB elimination/substitution attack as EPR is hidden in first and second ISB planes, unlike the latter wherein LSB has also been used for embedding. We have tested our scheme to various attacks like addition of noise, filtering using different filters, JPEG compression and rotation of varied angles etc. Experimental studies show that the proposed scheme could detect tamper caused due to all attacks carried out on it.

References

1. Mersini, P., Evangelos, S., Efrosini, S., & Athanasios, T. (2013). Health internet of things: Metrics and methods for efficient data transfer. *Simulation Modelling Practice and Theory, 34,* 186–199.
2. Atzori, L., Lera, A., & Morabito, G. (2010). The internet of things: A survey. *Computer Networks* 54, 2787–2805.
3. Parah, S. A., Sheikh, J. A., Ahad, F., Loan, N. A., & Bhat, G. M. (2015). Information hiding in medical images: A robust medical image watermarking system for E-healthcare. *Multimed Tools and Applications*. doi:10.1007/s11042-015-3127-y
4. Chakraborty, S., Maji, P., Pal, A. K., Biswas, D., & Dey, N. (2014, January). Reversible color image watermarking using trigonometric functions. In *2014 International Conference on Electronic Systems, Signal Processing and Computing Technologies (ICESC)*, pp. 105–110.
5. Boyi, X., Xu, L., Hongming, C., Cheng, X., Jingyuan, H., & Fenglin, B. (2014). Ubiquitous data accessing method in IoT-based information system for emergency medical services. *IEEE Transactions On Industrial Informatics, 10*, 3131–3143.
6. Zhao, K., & Lina, G. (2013). A survey on the internet of things security. In *2013 Ninth International Conference on Computational Intelligence and Security*, pp. 663–667.
7. Haux, R. (2010). Medical informatics: Past, present, future. *International Journal of Medical Informatics*, *79*, 599–610.
8. Parah, S. A., Sheikh, J. A., Hafiz, A. M., & Bhat G. M. (2014). A secure and robust information hiding technique for covert communication. *International Journal of Electronics*. doi:10.1080/00207217.2014.954635
9. Dey, N., Samanta, S., Chakraborty, S., Das, A., Chaudhuri, S. S., & Suri, J. S. (2014). Firefly algorithm for optimization of scaling factors during embedding of manifold medical information: An application in ophthalmology imaging. *Journal of Medical Imaging and Health Informatics, 4*(3), 384–394.
10. Shabir, A. P., Javaid, A. S., & Bhat, G. M. (2014). Data hiding in scrambled images: A new double layer security data hiding technique. *Computers and Electrical Engineering, 40*(1), 70–82. Elsevier.
11. Shabir, A. P., Javaid, A. S., Nazir, A. L., & Bhat, G. M. (2016). Robust and blind watermarking technique in DCT domain using inter-block coefficient differencing. *Digital Signal Processing*. Elsevier. doi:10.1016/j.dsp.2016.02.005
12. Dey, N., Maji, P., Das, P., Biswas, S., Das, A., & Chaudhuri, S. S. (2013). An edge based blind watermarking technique of medical images without devalorizing diagnostic parameters. In *2013 International Conference on Advances in Technology and Engineering* (ICATE), pp. 1–5.
13. Shabir, A. P., Javaid, A. S., & Bhat, G. M. (2013). On the realization of a spatial domain data hiding technique based on intermediate significant bit plane embedding (ISBPE) and post embedding pixel adjustment (PEPA). In *Proceedings of IEEE International Conference on Multimedia Signal Processing and Communication Technologies-IMPACT 2013*. AMU, Aligargh. November 23–25, 2013, pp. 51–55.
14. Dey, N., Ashour, A., Chakraborty, S., Banerjee, S., Gospodinova, E., Gospodinov, M., et al. (2017). Watermarking in bio-medical signal processing. *Intelligent Techniques in Signal Processing for Multimedia Security, 660,* 345–370.
15. Parah, S,. Ahad, F., Sheikh, J., Loan, N., & Bhat, G. M. (2017). Pixel repetition technique: A high capacity and reversible data hiding method for e-healthcare applications. *Intelligent Techniques in Signal Processing for Multimedia Security*, *660*, 371–400.
16. Chakraborty, S., Chatterjee, S., Dey, N., Ashour, A., & Hassanien, A. E. (2017). Comparative approach between singular value decomposition and randomized singular value decomposition based watermarking. *Intelligent Techniques in Signal Processing for Multimedia Security, 660,* 133–150.

17. Parah, S., Sheikh, J., & Bhat, G. M. (2017). StegNmark: A joint stego-watermark approach for early tamper detection. *Intelligent Techniques in Signal Processing for Multimedia Security*, *660*, 371-4427-452.
18. Ahad, F., Sheikh, J., Bhat, G. M., Parah, S. (2017). Hiding clinical information in medical images: A new high capacity and reversible data hiding technique. *Journal of Biomedical Informatics*. doi:10.1016/j.jbi.2017.01.006 (SCI IF, 2.98).
19. Parah, S., Ahad, F., Sheikh, J., Loan, N., & Bhat, G. M. (2016). Reversible and high capacity data hiding technique for E-healthcare applications. *Multimedia Tools and Applications*. Springer. doi:10.1007/s11042-016-4196-2
20. Akhoon, J., Parah, S,. Sheikh, J., Loan, N., & Bhat, G. M. (2016). Information hiding in edges: A high capacity information hiding technique using hybrid edge detection. *Multimedia Tools and Applications*. Springer. doi:10.1007/s11042-016-4253-x
21. Assad, U., Parah, S., Sheikh, J., & Bhat, G. M. (2016). Realization and robustness evaluation of a blind spatial domain watermarking technique. *International Journal of Electronics*. doi:10.1080/00207217.2016.1242162
22. Parah, S., Sheikh, J., & Bhat, G. M. (2013). Data hiding in color images: A high capacity data hiding technique for covert communication. *Computer Engineering and Intelligent Systems, 4*(13), 107.
23. Sukanya, B., Chakraborty, S., Dey, N., & Pal, A. (2015). High payload watermarking using residue number system. *International Journal of Image, Graphics and Signal Processing, 3,* 1–8.
24. Shabir, P., Sheikh, J., & Bhat, G. M. (2013). On the realization of a secure, high capacity data embedding technique using joint top-down and down-top embedding approach. *Elixir Computer Science & Engineering, 49,* 10141–10146.
25. Shabir, P., Sheikh, J., & Bhat, G. M. (2013). High capacity data embedding using joint intermediate significant bit and least significant technique. *International Journal of Information Engineering and Applications, 2*(11), 1–11.
26. Shazia, A., Ayash, A,. & Parah, S. (2015). Robustness analysis of a digital image watermarking technique for various frequency bands in DCT domain. *2015 IEEE International Symposium on Nanoelectronic and Information Systems*, pp. 57–62. doi:10.1109/iNIS.2015.41
27. Subrata, N., Satyabrata, R., Jayanti, D., Wahiba, B., Shatadru, R., Sayan, C., et al. (2014). Cellular automata based encrypted ECG-hash code generation: An application in inter-human biometric authentication system. *International Journal of Computer Network and Information Security, 11,* 1–12.
28. Akhoon, J., Parah, S,. Sheikh, J., Loan, N., & Bhat, G. M. (2015). A high capacity data hiding scheme based on edge detection and even-odd plane separation. In *2015 Annual IEEE India Conference (INDICON)*. doi:10.1109/INDICON.2015.7443595
29. Dharavath, K., Talukdar, F., Laskar, R., & Dey, N. (2017). Face recognition under dry and wet face conditions. *Intelligent Techniques in Signal Processing for Multimedia Security, 660,* 253–272.
30. Borra, S., Kanchan, J., Raju, S., & Dey, N. (2017). Attendance recording system using partial face recognition algorithm. *Intelligent Techniques in Signal Processing for Multimedia Security, 660,* 293–320.
31. Ahad, F., Parah, S., Sheikh, J., & Bhat, G. M. (2015). On the realization of robust watermarking system for medical images. In *2015 Annual IEEE India Conference (INDICON)*, 978-1-4673-6540-6/15, 2015 IEEE. doi:10.1109/INDICON.2015.7443363
32. Parah, S., Sheikh, J., & Bhat, G. M. (2011). On the design and realization of non-synchronous frequency hopping spread spectrum for securing data over mobile and computer networks. *Global trends in Computing and Communication Systems*, ObCom-2011. Springer, Proceedings Part-1, Vol. 269, pp. 227–236.
33. Sheikh, J., Parah, S., & Bhat, G. M. (2009). On the realization and design of chaotic spread spectrum modulation technique for secure data transmission. In *Proceedings of First IEEE*

Sponsored International Conference IMPACT-2009 at AMU Aligarh India, pp 241-244. 978-1-4244-3604-0/09/2009.
34. Parah, S. A., Sheikh, J. A., & Mohiuddin Bhat, G. (2014). A secure and efficient spatial domain data hiding technique based on pixel adjustment. *American Journal of Engineering and Technology Research, US Library Congress, (USA), 14*(2), 38–44.
35. Hartung, F., & Kutter, M. (1999). Multimedia watermarking techniques. *Proceedings of the IEEE, 87,* 1079–1107.
36. Wu, D. C., & Tsai, W. H. (2003). A stegnographic method for images by pixel value differencing. *Pattern Recognition Letters*, *24*, 1613–1626.
37. Frith, D. (2007). Steganography approaches, options, and implications. *Network Security, 8,* 4–7.
38. Wang, H., & Wang, S. (2004). Cyber warfare: Steganography vs. steganalysis. *Communications of the ACM*, *47*, 10 (October issue).
39. Petitcolas, F. P., Anderson, R. J., & Kuhn, N. G. (1999). Information hiding—A survey. *Proceedings of the IEEE, 87*(7), 1062–1078.
40. Artz, D. (2001). Digital steganography: Hiding data within data. *IEEE Internet Computing*, *5*, 75–80.
41. Lin, E. T., & Delp, E. J. (1999). A review of data hiding in digital images. In *Proceedings of the Image Processing, Image Quality, Image Capture Systems Conference, PICS '99'*, pp. 274–278.
42. Chan, C. K., & Cheng, L. M. (2004). Hiding data in images by simple LSB substitution. *Pattern Recognition, 37,* 469–474.
43. Baiying, L., Ee-Leng, T., Siping, C., Dong, N., Tianfu, W., & Haijun, L. (2014). Reversible watermarking scheme for medical image based on differential evolution. *Expert Systems with Applications, 41,* 3178–3188. Elsevier.
44. Arijit, K. P., Nilanjan, D., Sourav, S., Achintya, D., & Sheli, S. C. (2013). A hybrid reversible watermarking technique for color biomedical images. In *IEEE International Conference on Computational Intelligence and Computing Research*, IEEE, 978-1-4799-1597-2/13.
45. Chun, K. T., Jason, C. N., Xiaotian, X., Chueh, L. P., Yong, L. G., & Kenneth, S. (2011). Security protection of DICOM medical images using dual-layer reversible watermarking with tamper detection capability. *Journal of Digital Imaging, 24,* 528–540.
46. Rahimi, F., & Rabbani, H. (2011). A dual adaptive watermarking scheme in contourlet domain for DICOM images. Biomedical Engineering Online.
47. Malay, K. K., & Sudeb, D. (2010). Lossless ROI medical image watermarking technique with enhanced security and high payload embedding. *IEEE International Conference on Pattern Recognition*, pp. 1457–1460.
48. Shohidul, I., Kim, C. H., & Kim, J. (2015). A GPU-based (8, 4) Hamming decoder for secure transmission of watermarked medical images. *Cluster Computing, 18,* 333–341.
49. Solanki, N., & Malik, S. K. (2014). ROI based medical image watermarking with zero distortion and enhanced security. *International Journal of Education and Computer Science, 10,* 40–48.
50. Jianfeng, L., Meng, W., Junping, D., Qianru, H., Li, L., & Chang, C. (2015). Multiple watermark scheme based on DWT-DCT quantization for medical images. *Journal of Information Hiding and Multimedia Signal Processing*, *6*, 458–472.
51. Das, S., & Kundu, M. K. (2012). Effective management of medical information through a novel blind watermarking technique. *Journal of Medical Systems, 37*, 663–675.
52. Lee, C., & Huang, Y. (2012). An efficient image interpolation increasing payload in reversible data hiding. *Expert Systems with Applications, 39,* 6712–6719. Elsevier.
53. Talat, N., Imran, U., Tariq, M. K., Amir, H. D. & Muhammad, F. S. (2013). Intelligent reversible watermarking technique in medical images using GA and PSO. Optik (Elsevier). doi:10.1016/j.ijleo.2013.10.2014
54. Shabir, A. P., Javaid, A. S., & Bhat, G. M. (2015). Hiding in encrypted images: A three tier security data hiding system. *Multidimensional Systems and Signal Processing* (Springer), September, 2015. doi:10.1007/s11045-015-0358-z

55. Shabir, A. P., Javaid, A. S., & Bhat, G. M. (2012). Data hiding in ISB planes: A high capacity blind stenographic technique. In *Proceedings of IEEE Sponsored International Conference INCOSET-2012*, pp. 192–197, Tiruchirappalli, Tamilnadu, India.
56. Schneier, B. (1994). *Applied cryptography, protocols, algorithms, and source code in C*. New York: Wiley.
57. Bhat, G. M., Parah, S. A., Sheikh, J. A. (2010). FPGA implementation of novel complex PN code generator based data scrambler and descrambler. Maejo International Journal of Science and Technology, *4*(01), 125–135.
58. Bhat, G. M., Parah, S. A., & Sheikh, J. A. (2009). VHDL modeling and simulation of data scrambler and descrambler for secure data communication. *Indian Journal of Science and Technology*, *2*, 41–43.
59. Silman, J. (2001). Steganography and steganalysis: An overview. *SANS Institute*, *3*, 61.
60. Sharadqeh, A. A. M. (2012). Linear model of resolution and quality of digital images. *Contemporary Engineering Sciences*, *5*, 273–279.
61. Parah, S. A., Sheikh, J. A., Loan, N., & Bhat, G. M. (2017). Utilizing neighborhood coefficient correlation: A new image watermarking technique robust to singular and hybrid attacks. *Multidimentional Systems and Signal Processing*. DOI 10.1007/s11045-017-0490-z
62. Parah, S. A., Sheikh, J. A., Loan, N. A., & Bhat, G. M. (2016). A Robust and Computationally Efficient Digital Watermarking Technique Using Inter Block Pixel Differencing. *Multimedia Forensics and Security,* Springer, Vol: 115. DOI: 10.1007/978-3-319-44270-9_10

Practical Techniques for Securing the Internet of Things (IoT) Against Side Channel Attacks

Hippolyte Djonon Tsague and Bheki Twala

Abstract As a global infrastructure with the aim of enabling objects to communicate with each other, the Internet of Things (IoT) is being widely used and applied to many critical applications. While that is true, it should be pointed out that the introduction of IoT could also expose Information Communication and Technology (ICT) environments to new security threats such as side channel attacks due to increased openness. Side-channel analysis is known to be a serious threat to embedded devices. Side-channel analysis or power analysis attempts to expose devices cryptographic keys through the evaluation of leakage information that emanates from a physical implementation. In the work presented herein, it is shown that a skilful attacker can take advantage of side channel analysis to break a 3DES implementation on an FPGA platform. Because of the threats posed by side channel analysis to ICT systems in general and IoT in particular, counter attack mechanisms in the form of leakage reduction techniques applicable to CMOS devices are proposed and evaluated. The modelling results revealed that building CMOS devices with high-κ dielectrics or adding strain in silicon during the device fabrication could help drastically reduce leakages in CMOS devices and therefore assist in designing more effective countermeasures for side channel analysis.

Keywords Side channel · IoT · Cryptography · Security · Power consumption · Machine learning · Dimensionality reduction

H.D. Tsague (✉)
Smart Token Research Group, Modelling and Digital Science (MDS)
Council for Scientific and Industrial Research (CSIR), Pretoria, South Africa
e-mail: hdjonontsague@csir.co.za

B. Twala
Department of Electrical and Electronic Engineering Science,
Faculty of Engineering, Institute for Intelligent Systems,
University of Johannesburg (UJ), Johannesburg, South Africa

N. Dey et al. (eds.), *Internet of Things and Big Data Analytics Toward Next-Generation Intelligence*, Studies in Big Data 30,
DOI 10.1007/978-3-319-60435-0_18

1 Introduction

The Internet of things (IoT) which is defined as the inter-networking of embedded devices, sensors and computers that collect and distribute large amount of data is thoroughly on its way after more than 10 years of discussion and apprehension. The IoT promises to create opportunities to expand our online world in countless of ways. Integrating the physical world as well as people into computer based systems, and culminating into greater efficiency, accuracy and financial benefits. Most researchers argue that "the Internet of Things is a misnomer. The IoT is not a unified collection of connected devices, but rather a grouping of various sensors and technologies that can be put to work in coordination together at the service and to the ultimate benefit of people in both developed and under developed countries" [1]. "This combination and interaction of IoT objects is accomplishing a vision of an interconnected, embedded, automated set of devices communicating automatically on a regular basis" [2]. However, connecting up devices and objects is only a means to an end; the most exciting and challenging question is: what can be done with the recorded data and how can we ensure that it is handled securely for improving our future? "There are many difficult trades-offs involved; only some of which are technological (for example, the trade-off between robustness and reliability, and the sophisticated functionality of sensors on a water pump). Other trade-offs enter into broader issues (for example, gaps between technical security and users' perceptions of security and trust, or the detailed information yielded by geo-localization technologies). Moreover, the purpose for which technology and applications are developed does not always end up as the sole—or even major—purpose for which they are actually used" [1].

While it is widely acknowledged that technology has the capacity as well as the capability to transform people lives, yet throughout the world and especially in developing and undeveloped regions of the world, "populations still go through difficulties ranging from climate change, poverty, environmental degradation, to the lack of access to quality of education, and communicable disease. While these big issues might look and appear insurmountable, it is believed that technology can play a critical role in addressing many of these challenges, while creating a host of opportunities" [1]. Today, IoT is a significant role player in the improvement of the live of citizens worldwide. In both developed and developing countries traffic monitoring sensors positioned at strategic locations record and provide cities traffic control officials with important traffic patterns that assist them on improving transportation operations and traffic flow. "Likewise, effective data management is improving service in hospitals and healthcare systems, education and basic government services such as safety, fire, and utilities" [2]. Actuators and sensors in manufacturing plants, mining operations, and oil fields are also playing an important role in raising the production, lowering costs and increasing safety. In healthcare facilities in underdeveloped countries, "the Internet of Things is also helping monitor critical vaccines through the use of IP-connected thermometers. Various sensors such as moisture sensors in agricultural fields help alert farmers to the exact needs of food crops. Similarly acoustic sensors in protected rainforests are helping to curb illegal logging. For the IoT to have an even greater impact there is still more to be done to improve the deployment of these technologies in developing countries. Network deployment, power requirements, reliability and durability are all uneven, at best, and policy considerations concerning access to data, legacy regulatory models, standards, interoperability and most importantly security, and privacy need to be addressed" [1].

A good proportion of systems rely on the tamper resistance capability of microcontroller devices to ensure privacy [3, 4]. The microelectronics industry which is at the heart of IoT has improved enormously over the last few years. It has quickly evolved to occupy a key position in modern electronic system design. This rapid growth is mostly due to its capacity to always make Complementary Metal Oxide Silicon (CMOS) device's performance better while at the same time keeping costs low. Such improvements have resulted in the development of smaller devices that consume much less power and produce shorter delays as predicted by Moore's law. The law states that transistor density on integrated circuits doubles every 18 months. It cannot be argued that the major reason for the success of CMOS transistors is based on the fact that it can be continually scaled down to smaller sizes. "This ability to improve performance while reducing the device power consumption has resulted in CMOS becoming the dominant technology for Very Large Scale Integration (VLSI) circuits" [2].

As a direct result of the CMOS improvement, an increasing number of embedded systems have gained entrance to our everyday life. From consumer electronics, like coffee machines, smart television, digital versatile disc (DVD) and Blu-ray players, refrigerators, washing machines to name but a few over smart home automation, e.g., thermostats, light control, cameras, garage door openers and car electronics. Almost every modern electronic device is equipped with an embedded CMOS based microcontroller (μC) that is monitoring a system. In the context of IoT, an increasing number of devices are getting "smarter", incorporating sensors to monitor defined situations and μCs to react automatically, while simultaneously networking with other devices. Microcontrollers are perfectly suited for such applications and environments, as they combine a number of multi-purpose modules in a chip, i.e., a central processing unit, clock generation and peripherals as well as program and data memory. For that reason, they are often referred to as self-contained computing systems.

The increasing need for smart devices is reflected by the rapidly rising number of microcontroller units (MCUs) sold in the last years. According to [5], more than 14 billion micro-controller units were sold in 2010. The unit shipments are forecast to double by 2017, while at the same time the average selling price per unit is expected to drop by about half of the current value. Besides the everyday use of μCs in non-critical applications, a steadily increasing number is assigned to manage security-critical tasks. The banking and access control industries are good examples where both contact and contactless smart cards are extensively utilised for secured banking transactions and access control. This trend is increasingly seen in the automobile industry with electronic control units (ECUs) used for services such as software download or feature activation. The increasing trend to get everything networked and inter-connected will continually motivate for the need for embedded applications with strong security features [6–10]. It is therefore imperative for security experts, design engineers and users to further interrogate the suitability of μCs as security devices. For many years it was believed that embedding programs into microcontrollers make them more resistant to fraud and attacks, however, over the last decade, microcontrollers have been found to be vulnerable to electromagnetic and side-channel attacks. Whenever microcontrollers execute programs, their power consumption can be used to reveal the content of their program and/or data memory content. By using the correlation between power consumption and microcontroller activity attackers can successfully recover cryptographic keys, reveal hardware faults, create a covert channel or reverse engineer the code executed by the processor. This is highly worrying, given the exponential demand and dependability on microcontrollers in protected applications.

There is a need for privacy and security implementation in embedded systems and the best way to provide for such a need is through an effective and efficient use of cryptographic protocols. Yet most cryptographic algorithms are vulnerable to attacks known as side channel analysis. Skilful attackers can easily take full advantage of emanations from those devices to breach their security. The attacking mechanism is simple; once the attackers can record the physical properties of the device while it is performing a cryptographic operation, he/she can apply statistical methods on the recorded data to retrieve the encryption key stored within the device. "In a side channel attack, the intruder eavesdrops on the device's side channel emissions and takes note when an encryption key is used to access the device. This tiny amount of information can then be used to duplicate the key" [11]. Even better an attacker can use the technique to reverse engineer contents of CMOS devices without the user being aware of such attacks. Clearly, side channel attacks are among the most dangerous attacks in cryptanalysis. IoT devices are mostly CMOS based and therefore highly vulnerable to such attacks. One of the limitations of IoT devices is that they lack the power and processor resources for malware detection. This is true with supervisory control and data acquisition or SCADA systems as well. Of the scores of papers and research reports published, it is believed that one of the most often overlooked factors behind hacking is not only technological vulnerabilities but also economic ones and that is the reason why a lot of this hacking happens in developing countries as well as those that are economically not that well off. "IoT devices collect data and connect to the Internet. In doing so, they emit signals known as side channels. These signals show levels of power consumption at any given moment, as well as electromagnetic and acoustic emissions. An attacker can overcome the encryption protecting an IoT device by executing a side channel attack. Compared to attacking the mathematical properties of encryption, side-channel attacks require much less time and trouble to correctly guess a secret key" [11]. Instead of seeking for mathematical weaknesses of a given cipher, an adversary attacks its physical implementation by analysing the side-channel leakages, e.g., power consumption, electromagnetic emanations, timing behaviour, etc. These attacks often only make use of readily available off-the shelf components that executes quickly i.e. within a few minutes and almost impossible to detect.

After the first side channel publication by Kocher et al. in 1996, a community emerged that focused on protecting cipher implementations against this type of attacks [12, 13]. Common countermeasures to complicate side-channel attacks are mainly so-called masking and hiding methods, which either try to reduce the dependency between processed data and side-channel observations or add noise in the time dimension to complicate a proper alignment of the measurements. Although in many instances such techniques result in an attacker doing more work to successfully attack an implementation, they are still not capable of preventing them completely. Furthermore, it was noticed that most authors present new side channel attack without much information on their attack implementation or data to verify the claimed results.

Also, As IoT expands; "a myriad of cheap, low-power devices will record data and interact with humans and companies across the Internet in countless ways. As IoT devices and systems become cheaper and more ubiquitous, device manufacturers likely won't have much incentive to invest in complex security mechanisms. Cyber criminals will take advantage using well-understood hacking techniques, such as side channel attacks" [14]. In this chapter, there is a need to explore what can be achieved in terms of cryptanalysis of embedded devices in IoT environments using power analysis techniques. This research area has only been sparsely discussed in the available literature. The chapter's major contribution is in the use of machine learning techniques in the statistical analysis of power consumption data. Furthermore, to the authors' knowledge, this is the first documented work that investigates and discusses side channel implementations together with countermeasures for IoT systems in under developed economies.

2 Problem Formulation

A substantial number of systems rely upon tamper resistant microprocessor devices. This includes an important ratio of devices utilized in IoT systems in both developed and underdeveloped countries. "The IoT finds applications in all domains. Currently, the IoT has been deployed in smart environments at homes and cities, transportation and logistics with the prospect of driverless vehicles, healthcare of tracking and monitoring, and social domain on personal services" [15]. The Oxford Dictionary has added the following warning to their definition of IoT: "If one thing can prevent the Internet of Things from transforming the way we live and work, it will be a breakdown in security." Clearly Ubiquitous connectivity is most often accompanied by pervasive attacks. Yet it is interesting to note that discussions around designing secure devices for IoT applications have been rather sparse in the available literature. "Domain experts, not security engineers, will develop specialized cyber-physical control and critical infrastructure systems based on the IoT. Without security mechanisms being included in the research and development stage of the IoT, the nightmare of cyber threats that we are experiencing now on the Internet will repeat" [15]. For example, it has now become well known that microprocessor devices leak information about the activity being performed through side-channels. Side-channel or power analysis attacks "exploit correlations between the internal sensitive information and unintentionally externally available information such as time and power consumption, optical, acoustic or electromagnetic radiation. These attacks can be invasive or non-invasive such that no traces are left behind" [16].

Power analysis or side channel attacks are not new. It is reported that military and government organizations have used them for some times [5]. Such attacks were reported to the public for the first time in 1985 when the researchers of [17] showed how to eavesdrop on video display units from a considerable distance via emitted magnetic radiations. In 1996 authors of [18] published their ground-breaking work on side channel analysis. Immediately after that publication, Simple Power Analysis (SPA) and Differential Power Analysis (DPA) received some attention from, among others, the banking industry, and countermeasures were publicly announced. Power analysis has since then received a lot of attention, with many proposed and improved attacks and countermeasures.

Due to the continuing reduction in transistor size (Moore's law), microprocessor technology continues to shrink in size yet increase in power. This development has made possible devices such as smart phones and smart cards. However, as we become more dependent upon these devices and use them increasingly for sensitive information, protecting the device and the information it contains be-comes ever more important. The importance of understanding side-channel attacks is therefore higher than ever especially in IoT, as there is no complete or supreme protection. Each attack has to be studied and scrutinized in its own context as the solutions that appeared to be state of the art a couple of years ago might be worthless today. An ongoing effort to grasp what is deducible from side-channel information is of paramount importance as vulnerabilities, previously found infeasible to exploit may be within the adversaries reach today.

This chapter intends to provide a detailed overview of side channel attacks as it pertains to devices used in IoT environments in developing countries; the focus will be to raise awareness of this type of attacks and how they can be circumvented. Simple attacks scenarios will be outlined and effective countermeasures to this type of attacks proposed and simulated. Further steps will be taken in illustrating the need and importance of statistical tools for the analysis of power consumption data using machine learning techniques. Many publications have already described and presented the role of machine learning techniques in cryptanalysis [16]. The work presented in this chapter capitalizes on two areas of research of side channel analysis that have not yet been thoroughly explored. These are dimension reduction and model selection. A good understanding of these two areas of research will certainly help enhance the understanding of side channel attacks on IoT devices so as to assist in designing better and more effective counter-measures to such attacks.

3 Side Channel Overview

Although there is a varied mixture of cypher tools capable of providing very high levels of security for the IoT and other embedded systems, their implementations in such systems can easily be circumvented using side-channel analysis (SCA). Instead of concentrating on the mathematical characteristics of the cryptographic algorithm, a SCA implementation takes advantage of emitted secondary effects occurring while the device performs its normal operations. Such unintentional emissions can be used as a way to break into systems or disrupt their functionality. Besides power consumption, a few more leakage types have emerged including electromagnetic emanations, acoustic emissions and timing behaviour. The most common side-channel attacks mechanisms are explored next.

3.1 Timing Attack

In the early 1986 it was discovered that "Implementations of cryptographic algorithms often perform computations in non-constant time, due to performance optimizations. If such operations involve secret parameters, these timing variations can leak information and provided enough knowledge of the implementation is at hand, a careful statistical analysis could even lead to the total recovery of these secret parameters" [19]. Timing attacks involves measuring the amount of time it takes to complete certain cryptographic operations with the chief aim of taking advantage of such timing variances to "steal" the user's private data such as Personal Identification Number (PIN) or encryption keys. Following are key assumptions made while implementing timing attacks:

- The run time of a cryptographic operation depends to some extent on the key. With present hardware this is likely to be the case, but note that there are various efficient hardware based proposals to make the timing attack less feasible through 'noise injection'. Software approaches to make the timing attack infeasible are based on the idea that the computations in two branches of a conditional should take the same amount of time ('branch equalization').
- A sufficiently large number of encryptions can be carried out, during which time the key does not change. A challenge response protocol is ideal for timing attacks.
- Time can be measured with known error. The smaller the error, the fewer time measurements are required.

Performance-optimized implementations of cryptographic algorithms are usually non-constant in time. Primarily caused by conditional branches during the computation, the timing behaviour of the implementation becomes dependent on the processed data. If a secret key is involved in the timing variations, it might be sufficient to measure the elapsed time of the execution and use statistical analysis to extract details of the key. Typical targets for timing attacks are implementations of an RSA exponentiation using the square-and-multiply algorithm or the Chinese remainder theorem on smartcards or embedded devices [20].

3.2 Fault Injection Attacks

Generally, an attacker familiar with power analysis can easily separate active and passive power analysis attacks. "While passive attacks measure quantities such as the timing behaviour or the power consumption of a device, active attacks like fault injection attacks interfere directly with the encryption algorithm to trigger a device malfunction during computation" [20]. In some instances such as when RSA signature is being generated using the Chinese remainder theorem, an incorrect computation may leak secret information about the key. The following are possible sources of fault generation: (i) laser high temperature, (ii) overclocking, (iii) X-rays, and (iv) ion beam. It should be noted that in the process of generating a device malfunction, these manipulations can irreversibly damage the device. Another disadvantage of this method relates to the amount of effort required to successfully induce device faults.

3.3 Electromagnetic Analysis (EMA)

Devices such as cellular phones and desktop computer often emit electromagnetic radiation during their day to day operations. This is also true for any other CMOS based devices. Anyone with sufficient knowledge on how to manipulate and exploit such emanations can pull out a massive amount of valuable information from such devices. It is not new knowledge that IoT devices store and manipulate user's private information; clearly such attacks if carried out in an IoT based network will have serious consequences. "The potential of exploiting electromagnetic emanations has been known in military circles for a long time. The recently declassified TEMPEST document by the National Security Agency reports on investigations of compromising emanations including electromagnetic radiation, line conduction, and acoustic emissions" [21]. EMA takes advantage of local information. Though noisier, measurements can be carried out from a distance. This fact broadens the range of targets to which side channel attacks can be applied. Of concern are not only smart cards and similar tokens but also Secure Sockets Layer (SSL) accelerators and many other devices used in IoT systems. "Two types of EM radiations exit: intentional and unintentional. The first type results from direct current flows. The second type is caused by various couplings, modulations (AM and FM), etc. The real advantage of EMA over other side channel attacks lies in exploring unintentional radiations" [22, 23]. More precisely, EMA leakage consists of multiple channels. Therefore, EMA seem to succeed even in the presence DPA countermeasures.

3.4 *Power Analysis*

Power analysis hinges on the assumption that the power consumption of an integrated circuit (IC) is not constant, but data dependent. This is truly the case for complementary metal oxide semiconductor (CMOS) circuits. For instance, if a state of a CMOS inverter stays constant, only a small static leakage current can be observed. Instead, switching the state of the inverter results in high power consumption for a short moment in time. Power analysis can be divided in 3 components namely: Simple Power Analysis (SPA), Differential Power Analysis (DPA) and Correlation Power Analysis (CPA) as detailed below.

3.4.1 Simple Power Analysis (SPA)

The Simple Power Analysis was first introduced in the ground-breaking paper by Kocher [19]. Basically, SPA is a technique that directly interprets power consumption traces that are measured while a cryptographic algorithm is performed. The power consumption highly depends on the switching behaviour of the cells. Hence, the instructions that are executed on the device and especially the intermediate values that are produced during the encryption process determine the shape of the power consumption trace. Large features, like encryption rounds, can therefore be identified by inspecting plots of power consumption traces. Additionally, it is possible to identify even small differences by inspecting a visual representation of just a small part of the trace. An attacker's intention is to find such differences. More precisely, he searches for key dependent varieties in order to reveal the secret key bit by bit. As one can imagine, detailed knowledge about the implementation of the algorithm is usually required to mount an attack. However, if only single or just a small number of power traces can be recorded, e.g. when the device is only available for a short period of time, SPA is quite useful in practice. On the other hand, SPA is a great tool for reverse engineering, since it can be used to analyse the implications on the power consumption for different code sequences or values [24].

3.4.2 Differential Power Analysis (DPA)

Differential Power Analysis, also published in [16], is the most popular power analysis method. Compared to SPA, it has some important advantages. First, this type of analysis does not require any implementation details. In fact, it is sufficient to know the implemented algorithm. Due to Kirchhoff's principle this can generally be taken for granted. Second, DPA can even be used in extremely noisy environments [24]. On the other hand, the need for a large number of power traces has to be mentioned. This is because traces are not analysed visually but by means of statistics.

3.4.3 Correlation Coefficient and Power Models

A way to improve the DPA by Kocher is to use correlation coefficients instead of difference of means to estimate the correct key. This method was first published by [13] and is described as follows. At first, intermediate values are calculated as with the DPA by Kocher. Subsequently, these values are mapped to hypothetical power consumptions based on a destined power model. The correct key can then be found by correlating the recorded power consumption traces to the hypothetical power consumptions. To understand this attack in detail, it is important to have a basic knowledge of power models. As already mentioned, an attacker tries to simulate the power consumed by intermediate values. However, his knowledge about the hardware is generally limited. As a result, only simplistic power models can be used for this purpose. One basic model is the Hamming-Weight model. This model assumes that the power consumption of a device is proportional to the number of bits set in a chosen intermediate value and can thus be estimated by this number.

4 Mathematical and Statistical Introduction

Throughout this chapter the reader will be faced with mathematical problems. While trying to keep them as simple as possible, solving these problems demands some basic knowledge in statistics and algebra. In this chapter, formulas and descriptions that cover most of the mathematical back-ground needed to understand the techniques used in this work are provided. They are divided into two blocks: Statistics is not only fundamental for side-channel attacks but can also be helpful in analysing time series data for identifying μC instructions. The second part involves matrix algebra. It describes the use of eigenvectors and eigenvalues which are important properties of dimensionality reduction algorithms such as Fisher's linear discriminant. For a more comprehensive introduction to statistics refer to [18].

4.1 Probability Space

The probability space in statistics is a mathematical build-up used for the modelling of real-world processes or experiment. This quantity often define random event and is represented using the triplet $(\Omega, \mathcal{F}, \wp)$

- Ω Denotes a collection of elementary and random events.
- $\mathcal{F}$ Denotes a subset of the sample space Ω that possesses the following properties:
  - $\Omega \in \mathcal{F}$
  - $\mathrm{C} \in \mathcal{F} \Rightarrow (\Omega \backslash \mathrm{C}) \in \mathcal{F}$
  - $\mathrm{C}_1, \mathrm{C}_2, \ldots \in \mathcal{F} \Rightarrow \bigcup_{i=1}^{\infty} \mathrm{C}_i \in \mathcal{F}$

  Here C represents a group of random events.

- $\wp : \mathcal{F} \rightarrow \mathbb{R}$ represents a group of event in a probability space and satisfies the following postulate:
  - $P(\mathrm{C}) \geq 0$,
  - $(\mathrm{C}_1 \cup \mathrm{C}_2 \cup \cdots) = P(\mathrm{C}_1) + P(\mathrm{C}_2) + \cdots$ for $\mathrm{C}_i \cap \mathrm{C}_j = \varnothing$ *if* $i \neq j$
  - $P(\Omega) = 1$

4.2 *Discrete Random Variable*

Discrete random variable can only hold a defined number of values. A good example includes throwing of a dice where the number of possible values is only 6 (1, 2, 3, 4, 5, 6) and therefore the variable is discrete. Variables are defined as random when their probabilities sum up to 1 i.e.:

$$p_i = P(\mathrm{X} = x_i) \quad \text{with} \sum_i p_i = 1 \tag{1}$$

4.3 *Expected Value*

This quantity is analogous to the mean. This is perhaps the reason why those quantities are often confused. In general, experimental and predicted appearance of the values helps distinguish them. When the average is computed from past experimental data, the result obtained defines the mean μ which describes as follows:

$$\mu = AM(\mathrm{X}) = \frac{1}{n}\sum_{i=1}^{n} x_i \tag{2}$$

X represents a random variable, x_i represents the sub-components of X_i, *n* represents the total number of sub-component of X.

In a similar fashion, when the average is worked out from future experimental data by taking into consideration the probability of the occurrence of such events, the expected value is obtained. In general, the expected value can be calculated using the following formula:

$$E(\mathrm{X}) = \sum_{i=1}^{n} x_i p(x_i) \tag{3}$$

4.4 Standard Deviation and Variance

The variance is mathematically defined as the average squared deviation of the expected value:

$$\begin{aligned} Var(X) &= \sigma^2 = E\left((X - E(X))^2\right) \\ &= E\left(X^2\right) - E(X)^2 \end{aligned} \tag{4}$$

The variance is a quantity usually utilised to assist with estimating the deviation of a discreet random variable X from the expected value. By taking the square root of the variance, the standard deviation σ is obtained. A typical standard deviation curve for a normally distributed random variable is shown in Fig. 1.

The curve in Fig. 1 can also be called a Bell Curve since its curve is analogous to the density function of a standard probability distribution. The standard normal distribution is defined by two distinct variables namely the mean and the standard deviation.

4.5 Covariance and Correlation Coefficient

The covariance is an estimate of the joint variability between two discreet random variables. In other words, the covariance measures the relationship among two discreet random variables say X and Y. The covariance matrix is computed as follows:

$$\begin{aligned} Cov(X, Y) &= E((X - E(X)) \cdot (Y - E(Y)) \\ &= E(X \cdot Y) - E(X) \cdot E(Y) \end{aligned} \tag{5}$$

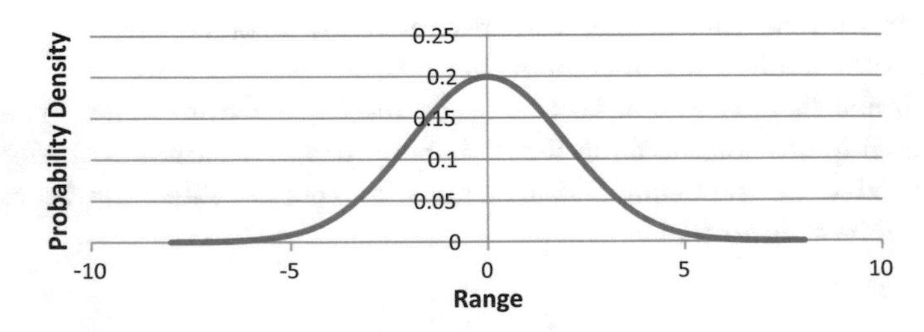

Fig. 1 Standard normal distribution

In statistics and probability theory, the covariance is a general form of variance and can be written as:

$$\begin{aligned} Cov(X,X) &= E((X-E(X))\cdot(X-E(X)) \\ &= Var(X) \end{aligned} \quad (6)$$

The outcome of a computed covariance always indicates one of the three scenarios shown below:

1. When the covariance is positive one can safely assume a positive linear relationship between variables.
2. When the covariance is negative however, a negative linear relationship exists between the variables of interest.
3. When the calculated covariance is 0, the variables are totally uncorrelated.

"When the covariance is divided by the product of the standard deviation of the two random variables we get a more accurate estimate of their interdependency" [18]. This normalization yields the correlation coefficient which is a quantity that is between -1 and 1, and is computed as:

$$\begin{aligned} \varrho(X,Y) = r_{XY} &= \frac{Cov(X,Y)}{\sigma_X\sigma_Y} \\ &= \frac{E(X-E(X))(Y-E(Y))}{\sqrt{Var(X)}\cdot\sqrt{Var(Y)}} \end{aligned} \quad (7)$$

Similar to the covariance defined earlier, a large value of the correlation coefficient depicts a positive linear relationship between the random variables X and Y. While negative linear relationship between the variables is assumed for values close to -1. When the correlation coefficient is 0 or very close to 0, no linear interdependency exists; this should not be confused with the absence relationship between the random variables X and Y.

As it pertains to power consumption or side channel traces, in order to correlate two measured traces A and B, with values $a_1, a_2, \ldots a_n$ and $b_1, b_2, \ldots b_n$, a term known as the "Pearson product-moment correlation coefficient" can be utilised and is given below:

$$\varrho(A,B) = r_{AB} = \frac{\sum_{i=1}^{n}(a_i-\bar{a})(b_i-\bar{b})}{\sqrt{\sum_{i=1}^{n}(a_i-\bar{a})^2\sum_{i=1}^{n}(b_i-\bar{b})^2}} \quad (8)$$

where $\bar{a} = \mu_A$ and $\bar{b} = \mu_B$ are the mean values of A and B.

4.6 Covariance Matrix

The covariance is always computed between two dimensions. In cases where more than two dimensions exist, the covariance values can be represented in a matrix form. For example, considering a random vector X consisting of three dimensions x_1, x_2 and x_3, the covariance matrix is defined as:

$$Cov(X) = \begin{pmatrix} Cov(x_1,x_1) & Cov(x_1,x_2) & Cov(x_1,x_3) \\ Cov(x_2,x_1) & Cov(x_2,x_2) & Cov(x_2,x_3) \\ Cov(x_3,x_1) & Cov(x_3,x_2) & Cov(x_3,x_3) \end{pmatrix} \tag{9}$$

The entries of the main diagonal matrix contain the covariance between one dimension and itself which is equivalent to the variance. For this reason the covariance matrix is also known as variance-covariance matrix. Another characteristic of the matrix is its symmetry, since $\mathrm{Cov}(x_i, x_j) = \mathrm{Cov}(x_j, x_i)$.

4.7 Matrix Algebra

Matrix algebra is a very comprehensive term which includes basic operations like matrix addition, scalar multiplication, transposition, and so on. In order to keep this section (and also this thesis) short and manageable, it is assumed that many of the basic operations are known by the reader and concentrate only on Eigenvectors and Eigenvalues, as these are two fundamentals required for principal component analysis (PCA) and linear discriminant analysis (LDA).

4.7.1 Eigenvalues and Eigenvectors

Let us consider the following example:

A square matrix $A \in \mathbb{R}^{2\times2}$ and two vectors v_1 and $v_2 \in \mathbb{R}^2$ are given by the equation:

$$A = \begin{pmatrix} 2 & 0 \\ -1 & 8 \end{pmatrix},\ v_1 = \begin{pmatrix} 2 \\ 1 \end{pmatrix}\ and\ v_2 = \begin{pmatrix} 6 \\ 1 \end{pmatrix}$$

$$Av_1 = \begin{pmatrix} 2 & 0 \\ -1 & 8 \end{pmatrix}\begin{pmatrix} 2 \\ 1 \end{pmatrix} = \begin{pmatrix} 4 \\ 6 \end{pmatrix}$$

$$Av_2 = \begin{pmatrix} 2 & 0 \\ -1 & 8 \end{pmatrix}\begin{pmatrix} 6 \\ 1 \end{pmatrix} = \begin{pmatrix} 12 \\ 2 \end{pmatrix}$$

In most cases, a multiplication between a matrix and a vector results in a vector with no special properties, as can be seen in the first example. In contrast, in the second example the newly derived vector Av_2 is 2 times the original vector v_2. It holds that:

$$Av_2 = \alpha v_2 \tag{10}$$

While Av_1 and v_1 point to different directions, Av_2 and v_2 are exactly parallel, with Av_2 being twice as long as v_2. A vector that fulfils this special property is called an eigenvector of the matrix A.

The scalar α is the corresponding eigenvalue. Eigenvalues are closely related to the direction of the eigenvectors, but not to their lengths, i.e. multiplying an eigenvector with a scalar does not change the eigenvalue. For instance,

$$v_3 = 4 \cdot v_2 = \begin{pmatrix} 24 \\ 4 \end{pmatrix}$$

$$Av_3 = \begin{pmatrix} 2 & 0 \\ -1 & 8 \end{pmatrix} \begin{pmatrix} 24 \\ 4 \end{pmatrix} = \begin{pmatrix} 48 \\ 8 \end{pmatrix} = 2 \cdot v_3$$

Hence, it is preferable to specify eigenvectors in its normalized form, that is, with a length of 1.

4.7.2 Computation of Eigenvalues

For a given $A \in \mathbb{R}^{n \times n}$ how are eigenvalues computed?

Let v be an eigenvector of A. By transposing (Eq. 1), we obtain a homogeneous system of linear equations:

$$\begin{aligned} & Av = \alpha v \\ & \Leftrightarrow Av - \alpha v = 0 \\ & \Leftrightarrow (A - \alpha I)v = 0 \end{aligned} \tag{11}$$

where I is the identity matrix of size n. This system of linear equations has exactly one solution if and only if Det $(A - \alpha I) \neq 0$. However, since there is a need to exclude the zero vectors (which is by definition always a solution for homogeneous systems of linear equations), we are looking for a non-trivial solution for v, which only exists if:

$$Det(A - \alpha I) = 0 \tag{12}$$

By applying the formula for computing the determinant, a characteristic polynomial $p(\alpha) = \text{Det}(A - \alpha I)$ with degree n is obtained. The desired eigenvalues are obtained by solving the equation $p(\alpha) = 0$.

Example: Let A be the above mentioned matrix.

$$A = \begin{pmatrix} 2 & 0 \\ -1 & 8 \end{pmatrix}$$

The characteristic polynomial is given by the determinant of (A − αI):

$$\begin{aligned} p(\alpha) &= Det(A - \alpha I) \\ &= \begin{vmatrix} 2-\alpha & 0 \\ -1 & 8-\alpha \end{vmatrix} \\ &= (2-\alpha)(8-\alpha) \\ &= \alpha^2 - 10\alpha + 16 \end{aligned}$$

Solving p(α) = 0 leads to the 2 eigenvalues $\alpha_1 = 2$ and $\alpha_1 = 8$.

In theory, finding eigenvalues of a given matrix seems to be straightforward, which holds for small n. However, for n > 4, there exists no algebraic solution. Thus, in practice, approximation algorithms as, e.g., the (implicit) QR algorithm is used to compute the eigenvalues. Basically, the QR algorithm transforms the matrix into an upper triangular form while preserving the original eigenvalues as much as possible. Because the determinant $Det(A' - \alpha I)$ of a triangular matrix A' is given by:

$$Det(A' - \alpha I) = \begin{vmatrix} a_{1,1}-\alpha & a_{1,2} & a_{1,3} & \cdots & a_{1,n} \\ & & a_{2,3} & & \\ 0 & a_{2,2}-\alpha & & \cdots & a_{2,n} \\ & & \vdots & & \\ 0 & 0 & 0 & \cdots & a_{n,n}-\alpha \end{vmatrix}$$
$$= (a_{1,1}-\alpha)(a_{2,2}-\alpha)(a_{3,3}-\alpha)(a_{n,n}-\alpha)$$

The eigenvalues can be directly read out from the elements of the main diagonal. For more detailed information about the QR algorithm we refer to [25].

4.7.3 Computation of Eigenvectors

Once the eigenvalues are found, they can each be inserted (Eq. 11) to obtain a homogeneous system of linear equations. Solving these equations with the Gaussian elimination leads to a solution set of eigenvectors. The solution is not unique due to the above mentioned characteristic that multiplying an eigenvector with an arbitrary number c does not change the eigenvalue. It becomes even clearer if (Eq. 10) is considered: If a combination α and v that fulfills this equation is found, the value of v can be replaced with every new product c · v without the equation becoming invalid.

5 Related Work

Very few publications have investigated and documented side channel attacks in IoT although many researchers and security experts have designed attack scenarios for sensor based devices and embedded platforms. Authors of [26] showed that unrestricted use of cellular phone accelerometer information allowed a harmful application to successfully listen to and capture and decipher the vibrations emanating from the keystrokes of an attached keyboard. In a similar way, it was shown in [27] that accelerometer sensors could be assimilated to side channels in that they could be used to learn user inputs such as gestures when keying in their PIN to access their cellular phones. In a research carried by authors of [28], it was demonstrated that analog sensors are prone to injection attacks. The authors "analyse the feasibility of these attacks with intentionally generated electromagnetic interference on cardiac medical devices and microphones". In [29] the authors investigate the security of smart cards using power analysis. In their research, "the attacker guesses the data from power consumption of the smart card by measuring them several times and comparing to each other prior to analysing the results". To prevent the power analysis attack, a mechanism based on hiding is presented. Although the logic implemented in this research works well its major drawback is that the encryption algorithm used is very weak and can be broken fairly easily. In recent years, machined learning has gained momentum and is being used in many applications. The advantage of using machine learning techniques in cryptanalysis and side channel was briefly described in [29, 30] where a machine learning tools are effectively used to "recover information on printed characters of a printer by exploiting the information hidden into the acoustic noise". In [31] the authors investigated the application of a machine learning technique in side-channel analysis by applying the Least Squares Support Vector Machine. The authors used the device's power consumption as the side-channel while the target was an implementation of the Advanced Encryption Standard. Their results show that the choice of parameters of the machine learning technique strongly impacts the performance of the classification. They also found out that the number of power traces and time instants does not influence the results in the same proportion. In a similar way authors of [32] mounted an attack on an 8-bit smart card performing DES decryptions with a 56-bit key. The machine learning tool used to exploit the side channel leakage was Support Vector Machines (SVM). The authors claim a classification rate of close to 100% using features drawn from measurements during the DES key schedule, after training with at least 1000 samples out of our set of 11,000.

6 A Machine Learning Based Implementation of an Attack on 3DES Algorithm

Due to its popularity and simplicity, the 3DES algorithm has been extensively utilised for secure transactions in IoT implementations. In this book chapter the main goal is to demonstrate how an attacker familiar with side channel techniques can successfully break a 3DES cryptographic implementation built on CMOS devices such as the FPGA. In this scenario, the key objective is to put forward the relationship between the device power consumption and the cryptographic key used to secure or encrypt sensitive information stored in the device. We then proceed to compare, contrast and assess the efficiency of classification algorithm and dimension reduction tools in a real-world scenario. The results obtained confirm the importance of taking side channel based attacks seriously especially in IoT systems.

As stated earlier, DPA uses advanced statistical analysis by modelling the theoretic power consumption for each key. The likelihood of the observed power consumption for each model is then used to predict the key. A better approach exists and is known as Template Based DPA (TDPA). This technique “adds an additional step by estimating the conditional probability of the trace for each key” [33]. This method retrieves bits of information present in each trace and is regarded as one of the strongest technique available for side channel attacks. The method borrows a lot from Gaussian para-metric estimation techniques which in some instances has shown several limitations. One such limitation relates to the fact that, “the method can be ill-conditioned when the number of observed traces is smaller than the number of features used to describe the trace” [33]. Such behaviour is usually observed when the full set of measured traced or power consumption is taken into account.

In this approach, the intentions are to go a step further into the statistical analysis of power traces by making intelligent use of machine learning tools available. Many recent publications have outlined the used and importance of statistics and machine learning in cryptanalysis [34, 35]. This work takes a similar approach and demonstrates that machine learning implementations enhanced with a combination of dimension reduction and model selection is capable of outperforming conventional TPDA.

6.1 The TDPA Approach

Consider a cryptographic device in an IoT scenario that executes cryptographic operation with a binary key O_k. $k = 1, 2, 3, \ldots, K$.

Where $K = 2^B$ represents the entire set of realistic combination that the key can hold. The number of bits or the key size is represented by B if it is assumed that N power consumption traces of the device over a time interval n have been recorded. A trace is represented by a set of observations as follows:

$T_i^{(k)} \in \mathbb{R}^n$. $i = 1, 2, 3, \ldots, N$. linked to the k_{th} key.

The TDPA model "represents the dependency between the key and the trace by making use of a multivariate normal conditional density" [33].

$$P\left(T^{(k)}|O_k; \mu_k, \Sigma_k\right) = \frac{1}{\sqrt{(2\pi)^n |\Sigma_k|}} e^{-\frac{1}{2}\left(T^k - \mu_k\right)\Sigma_k^{-1}\left(T^k - \mu_k\right)^T} \quad (13)$$

In (13) $\mu_k \in \mathbb{R}^n$ is the expected variable while Σ_k, $k = 1, 2, 3, \ldots, K$, is the covariance matrix.

Whenever a group of N power consumption measurements $T_i^{(k)}$, $i = 1, 2, 3, \ldots, N$ are recorded for the corresponding key, the expected value μ_k and the covariance matrix Σ_k can be computed using the TDPA technique.

$$\widehat{\mu}_k = \frac{1}{N}\sum_{i=1}^{n} T_i^{(k)} \quad (14)$$

and the sample covariance

$$\widehat{\Sigma}_k = \frac{1}{N}\sum_{i=1}^{n}\left(T_i^{(k)} - \widehat{\mu}_k\right)^T\left(T_i^{(k)} - \widehat{\mu}_k\right) \quad (15)$$

A given trace is considered unclassified when the recorded cryptographic key associated with it is unknown. Whenever such unclassified item is observed, the approach returns the key that maximizes its likelihood:

$$\widehat{\kappa} = \arg\max_k \widehat{P}(T|O_k) = \widehat{P}\left(T|O_k; \widehat{\mu}_k, \widehat{\Sigma}_k\right) \quad (16)$$

Like other researchers, it is assumed in this that "the distribution of power traces for a particular key follows a parametric Gaussian distribution, where the number of parameters is given by: $(n^2 + 3n)/2$" [33].

It is important at this stage to point out that n can grow exponentially to become greater than N especially for moderate observation intervals.

6.2 *Proposed Approach*

The proposed approach relies on statistics and machine learning techniques to evaluate the conditional distribution $P(O_k|T)$ where $O_k \in \{0, 1\}^B$ from a set of recorded power traces.

In this work an implementation procedure based on three steps is put forward namely:

1. Breaking down the prediction task into distinct and independent classification tasks.
2. Dimension reduction.
3. Model selection.

Furthermore, 2 techniques for dimension reduction are utilised, namely:

- Principal component analysis (PCA). A technique extensively used in available literature to reduce the number of variables being manipulated with limited losses in precision.
- Linear discriminant analysis (LDA) which is a simple but effective selection tool that usually returns the maximal variant variables.

To measure the predictive effectiveness of the proposed classification models, the proposed technique borrows from a popular method commonly referred to as the "leave-one-out". The method is run over k stages. At every stage k − 1 recorded power traces are used to train the model. The unused power trace is used to assess generalization accuracy of the model. This process is repeated till all power traces available for testing have been utilised. The power trace maximises the value returned by the "leave-on-out" approach is selected as the best one. Although computationally expensive, this estimation technique is very accurate.

6.3 Experimental Method

The experimental setup is based on the attack of a 3DES block cipher running on an FPGA. The system is designed to be compatible with most FPGA boards (especially the Spartan 6 FPGA) as shown in Fig. 2. It is important to state here that the board has a ZTEX FPGA Module incorporated. "This board provides several features specific to side-channel analysis: two headers for mounting ADC or DAC boards, an AVR programmer, voltage-level translators for the target device, clock inputs, power for a differential probe and Low Noise Amplifier (LNA), external Phase Locked Loop (PLL) for clock recovery, and extension connectors for future improvements such as fault injection hardware. This board will be referred to as the Chip Whisperer Capture Rev2" [36].

Encryption and decryption procedures are implemented with randomly selected messages of constant length (in this case 64 bits). 3DES is a variant of the DES encryption algorithm. The 3DES key length is exactly 192 bytes due to the fact that it is made up of a combination of three 64-bit keys. "The procedure for encryption is exactly the same as regular DES, but it is repeated three times, hence the name

Fig. 2 Side channel equipment setup

Triple DES. The data is encrypted with the first key, decrypted with the second key, and finally encrypted again with the third key" [37]. Although a lot more secured than DES, it is also 3 times slower. "Like DES, data is encrypted and decrypted in 64-bit chunks. Although the input key for DES is 64 bits long, the actual key used by DES is only 56 bits in length. The least significant (right-most) bit in each byte is a parity bit, and should be set so that there are always an odd number of 1s in every byte. These parity bits are ignored, so only the seven most significant bits of each byte are used, resulting in a key length of 56 bits. This means that the effective key strength of 3DES is 168 bits because each of the three keys contains 8 parity bits that are not used during the encryption process" [37].

For simplicity sake, the attack is narrowed down to a single byte of the 3DES key set. In other words for this attack, a "target" value O_k capable of accommodating up to $2^7 = 128$ different values is selected. The proposed experimental setup is identical to the setup outlined in [33] and is constituted of four components as follows:

- The first component is the data generator component that runs on the Xilinx FPGA hardware. It uses a message and key combination to provide a measure of power consumption of the device throughout encryption/decryption operations.
- The second component represents the filter. This component is a feature selection algorithm with the main purpose of reducing the trace dimensionality thereby accelerating the attack.

- The third component is the model also known as the "an attack". It returns the encryption key used to encrypt data based on recorded traces.
- The last component is the verification function used to estimate and assess the quality of the attack. It can be a function of a number of variables including but not limited to the attack execution time or amount memory utilised.

To compare the quality of the model, one of the most basic and widely used attack i.e. Correlation Power Analysis (CPA) is used. The power traces are recorded using a Xilinx FPGA device running the 3DES implementation. Approximately 600 traces are recorded for every possible value of the key to a total of 50,000 power traces which are filtered by averaging to reduce the noise level as follows:

$$\widehat{\mu}_k = \frac{1}{600}\sum_{i=1}^{600} T_i^{(k)}$$

Two types of models are taken into consideration: support vector machine (SVM) and random forest (RF); six (6) test types of feature selection are used as follows:

- SVM with a ranking feature selection
- SVM with a nosel feature selection
- RF with ranking feature selection
- RF with nosel feature selection
- RF with self-organizing maps
- RF with PCA.

The results for various learning configurations at different bits are shown in Table 1.

It is interesting to note that the prediction accuracy for the first bit does not exceed 50%; the classification accuracy thereafter increases steadily to reach about 98% for the 7th bit. It can also be observed from the table that the most accurate classification is given by the PCA combined with random forest. Clearly, this classification combination should be preferred over the others when attempting to retrieve cryptographic keys of a DES implementation using side channel analysis.

Table 1 Classification result for various bits

	7th bit	6th bit	5th bit	4th bit	3th bit	2nd bit	1st bit
SVM_{Rank}	94.07	81.09	83.75	71.23	63.78	53.26	50.00
SVM_{Model}	95.14	81.14	78.76	74.80	68.56	54.64	50.00
RF_{Rank}	97.34	82.39	83.37	69.97	63.37	53.12	50.00
RF_{mosel}	98.16	84.80	77.10	72.65	70.56	55.32	50.00
RF_{SOM}	98.06	85.01	86.81	74.09	64.78	54.01	50.00
RF_{PCA}	98.56	83.52	80.92	74.29	62.43	53.87	50.00

7 Side Channel Counter Measures

As pointed out by researchers "Non-traditional endpoints of national networks, such as motor vehicles, medical devices, and home surveillance cameras, advance our society from the connected people of today to the new era of connected things" [15]. The introduction and emergence of IoT in undeveloped countries if not properly thought of in the point of view of security may bring with it cybercriminals whose aim is to attack smart devices to proliferate ransomware. As stated earlier, most devices present in IoT infrastructures are CMOS based and therefore vulnerable to side channel attacks. "Since the publication of side-channel attacks against cryptographic devices, numerous countermeasures have been devised. Countermeasures are available at both the hardware and the software levels but the sought-after goal is the same: to suppress the correlation between the side-channel information and the actual secret value being manipulated. A first class of countermeasures consists in reducing the available side-channel signal. This implies that the implementations should behave regularly. In particular, branching conditioned by secret values should be avoided" [38].

Noise has also emerged recently as a possible candidate for side channel countermeasures. "This includes power smoothing whose purpose is to render power traces smoother. The insertion of wait states (hardware level) or dummy cycles (software level) is another example. This technique is useful to make statistical attacks harder to implement as the signals should first be re-aligned (i.e., the introduced delays should be removed). One can also make use of an unstable clock to desynchronize the signals. At the software level, data masking is another popular countermeasure" [38]. For example, an RSA can equivalently be evaluated as:

$$S = \left[(\mu(m) + r_1 N)^{d + r_2 \phi(N)} mod(r_3 N)\right] mod\, N \tag{17}$$

Many factors exist when building or implementing cryptographic products. Efficiency is one of those factors; and so such implementation needs to be very efficient. "The implementations must also be resistant against side-channel attacks. To protect against the vast majority of known attacks, they must combine countermeasures of both classes described above" [38]. Moreover, experience shows that although the above described countermeasures work well in some cases, experienced attackers can still overcome them and so there is a need to device better and stronger counter measures as described next.

Most modern cryptographic devices are implemented using Complementary Metal-Oxide-Semiconductor (CMOS). During the data transition, electrons flow across the silicon substrate when power is applied to (or removed from) a transistor's gate; this results in power consumption and generation of electromagnetic radiation which can be used as a proper source of leakage that attackers can use to guess secret data stored on cryptographic devices. The transistor is one of the key building blocks of present day cryptographic devices. "The transistor is a semiconductor device used to amplify and switch electronic signals and electrical power" [39]. "The transistor is perhaps the key active component in practically all modern electronics. Many researchers and inventors consider it to be one of the greatest inventions of the 20th century" [40]. One of the main challenges associated with CMOS transistor is the reduction of device dimension. The main concern is to be able to very accurately predict the device performance and how the transistor works and behaves as its size is reduced.

Scaling CMOS technologies has contributed to a contraction in gate length and a comparable reduction of the device gate oxide thickness. For example not long ago the gate length was roughly 9000 nm and the gate oxide thickness was approximately 90 nm. However, by the end of 2011 the gate length had decreased to just about 20 nm while the device oxide thickness was dropped to about 1 nm. The major issue arising from decreasing the gate oxide thickness (ultra-thin SiO_2 gate oxide materials) is that it has heavily contributed to a dramatic increase in the gate leakage current that flows through the gate oxide material by a quantum mechanical tunnelling effect [41, 42]. Various improvement techniques aimed at reducing and suppressing the gate leakage current are currently being investigated. Such techniques include strained channel regions, high-κ gate dielectrics, dual orientation devices and new channel material (e.g. Germanium). While viewed as short term fixes, these CMOS improvement mechanisms are expected to allow the industry to continue moving forwards until the post CMOS era starts, and are therefore extremely important. The use of strained channel regions and high-κ dielectrics to replace the gate oxide have been most heavily investigated in recent years as they appear to be the most beneficial.

7.1 The Gate Biaxial Strained Silicon MOSFETs

Strained Silicon is a technology that pertains entails stretching or compressing the silicon crystal lattice through various techniques, which in turn increases carrier mobility and enhances the performance of the transistors without having to reduce their physical structures [43]. As the benefits associated with transistor device scaling continue to decrease, researchers have recently diverted their interest to using s-Si in CMOS devices [44]. Additionally, s-Si still retains its integrality throughout the manufacturing process contrary to any other semiconductor material. At the molecular level, the transistor channel looks like a crystal lattice. The basic idea behind this is, if silicon atoms can be forcibly pushed apart, then electrons flowing between the structure gates will be less impeded. Less impedance is equal to better flow and better flow translates to faster moving electrons, less power consumed, and less heat generated and thus less leakage. The problem with this is finding an economical way to stretch out these atoms. Crystalline structures have a tendency to line-up with each other when in contact with a similar element or compound. In this case, a SiGe layer is laid down; then a layer of pure silicon is deposited on top.

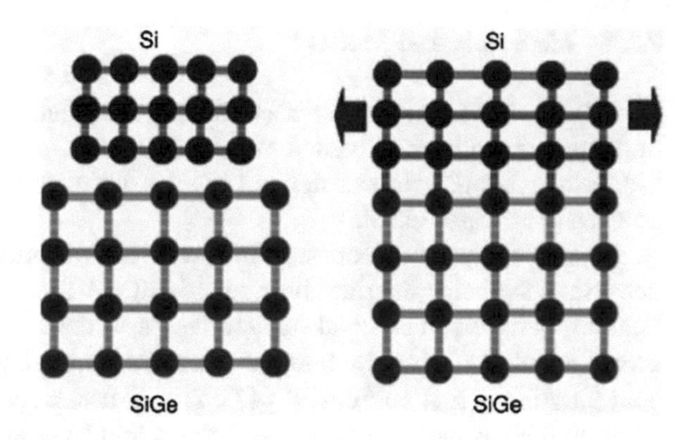

Fig. 3 Strained silicon layer process

The SiGe matrix has been shown to spread out much more than pure silicon only. When the new layer of silicon is deposited on to SiGe, the pure silicon lattice tries to be aligned with SiGe and therefore stretches itself as shown in Fig. 3. This strained layer of silicon becomes the preferred route for electrons to flow through between the metal gates because electrons move faster in strained silicon. It has been proven that electrons flow through strained silicon 68% faster than in conventional silicon, and strained chip designs can be up 33% faster than a standard design resulting in better chip performance and lower energy consumed [45].

The change in the carrier speed depends on the direction of strain as well as the type of channel under consideration. The two most researched types of induced strain are biaxial and uniaxial. "Biaxial strain has been shown to be able to create sheets of uniformly strained material, most commonly in the form of nano-membranes. These created nano-membranes provide flexible and transferable Silicon with increased electron mobility" [45]. Uniaxial strain on the other hand is the most recent alternative used to increase the CMOS carrier mobility as "It can offer benefits not obtained in biaxial strain such as creating a direct band gap in materials of certain orientation or a super-lattice of strain" [46]. The growing importance of size effects requires that efforts be put into understanding the distribution of strain on uniaxial semiconductor as well as its effect on the band structure.

7.2 *Design Simulation of Strained Silicon N-Channel MOSFET*

In this paragraph, the design flow of the structure of strained silicon n-channel MOSFET is discussed in details. Both of the strained silicon and conventional MOSFETs are fabricated virtually in TCAD tools. Process simulation is done in Silvaco's Athena for the virtual fabrication while device simulation is done in ATLAS for characterization of the transistor.

7.2.1 Materials and Method

The 90 nm NMOS transistor was fabricated virtually. The first step consisted in building the grid which had a width of $0.5\,\mu m$ and a depth of $0.15\,\mu m$. For x-direction, a finer grid was defined in the right region whereas for the y-direction, an even grid was defined.

For this design a silicon substrate with crystal orientation is selected. This is because of the better interface between Si/SiO_2. "This interface relates to the atomic bonding between silicon and oxygen atoms in the oxide layer which is thermally grown on silicon substrate. Since the substrate is a p-type, boron was doped with a concentration of $2 \times 10^{18}\,cm^{-3}$" [47]. The third step consisted in adding an Epitaxy layer. Epitaxy is the process of depositing a thin layer of single crystal material over a crystal substrate. The material used in epitaxy layer must be same as the substrate. The reason to grow epitaxy layer over a heavily doped substrate is to minimize the latch-up occurrence in VLSI design. This will allow better controllability of doping concentration and improve the device's performance. The next step consisted in Silicon and silicon Germanium deposition. Depositions was done on a layer by layer basis, i.e. silicon first then Silicon Germanium and lastly strained silicon. The next step was gate oxidation. An oxide layer is deposited to get ready for gate forming. The oxide is diffused onto the surface of the s strained silicon layer at a temperature of 930 °C and pressure of 1 atm. The thickness of oxide is then extracted to obtain an accurate value. "The next step was polysilicon deposition and pattering. Polysilicon gate is used in this simulation, instead of a metal gate. Firstly, polysilicon is deposited on the oxide layer. Then, the polysilicon and oxide are etched to a correct size from the left hand side. The process is followed by the polysilicon oxidation where another oxide layer is deposited on top of the gate by diffusion. Polysilicon is implanted with phosphorous at a concentration of $5 \times 10^{14}\,cm^{-3}$. The next step consisted of spacer oxide deposition and etching. This step is to deposit a further layer of oxide above the polysilicon gate and to etch the spacer oxide layer to provide an optimized thickness. This spacer oxide layer's function is to prevent ions from being implanted into the gate. With this oxide layer, ions will only be implanted into source/drain region. The next step is concerned with source/drain implantation and annealing. The source and drain of NMOS are created by ion implantation process. Phosphorus (N-type ion) is used for implantation at a concentration $1 \times 10^{16}\,cm^{-3}$. Ion implantation is a materials engineering process which consists of accelerating ions of a material in an electrical field through a solid structure" [47]. This process is effectively used to change the physical, chemical, or electrical properties of the solid.

Ion implant has slowly and effectively replaced thermal diffusion for doping a material in wafer fabrication because of its perceived advantages. The greatest advantage of ion implant over diffusion relies in its precise control for depositing dopant atoms into the substrate. As mentioned, every implanted ion goes into collision with several target atoms before it comes to a rest. Such collisions may involve the nucleus of the target atom structure or one of its electrons. The total power required to stop an ion S is the sum of the stopping power of the nucleus and the stopping power of the electron. Stopping power is described as the energy loss of the ion per unit path length of the ion.

It is important to note that "damages caused by atomic collisions during ion implantation change the electrical characteristics of the targeted structure. Many of the targeted atoms are displaced, creating deep electron and hole traps which neutralize mobile carriers and in that process increase resistivity in the structure" [47]. A process known as "annealing is therefore required to repair the lattice damage and put dopant atoms in substitutional sites where they can be electrically active again" [43].

The eight step consisted of Metallization and Contact Windows Patterning as shown in Fig. 4. "The metallization process refers to the metal layers or contacts that electrically interconnect the various device structures fabricated on the silicon substrate. Thin-film aluminium is the most widely used material for metallization, and is said to be the third major ingredient for IC fabrication, with the other two being silicon and silicon dioxide (SiO_2). Aluminium is very suitable for this purpose with its very low resistivity and its adhesion compatibility with SiO_2" [48].

A thin layer of aluminium is then deposited on the surface, and thereafter etched away except the one above source/drain region, to form the device electrodes. The last step in the process was structure reflection. From the first step, only the left hand side of the structure was being fabricated. Since the left hand side is a pure reflection of the right hand side, the structure is reflected to obtain the right hand side structure to complete the fabrication process. Lastly, the device is labelled with electrode name for source, drain, gate and substrate. The final structure is shown in Fig. 5.

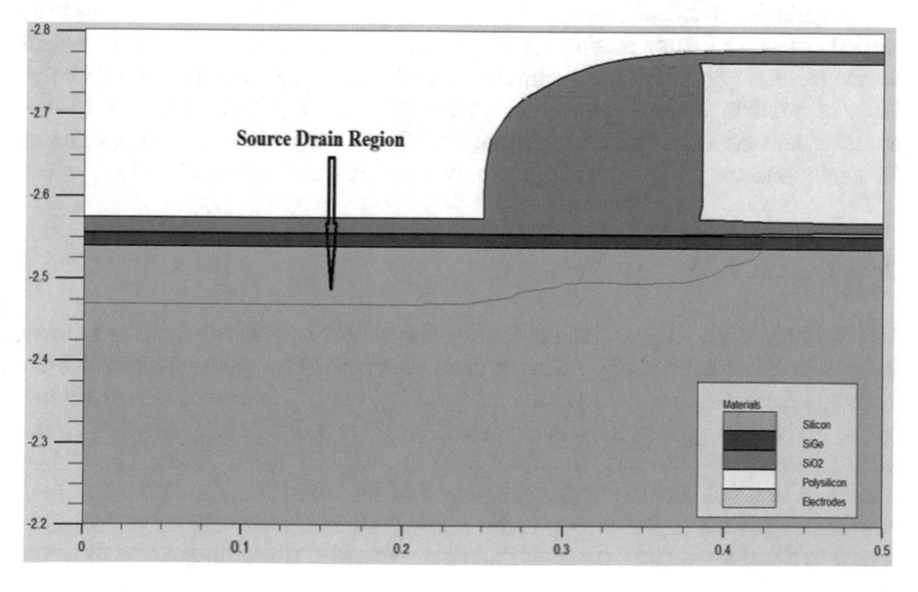

Fig. 4 Metallization

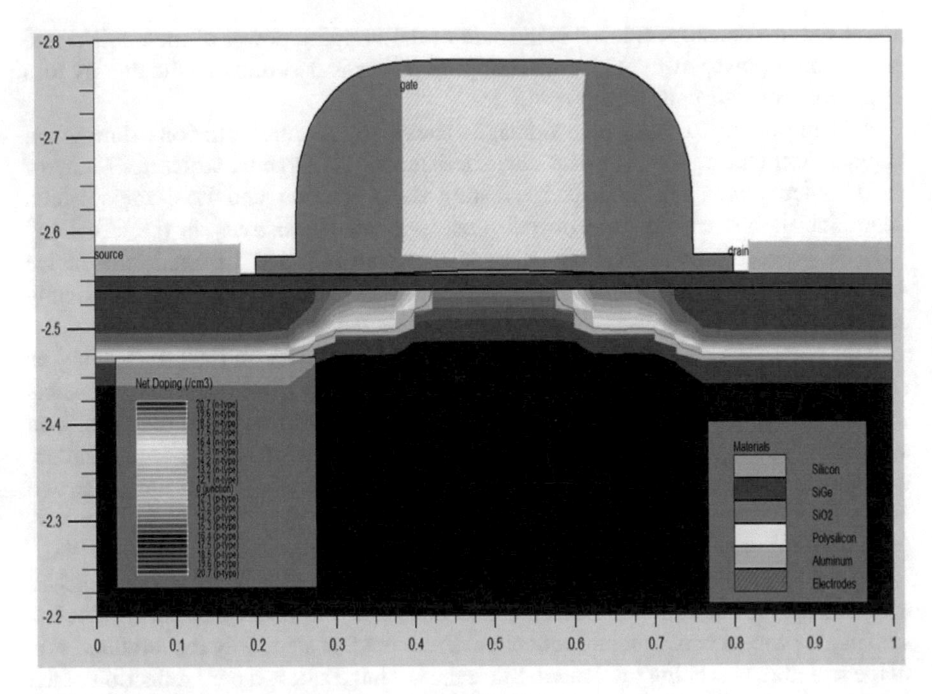

Fig. 5 Complete structure of biaxial strain NMOS silicon

7.2.2 Results and Discussion

To compute the electrical characteristics of the built model device, simulations were carried out on the Atlas module of the Silvaco simulation tool. "The tool enables device technology engineers to simulate the electrical, optical, and thermal behaviour of semiconductor devices. ATLAS provides a physics-based, easy to use, modular, and extensible platform to analyse DC, AC, and time domain responses of all semiconductor based technologies in two and three dimensions" [47].

Drain Current Versus Gate Voltage (I_d Vs. V_{gs})

To plot the I_d versus V_{gs} graph, the drain voltage must be constant in order in order to have a direct current (DC) bias at drain electrode. The gate voltage is slowly increased from zero to a final value in steps. Also, the source electrode is grounded. In this project, the gate voltage is increased from 0 to 3.0 V in steps of +0.1 V. The drain is biased at two critical values namely, 0.1 and 1.0 V. These two values indicate a low voltage and high voltage bias of the transistor. Clearly, both of the NMOS devices are biased with positive value. This is because electrons flow from source to drain terminal to produce drain current which flows in opposite direction

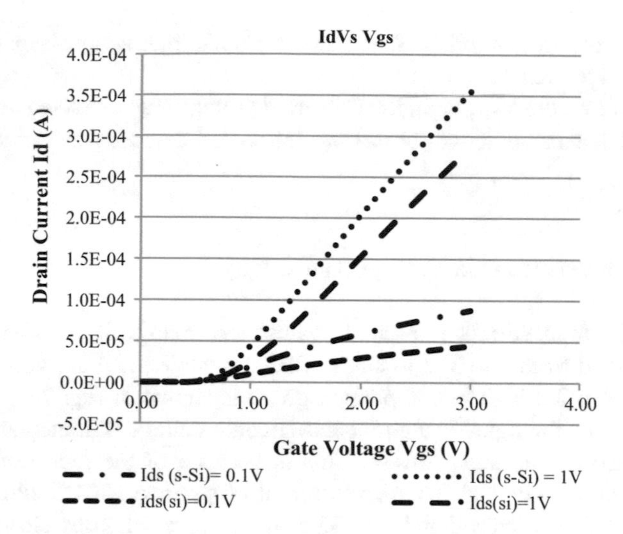

Fig. 6 Drain current versus gate voltage (I_d vs. V_{gs})

Table 2 Threshold voltage at different channel lengths

Channel length	Drain voltage	Threshold voltage	
		Strained silicon NMOS	Conventional NMOS
90	0.1	0.5671	0.6325
	1.0	0.2782	0.4679
150	0.1	0.5738	0.7790
	1.0	0.3516	0.4880
300	0.1	0.6023	0.8122
	1.0	0.4109	0.6787

with the electrons flow. The I_d versus V_{gs} characteristics of the fabricated device are shown in Fig. 6.

From the I_d versus V_{gs} graph in Fig. 6, it can be observed that the drain current of strained silicon NMOS are higher than the conventional one for both 0.1 and 1.0 V drain bias. This clearly means the current flows faster in strained silicon NMOS. Furthermore, electron mobility also is increased as current is directly proportional to mobility. It is important to note that when the drain voltage increases the drain current increases as well. These facts are in support of (1) below for the relationship between current and mobility, as well as drain bias increment, when the transistor operates in linear mode.

$$I_d(\mathrm{lin}) = \frac{\mu W C_{ox}}{L}\left(V_g - V_t - \frac{V_d}{2}\right)V_d \tag{18}$$

where μ is the electron mobility, W is transistor's width, L is transistor's length, C_{ox} is the oxide capacitance.

From the I_d versus V_{gs} graph, the threshold voltage was extracted as can be seen in Table 2. The channel length of the two fabricated devices was varied from 90 to 300 nm.

Drain Current Versus Drain Voltage (I_d Vs. V_{ds})

At this stage, drain current is plotted against the drain voltage. The device gate voltage is varied from 1 to 3 V in steps of 0.1 V; while the drain voltage is slowly varied from 0 to 3.3 V. The comparison graph is shown in Fig. 7.

It is clear from the figure that an increase in gate voltage is immediately followed by an increase in the drain current. This is because of the increased number of electrons along the channel. The drive current of strained NMOS silicon device is higher than that of conventional NMOS and causes a vigorous electron mobility enhancement along the channel. For V_{gs} = 3 V, the percentage of increment of current drive in strained silicon NMOS (compared with normal NMOS at V_{gs} = 3 V) is around 35.7%. This clearly demonstrates an enhancement of electron mobility in the channel.

Clearly, the transistors are operating in their linear region as the current value is not constant. In the linear region of operation, the channel is induced from source to drain since $V_{gs} > V_t$. However, the drain bias must be kept small enough so that the channel is continuous and not pinched off at the drain end.

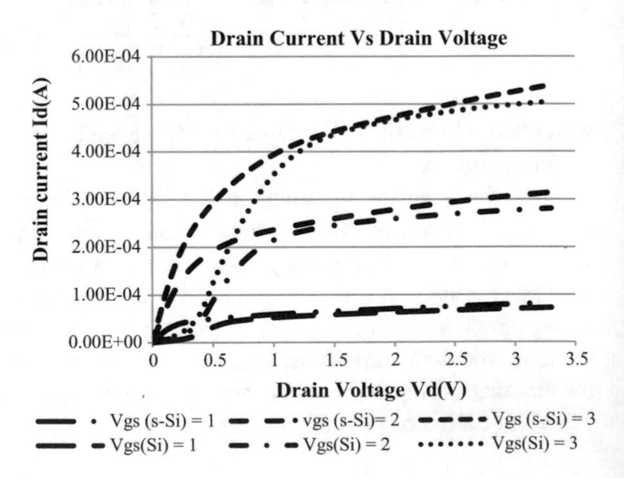

Fig. 7 I_d versus V_{ds} in strain NMOS silicon devices

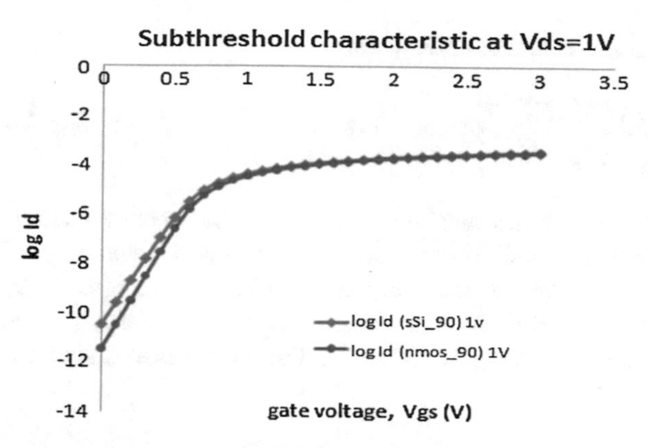

Fig. 8 Subthreshold characteristics at 0.1 V

Table 3 Sub-threshold swing values

Drain biased V_d (V)	Sub-threshold swing in mV/dec	
	Strained silicon NMOS	Conventional NMOS
0.1	112.03	117.1
0.99	111.78	116.2

Sub-threshold Characteristics

Based on the I_d/V_{gs} graph, the device's sub-threshold characteristics can be worked out using the I_d vs. V_{gs} plot as shown in Fig. 8 for a $V_{ds} = 0.1$ V.

The slope of the lines in these graphs is known as the sub-threshold slope. In the other hand, the inverse of the slope referred to above is known as the sub-threshold swing, (S) and is given in units (mV/dec). "The Sub-threshold swing can be interpreted as the voltage required to increase or decrease I_d by one decade. Sub-threshold swing is one of the most critical performance quantities of MOSFET devices" [47]. The computed sub-threshold quantities for this project are shown in Table 3.

The sub-threshold swing values of the strained silicon NMOS are slightly lower compared to conventional NMOS in both $V_{ds} = 0.1$ V and $V_{ds} = 1$ V. Clearly, when V_{ds} increases, the sub-threshold swing decreases accordingly. This indicates that strained silicon NMOS has better transition performance in switching application. The sub-threshold slope of the device can be expressed as:

$$\log I_{ds} = \log\left(\frac{W}{L} I'\right) + \frac{q}{kT}\left(\frac{V_{gs} - V_t}{n}\right) \tag{19}$$

where l′ and n are constant defined as:

$$l' = \frac{\mu\sqrt{2q\varepsilon_s N_A}}{2\sqrt{2\varnothing_F + V_{SB}}}\varnothing^2, n = 1 + \frac{\gamma}{2\sqrt{2\varnothing_F + V_{SB}}} \text{ and slope } \frac{1}{s} = \frac{1}{n}\frac{q}{nkT}$$

Researches have shown that "it is highly desirable to have a sub-threshold swing that is as small as possible and still get large current variations. This is a key parameter that determines the amount of voltage swing necessary to switch a MOSFET ON and OFF. It is especially important for modern MOSFETs with supply voltage approaching 1.0 V" [47]. The sub-threshold swing expression can be given by:

$$S = \frac{kT}{q}\left(1 + \frac{C'_{dep} + C'_{it}}{C'_{ox}}\right)\ln(10) \tag{20}$$

where C'_{dep} defines "the depletion region capacitance per unit area of MOS gate is determined by the doping density in channel region, and C'_{it} is the interface trap capacitance" [49].

It was demonstrated that "lower channel doping densities yield wider depletion region widths and hence smaller C'_{dep}. Another critical parameter is the gate oxide thickness, t_{ox} which determines C'_{ox}. To minimize sub-threshold swing, the thinnest possible of oxide must be used" [49]. The typical value of sub-threshold swing for MOSFET is in the range of 60–100 mV/dec. However, there is a limit for MOSFET to go below 60 mV/dec for sub-threshold swing [35]. Finally, the sub-threshold characteristics for strained silicon for the fabricated device are shown Fig. 9.

It is clear from Fig. 9 that as the channel length is shortened or reduced; it is observed that the sub-threshold swing increases. It is also interesting to observe that as soon as the channel length exceeds the 1 μm mark, the decrease on the sub-threshold swing is less pronounced; in fact the decease rate is slower and almost reaches zero when the channel length gets closer to 2 μm. Clearly, the drain voltage effects tend to become insignificant at such length and the device behaviour is now comparable to that of a long channel.

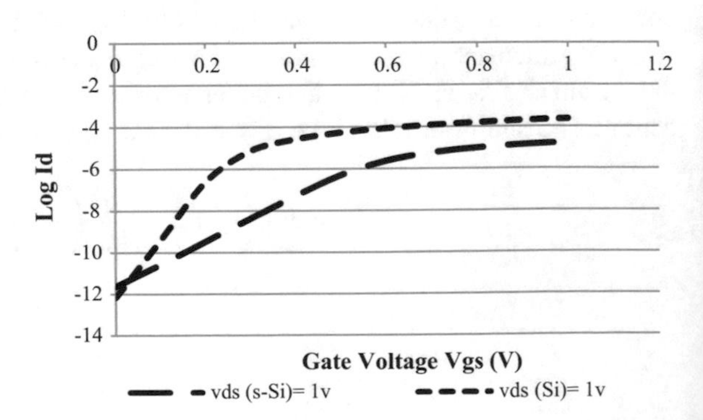

Fig. 9 Sub-threshold characteristic at 1 V

Table 4 Comparison between strained silicon PMOS done by other researches

Characteristics	Strained silicon PMOS (uniaxial)	Strained silicon NMOS (biaxial)
Threshold voltage (Vd = 0.1 V)	−0.596894 V (100 nm) [5] −0.511299 V (71 nm) [42]	0.571733 V
Sub-threshold Swing (mV/dec)	186.153 [41]	112.8
DIBL (mV/V)	693.564 [5]	354
Mobility enhancement at V_{gs} = 3 V (%)	25.65% hole mobility enhancement [5]	35.7% electron mobility enhancement

Drain Induced Barrier Lowering

Drain induced barrier lowering (DIBL) is another highly important factor to be considered in MOSFET design. Obtaining a large ON/OFF rate during simulation is of paramount importance. This factor is without any doubt one of the greatest limitations affecting the short channel MOSFET. Variations observed with the threshold voltage are the result of changes in the drain voltage (ΔV_t), and is computed as an index of the DIBL. Table 4 shows the DIBL value calculated from the previous threshold voltage and drain voltage.

7.3 *High-κ Gate Dielectrics*

CMOS technology was invented at Fairchild Semiconductor in 1963 by Wanlanss Frank. The original idea behind the technology was to design a low power alternative to Transistor–Transistor Logic (TTL). The early adopters of the technology were watch designers who realized the importance of low battery power consumption over the device processing capabilities for electronic circuits. Nowadays, the CMOS technology has grown substantially to become the dominant technology in integrated circuit design as well as IoT. However, off State leakages have been identified by semiconductor manufacturers as a major hindrance for future microcontroller integration. “Off-state leakage is a static power current that leaks through transistors even when such devices are turned off. It is the second major cause of power dissipation in present day microcontrollers. The other source of leakage dynamic power is caused by the repeated capacitance charge and discharge on the output of the hundreds of millions of gates in today’s microcontrollers. Until recently, only dynamic power was identified as a significant source of power consumption, and Moore’s law has helped to control it through shrinking processor technology. Dynamic power is proportional to the square of the supply voltage; therefore reducing the voltage significantly reduces the device’s power consumption” [48].

Unfortunately transistor shrinking has aggravated leakage current in CMOS devices; it has allowed the static component to dominate the power consumption in those devices. “This is essentially because area occupation, operating speed, energy efficiency and manufacturing costs have benefited and continue to benefit from the geometric downsizing that comes with every new generation of semiconductor manufacturing processes. In addition, the simplicity and the low power dissipation of CMOS circuits have allowed for integration densities not possible in similar techniques such as bipolar junction transistors (BJT)” [50]. Despite its many benefits, authors of [39] argue that the power consumption of such devices has increased exponentially with the improvement of device speed and chip density. This has so triggered the rise in the number of attacks on cryptographic devices.

7.3.1 Power Fundamentals

Five mathematical equations govern the power performance in the CMOS logic circuits according to [39]. In this work, they are presented in a way that addresses the basics of physics and logic circuitry design. The first mathematical equations related to CMOS power fundamentals are the basics of low power consumption [12] while the last two equations are more concerned with sub-threshold and gate-oxide leakage modelling in CMOS technologies.

Investigation of Frequency and Voltage Relationships

Equation (21) below depicts the supply voltage dependency of the operating frequency of the device as computed in:

$$f \propto (V - V_{th})^{\alpha/V} \tag{21}$$

In this equation, V represents the transistor’s supply voltage while V_{th} is the device’s voltage. “The exponent α is an experimentally derived constant with a value of 1.3 approximately” [48]. Dynamic voltage scaling in CMOS devices is used to control switching power dissipation in battery operated systems. Also, power consumption minimization techniques rely on low voltage modes and lowered clock frequencies. In [50] authors have used the relation derived in (Eq. 21) to compute an equation that depicts the relationship between frequency and supply voltage. The derivation begins with the selection of the device’s working voltage and frequency defined as V_{norm} and f_{norm} respectively. The quantities selected are normalized entities depicting the relationship between the largest possible device’s operating voltage V_{max} and frequency f_{max}. This relationship is shown in (Eq. 22):

$$V_{norm} = \beta_1 + \beta_2 \cdot f_{norm} = \frac{V_{th}}{V_{max}} + \left(1 - \frac{V_{th}}{V_{max}}\right) \cdot f_{norm} \tag{22}$$

From (Eq. 20) it is evident that if f = 0, then (Eq. 21) becomes:

$$\mathrm{V_{norm}} = \beta_1 = \frac{\mathrm{V_{th}}}{\mathrm{V_{max}}} \tag{23}$$

The value of V_{norm} can safely be approximated to 0.37. That approximation closely matches present day's industrial data [39]. It is also worth mentioning at this stage that f_{max} is proportional to V_{max} and that the frequency will drop to zero if V is equal to V_{th}, as clearly shown in (Eq. 23).

CMOS Power Dissipation

Many origins of power consumption in CMOS devices exist in CMOS devices and that what the third equation attempts to clarify. The CMOS power dissipation can be computed as the sum of dynamic and static power as shown in (Eq. 24):

$$\mathrm{P} = \mathrm{P_{dynamic}} + \mathrm{P_{static}} \tag{24}$$

The first term of (Eq. 24) can be broken off into two distinct entities namely P_{short} and P_{switch}. The first component P_{short} is the power dissipated during gate voltage transient time while the second component P_{switch} comes as a result of the many charging and discharging of capacitances in the device. The last term P_{static} represents the power generated when the transistor is not in the process of switching; (Eq. 24) can be rewritten as:

$$\mathrm{P} = (\mathrm{P_{short}} + \mathrm{P_{switch}}) + \mathrm{P_{static}} = \mathrm{ACV^2f} + \mathrm{VI_{leak}} \tag{25}$$

where: A denotes the number of switching bits and C is the combination of the device's load and internal capacitance.

It is worth mentioning at this stage that for ease of simplicity, the power lost to spasmodic short circuit at the gate's output has been neglected.

From (Eq. 25) it is evident that dropping the supply voltage leads to an important decrease in the device power consumption. Mathematically speaking, dividing the supply voltage by 2 or halving it will cut down power consumption by a factor of 4. "The main drawback of that proposition is that it will shrink the processor's top operating frequency by more than half" [50]. A better approach suggested in [50] relies on the use parallel or pipelined techniques to compensate for the performance losses due to supply voltage reduction.

Computing Leakage Current

Parallelism and pipelining techniques for power reduction were first proposed by Zhang et al. [46]. Since then researchers have conducted studies aimed at optimizing the pipelining depth for dissipated power reduction in CMOS devices. Furthermore, researches have been conducted at a functional block level to compare the performances of pipelining and parallelism to find out which technique performs best when it comes to minimizing total switching power. In (Eq. 25) it was shown that adding the subthreshold current component to the gate-oxide leakage current produces the leakage current i.e.:

$$I_{leak} = I_{sub} + I_{ox} \tag{26}$$

To derive the sub-threshold leakage, authors of [51] presented an equation representing the direct relationship between a CMOS device threshold voltage, its subthreshold leakage current and the device supply voltage as follows:

$$I_{sub} = K_1 W e^{-V_{th}/nV_0\left(1-e^{-V/V_\theta}\right)} \tag{27}$$

In Eq. (27), K_1 and n are normally derived experimentally, W represents the device's gate width, and V_θ is its thermal voltage. The quantity V_θ can safely be approximated to 25 mV at room temperature (20 °C). If I_{sub} rises enough to generate, V_θ will rise as well and in that process cause an increase in I_{sub} and this may result in thermal runaway. From (Eq. 26) it becomes clear that two ways exist for reducing I_{sub} which are:

i. Turning off the supply voltage and
ii. Stepping-up the threshold voltage.

In [50] it is argued that "since this quantity shows up as a negative exponent, increasing that value could have a dramatic effect in even small increments. On the other hand, it is evident from (Eq. 25) that increasing V_{th} automatically creates a reduction in speed. The obvious problem with the first approach is loss of state; as for the second option, its major inconvenience relates to the loss of performance". The device's gate width W, its gate length Lg, the device's oxide thickness Tox, and doping concentration Npocket are other major contributors to subthreshold leakage in CMOS based technologies. Processor designers often optimize one or a few of those leakage components as a convenient technique to reduce subthreshold leakage as will be seen in subsequent paragraphs.

Subthreshold Power Leakage: Gate leakage mechanisms, such as tunnelling across thin gate oxide leading to gate oxide leakage current become significant at the 90 nm node and smaller. The gate oxide leakage is a rather poorly understood quantity as opposed to the subthreshold leakage. For the purpose of this research, a simplification of equations as proposed by [5] produces:

$$I_{ox} = K_2 W \left(\frac{V}{T_{ox}} \right)^2 e^{-\alpha T_{ox}/V} \tag{28}$$

where K_2 and α are derived experimentally. In (Eq. 28) the attention is drawn to the oxide thickness, T_{ox} component of the equation. "Increasing T_{ox} will reduce the gate leakage. However, it also negatively affect the transistor's efficiency since T_{ox} must decrease proportionately with process scaling to avoid short channel effects. Therefore, increasing T_{ox} is not a viable option" [39]. A better approach to this problem consists in the adoption of high-κ dielectric gate insulators. This approach is currently under heavy investigation by the research community.

Reducing Static Power Consumption

Power models to scale down the static power dissipation of embedded devices have been developed by many researchers. Among them, power gating [52, 53] is slowly becoming a very popular design technique for decreasing leakage currents. "Although effective in reducing static power consumption in many instances, its major drawback lies in its tendency to introduce delays by adding extra circuitry and wires and also uses extra area and power" [48].

It should be noted that other techniques for reducing static power consumption exist and are based on the use of threshold voltage techniques as shown in [48]. "Present day's processes typically offer two threshold voltages. Microprocessor designers assign a low threshold voltage to some of the few identified performance-critical transistors and a high threshold voltage to the majority of less time critical transistors. This approach has the tendency to incur a high subthreshold leakage current for the performance-critical transistors, but can significantly reduce the overall leakage" [50]. "Other techniques for decreasing subthreshold leakage are closely related to gate tunnelling current, however, their effects are still under investigations. Gate-oxide leakage has a negligible dependence on temperature" [50]. Therefore, as it subsides with drops in temperature, gate-oxide related current leakage become important.

7.3.2 Fabrication Method

The fabrication method is identical to [48]. The device electrical characteristics and performance are fabricated and simulated using the SILVACO simulator from Athena. "The sample specifications for the experiment is a p-type boron doped silicon substrate at a concentration of $1.5e^{15}$ atoms cm^{-3} and $\langle 100 \rangle$ orientation" [48]. A P-well is then developed by growing a 800 Å oxide screen on top of bulk silicon. The technique utilises dry oxygen at a temperature of about 800 °C which is followed by the development of boron as dopant at a concentration of $3.75e^{13}$ atoms cm^{-3}. The next stage consists in depositing an oxide layer that is later etched and annealed to spread all boron atoms uniformly at a temperature of 900 °C using nitrogen. In the final stage the temperature is increased to 950 °C using dry oxygen. Neighbouring transistors are then isolated by creating a shallow trench isolator with a thicknesses of 130 Å. The wafer is then oxidized using dry oxygen for 25 min at 1000 °C as in [25, 54–56]. "It shall be noted that the deposition of high-κ dielectric process with gate oxide thickness is selected so that they have the same equivalent oxide thickness as SiO_2. Furthermore, the length of the high-κ material was scaled so as to get the equivalent 22 nm gate length of the transistor" [48]. The simulation of the model was done in the Atlas subcomponent of the SIVACO simulator.

7.3.3 Results and Discussions

The complete NMOS structure is shown in Fig. 10. The fabrication process is the same for all high-κ devices fabricated for this research except that the dielectric materials were varied.

In Fig. 11, we show a doping profile fabricated using an NMOS device with a gate length of 22 nm. The electrical characteristics of that device are depicted in Fig. 12.

The plots are also known as "V_t Curves", because devices designers use them extensively to extract the threshold voltage (V_t), which defines an approximation of when a transistor is "on" and allows current to flow across the channel. For this research, Fig. 12 presents the plot of I_d which is the drain current versus V_{GS} the gate voltage for Si_3N_4 (k ~ 29), HfO_2 (k ~ 21) and a conventional device made of SiO_2. Typically, the fabricated device's drain voltage V_{DS} was fixed when (I_d)

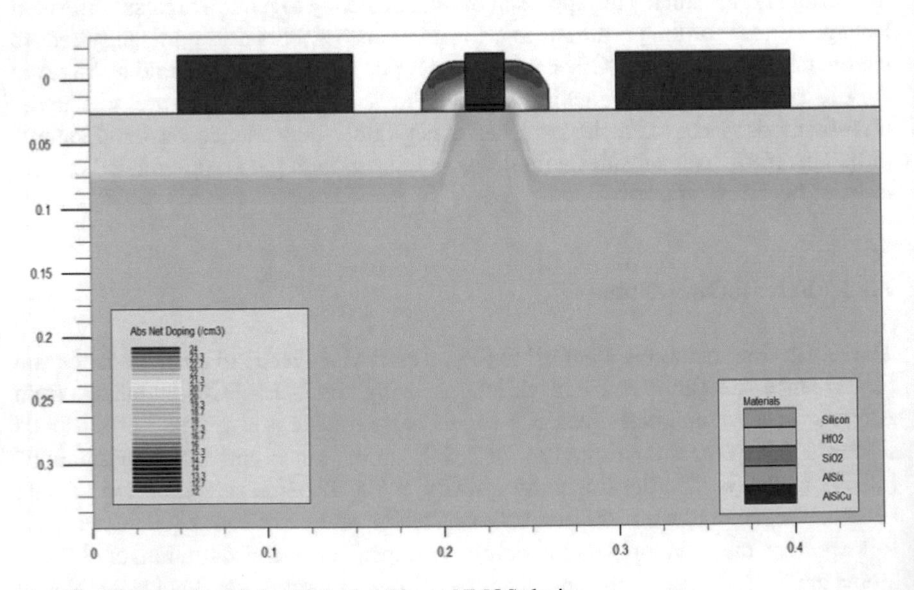

Fig. 10 Cross section of the fabricated 22 nm NMOS device

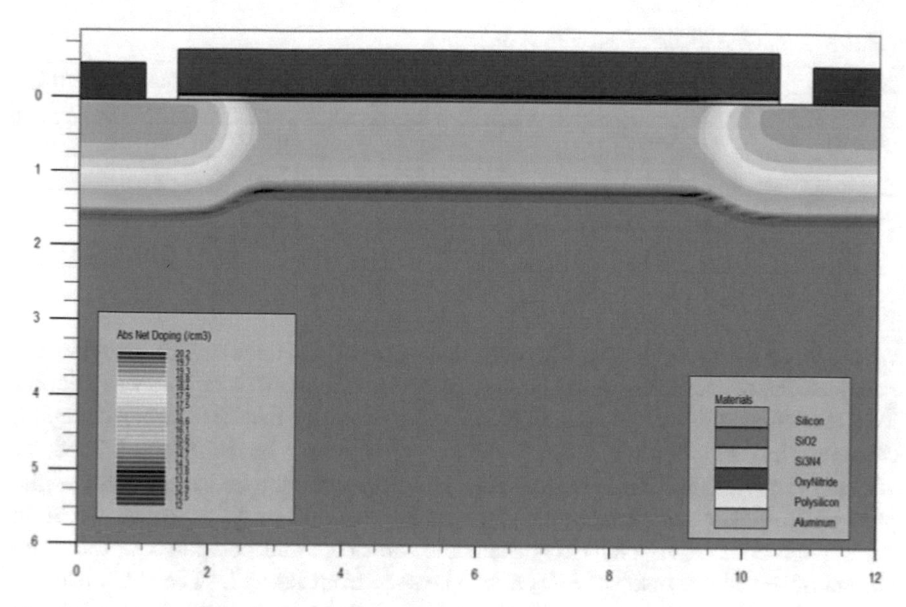

Fig. 11 Doping profile of the fabricated device

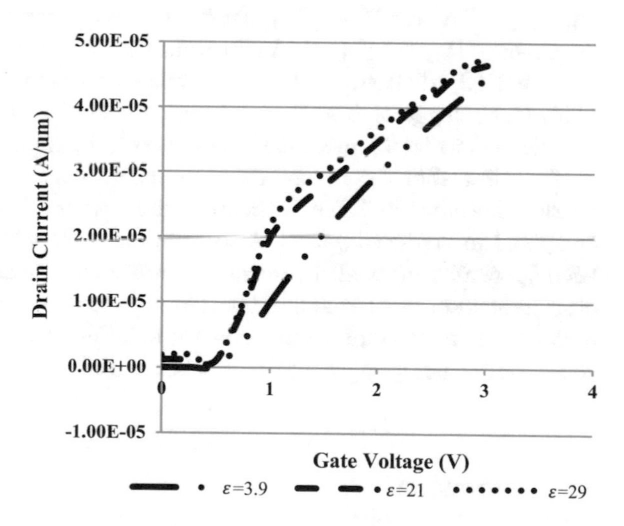

Fig. 12 Drain current versus gate voltage curves (threshold curves)

versus (V_{GS}) was plotted. “The threshold voltage (V_{th}), state on current (I_{on}) and state off current (I_{off}) can be extracted from the (I_d) versus (V_{GS}) curve” [48].

A good doping concentration is crucial in ensuring that the transistor works well and emits fewer leakage currents so as to improve gate control [57]. “Four distinct factors heavily affect the threshold voltage. They are: (1) threshold voltage adjustment implant, (2) halo implant, (3) channel implant and (4) compensation implant” [48]. However for the purpose of this work where the main requirement

Table 5 Simulated results of the utilised dielectrics

Studied parameters	Dielectric 1 (Si_3N_4)	Dielectric 2 (HfO_2)	Prediction from ITRS (2011)
V_{th} (V)	0.30226	0.30226	0.302
I_{on}	3.03345^{-4}	2.63345^{-4}	1.03^{-7}
I_{off}	2.83345^{-15}	1.03345^{-15}	1.495^{-6}
$\frac{I_{on}}{I_{off}}$	1.071^{11}	2.541^{11}	6.6071^{-2}

was to investigate the leakage current emissions of the device while varying the gate material used, the threshold voltage adjustment implant technique was utilised. To get a threshold voltage of 0.302651 as stipulated by the ITRS the best doping concentration with boron was set to $8.5763 \times 10^{13}\,cm^{-2}$ for HfO_2 and $9.73654 \times 10^{13}\,cm^{-2}$ for Si_3N_4. This doping concentration variation is to account for the physically thicker and stronger dielectric materials utilised. As expected, the drain current for both Si_3N_4 and HfO_2 dielectric are decreased compared to the drain current of the device made of SiO_2 as can be seen from Fig. 12. "The Drain leakage current (I_{off}) or sub-threshold leakage current occurs when the gate voltage (V_{GS}) is lower than the threshold voltage (V_{th}). In ideal case, when the transistor is turned off, $V_{GS} = 0$ V and $V_{DS} = V_{DD}$ (voltage supply), there is no current flow through the channel ($I_{off} = 0$)" [48]. Again from Fig. 3 it is clear that the leakage current through HfO_2 dielectric is lowest compared to both Si_3N_4 and SiO_2 dielectrics. "This result suggests that HfO_2 dielectric material is more compatible with silicon and appears to be the most stable oxide with the highest heat of formation" [55].

The simulation results of the various dielectric materials simulated in this research is shown in Table 5. Clearly, the I_{on} values from the simulation are larger compared to predicted values. Also from the same table it can be seen that simulated I_{off} results are smaller compared to predicted values from ITRS-2011. This is a clear indication that the above selected high-κ dielectric materials combined with the metal gate are better alternatives for building low leakage transistors. Again as stated earlier HfO_2 appears to be the better option.

8 Conclusion

As a global infrastructure with the aim of enabling objects to communicate with each other, the Internet of Things (IoT) is being widely used and applied to many critical applications. This openness however exposes devices participating in IoT implementations to all sorts of attacks including side channel analysis attacks perpetrates by groups whose main goal is to "steal" users confidential information or the cryptographic keys used to protect them. Side Channels particularly are of concern because those attacks can be set up very fast only using cheap and readily available off-the-shelf equipment and produce devastating results; especially in undeveloped and underdeveloped countries. The amount of time needed by a skilful attacker to succeed in breaching the security of a microcontroller or smart card is in the range of a few seconds. Side-channel analysis is undoubtedly one of the most powerful tools when targeting cipher implementations. In this chapter, we have given a small insight into further applications of side channel analysis with focus on IoT environments. As an example, we have picked a subject that has not been thoroughly discussed so far in terms of embedded μCs: breaching the security of a secure FPGA device. Because of their versatility and power FPGA are very good candidates for IoT implementations. We have shown that it is possible to reveal the encryption key in such a setup only by observing and analysing the side-channel leakages that emerge during the execution of a code. We have then gone a step ahead by designing and simulating countermeasures to such attacks. It was demonstrated that design techniques such as high-k dielectrics or strain silicon based design are highly effective in reducing leakage currents observed with CMOS devices and are therefore the best known counter measures to such threats.

References

1. Biggs, P., Garrity, J., Lasalle, C., & Polomska, A. (2015). Harnessing the internet of things for global development: ITU/UNESCO broadband commission for sustainable development.
2. Houlin, Z. (2016). Harnessing the Internet of things for global development. *White paper*. Available from http://theinternetofthings.report/view-resource.aspx?id=2574
3. Subrata, N., et al. (2014). Cellular automata based encrypted ECG-hash code generation: An application in inter-human biometric authentication system. *International Journal of Computer Network and Information Security*.
4. Shubhendu, B., et al. (2015). High Payload watermarking using residue number system. *International Journal of Computer Network and Information Security*.
5. Wright, D. (1987). *Spy catcher*. Viking Penguin Inc.
6. Nilanjan, D., et al. (2017). *Watermarking in biomedical signal processing: Intelligent techniques in signal processing for multimedia security*. New York: Springer.
7. Chakraborty, S., et al. (2017). Comparative approach between singular value decomposition and randomized singular value decomposition-based watermarking. In *Intelligent techniques in signal processing for multimedia security*. New York: Springer.
8. Dharavath, K., et al. (2017). Face recognition under dry and wet face conditions. In *Intelligent techniques in signal processing for multimedia security*. New York, NY: Springer.
9. Surekha, B., et al. (2017). Attendance recording system using partial face recognition algorithm. In *Intelligent techniques in signal processing for multimedia security*. New York, NY: Springer.
10. Rajeswari, P. (2017). Multi-fingerprint unimodal-based biometric authentication supporting cloud computing. In *Intelligent techniques in signal processing for multimedia security*. New York, NY: Springer.
11. Anderson, M. (2016). Vulnerable smart devices make an internet of insecure things: IEEE spectrum. http://spectrum.ieee.org/riskfactor/computing/networks/vulnerable-smart-devices-make-an-internet-of-insecure-things
12. Coron, J., & Goubin, L. (2009). On Boolean and arithmetic masking against differential power analysis. In *Cetin Kaya Koc and Paar* (pp. 231–237).

13. Clavier, C., Isorez, Q., & Wurcker, A. (2013), Complete SCARE of AES-like block ciphers by chosen plaintext collision power analysis: In G. Paul & S. Vaudenay (Eds.), *INDOCRYPT* (Vol. 8250 of Lecture Notes in Computer Science, pp. 116–135). Berlin: Springer.
14. Byron, A. (2017) Securing the internet of things: Side channel attacks expose sensitive data collected by IoT devices. http://thirdcertainty.com/featured-story/securing-the-internet-of-things-side-channel-attacks-expose-sensitive-data-collected-by-iot-devices/. Accessed January 12, 2017.
15. Crossman, M. A., & Hong, L. (2015). Study of authentication with IoT testbed. In *IEEE International Symposium on Technologies for Homeland Security (HST).*
16. Mangard, S., Oswald, E., & Popp, T. (2007). *Power analysis attack—Revealing the secret of smart cards*. Heidelberg: Springer.
17. Van Eck, W. (1985). Electromagnetic radiation from video display units: An eavesdropping risk. *Computers and Security, 4,* 269–286.
18. Richard, J. L., & Morris, L. M. (2005). *An introduction to mathematical statistics and its applications* (4th ed.). Boston: Prentice Hall.
19. Kocher, P. (1996). Timing attacks on implementations of Diffie-Hellmann, RSA, DSS and other systems. In *CRYPTO'96, LNCS 1109* (pp. 104–113).
20. Daehyun, S. (2014). *Novel application for side-channel analyses of embedded microcontrollers*. PhD thesis, Ruhr-Universitat Bochum, Germany.
21. Quisquater, J. J., & Samyde, D. (2001). Electromagnetic analysis (EMA): Measures and countermeasures for smart cards. *E-smart: LNCS 2140* (pp. 200–210).
22. Agrawal, D., Archambeault, B., Rao, J. R., & Rohatgi, P. (2002). The EM side-channel(s): Attacks and assessment methodologies: In B. S. Kaliski Jr., Ç. K. Koç, & C. Paar, (Eds.), *Proceedings of the 4th International Workshop on Cryptographic Hardware and Embedded Systems (CHES)* (Vol. 2523 of LNCS, pp. 29–45). Berlin: Springer.
23. Agrawal, D., Archambeault, B., Chari, S., Rao, J. R., & Rohatgi, P. (2003). Advances in side-channel cryptanalysis. *RSA Laboratories Cryptobytes, 6*(1), 20–32.
24. Goldack, M. (2008). *Side-channel based reverse engineering for microcontrollers*. Bochum: Ruhr-University.
25. Gene, H. G., & Charles, F. L. (1996). *Matrix computations* (3rd ed.). Baltimore: The Johns Hopkins University Press.
26. Aviv, A. J., et.al. (2012). Practicality of accelerometer side channels on smartphones. In *Proc. of 28th ACM ACSAC.*
27. Rouf, I., et.al. (2010). Security and privacy vulnerabilities of in-car wireless networks: A tire pressure monitoring system case study. In *Proc. of the USENIX Security Symposium* (pp. 323–338).
28. Foo, K. D. (2013). Ghost talk: Mitigating EMI signal injection attacks against analog sensors. In *Proceedings of the IEEE Symposium on Security and Privacy.*
29. Backes, M., Dürmuth, M., Gerling, S., Pinkal, M., & Sporleder, C. (2010). Acoustic side-channel attacks on printers. In *Proceedings of the 19th USENIX Security Symposium.* Washington, DC, USA.
30. Rivest, R. L. (1993). *Cryptography and machine learning*. Cambridge: Laboratory for Computer Science, Massachusetts Institute of Technology.
31. Gabriel, H., et al. (2011). Machine learning in side-channel analysis: A first study. *Journal of Cryptographic Engineering, 1*(4), 293–302.
32. Hera, H., Josh, J., & Long, Z. (2012). *Side channel cryptanalysis using machine learning using an SVM to recover DES keys from a smart card*. Stanford University.
33. Lerman, L., Bontempi, G., & Markowitch, O. (2011). Side channel attack: An approach based on machine learning. In *COSADE, Second International Workshop on Constructive Side-Channel Analysis and Secure Design, 2011.*
34. Hastie, T., Tibshirani, R., & Friedman, J. (2009). *The elements of statistical learning: Data mining, inference, and prediction* (2nd ed.). New York: Springer.
35. Rivest, R. L. (1993). *Cryptography and machine learning: Laboratory for computer science*. Cambridge: Massachusetts Institute of Technology.

36. Colin, O., & Zhizhang, D. (2014). *Chip whisperer an open-source platform for hardware embedded security research*. Halifax: Dalhousie University.
37. Dough, S. (2002). Triple DES and encrypting PIN pad technology on triton ATMs: Triton systems of Delaware, Inc. ATMdepot.
38. Marc J (2009) *Basics of side-channel analysis: Cryptographic engineering.*
39. Eason, G., Noble, B., & Sneddon, I. N. (1955). On certain integrals of Lipschitz-Hankel type involving products of Bessel functions. *Philosophical Transactions of the Royal Society of London, A247,* 529–551.
40. Price, W. R. (2004). *Roadmap to entrepreneurial success: AMACOM div* (p. 42). American Management Assocation. ISBN 978-0-8144-7190-6.
41. Depas, M., Vermeire, B., Mertens, P. W., Van Meirhaeghe, R. L., & Heyns, M. M. (2012). Determination of tunnelling parameters in ultra-thin oxide layer poly-Si/SiO_2/Si structures. *Solid-State Electronics, 38,* 1465.
42. Lo, S. H., Buchanan, D. A., Taur, Y., & Wang, W. (2009). Quantum-mechanical modelling of electron tunnelling current from the inversion layer of ultra-thin-oxide nMOSFET's. *IEEE Electron Device Letters, 18,* 209.
43. Wong, Y. J., Saad, I., & Ismail, R. (2006). Characterisation of strained silicon MOSFET using semiconductor TCAD tools. In *ICSE2006 Proc*, Kuala Lumpur.
44. Iwai, H., & Ohmi, S. (2002). Silicon integrated circuit technology from past to future. *Microelectronics Reliability, 42,* 465–491.
45. Acosta, T., & Sood, S. (2006). Engineering strained silicon-looking back and into the future. *IEEE Potentials, 25*(4), 31–34.
46. Zhang, F., Crispi, V. H., & Zhang, P. (2009). Prediction that uniaxial tension along $\langle 111 \rangle$ produces a direct band gap in germanium. *Physical Review Letters, 102*(15), 156401.
47. Ngei, L. O. (2010). *Design and characterization of biaxial strained silicon N-Mosfet*. Master Thesis, Faculty of Electrical Engineering Universiti Teknologi Malaysia.
48. Djonon Tsague, H., & Twala, B. (2015). First principle leakage current reduction technique for CMOS devices. In *IEEE International Conference on Computing, Communication and Security (ICCCS), Mauritius.*
49. Misra, V. (2005), Field effect transistors: The electrical engineering handbook.
50. Chattererjee, S., Kuo, Y., Lu, J., Tewg, J., & Majhi, P. (2012). Electrical reliability aspects of HfO_2 high-K gate dielectric with TaN metal gate electrodes under constant voltage stress. *Microelectronics Reliability, 46,* 69–76.
51. Ganymede. (2015). *Complementary Metal Oxide Semiconductor (CMOS)* (online). Accessed on October 20, 2015.
52. Shin, Y., Seomun, J., Choi, K. M., & Sakurai, T. (2010). Power gating: Circuits, design methodologies, and best practice for standard-cell VLSI designs. *ACM Transactions on Design Automation of Electronic Systems, 15*(4), 28:1–28:37.
53. Greer, J., Korkin, A., & Lebanowsky, J. (2003). *Nano and Giga challenges in microelectronics: Molecular and nano electronics: Analysis, design and simulation* (1st ed).
54. Elgomati, H. A., Majlis, B. Y., Ahmad, I., Salahuddin, F., Hamid, F. A., Zaharim, A., et al. (2011). Investigation of the effect for 32 nm PMOS transistor and optimizing using Taguchi method. *Asian Journal of Applied Science.*
55. Chen, Y., et al. (2014). Using simulation to characterize high-performance 65 nm node planar. In *International Symposium on Nano-Science and Technology, Taiwan.*
56. Wong, H., & Iwai, H. (2013). On the scaling issues and high-k replacement of ultrathin gate dielectric for nanoscale MOS transistor. *Microelectronic Engineering, 83*(10), 1867–1904.
57. He, G., & Sun, Z. (2012). *High-k dielectrics for CMOS technologies*. New York: Wiley.

Framework of Temporal Data Stream Mining by Using Incrementally Optimized Very Fast Decision Forest

Simon Fong, Wei Song, Raymond Wong, Chintan Bhatt and Dmitry Korzun

Abstract Incrementally Optimized Very Fast Decision Tree (iOVFDT) is a new data stream mining model that optimizes a balance of compact tree size and prediction accuracy. The iOVFDT was developed into open source on Massive Online Analysis as a prior art. In this book chapter, we review related techniques and extend iOVFDT into iOVFDF ('F' for forest of Trees) for temporal data stream mining. A framework for follow-up research is reported in this article. A major issue to the current temporal data mining algorithms is due to the inherent limitation of batch learning. But in real-life, the hidden concepts of data streams may change rapidly, and the data may amount to infinity. In the big Data era, incremental learning is attractive since it does not require processing the full volume of dataset. Under this framework we propose to research and develop a new breed of temporal data stream algorithms—iOVFDF. We integrate for a "meta-classifier" called iOVFD Forest over a collection of iOVFDT classifiers. The new iOVFD Forest can

S. Fong (✉)
Department of Computer and Information Science, University of Macau, Taipa, Macau SAR, China
e-mail: ccfong@umac.mo

W. Song
School of Computer Science and Technology, North China University of Technology, Beijing, China
e-mail: sw@ncut.edu.cn

R. Wong
School of Computer Science and Engineering, University of New South Wales, Sydney, NSW 2052, Australia
e-mail: wong@cse.unsw.edu.au

C. Bhatt
Patel Institute of Technology, Charotar University of Science and Technology (CHARUSAT), Changa 388421, Gujarat, India
e-mail: chintanbhatt.ce@charusat.ac.in

D. Korzun
Department of Computer Science, Faculty of Mathematics, Petrozavodsk State University, Lenin St., 33, Petrozavodsk, Republic of Karelia, 185910, Russian Federation
e-mail: dkorzun@cs.karelia.ru

N. Dey et al. (eds.), *Internet of Things and Big Data Analytics Toward Next-Generation Intelligence*, Studies in Big Data 30,
DOI 10.1007/978-3-319-60435-0_19

incrementally learn temporal associations across multiple time-series in real-time, while each underlying individual iOVFDTree learns and recognizes sub-sequence patterns dynamically.

Keywords Data stream mining · Decision trees · Meta-classifiers · Big data

1 Introduction

Data stream mining is a relatively new branch of machine learning. It works about analyzing, inferring and building a prediction or classification model over data stream(s) that feed in continuously as time-series. There are many possible application scenarios ranging from data mining over bio-signals, sensor data feeds, IoT sequences, social tweets and live trajectories, etc., just to name a few.

In the past, traditional temporal data mining algorithms have existed, but they all belong to classical batch-mode machine learning, that needs all the data to be loaded and processed for model refresh. In this review study, we focus on incremental learning, which makes real-time temporal data stream mining possible for the new breed of applications that fed on continuous data streams in nature. To the best of our knowledge, data stream mining on temporal pattern is a relatively new and popular research domain, because of the prevalent ubiquitous computing, data collection and big Data applications. The objective of this article is to narrate a framework under which relevant techniques are reviewed, possibilities of combining several data stream mining models together, known as Incrementally Optimized Very Fast Decision Forest (iOVFDF) is pondered on. The significance of the new model iOVFDForest is to provide possibilities of stream mining temporal patterns in real-time.

The framework of iOVFDF is based on the direction of the work [1]. In [1] the major direction is to analyze and make sense of data streams in an example of physiological patient's vital feeds, in the name of temporal data stream mining. Their proposed method is mainly for detecting anomalies and association of irregular events on a singular data stream. The core learner is k-NN which is quick but lack of comprehensive knowledge representation (unlike rules from decision trees). Our proposed model, based on iOVFDT which has proven its efficiency (high accuracy, compact tree size, and decision rules availability), is to work on similar scenario, multiple temporal data stream mining. A two-level forest of iOVFDT called iOVFDF is proposed that is comprised of individual learners at the low level, doing pattern recognition over on-going individual data stream; at the high level, a meta-classifier learns the full picture based on the information learnt from the individual iOVFDT's. The main concept of the iOVFDF hence is to tap on the collective power of multiple iOVFDT's for understanding the macro meaning of a complex problem.

As an analogy, it is like training an expert in playing jigsaw puzzles. Two cognition levels exist in our brain—one is to coordinate with hand and eyes to recognize the shape and colour of each individual piece, see what it is, where it does fit (border piece or centre piece, sky or earth etc.). On a higher level, meta-knowledge is constructed from knowledge of each individual piece that has been examined so far, and the meta-knowledge learner tries to make sense from the full picture. Individually each piece of knowledge learnt from iOVFDT often may be incomplete. In the example given in Fig. 1, the micro view (a) of an iOVFDT predicts about there may be a camomile or two on a hazel color background. However, only when the outputs of all the individual iOVFDTs piece together, one can see a macro view (b) of the full picture—that is a rabbit sitting on meadow.

Now imagine the puzzle picture is not static, but dynamic. The content of each one of these pieces may change in time, just like temporal data streams. E.g. tweets, online opinions, traffic conditions, battlefield information etc. And because the data streams are temporal, the concepts may drift, and the patterns of movement such as interval, frequency, regularity, etc. (as well as sub-patterns) can change over time. This adds another dimension of longitudinal analysis which has been relatively less

(a) Macro view from iOVFDF

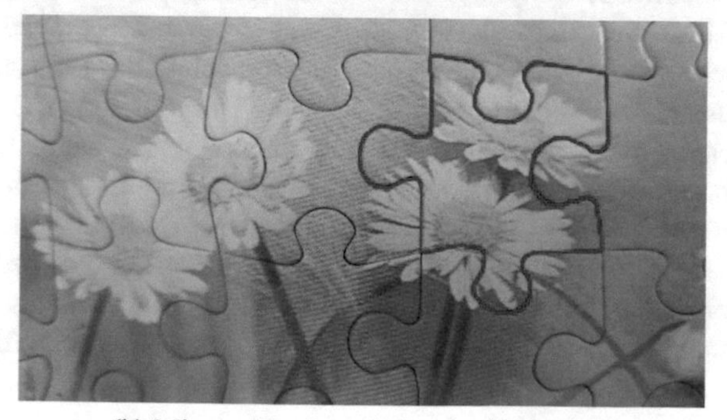

(b) Micro view from individual iOVFDT

Fig. 1 Different views between iOVFDF and iOVFDT

explored in the literature. Given the longitudinal study, knowledge of both long-term and dynamic instances, future predictions and causality in relations could be made.

In this above puzzle analogy, the individual piece is handled by individual iOVFDT (data stream learning tree). iOVFDT was designed to be adaptive in nature, being able to deal with concept drifts which are very common nowadays in dynamic situations hence dynamic data (not only they vary in shape in the temporal domain, but also vary on what the concepts they represent). The meta classifier is the central brain that receives information (knowledge, in the form of decision rules) from each individual iOVFDT. In ecology, a forest is also considered as a united entity, when some trees are sick or on draught, the whole forest is affected. The efficacy of a forest learner depends on the performance of each individual decision tree. The meta classifier is called iOVFDF (forest) which learns on the full picture from a macro view, from individual trees.

Back to the paper [1], physiological data streams are not individual, they are related and correlated. Individually the current method by IBM can detect and monitor irregular heartbeats, for example, on an ECG pattern. But in ICU, often several vital feeds are measured and put together in consideration, as they are known to have cross-effects, e.g. oxygen level in blood and respiration rate. The same goes on for dynamic stock market data where multiple factors are influencing one another, e.g. property market price and bank interest rate.

The current research on temporal data stream mining lacks of lightweight learner with good balance of accuracy, speed and size, iOVFDT was designed for that, as a fundamental component. This is exactly why this research framework needs to be founded for combining the powers of incremental decision trees, in the hope of rolling up the research momentum for developing iOVFDF. Two previous publications [2] and [3] show that iOVFDT is a reality, not only theoretical, but has already been implemented as an open source Java tool, available for download at here: http://moa.cms.waikato.ac.nz/moa-extensions (under data stream mining extension tool, classification and regression). The following two journal papers [4] and [5] have shown that applying iOVFDT for mining physiological data streams are possible. The experiments show that iOVFDT outperformed the current algorithms. The methods are however limited to a single classifier for a single temporal data stream, the same limitation is found from [1] by IBM too. The new forest design is meant to build on top of incremental associate rule learning, a meta-classifier that receive knowledge input from association rules from (ARM) [6] and from multiple iOVFDT's as well. The knowledge will be time-stamped and source identified, as well as proximity and sources ontology mapped. In the latter section, it would be shown clearly why and how in mining temporal data patterns, we need to cross-mine them in order to better make sense of the whole situation.

2 Motivation

Data stream mining on temporal pattern is a relatively new but grounded on a popular research domain, because of the prevalent ubiquitous computing, data collection and big data explosion and data-intensive applications, advances in data collections in bioinformatics large sequences, genome datasets etc., waiting for efficient classifiers to extract insights out of them.

iOVFDT is already proven good and effective as observed from an extensive record of publications, this newly proposed iOVFDF is supposed to make another impact in the research community, when it is tailored to do temporal data stream mining. In addition to building iOVFDF, there exists possibilities to demonstrate its efficacy by trying on the fMRI, EEG, Disease Pathway microarray datasets. Some major and significant discovery may be made in life science through the tools developed by the design of iOVFDF.

As a matter of fact, when one searches on Google Scholar with keywords of "temporal data mining", more than 1 million results are retrieved. But on the first page of results, most of them dated back to a decade ago, implying there has been some sort of stagnancy in research momentum waiting for an innovative breakthrough.

http://scholar.google.com.hk/scholar?q=temporal+data+mining

For strengthening up the framework of iOVFDF, a theoretical ground-breaking method, based on data stream classifiers should be in place. Quoted from a recent and highly cited survey paper on temporal data mining [7], titled "Temporal Data Mining: An Overview". Page 1, by Antunes and Oliveria, from IST/INESC, Lisboa, Portugal:

> One of the main unresolved problems that arise during the data mining process is treating data that contains temporal information. In this case, a complete understanding of the entire phenomenon requires that the data should be viewed as a sequence of events. Temporal sequences appear in a vast range of domains, from engineering to medicine and finance, and the ability to model and extract information from them is crucial for the advance of the information society.

The statements are credible as the survey paper has been cited more than 302 times. So this is what the new generation of data analytics tool we need—data stream mining, for today's and tomorrow's digital societies where our lives will be engulfed by ubiquitous devices and information. big data and deep learning, will need to be mined by new and light-weighted tools.

In order to reach this research objective, other researches mainly based on developing their tools to do pattern recognition and detection, mostly by Association Rule Mining methods. Classification methods are important but relatively rare, as quoted from the same survey paper, page 10, it commented the current shortage of classification methods for Temporal Data Mining.

Classification is one of the most typical operations in supervised learning, but it hasn't deserved much attention in temporal data mining. In fact, a comprehensive search of applications based on classification has returned relatively few instances of actual uses of temporal information.

While it is true that most of the research focus shifted towards association rules and sequential pattern discovery, however, classification has relatively been neglected for temporal data mining because of its traditional supervised learning that relies on batch-mode data induction, and the presumably structured data format with attributes and instances rigidly formatted as a fixed data matrix. The differences between the batch-mode traditional classifiers and the stream-based classifiers have been studied and reported in [8].

Stream-based classifiers will have enormous application potentials, as it has learning, predicting and discovery abilities. Data streams contain very rich information, ranging from temporal information, events information and regularity in the time domain. There are potentially niche rooms in inventing and extending related algorithms like classification-by-association, by fundamentally revamping the underlying greedy-search learning structure to incremental learning. Again, iOVFDT is a precedent success example, but it is limited by only learning a single thread of data stream. Another layer of meta-learner that should be designed with a compatible incremental learning structure is needed to piece the information from different sub-sequences together. It would be designed to predict evolving time-series at the level of sub-sequences and meta-level of multiple series. The result of this research has a good potential for developing a useful stream mining tool, suitable for a wide range of modern applications.

With the emergence of IoT technology, the need of temporal data stream mining is essentially increased. In particular, eHealth applications adopt continuous monitoring of the critical health parameters of the patients based on wireless medical devices (wearable or implantable). Multiple online data flows are simultaneously going through the patient's personal mobile device (e.g., smartphone), which acts as a preliminary processor. Certain crucial healthcare decisions can be made based directly on the analysis of these physiological data [9]. Let us mention the following examples of such real-time analysis problems.

1. Analysis of the individual behavior of a real-time monitored data flow. A change of some characteristics of the data flow can indicate a qualitative change in the patient's health status.
2. Analysis of joint behavior of multiple real-time monitored data flows. A change in correlation characteristics of these data flows provides higher-level semantic sign on changes in the patient's health status.
3. In addition to the monitored physiological data, there can be data flows coming from non-medical sensors (e.g., surrounding light level or temperature). The sensors are represented as body-area devices or surrounding devices of the physical environment or even nearby people. Joint analysis of such data flows can provide new insights on the recent patient's health status in the recent situation.

In general, this IoT-enabled application of temporal data stream mining provides a promising way to construct advanced (smart) services as solving a puzzle over multiple heterogeneous data flows, when some information fragments are discovered in the flows and the fragments then are assembled into appropriate situational knowledge that the user needs at any moment [10, 11].

3 Discussion on Developing iOVFDF

iOVFDF would be devised in both conceptual or theoretical level as well as carried out on a software development level. In particular, there will be software development, computer simulation experiment, data processing, computational performance evaluations, and results visualization. They help to vigorously evaluate iOVFDF for intellectual inquiry and practical necessity.

The general methodology of the developing iOVFDF follows that of typical software engineering. In general, it includes in-depth study of the current problems, literature review, problem and solution formulation, algorithm designs, model system design, followed by computer simulation experiments as well as applying the conceptual models in several case studies. Results are then collated, analyzed, documented and disseminated.

Several challenges are identified at the beginning of the development. Currently many data mining methods neglect the temporal aspects during the typical knowledge discovery process—data are treated as unordered instances, lumped as an archive, and batch learning algorithms thereafter developed for prediction model induction. Data however have temporal dependencies, and the temporal attributes contain very rich information to be analyzed. Most of the data mining tools, including even some of the data stream methods tend to treat temporal data as an unordered collection of events, missing out the useful temporal information.

The related problems and issues associated with the real-life applications will be looked into. Then the development moves on to the tasks of representing temporal sequences used by different researchers and techniques. The discovery of relations among multiple sequences and their sub-sequences as well as the important events, takes three major steps—(1) representation and modeling of the data sequence in some format suitable for data stream mining, (2) defining some appropriate similarity measures between sequences. There is an article that is about using data mining methods to enumerate similarity measure according to the predicted class [12]. The method, though is based on data-mining and geared for unordered rule induction, shall be extended to cover the temporal aspects as the key important attributes.

Representation of temporal sequences is a part of pre-processing, and it needs to be meticulously defined, in both mathematical means and data structure in programming. It is especially important when it comes to data stream mining because time series are being dealt with, finding the right data attributes or features that describe these data which involve manipulation of continuous and high-dimensional data is essential and challenging. Under the proposed framework the model formulation takes several approaches. One is to manipulate the time-series with minimal transformation by representing its original form as univariate data or windowing the sequence with piecewise linear approximation segmenting them into sub-sequences. The univariate in the first approach would be mapped to hyperspace which would be either in the time-domain or discrete domain. Generative model be tried too as we have succeeded in applying data mining classification on statistical model which is derived from the original data for doing complex voice classification, as reported in a journal paper [13].

This work nevertheless is limited to a single time-series in the full processing window from the beginning to the end in voice classification. In this research framework, as a part of data representation and pre-processing, multi-dimensional transaction bases with timing information would be explored as option. Regardless of which representation methods which will be finally adopted, for data stream mining with multiple data streams, it is necessary to meet the objective of discovery and representation of the subsequences of a sequence, in single or in parallel, for the meta-classification. Both time-domain continuous representations like windowing, segmenting, discretization based representation by Shape Definition Language, constraint-based pattern specification language, Self-organizing Map, Generative Models, and transformation types of representations such as Discrete Fourier Transform, Discrete Wavelet Transform would be tested.

Before moving on to the redesign of iOVFDT and subsequent iOVFDF, several types of similar measures for sequences would needed to be tested out. They would be tested according to the use of detecting outlying points (or outliers), episodes, events, amplitude differences, time scale distortion etc. This step and the one mentioned above are related, how the sequences of data being represented would have certain influences on the efficacy of similar measures. The popular Dynamic Time Warping technique and its variants would be tested out in defining varying lengths of warping paths, their embedding probabilistic models that constitute to the global shape sequence with local features with different degree of deformation, stretching and elasticity would be investigated, prior to data stream mining. There are others like deterministic grammars, graph-based mining models, stochastic generative models (Markov chains and hybrids of heuristic similarity measures) to be tried out as well, especially for dealing with different applications and priority of data sequences.

When it comes to temporal data mining, it is known that the most common approaches widely published by many others are association rule mining by Aprior method or its variants. Frequent pattern mining has been tried by many researchers too. In particular, the challenging task here in this project is the transformation of univariate time series and transactional itemsets to multi-variate datasets coupled with some representation methods as above-mentioned, for iOVFDT and iOVFDF.

The main features of iOVFDT: (1) Loss function, replaced Hoeffding Tree for higher accuracy. (2) Auxiliary classifier, solved imperfect data stream problems. (3) ROC optimizer, relieved concept drift problems.

The core algorithm of iOVFDT would have to be modified with the addition of a two-step data stream mining algorithm. It would consist of two classifiers, one serves like a cache (for accommodating the distortion and possibly latency when multiple data streams are considered) and the main classifier. The cache-based classifier is embedded within a sliding-window that updates the instances and builds a decision table classifier by using the locally cached data in the window. Here it is called decision table cache-based classifier (DTC). The DTC serves like a tester which verifies the quality of the incoming data; if the data are clear they would be passed to the main classifier, otherwise the induction at the main classifier will pause. The sliding window size should be chosen corresponding to the rate of data change. And its formation should be adaptive depending on the representation of the sequences or subsequence would adopt as fore-mentioned. The sorting algorithm would be exploited from the family of 'incremental learning' or 'lazy leaner' which only updates whenever an instance comes, which contrasts with greedy-search that needs the full dataset to be loaded. The stream mining algorithm has been tested and some good progress is reported in [14].

Under this framework, other incremental algorithms should be attempted to be used as the core mechanism, as we believe different algorithms would suit different sequence representation and different application scenarios. The size of window W is defined as the maximum number of instances that a window can cache. The window is updated in interval W. Suppose $|D_t|$ is the amount of instances that has been cached in a window at timestamp t. The way to cache instances in the sliding-window is carried out by: (1) Adding it to the window until cache $|D_t| = W$; (2) Building cache classifier (DTC) by decision table; (3) Resetting the cache. The speed is much faster than that of decision tree. However, one of the greatest disadvantages of decision table is it does not depict the flow by logic of a solution to a given problem. We intend to try out some more adaptive kinds of incremental learning algorithms that can map very non-linear relations in lieu of decision table, e.g. Updatable Naïve Bayes would be a suitable candidate, as the aprior probabilities would be updated in real-time, and all the conditional dependences of the data (that represent the sequences) would be catered for; most importantly it is light-weighted and fast in computation.

The main tree classifier (MTC) as the core of iOVFDT is responsible for classifying samples into target labels. A new challenge in this research framework is to enhance from the existing "elastic divider" that allows certain fuzzy or blurry accommodation for slightly deviation when it comes to aligning or measuring multiple time sequences. It is explained in details; to start with the currently known function, which is called a loss function which dynamically controls the size of the statistic bound for node-splitting, in lieu of *Hoeffding Bound* (HB) which was originally adopted by VFDT in data stream mining.

It is static and it doesn't offer desirable accuracy on par with iOVFDT. MTC is built by reading through all the streaming instances; hence it should have a global perspective of all the data. DTC is the cache-based classifier that is built with the scope of seeing instances that are locally cached in the current load of a sliding-window. DTC is rebuilt in every phase of the window. When a new instance

(X, y_k) arrives, we can use MTC and DTC to predict the result separately. The numbers of true/false prediction are computed for both classifiers as follow:

$$MTC(X) \rightarrow \widehat{y_k} \begin{cases} = y_k \Rightarrow T_k^{MTC} = T_k^{MTC} + 1 \\ \neq y_k \Rightarrow F_k^{MTC} = F_k^{MTC} + 1 \end{cases} \quad (1)$$

$$DTC(X) \rightarrow \widehat{y_k} \begin{cases} = y_k \Rightarrow T_k^{DTC} = T_k^{DTC} + 1 \\ \neq y_k \Rightarrow F_k^{DTC} = F_k^{DTC} + 1 \end{cases} \quad (2)$$

After computing T_k and F_k of both MTC and DTC, we define the loss function of accuracy:

$$Loss_k^{MTC} = T_k^{MTC} - F_k^{MTC}\, Loss_k^{MTC} = T_k^{MTC} - F_k^{MTC} \quad (3)$$

In order to measure the performance of current DTC and the overall MTC for a certain class y_k, a coefficient P_k is proposed in this paper. Suppose N is the total number of instances by which MTC has been trained, and n_{ijk} is the sufficient statistics value that counts the number of instances belonging to class y_k within a window. When a DTC has been updated and a new DTC of the new window is initialized, $Loss_k^{DTC}$ is reset to zero, hence the coefficient P_k is:

$$P_k = \frac{Loss_k^{MTC} \times \sum_{i=1}^{I} \sum_{j=1}^{J} n_{ijk}}{Loss_k^{DTC} \times N + 1} \quad (4)$$

For any value of P_k, the smaller it is, the better overall accuracy should be. To this end, we remember the values of the currently P_k^t and its previously P_k^{t-1} respectively, and we keep them being updated whenever new instance arrives. If $P_k^t > P_k^{t-1}$, it means the accuracy is degrading by the newly arrived data. There is a bound $\in_k$ for these statistics, with a confidence δ. This bound for a partition P_k^t is computed as follows:

$$m_k = \frac{1}{\frac{1}{P_k^t} + \frac{1}{P_k^{t-1}}}; \delta' = \frac{\delta}{P_k^t + P_k^{t-1}}; \in_k = \sqrt{\frac{1}{2mk} \times \ln(\frac{4}{\delta'})} \quad (5)$$

where m_k is the harmonic mean of the currently P_k^t and its previously P_k^{t-1}. Suppose $H(\cdot)$ is the heuristic function using Information Gain. It is used to choose one attribute of the data that most effectively splits its set of sample into subsets enriched in one class or the other. Let y_{k_a} and y_{k_b} be the classes belong to the highest heuristic value x_{ia} and the second highest heuristic values x_{ib} respectively. *HB* is proposed for the case of *n* independent observations of $H(\cdot)$ whose range is $\log_2 n_k$ that with a confidence level of $1 - \delta$, the true mean of $H_{true}(\cdot)$ is at least $H_{obs}(\cdot) - HB$, where $H_{obs}(\cdot)$ is the observed mean of heuristic value. The original condition for node-splitting in iOVFDT is: *If* $H(x_{ia}) - H(x_{ib}) > HB$, *then the attribute* x_{Best} *becomes a splitting-node.* Here *HB* only considers the bound between those two values of heuristic function at a tree leaf. However, it does not reflect the overall accuracy of the main tree. The accuracy keeps on degrading as long as concept-drift persists. To handle this problem and to balance the performance with accuracy and size, we propose using $\in_k$ instead of *HB*, where *k* is the index of class y_{k_b}.

While the core classification mechanism has been implemented in Java, the other challenge at the integration level is the design of the meta-classifier which is based on the design of the above-shown learning algorithm, except that there are additional issues to be considered—synchronization, incomplete information, scalability, distributed operation, etc. The input data would be the output of the individual iOVFDT, usually in a designated format of knowledge. Extensive experiments would be carried out, with different variables in parameters, and with benchmark datasets. When the performance is validated to be satisfactory, it will then move on to testing the iOVFDForest model in different case study applications. Parameters tuning will be a key step too, as it is foreseen that the ultimate model will be working with extension of multiple univariate time series being transformed and/or data-mined into multivariate time series with temporal and sequencing information, the complication of parameter tuning would depend on the sophistication of sequence modeling in the actual application.

4 Related Work

The related work done by researchers which have progressed so far along this direction are shown as follow. Each one of these are potential candidate for implementing a service block under the iOVFDF framework. These works should be extended, then integrated as a fully working system for doing Temporal Pattern Mining at high speed and within the meta-learning context. The unique feature of iOVFDF is that none of other works being reviewed here so far produce decision rules, unless substantial post-processing is done. The possibilities and the challenges involved in extending these current state-of-arts to iOVFDF are discussed.

The definition of Temporal Data Mining (TDM) is about any data mining task that involves analyzing data with time dimension. A lot of existing techniques, they include temporal association rules, evolutionary clustering, trajectory clustering, but very few are on classification especially data-stream kind of classification. Most common one is time-series data mining, which has had a lot on time-series forecasting, stock market prediction, etc. But again, few is on classification based on multi-variate data. TDM however is important in areas like anomaly detection, real-time fault diagnosis, special event prediction etc.

Three main tasks of TDM exist, and form the scope of the proposed research framework here. The new iOVFDF miner should answer these questions in this research task: (1) The Data Models—how temporal data in stream form is represented? Time points, or time interval? (2) Temporal Concepts—what is the target semantic? Ordered data stream and/or concurrent data, dependences? For what applications and what mining objectives? (3) Temporal Operators—how the data elements in the data streams compared to and combined by different algorithms?

The above research questions form a hierarchy of multi-level taxonomy for the temporal data streams that we are going to work with: At the top level called Time Points, models are constructed that embrace observation over associated data sequence with specific point in time. Most time-series forecasting applications like ARMIA, Regression, etc., are on the Univariate domain. But when multiple data streams are considered, there exist multiple dimensions of observation. The operations of data streams and their elements moving in parallel are treated. In the literature, most commonly mined concepts are order or the less strict concurrency. See Fig. 2. Through data transformation, and data model representation, raw time points can be transformed to Time Intervals, as shown in Fig. 2.

Now when time points are marked into time interval, which could be modeled by sliding window as in the cache part of iOVFDT, we can have different concepts to work alone with: Order, Coincidence, and Synchronicity. Currently in the literature, there are four main schools of research tools for revealing these concepts; they are Allen, Freksa, UTG and TSKR respectively. A very good review paper can be found in [15]. They are known as Allen's interval relations, Freksa's semi-interval relations, Reichs's interval point relations, Roddick's Midpoint interval relations and so on. Depends on the applications, different data models can be used. Below in Fig. 3 shows some possibilities of how different applications would work with different data models. In our research framework, iOVFDF should be designed as universal as possible, as the formats of the data model can be adaptively customized at the cache classifier level of the individual iOVFDT. In other words, different iOVFDT trees can be configured differently for different data stream type for different applications. This will be flexible in regards of working with data that are streaming from different sources. Figure 3 shows the importance of data preprocessing too. From the crude data, let say from sensor device, a numeric time series ordered with timestamps can be transformed to different formats, with different level of complication and compatibilities to the natures of the applications. E.g. shopping baskets format is usually for marketing and business intelligence where association between sales items are of concern. The design of iOVFDF should be made flexible to accommodate such various types. This is technical possible in building different data transformation in the pre-processing part/stage of the mining process.

As in Fig. 4, it depends on the length of the sequence that the users are interested in. The benefit of data stream mining is that not ALL the data are needed to be loaded. Subsequently rescanning the whole database when new data come is unnecessary, for refreshing the data mining model. With relatively shorter sequences, or even sub-sequences, we can examine more for interesting properties for a wide range of applications. Figure 4 shows some possibilities. The long-term and short-term dependency relations across multiple sequences can be examined too by data stream mining models if meta-learner is available that can work and learn from multiple data sources. This topic has been touched on before and some

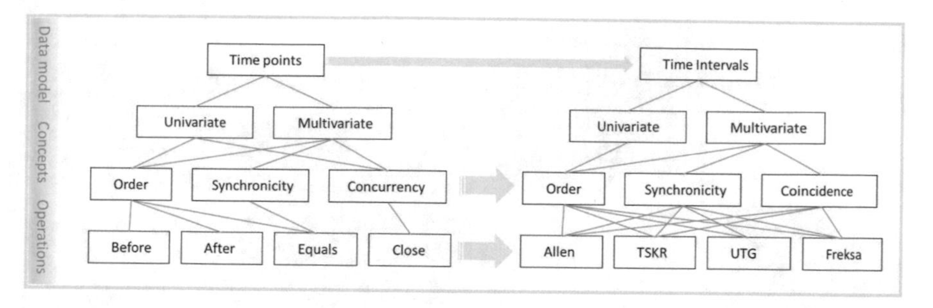

Fig. 2 Multi-level model of time points, concepts and operator, before and after transformation

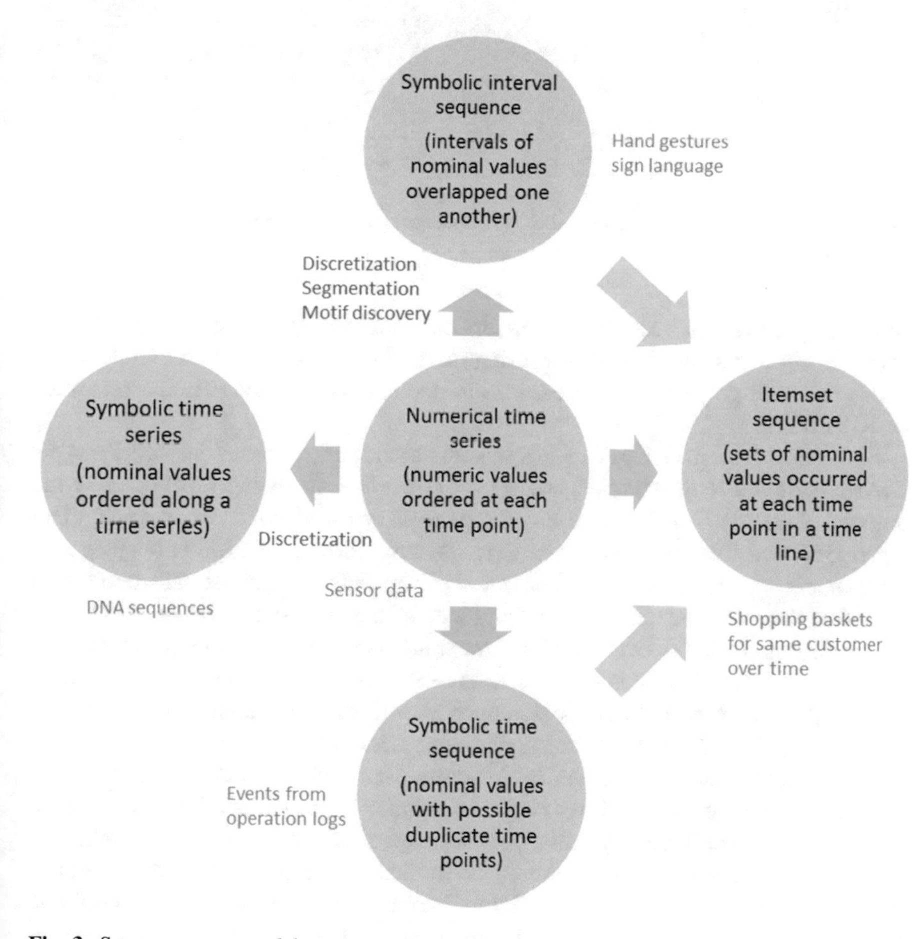

Fig. 3 Some common models and transformations

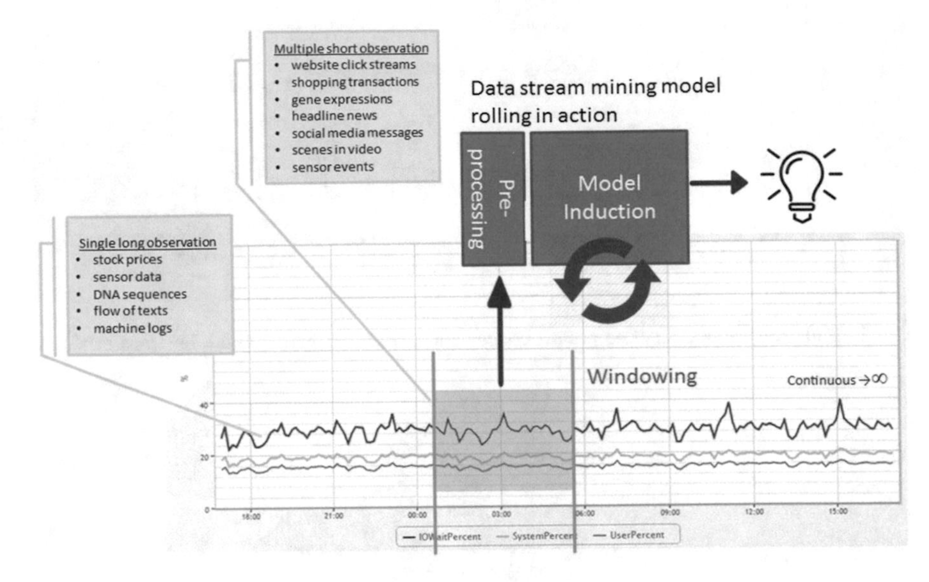

Fig. 4 Transforming long sequence into subsequences, more of these should be done for meta-learning

preliminary work has been done. Readers can refer to [16] for more information. There are some distance-based measuring methods that cater for multivariate more. In this research project, we will explore in depth, and formulate similar measures for iOVFDT/F.

As it can be seen in Fig. 5, for just some examples, different concepts can be derived from subsequences, depending on how they are being positioned in the temporal domain. Usually there are six general types of concepts, we call them "operators" and they can be defined even as some formal descriptive language. In our projects, our classifiers that are trained from data streams, should be able to recognize these six operators—(1) Duration, is the persistence of an event over several time points; (2) Order is the sequential occurrence of time points or time intervals; (3) Concurrency is the closeness of two or more temporal events in time without particular order; (4) Coincidence describes the intersection of several intervals; (5) Synchronicity is the synchronous occurrence of two temporal events, and (6) Periodicities he repetition of the same even with a constant period.

With the operators defined, we can have another 7 interval relations as shown in Figs. 6 and 7 respectively, by Allen's[1] and six semi-interval relations by Freksa's.[2] That once again, our classifiers should be trained to recognize from the inputs of multiple, two or more data streams. Combinations of Allen's and Freksa's could be made programmatically possible, depending on the application needs. Our design

[1]https://www.ics.uci.edu/~alspaugh/cls/shr/allen.html.

[2]https://arxiv.org/pdf/1603.09012.

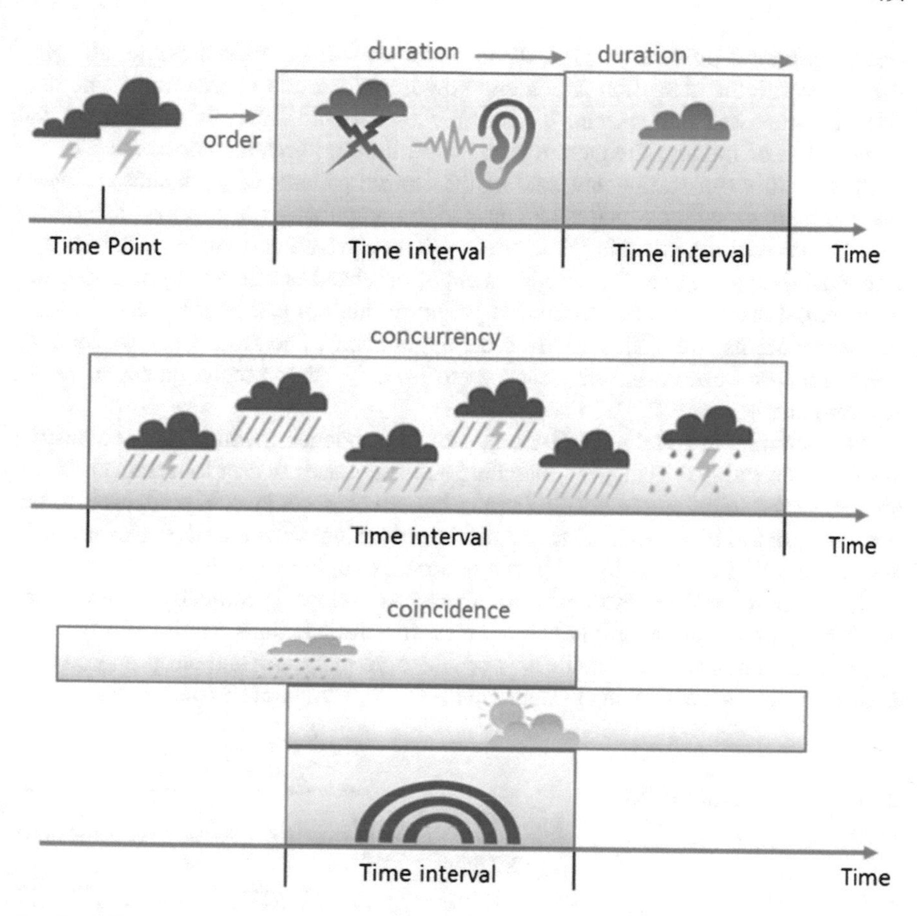

Fig. 5 Different concepts can be derived from subsequences, with a case of weather conditions

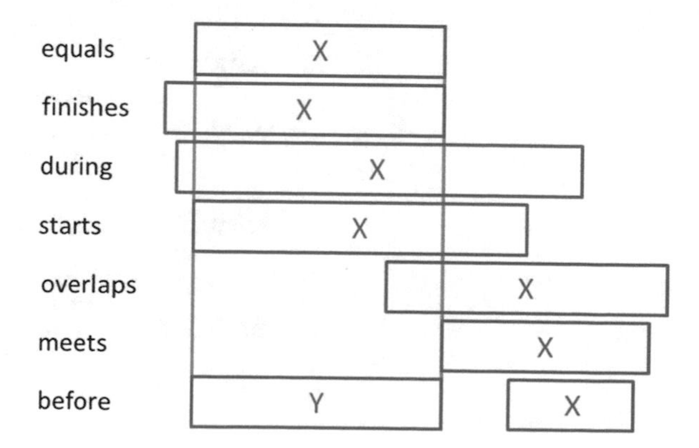

Fig. 6 Allen's six interval relations

on the initiation process of iOVFDF classifier should be made flexible, allowing users to configure at will on which and how many relation operators that the user wish to empower our system. Technically this would have to go back to the formulation of training the pattern recognition that requires the cooperation of the cache classifier (for cutting and holding the desired patterns of input subsequences) and the main classifier which is responsible in recognizing the required occurrence of patterns from the incoming data streams. There are other complex relations too, like overlapping, degree of overlaps, extent of far ahead or fall behind, multiple and geometrical intersection etc. Instead of programming our classifier to cater for each of these patterns, we shall start from the applications, and from there we identify what relation operators are needed to satisfy the prediction/classification requirements.

For example, in matching DNA sequence, or sequencing which is an exhaustive computation on identifying and matching sub-sequences over a long sequence, we shall have the following interval relations, known as X contains Y contains Z in the top example in Fig. 8; X, Y, Z are approximately equal as in the middle example in Fig. 8, and X, Y, Z coincide as in the bottom example in Fig. 8.

Some more realistic examples are shown as follow, specifically designed for substring pattern matching. These are typical matching actions required for genome sequencing. The major difference between the current system and our proposed one, is that the current sequencing system requires a huge amount of computation power

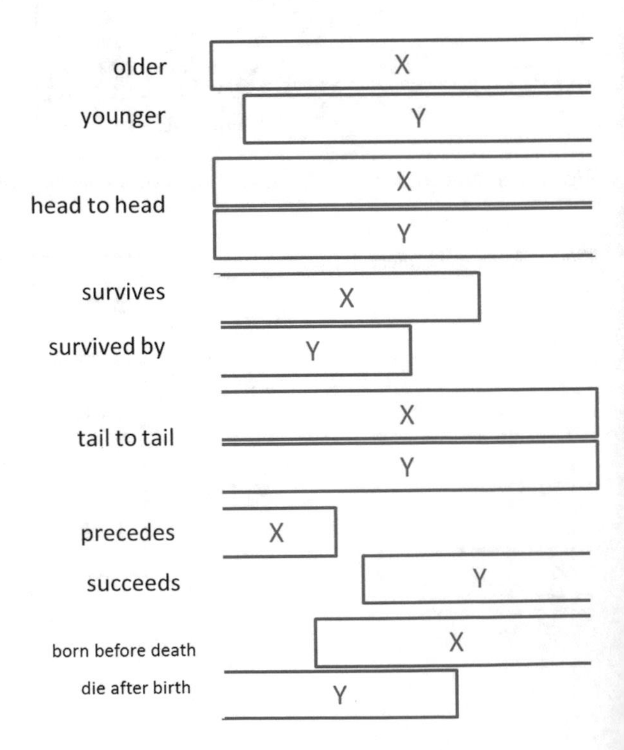

Fig. 7 Freksa's semi-interval relation

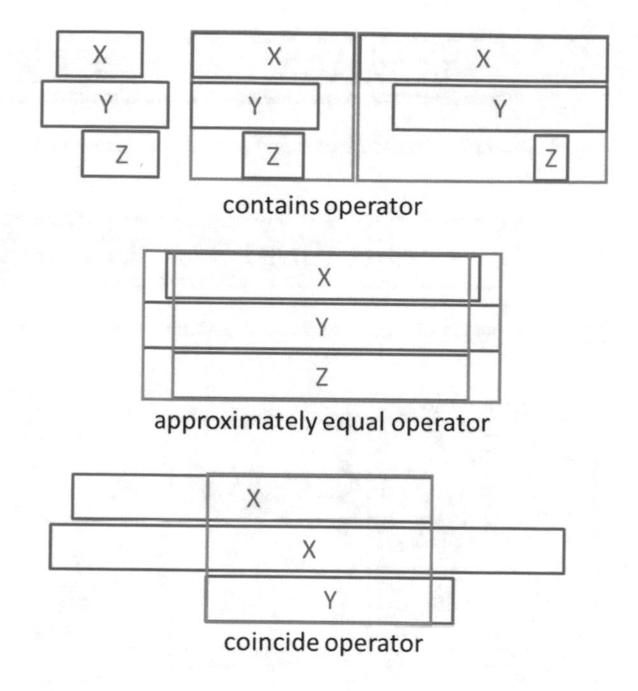

Fig. 8 Contains operator, approximately equal operator and coincide operator

and memory space in scanning through millions of billions of DNA symbols, for decoding some sensible information. In our system, it is light-weighted, incremental, consequently this translates to the advantage of 'read and forget' without remembering the whole or large portion of data. The detection can be done on the fly; prediction is also possible by crunching over the historical data to infer what would be the likely outcome. Our proposed iOVFDF system is meant to be fast, efficient and light in computational demand. Example 1—there is recognition of substring patterns. Sequences of symbols are scanned for without gaps. It expresses the concept of "order", or sequence, for example, B → B → B (Fig. 9).

Example 2, it scans for regular expression patterns. Some extensions are allowed for gaps, it works like inserting a wildcard, negation, repetitions and so on. Example, B → ^C → A|B (Fig. 10).

The next diagram shows when several time-series are being processed together, our meta-classifier in the iOVFDF is supposed to recognize the interaction of them, therefore to conclude some associative outcomes (Fig. 11).

In this case, each iOVFDT will take care of each individual data stream in the pre-processing step 1. They are to be represented in different formats, model, windowing variation etc. From step 2 onwards, the individual iOVFDT is supposed to derive certain higher level information, e.g. the order, the intervals, etc. In steps 3, 4 and 5, meta-classifier is deployed to recognize the interaction, conjunction and alignment of these sub-sequences, such as duration, coincidence and partial order plus those interval relations as in Figs. 6, 7 and 8.

ABCBDADBBCBAAABBCBDBABDABA

Fig. 9 Example of regular expression pattern—scanning without gaps

Fig. 10 Example of regular expression pattern—with gap allowance

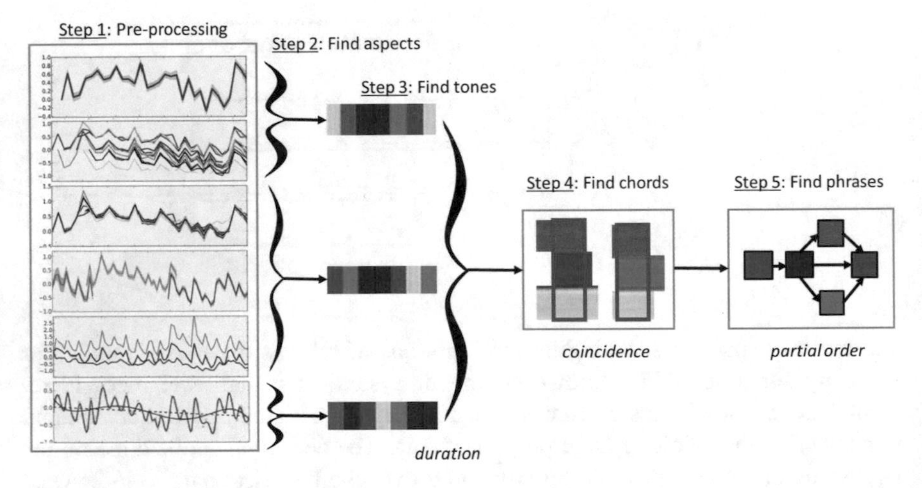

Fig. 11 Time-series multiple mining

The main advantage is the adaptability due to the nature of incremental pattern learning abilities that has already been researched maturely in the past. Other advantages are of course inherent from data stream mining algorithms, fast, reasonably accurate and configurable. It can learn (train) from incoming data streams on the fly, the challenge to be tackled in this reach is the preprocessing step, which is to convert data sequences into the desired format and symbolic temporal data model. The output of individual iOVFDT must be compatible at least in format and semantics to the input of the meta-classifiers. Certain research efforts are needed and expected to address this integration and ontological issues. The proposed data stream mining tool is supposed to extract interestingness, pattern significance, custom pattern detection; anomaly detection by which the classifier learns to distinguish normal sequence from abnormal symbolic temporal data based on patterns. E.g. if my pattern during weekday is 9:00 am reaching office, 7:00 pm leaving office, it is considered normal. Otherwise, if I show up in office 3:00 am and leaving at 10:00 pm, it may become suspicious. A physical human movement security model was published for physical human movement, anomaly detection about a time-based physical security system [17]; however hard-coded rules were defined to discriminate what are suspicious and otherwise. In this research framework, it is aimed at developing a flexible solution that can be configured and it learns and predicts from data streams, even they are changing in context. Micro- and macro level of resolution and contexts could be inferred from this iOVFDF model.

5 Conclusions

Data stream mining is a new and promising branch of machine learning suitable for analyzing, discovering and extracting insights from dynamic data streams. It is an anticipated approach for big data analytics where the analysis is usually being done in real-time and in runtime memory. Though there are many tools around, individually they learn the data stream in incremental manner. Some successful examples include but not limited to iOVFDT which can optimize [18,19] the decision tree size and other performance aspects in real-time. It is believed that combining multiple iOVFDT decision trees we can achieve a greater learning tool which can cater for meta-learning, solving the big data problems. In this article, a review of relevant techniques which can potentially be assembled and extended to construct a meta-learning system is proposed. The meta-learning system functions like a 'forest' of trees, hence the name, iOVFDF. The full name is incremental optimized very fast decision forest. A framework is presented in this article which can serve as inspiration and even blue print for fellow researchers for considering building multiple data stream mining classifiers to be bundled, and function cooperatively for a greater level of performance and wider applications. The prospects, possibilities and challenges by the concept of iOVFDF are discussed.

Acknowledgements The authors are thankful for the financial support from the Research Grant called "Temporal Data Stream Mining by Using Incrementally Optimized Very Fast Decision Forest (iOVFDF)", Grant no. MYRG2015-00128-FST, offered by the University of Macau, and Macau SAR government. The work of D. Korzun is financially supported by Russian Fund for Basic Research (RFBR) according to research project # 14-07-00252.

References

1. Sun, J., et al. (2010). A system for mining temporal physiological data streams for advanced prognostic decision support. *IEEE Data Mining (ICDM)*. doi:10.1109/ICDM.2010.102
2. Yang, H., & Fong, S. (2014). Countering the concept-drift problems in big data by an incrementally optimized stream mining model. *Journal of Systems and Software, 102*, 158–166.
3. Yang, H., & Fong, S. (2013). Incremental optimization mechanism for constructing a decision tree in data stream mining. *Mathematical Problems in Engineering*. doi:10.1155/2013/580397

4. Fong, S., et al. (2013). Evaluation of stream mining classifiers for real-time clinical decision support system: A case study of blood glucose prediction in diabetes therapy. *BioMed Research International*. doi:10.1155/2013/274193
5. Fong, S., et al. (2013). Using causality modeling and fuzzy lattice reasoning algorithm for predicting blood glucose. *Expert Systems with Applications, 40*(18), 7354–7366.
6. Naqvi, M., Hussain, K., Asghar, S., & Fong S. (2011). Mining temporal association rules with incremental standing for segment progressive filter. In *Proceedings of the Third International Conference on Networked Digital Technologies* (NDT 2011) (pp. 373–382). Macau: Springer CCIS.
7. Antunes, C., & Oliveira, A. L. (2001). Temporal data mining: An overview. *KDD Workshop on Temporal Data Mining, 1,* 1–13.
8. Fong, S., Hang, Y., Mohammed, S., & Fiaidhi, J. (2011). Stream-based biomedical classification algorithms for analyzing biosignals. *Journal of Information Processing Systems, 7,* 717–732.
9. Korzun D., Nikolaevskiy I., & Gurtov A. (2015). Service intelligence support for medical sensor networks in personalized mobile health systems. In *Proceedings of the 8th conference on Internet of Things and Smart Spaces* (ruSMART 2015). Switzerland: Springer International Publishing.
10. Acampora, G., Cook, D. J., Rashidi, P., & Vasilakos, A. V. (2013). A survey on ambient intelligence in healthcare. *Proceedings of the IEEE, 101,* 2470–2494.
11. Korzun, D., Borodin, A., Paramonov, I., Vasyliev, A., & Balandin, S. (2015). Smart spaces enabled mobile healthcare services in internet of things environments. *International Journal of Embedded and Real-Time Communication Systems (IJERTCS), 6*(1), 1–27.
12. Fong, S., et al. (2014). Measuring similarity by prediction class between biomedical datasets via Fuzzy unordered rule induction. *International Journal of Bio-Science and Bio-Technology, 6*(2), 159–168.
13. Lan, K. (2013). Classifying human voices by using hybrid SFX time-series pre-processing and ensemble feature selection. *BioMed Research International*. doi:10.1155/2013/720834
14. Yang, H., & Fong S. (2013). Improving the accuracy of incremental decision tree learning algorithm via loss function. In *Proceedings of the 2nd International Conference on Big Data Science and Engineering* (*BDSE 2013*), Sydney, Australia.
15. Mörchen, F. (2006). Unsupervised pattern mining from symbolic temporal data. *SIGKDD Explorations, 9,* 41–55.
16. Fong, S., Luo, Z., Yap, BW., & Deb, S. (2014). Incre-mental methods for detecting outliers from multivariate data stream. In *Proceedings of the 13th IASTED International Conference on Artificial Intelligence and Applications* (AIA 2014). Austria: Innsbruck.
17. Biuk-Aghai, R. P., Yain-Whar, S. I., Fong, S., & Yan, P. F. (2010). Security in physical environments: Algorithms and system for automated detection of suspicious activity. In *International Workshop on Behavior Informatics* (*BI 2010*) *in Conjunction with the 14th Pacific-Asia Conference on Knowledge Discovery and Data Mining* (PAKDD 2010). Hyderabad.
18. Kausar, N., Palaniappan, S., Samir, B. B., Abdullah, A., & Dey, N. (2015). *Systematic analysis of applied data mining based optimization algorithms in clinical attribute extraction and classification for diagnosis of cardiac patients, Applications of Intelligent Optimization in Biology and Medicine* (pp. 217–231). Switzerland: Springer.
19. Yang, H, & Fong, S. (2013). Countering the concept-drift problem in big data using Iovfdt. In *Proceedings of the IEEE 2nd International Congress on Big Data*. Santa Clara Marriott: BigData 2013. June 27–July 2, 2013.

Sentiment Analysis and Mining of Opinions

Surbhi Bhatia, Manisha Sharma and Komal Kumar Bhatia

Abstract Now days the way of expressing opinions on certain products that people purchase and the services that they receive in the various industries has been transformed considerably because of World Wide Web. Social Networking sites fascinate people to post feedbacks and reviews online on blogs, Internet forums, review portals and much more. These opinions play a very important role for customers and product manufacturers as they tend to give better knowledge of buying and selling by setting positive and negative comments on products and other information which can improve their decision making policies. Mining of such opinions have focused the researchers to pay a keen intention in developing such a system which can not only collect useful and relevant reviews online in a ranked manner and also produce an effective summary of such reviews collected on different products according to their respective domains. However, there is little evidence that researchers have approached this issue in opinion mining with the intent of developing such a system. Our work will focus on what opinion mining is the existing works on opinion mining, the challenges in the existing techniques and the workflow of mining opinions. Consequently, the aim of this chapter is to discuss the overall novel architecture of developing an opinion system that will address the remaining challenges and provide an overview of how to mine opinions. Existing research in sentiment analysis tend to focus on finding out how to classify the opinions and produce a collaborative summary in their respective domains, despite an increase in the field of opinion mining and its research, many challenges remain in designing a more comprehensive way of building a system to mine opinions. This chapter addresses the problem of how to classify sentiments and develop the opinion system by combining theories of supervised learning.

S. Bhatia (✉) · M. Sharma
CS department, Banasthali University, Vanasthali, India
e-mail: surbhibhatia1988@yahoo.com

M. Sharma
e-mail: manishasharma8@gmail.com

K.K. Bhatia
CSE Department, YMCA University of Science and Technology, Faridabad, India
e-mail: komal_bhatia1@rediffmail.com

N. Dey et al. (eds.), *Internet of Things and Big Data Analytics Toward Next-Generation Intelligence*, Studies in Big Data 30,
DOI 10.1007/978-3-319-60435-0_20

Keywords Mining · World Wide Web · Opinion · Sentiment analysis · Supervised learning

1 Introduction

Opinion Mining is closely associated with the Mining area. We can consider opinions to be a feeling, review, sentiment or assessment of an object, product or entity [1]. Mining refers to as extracting some useful information and representing it in the form of knowledge in an understandable structure as patterns, graphs etc. [2]. The two important aspects of Opinion Mining are Information Extraction and Question Answer [3]. Opinion mining is a technique of extracting knowledge automatically by taking opinions from others on the specific topic or problem [4]. Sentiment analysis can be used as a synonym for opinion mining. The increase in the growth of Social media has made the people dependent on search engines and Information retrieval systems. This dependency has created the need to develop such a system that will help people in making the correct decision of purchasing products by analyzing the positivity and negativity score of the opinions respective to the particular domain.

Opinion mining has numerous applications in purchasing and marketing products and services, setting up policies and making decisions, marketing and trading etc. We will explore the trends in the mining industry; in particular, how online reviews are reflecting the major shift in the decision making policies of customers. The various researchers like [5–8] have discussed briefly the meaning of opinion mining, its techniques, tasks, sentiment analysis and other challenging aspects in summarization of opinions. Penalver-Martinez discussed opinion mining in semantic web technologies [9]. A novel technique of assessing product reviews is presented by [10]. The aspect based classification and summarization of reviews is depicted by proposing a novel model which is represented with a generic tool which does not require any seed word or pre requisite knowledge about the product or service [11]. Mfenyana has proposed a tool to collect and visualize data from face book [12]. Features based summarization is discussed in detail by Bafna [13]. Sentiment analysis done by using learning techniques is evaluated and discussed diagrammatically [13]. The aspect based opinion mining by combining hybrid theories are discussed and results are evaluated by effective measures using search strategies [14]. Pang worked out the effectiveness of sentiment analysis problem with the application of machine learning algorithms [15]. Hemalatha developed the sentiment analysis tool that focuses on analyzing tweets from social media [16]. McDonald identified sentiments at all the levels with its subcomponents [17]. Liu and Chen presented a detailed analysis of twitter messages by making use of three metrics [18]. Bhatia presented the design and implementation of opinion crawler [19]. We will therefore design a novel system that is dynamic in nature containing less complex algorithms and promises to generate more effective and structured opinions considering its aspects.

Most previous research in opinion mining and sentiment analysis focuses on a specific content. Thus, there is little emphasis on a coherent structure—no integrative framework to mine opinions is proposed. Moreover, there is little attention given to conceptualize the structure of mining opinions. It has also been observed that many institutes are consistently doing research in Opinion mining area which has been listed as one of the prime means of creating enormous amount of online reviews and feedbacks of different products. The following section focus on the systematic literature review with the basics of Opinion mining.

The goals of the chapter are as follows:

(1) To define and explain Opinion Mining.
(2) To find the gaps in the existing definitions and frameworks.
(3) To explicate the need of information retrieval in Opinion Mining.
(4) To propose the structure using Opinion Mining.

The chapter has been structured as follows. The definition, existing challenges, literature review and the background has been explained in Sect. 2. Section 3 examines the framework design and explains all the modules in determining opinions. The analysis of the proposed work and results are evaluated in Sect. 4. The summarization is discussed in Sect. 5 and the scope for the future is concluded.

2 Literature Review

The Opinion mining is the crucial part of the Data mining, a significant part of our paper addresses the subject. In the previous work done, the various researchers examined it from a narrow scope. An in-depth study is required to present an overview of what has been published on opinion mining. So a need to determine and identify the areas in which detailed research is required in order to meet the demand required by the present and future day systems being built is explored in our work. A systematic mapping involves going into the depth of the literature and examining the nature, content and quality of the published papers in this domain. The grouping of primary studies is done to produce a view of the specific research area by the systematic mapping of the related studies. The systematic literature review process is necessary to ensure that the further innovative research can be carried out by the various researchers, practitioners and also to aim at gaining an overall idea of the distinct researchers in this field.

The present study intends to summarize the existing evidence in the opinion mining area. The review is undertaken in accordance with the predefined search strategy as proposed by Kitchenham [20]. The established of such a systematic review helps in minimizing the researcher bias. Few research questions are listed which seem to be a part of discussion in the literature review section given below. Section 1.2.2 defines the paper literature extraction strategy containing resource archives.

2.1 Research Questions

The main target of this assessment is to report the research done in the field of opinion mining. The following questions are discussed which reflect our purpose as under:

RQ1: How is the term 'opinion mining defined'?
RQ2: What types of studies are published by the abundant researchers?
RQ3: How components are inter-related and what are the different techniques used in classifying sentiments?
RQ4: How strong is the association between opinion mining and social media?

The relevant literature is a twofold process, i.e. searching through a keyword in the scientific libraries and second is the collecting papers from the workshop series. It consists of the following steps.

- Gathering information about the literature: This is the combination of the digital library and distinct sources such as ACM, Springer, and IEEE, Science Direct etc. which includes the search based on titles, keywords, and abstracts. The peer review process is considered for study selection in order to attempt for a broader prospective, so the restriction on the primary studies must be avoided.
- Applying the selection/rejection method: The paper that explores the relevant study regarding Opinion mining is taken into account to be selected. The papers that are not related i.e. other than mining opinions are rejected.
- Verify the rejected papers: The combination of the different keywords must be avoided. Only the terms "mining" and "opinion(s)" in the field's title, abstract, keywords are evaluated in the literature.
- Verify included papers: This is manually done by reading the abstracts and conclusions. The selected literature resulted to produce outputs that are directly or indirectly related to mining opinions.

2.2 Analysis

Based on the above mentioned research questions and the analogies extracted for RQ1, RQ2 and RQ3 with respect to RQ4, thorough analysis has been accomplished to achieve the goal.

The thorough literature survey has been done in order to present the in-depth study and the results are presented in the following section according to the questions listed above.

2.2.1 Defining Opinion Mining

Human feeling and emotions can be described and entities, events and properties are evaluated using a subjective approach that can be taken as an Opinion. Opinion mining can be considered as a process of collecting reviews that are posted online on various Social Networking sites. This definition, which is most prevalent, has been used by a number of researchers. This definition considers mining as a bigger term which contains many sub definitions.

According to another definition, to identify and abstract subjective facts and material from text documents can be accomplished by applying Opinion mining technique. There are various terms which can be used exchange ably with Opinion Mining like Review mining, Sentiment analysis, Appraisal Extraction as suggested in paper [21]. Considering the hierarchical structure, Web content mining list Opinion mining as part of it which basically comes under the category of Web Mining [22]. Data mining and World Wide Web lies in close proximity with one another. Several research communities consider web mining research an important resource of research since it lies at its junction. Data mining and World Wide Web lies in close proximity with one another. Several research communities consider web mining research an important resource of research since it lies at its junction. Each term is explained in detail as follows:

Data Mining (DM): The other name of DM is Knowledge Discovery. It is an important process which deals with analysis of data from various perspectives and summarizes it into valuable information that is useful for companies and individuals.

Text Mining (TM): It is analysis of data present in natural language text. It aims at getting fine information from text through means such as statistical pattern learning.

Web Mining (WM): It can be stated as discovering and analyzing useful information from World Wide Web. It is basically categorized into three types as Web usage mining, Web content mining and Web structure mining.

Web Usage mining: It is the process of discovering what users want to see on the Internet. The data can be of the textual form or in multimedia form by making use of user logs.

Web content mining: It targets the knowledge discovery including multimedia documents like images, video and audio. It scans the contents of web page deeply and extracts useful information.

Web Structure mining: It focuses on analysis of the link structure of the web and generate structural summary about the Web site and Web page.

The evidence of the above is database, information retrieval, and Artificial Intelligence [23]. The complete hierarchical model of data mining is diagrammatically explained in Fig. 1.

We can define Opinion Mining as a set of text documents (T), entity which has opinions, further determining attributes of the entity in text $t \in T$ and comments to be distinguished in the form of positive, negative or neutral classes [24]. It is one the most appropriate definitions as it considers different components evolved and

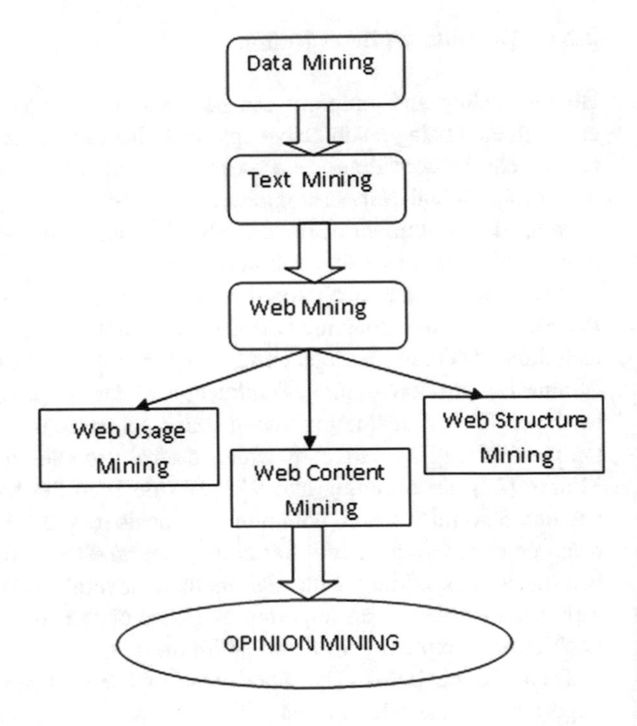

Fig. 1 Simplified model of data mining

their interactions with each other. The resonance of this definition with Information Retrieval is hard to ignore. In fact, the most important attribute of the definition of Opinion mining should be its ability to encompass varied components developed for the same product, working together for a common task. The definition should be open to ascertain the effect of components on the target audience and vice versa.

According to Bing Liu [6], the area that helps in analyzing person's opinion, emotions, attitudes, feelings towards various entities including products, services etc. is referred to as Opinion mining.

As discussed in his earlier works, Opinion Mining has been categorized into three levels;

- Document Level: Distinction is done taking into consideration the whole opinion document into positive category or negative category [15, 25].
- Sentence Level: This level goes into a deeper level than document, i.e. it distinct the complete sentence into positive, negative or neutral category. No opinion comes under neutral category. It is also related to subjectivity classification [26].
- Feature Level: A more detailed classification is aspect level performs advanced and more quality based analysis. Feature level is a synonym of Aspect level [27]. The features or views are taken into account at a deeper level than considering documents, paragraphs, sentences, clauses or phrases which come under language constructs. Opinions will be based on the target or object which will be either positive or negative.

According to our definition, Opinion Mining is a collaboration of reviews working as a prime component and interaction with the Public media for advertising and selling their goods and services. Maintenance of the relationship with the components is required which is underlined by the common platform by inter-changing information, resources or artifacts.

It may be noted that this definition takes into consideration almost all the factors mentioned in the previous definitions. However, with respect to the E-Commerce, the author sees Opinion mining as a set of businesses which is somewhat true. Businesses run and interact with the social media with a sole purpose of making profit. If all the components of a particular Opinion Mining compete for individual profits, then the definition will be correct. However, in our view this is not the case. The components of a Opinion mining perform the task they are required to perform. They might not be intending to make the profit but just contributing towards the goal of the overall mining system. In this sense, the Information Retrieval analogy would be more appropriate than the businesses one.

2.2.2 Studies by Various Researchers

This section gives a brief of works given by various researchers. The study of Opinion mining is gaining importance, from its inception in 2012. It is evident from the increase in the number of papers published. It, therefore, becomes important to understand what exactly Opinion mining is. For any evolving subject a divergence of views is necessary and is healthy for evolution. It is not essential to follow the previous definitions although they might show the way. The way might have much segregation and there is no reason why perspective segregation should be left.

The paper [28] discusses the properties of social networks associated with opinion mining. It explores a new opinion classification method which a variant of semantic orientation and presents the results by testing the algorithm for accuracy on data sets from real world. The future work concentrates on the response of opinions in text and improving the performance of the algorithm by creating an active learning strategy.

The paper [29] presents an outline of opinion mining and sentiment analysis clearly with the data sources, components and tools. The levels of sentiment analysis with its techniques are discussed with examples. The paper [30] discusses how link based graphs can be applied on the newsgroups which helps in classifying objects in different categories and hence are considered more accurate. The future work concentrates on text information by inculcating linguistic analysis for improving the accuracy. The paper [31] discusses the importance and functionalities of different types of mining areas in social networks. It emphasizes on how opinion mining and sentiment analysis can be studied to gather knowledge which can promote business, political scenarios and other areas. It promises to deploy the techniques which can be applied in real world applications. The paper [32] presents a model based on opinion based graph which focuses towards content oriented domain. It allows observing the evolution of the opinion inside a discussion.

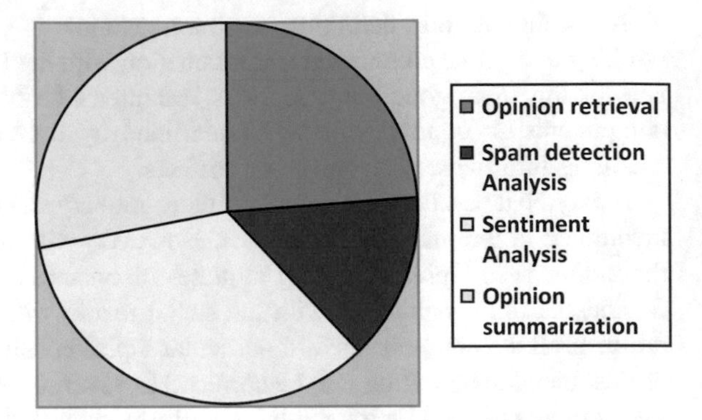

Fig. 2 Distributions of primary studies by research type

The comparison between the existing user based graph approach and the proposed opinion based graph is illustrated. The future work gives a summarized picture of combining user based graph and opinion based graph for analysis. It also explores how opinion changes over time. The paper [33] focuses on achievement of the opinion mining levels. The problems faced in sentiment analysis for reviewing a product are discussed in order to provide a summarized framework of opinions. Set of reviews and set of known expressions related to features which are taken from the reviews are provided as an input for Opinion Mining. The paper [5] presents a systematic literature review of opinion mining techniques and methodologies and also explores the existing gaps in the field. The author promises to improve the performance of the model and work on issues in sentiment analysis. The major limitations that need to be considered are language issues, dealing with negative terms and finally summarize the opinions on the basis of aspects etc. The paper [34] proposes a new framework for opinion mining with respect to content analysis in social media. It discusses the three prime components to track opinion data online. The further research focuses on the policy making issues through social media sources. Precise learning algorithms related to quantification perform sentiment analysis by conducting experiments on diverse tweets datasets. A naive way to tackle quantification is done by classifying each unlabelled item independently and computing the fraction of such items in the class is explored in the paper [35].

After going through the keen analysis of the various papers on Opinion mining, we classify the papers according to the study type. The major research has been done under the category of Sentiment analysis, followed by summarization, followed by Opinion retrieval and last is spam detection. Figure 2 shows the frequency of major studies by research type. Chiefly the distributions are listed.

2.2.3 Components, Techniques, Applications of Opinion Mining

Automatically identifying the opinions from documents and presenting it to the user is the prime core task of opinion mining [36, 37]. The task has been defined in terms of extracting public's opinion based on the features of products. These components are distinguished in the below listed three categories under Opinion Mining:

- Opinion Holder
- Object (Entity/Target)
- View.

Considering the example:

I purchased LG Television three days ago. I loved watching it. The picture quality was great. It contained smart television feature also. The wi-fi slot enables me to watch my choice movies. However my brother was angry as it did not have the 3-D feature. He also thought that the TV was expensive, so I should return it.

There are several opinions in this review. Talking about the view: Sentences 2, 3, 4 and 5 convey optimistic response and sentences 6 and 7 convey pessimistic response. There are various targets in the above example. The target is phone in sentence 1, picture quality in sentence 3, wifi slot in sentence 5 and price in the last sentence. The holder is author in sentences 1, 2, 3, 4 and 5. But in the last two sentences i.e. 6 and 7, the holder is brother.

Policy making which is a joint decision can be easily obtained by making use of Opinion mining applications. It is useful in making sense of thousands of interventions [38]. Search engines, Question answering systems and other practical applications are making extensive use of Opinion mining. The interaction between humans and machines has also been improved with it [3]. The primary areas where Opinion mining can set a major impact are:

- Purchasing and Quality improvements in products and services by placing ads in user generated content.
- Decision making and Policy setters by tracking political topics.
- Automated content analysis which provide general search for opinions.

In general, Opinions can be considered as an image of certainty as it is dependent on recognizing difficulties by paying attention through listening rather than by questioning. To classify opinions into categories like positive, negative and neutral, we have a large no. of sentiment detection approaches. Sentiment detection is closely associated with Natural Language Processing and Information Retrieval. Focusing on the uses of machine learning which mainly belong to supervised classification and text classification techniques in case of Sentiment Analysis. So it is called as supervised learning. On the other hand, unsupervised classification belongs to semantic orientation approach. Both of these two approaches necessitate working on a data set which involves training with pre worked examples. Thus, classification requires training the model [5].

Training data and Test data are involved in a machine learning based classification. Distinguishing features are required to be understood by the automatic classifier while working on training data set in text documents. The performance is validated using test data set of the automatic classifier.

Text categorization mainly involves supervised learning methods such as Naive Bayes, Maximum entropy, Support Vector Machines, Winnow classifier, N gram model etc. [39–43]. Unsupervised learning makes use of available lexical resources. In this approach, input given is a set of unknown labels (not known during training) and output comes as classified data. Clustering is a good example of the above technique. Word lexicons which appear in sentences containing opinions are compared and sentiment words are determined for classifying. Part of speech tagger, software which works by allocating parts of speech to every word that appear as an input in the text is widely used in determining sentiments. Corpus based or dictionary based techniques are used in this type of technique.

3 Proposed Work

An integrative framework is discussed in this section that draws on previous research, extracts, classifies and summarizes the opinions posted online on various social networking sites. We argue that such a framework is an organizing and unifying structure that accounts for mining opinions in a more general and easier way. Such an integrative framework is temporal independent and periodic in nature as the crawler that we will design to retrieve opinions regularly updates itself. Moreover, the work we will present in building a crawler, there exists no single theoretical perspective that can support such a platform that can extract relevant and quality reviews ignoring the spam reviews but also periodically update the sites by calculating the timestamp of the URL's. The work will also include to classify the sentiments by using different supervised learning techniques and finally to collate all the opinions to produce the one liner summary. The final ranking of the opinions is also done using different algorithms. As initial analysis was unable to find evidence suggesting that their exists such a unique system of summarizing opinions, we hypothesize that a set of different learning techniques will account for classifying opinions which will form the groundwork for explicitly consolidating them and to present the final opinion.

Our emphasis will be on describing the opinion system as a framework that can be used to present ranked opinions according to different domains and describe the flow of mining opinions. We begin by outlining the meaning of opinion mining, its advantages, information retrieval system, and sentiment analysis with a focus on the evaluation by using various search strategies. Then, we present a brief overview of the system, and discuss its modules and components one by one. The modules describe the overall flow of mining opinions under various domains that can be measured and evaluated in information retrieval search strategies. Hence, the combination of modules will be used to construct the framework and will provide the final summarization of ranked wise opinions, which in turn is expected to give the final results in the ranked manner in the form of opinions. More importantly, this system can be used to define the development of a complete architecture to mine opinions, providing a description of each component, including its definition

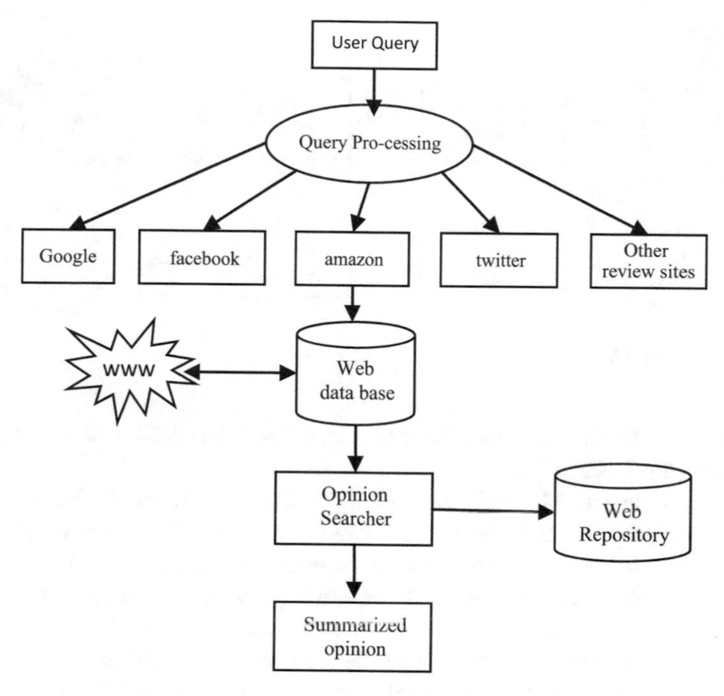

Fig. 3 Framework design

and behavior. Without a detailed description of every module, it would be difficult to design how the components would look and behave in mining opinions. On the other hand, with a set of definitions and behavior descriptions, more informed decisions can easily be made by people in whether to buy a certain product or not. The framework design is shown in Fig. 3.

The explanation of every module is given as follows:

- User Interface: It is basically an interface that enables users to write their queries for getting the output as opinions.
- Opinion Retriever: Opinion retriever contacts the World Wide Web to look into the Social Media, E-forums and other sites to collect reviews and feedbacks. The HTML source is parsed by checking the keywords such as comment, review summary, opinion in the div class or span class. Our earlier works develop such a crawler [19] that periodically updates reviews and present the most recent opinions with the date and time which fulfills our purpose of having relevant reviews which is time and temporal independent. The architecture for the Opinion Retriever is given in Fig. 4.
- Opinion Scheduler: This module will pick the URL from the queue of the URL database. If some URL's are specified as seed URL's, then it will prioritize and choose that URL from amongst all, otherwise it will follow the LIFO policy of

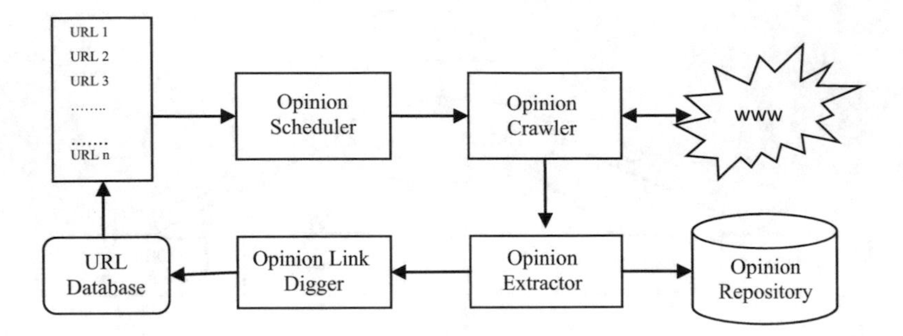

Fig. 4 Opinion retriever

the queue. It will first check the avail list of the URL buffer. If not empty, will traverse the buffer.

- Opinion Crawler: This module will fetch the URL from the buffer and will check whether the URL is of the review site or not. The code is taken and a structure is developed in memory which computer understands. HTML code is taken and relevant information like heading, paragraphs, links and hyperlinks in the page are taken to search for the opinions from the world Wide Web.
- Opinion Extractor: It will search for the specific tags from the HTML or XML pages to extract only the review content from the parsed HTML code by using basic string processing and index the opinions and store them in the repository.
- Opinion Repository: It is a storage area for the final relevant opinions found from the Opinion Extractor.
- Opinion Link Digger: This module will search the links and extract them from that particular web page in the depth first search manner and will add them in the URL queue associated with the URL Database.

The algorithm for the above is given in Table 1.

- Spam Detector: Spam can be referred to as an area of accomplishing search in text documents with the sole purpose to find irrelevant and fake opinions which can be regarded as useless. These reviews are created for the sole purpose of advertisement or deceiving the customers falsely. Machine learning approach has been widely studied in spam detection which has already been compared and studied [44]. This module will look for the spam reviews and filter out the relevant ones from the irrelevant ones. Since, words are repeated many a times in the fake reviews. Keyword stuffing is a technique which is used in our work by checking the review sentences with the keywords used by the spammers. Also, the threshold is set which checks to find if the length of review is more than expected.
- Tokenizer: This module will pick the relevant opinions one by one and will tokenize the opinions into individual sentences after removing the stop words and transforming the tokens from plurals to singulars. This will be done by using the automatic POS Tagger.

Table 1 Opinion retriever

char *text[], ch;
*text[] = {“list of URLs};
If(*text[i] ==NULL)
{
Printf(“no URL left”);
Exit(1);
}
While(ch ==gets(*text[])! = EOF)
{
If(found seedURL)
{
Prioritize URL;
}
else
Pick URL from *text[];
}
Fetch page;
Parse Page;
Resolve DNS;
Make threads for HTTP Socket;
If(!visible page content)
{
Break;
}
If(robots.txt exists)
{
If(.disallow statement exists)
{
Continue;
}
}
extract the opinion
If(meta tag found)
{
If(div class!! Span class)
{
If(review content!!comments found)
{
Extract sentence
}
}
}

(continued)

Table 1 (continued)

Store in opinions repository
}
Extract URLs;
Normalize URls;
If(already visited URL)
{
Remove from *text[])
}
Else
Add URL to *text[];
}

Table 2 Opinion categorizer

do
{
while (wo_i ! = NULL)
{
for(x = 1;x < n;x ++)
{
if (wo_i == T_d)
return +1 and −1
else
overall polarity = (log(+1)-log(−1))
}
}while(s_i ! = NULL);

- Opinions categorizer: Next is to categorize these opinions into positive and negative category by using the Naive Bayes. The well-known probabilistic classifier is Naïve Bayes Classifier. It is used to describe its application to text for incorporating unlabelled data. The learning of the model is conducted by including parameters which are predicted by using labeled training data. In the proposed algorithm, nouns which are taken as features are used to calculate probabilities of the positive and negative count [45].

The algorithm for Naive Bayes Classifier is given in Table 2:

Input: Sentences $\{s_1 + s_2 + s_3 + \ldots s_n\}$ divided into List of words (tokens) words = $\{wo_1 + wo_2 + wo_3 + \ldots wo_n\}$ where x = 1, 2, 3, …, n
Database: Naive Table T_d
Positive words: $\{pwo_1 + pwo_2 + pwo_3 + \ldots p\ wo_n\}$
Negative words: $\{nwo_1 + nwo_2 + nwo_3 + \ldots nwo_n\}$
Neutral words: $\{nuwo_1 + nuwo_2 + nuwo_3 + \ldots nuwo_n\}$
Positive Polarity =+1
Negative Polarity = −1

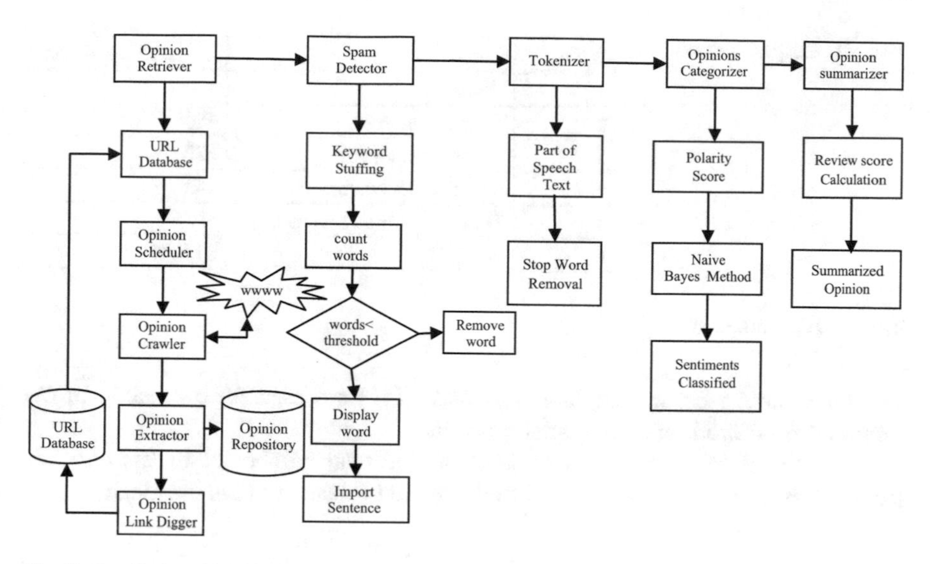

Fig. 5 Detailed architecture

Opinion Summarizer: After analyzing the sentiments and calculating the review score of the positive instances as well as negative instances, the summary is produced as to whether a product is recommended to purchase or not.

The complete model of the proposed work is given in Fig. 5.

4 Results and Analysis

The following section conducts the experiments on the sample data and analysis is done by calculating performance metrics.

4.1 Dataset Description

The customer review dataset of a product is used for our analysis. The tweets are collected from the http://www.Twitter.com. Opinions may contain complete sentences as reviews or may be rated as stars with date and time but tweets generally consists of short comments. LG television product reviews are used in our work. These opinions are categorized into individual sentences. The dataset used in the proposed system is shown in Table 3.

Table 3 Corpus details

S No.	Corpus	LG LED television
1	Opinions	150
2	Total sentences	460
3	Positive sentences	252
4	Negative sentences	143
5	Total opinion as sentences	395
6	Percentage	85.86%

4.2 Evaluation

The three basic measures explained as under [46] are used for evaluation of our proposed framework and comparison is done.

The ratio of relevant tweets retrieved to the total number of tweets retrieved (relevant and irrelevant tweets retrieved is called Precision. Mathematically,

$$\text{Precision} = \frac{\text{RTT}}{\text{RTT} + \text{RWT}} \tag{1}$$

where, RTT is the relevant tweets retrieved and RWT is the irrelevant tweets retrieved.

Recall is defined as the ratio of relevant tweets retrieved to the manually retrieved tweets by the classifier (relevant tweets retrieved and relevant tweets not retrieved). Mathematically,

$$\text{Recall} = \frac{\text{RTT}}{\text{RTT} + \text{RNT}} \tag{2}$$

where, RTT is number of relevant tweets retrieved and RNT are relevant tweets not retrieved.

F-Measure is explained mathematically [47].

$$\text{F-measure} = \frac{2 * \text{Precision} * \text{Recall}}{\text{Precision} + \text{Recall}} \tag{3}$$

The values are identified manually in the system to calculate the systems accuracy with these performance metrics.

4.3 Result

The test set used is LG Television which is annotated below using positive and negative polarities. On determining the true values, Opinions categorizer provides 92.36% of accuracy for the given dataset. The parameters for opinions categorizer are shown in the Fig. 6.

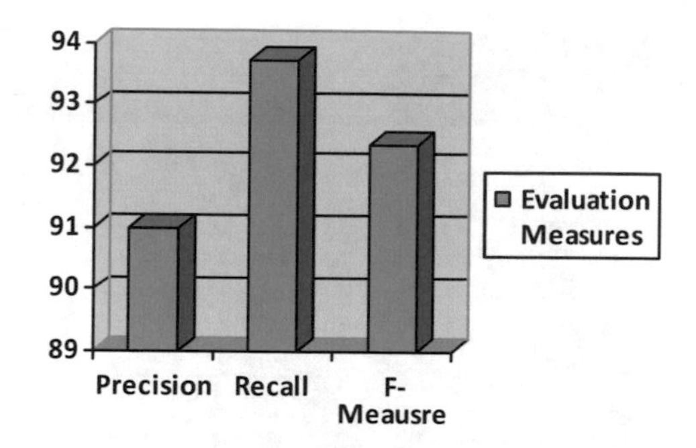

Fig. 6 Parameters for evaluation

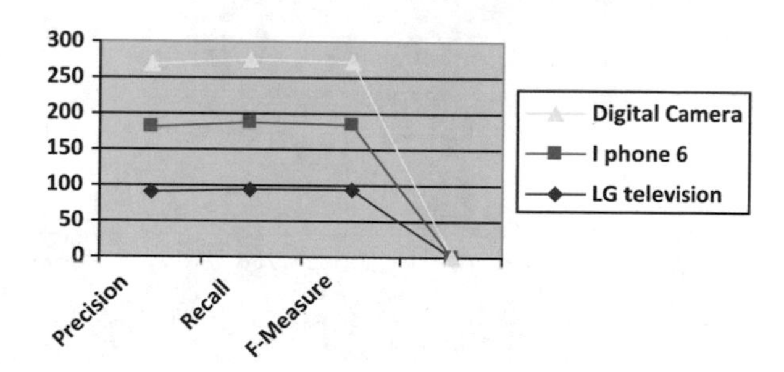

Fig. 7 Different products evaluation

Precision, recall and accuracy rates are calculated on the basis of features per sentence (Fig. 7). We did the same work for electronics and healthcare domain. Tests are done by collecting reviews also for I phone 6, Digital camera Sony and Bausch and Lomb Contact lens which are displayed in Fig. 8. Table 4 lists the details of the results.

The analysis of our work is demonstrated using WEKA[1] tool by using the data set "Bausch and Lomb contact lens.arff". We have taken first twenty four instances for analysis which consists of training and test data. The data consists of 14 positive and 9 negative samples. The snapshot of the Naive Bayes algorithm is shown in Fig. 8.

5 Future Trends and Conclusion

A novel framework is designed that provides relevant features and benefits in recommendation of a product by considering reviews. This chapter presents a research plan with an overarching goal to help ensure that the proposed opinion system is developed systematically with novel algorithms. The result of this plan will be a good approach to improve decision making policies of customers, made available to a wider community of academics and practitioners. The chapter concludes with a summarized opinion, a balanced assessment of the contribution of mining opinions and a roadmap for future directions. There is still a gap needed to

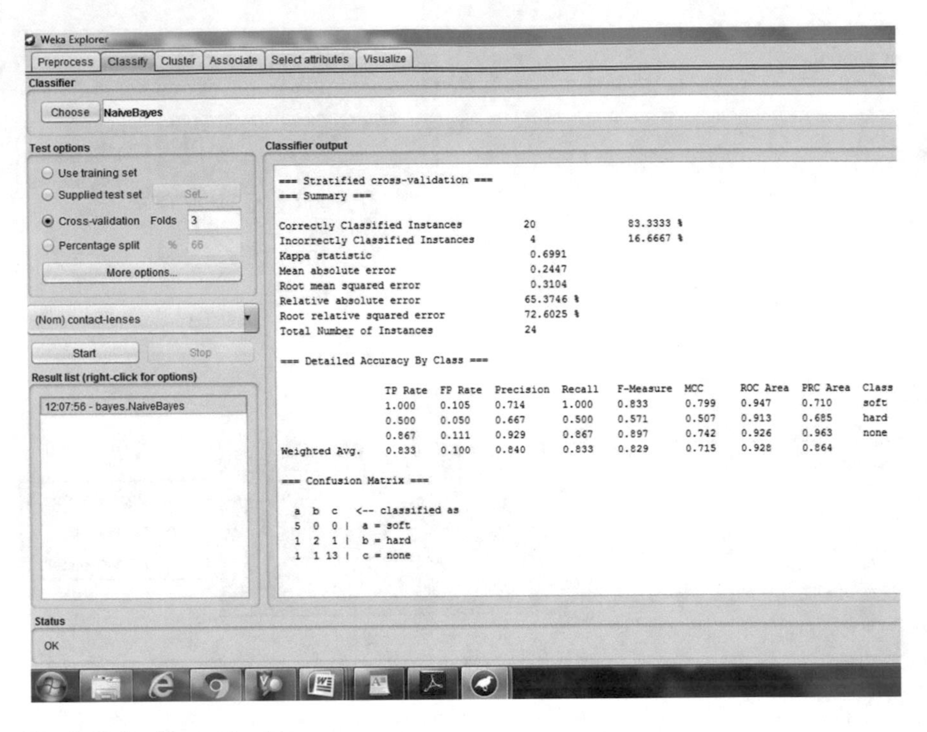

Fig. 8 Naïve Bayes algorithm

Table 4 Results

Product	Precision (%)	Recall (%)	F-measure (%)
LG television	91	93.73	92.36
I Phone 6	90	92.8	91.4
Digital camera sony	86.2	87.5	86.85

be filled to make our system more user friendly as compared with human annotation.

Our future work will include the analysis of various supervised, unsupervised and semi supervised learning techniques in detecting sentiments. The semantic (knowledge) data present in the review needs to be analyzed to detect sentiment by developing the opinion ontology.

5.1 Scope of Chapter

This chapter defines the scope of Internet of Things and big data, as covered in this book, discusses the need of mining opinions, developing information retrieval system, and outlines some of the benefits associated with the social interaction and social software on World Wide Web. In addition to the theoretical aspects of opinion mining, it covers various supervised learning algorithms used in the classification of opinions, and describes a typical coherent framework of how to mine opinions. The rapid evolution of Electronic media led to the growth of social media users to 127 million users in the year 2013 which will further increase substantially close to 283 million users by 2017. Therefore, the dependency of public on search engines and Information retrieval systems has become amplified. Opinion mining lies in close proximity with social media. As the time changes and E-business takes its place through Social media, demand of a new web world has awaken people to share views and opinions in order to make correct decisions by extracting useful content (aspects) followed with summarization of opinions with their polarity score. Interest of researchers towards the field of Opinion mining is a demanding task. Thus, a true need has been raised to mine opinions and develop an opinion mining system which will help people in taking better decisions of buying and selling products online considering their relevant aspects.

References

1. Bhatia, S., Sharma, M., & Bhatia, K. K. (2015). Strategies for mining opinions: A survey (2nd Edn.). In *International conference on computing for sustainable global development (INDIACom)* (pp 262–266). New Delhi: IEEE Xplore.
2. Abulaish, M., Doja, M. N., & Ahmad, T. (2009). Feature and opinion mining for customer review summarization. In *Pattern recognition and machine intelligence* (pp 219–224). Berlin: Springer.
3. Khan, K., Baharudin, B., Khan, A., & Ullah, A. (2014). Mining opinion components from unstructured reviews: A review. *Journal of King Saud University-Computer and Information Sciences, 26*(3), 258–275.
4. Seerat, B., & Azam, F. (2012). Opinion mining: Issues and challenges (a survey). *International Journal of Computer Applications* (0975–8887), *49*(9).
5. Vinodhini, G., & Chandrasekaran, R. M. (2012). Sentiment analysis and opinion mining: a survey. *International Journal, 2*(6).
6. Liu, B. (2012). Sentiment analysis and opinion mining. *Synthesis lectures on human language technologies, 5*(1), 1–67.
7. Lo, Y. W., & Potdar, V. (2009). A review of opinion mining and sentiment classification framework in social networks. In *2009 3rd IEEE international conference on digital ecosystems and technologies* (pp. 396–401). New York: IEEE.
8. Khan, F. H., Bashir, S., & Qamar, U. (2014). TOM: Twitter opinion mining framework using hybrid classification scheme. *Decision Support Systems, 31*(57), 245–257.
9. Penalver-Martinez, I., Garcia-Sanchez, F., Valencia-Garcia, R., Rodríguez-García, M. Á., Moreno, V., Fraga, A., et al. (2014). Feature-based opinion mining through ontologies. *Expert Systems with Applications, 41*(13), 5995–6008.
10. Song, Q., Ni, J., & Wang, G. (2013). A fast clustering-based feature subset selection algorithm for high-dimensional data. *IEEE Transactions on Knowledge and Data Engineering, 25*(1), 1–4.
11. Dongre, A. G., Dharurkar, S., Nagarkar, S., Shukla, R., & Pandita, V. (2016). A survey on aspect based opinion mining from product reviews. *International Journal of Innovative Research in Science, Engineering and Technology, 5*(2), 2319–8753.

12. Mfenyana, S. I., Moroosi, N., Thinyane, M., & Scott, S. M. (2013). Development of a facebook crawler for opinion trend monitoring and analysis purposes: case study of government service delivery in Dwesa. *International Journal of Computer Applications, 79* (17).
13. Zhang, S., Jia, W. J., Xia, Y. J., Meng, Y., & Yu, H. (2009). Opinion analysis of product reviews. In *Fuzzy Systems and Knowledge Discovery, 2009. FSKD'09*. Sixth International Conference on 2009 August 14 (Vol. 2, pp. 591–595). New York: IEEE.
14. Varghese, R., & Jayasree, M. (2013). Aspect based sentiment analysis using support vector machine classifier. In *Advances in computing, communications and informatics (ICACCI), international conference on IEEE.*
15. Pang, B., Lee, L., & Vaithyanathan, S. (2002). Thumbs up? Sentiment classification using machine learning techniques. In *Proceedings of the ACL-02 conference on Empirical methods in natural language processing-2002 July 6* (Vol. 10, pp. 79–86). Association for Computational Linguistics.
16. Hemalatha, I., Varma, D. G., & Govardhan, A. (2013). Sentiment analysis tool using machine learning algorithms. *International Journal of Emerging Trends & Technology in Computer Science (IJETTCS), 2*(2), 105–109.
17. McDonald, R., Hannan, K., Neylon, T., Wells, M., & Reynar, J. (2007). Structured models for fine-to-coarse sentiment analysis. In *Annual meeting-association for computational linguistics 2007 June 23* (Vol. 45, No. 1, p. 432).
18. Hu, M., & Liu, B. (2004). Mining opinion features in customer reviews. In *AAAI 2004 July 25* (Vol. 4, No. 4, pp. 755–760).
19. Bhatia, S., Sharma, M., & Bhatia, K. K. (2016). A novel approach for crawling the opinions from world wide web. *International Journal of Information Retrieval Research (IJIRR), 6*(2), 1–23.
20. Kitchenham, B. (2004). Procedures for performing systematic reviews. *Keele, UK, Keele University, 33*(2004), 1–26.
21. Ganesan, K. A., & Kim, H. D. (2008). Opinion mining—A Short Tutorial (Talk). University of Illinois at Urbana Champaign.
22. Sharma, N. R., & Chitre, V. D. (2014). Opinion mining, analysis and its challenges. *International Journal of Innovations & Advancement in Computer Science., 3*(1), 59–65.
23. Ma, Z. M. (2005). Databases modeling of engineering information. China: Northeastern University.
24. Seerat, B., & Azam, F. (2012). Opinion mining: Issues and challenges (a survey). *International Journal of Computer Applications, 49*(9).
25. Turney, P. D. (2002). Thumbs up or thumbs down? Semantic orientation applied to unsupervised classification of reviews. In *Proceedings of annual meeting of the association for computational linguistics* (ACL-2002).
26. Wiebe, J., Bruce, R. F., & O'Hara, T. P. (1999). Development and use of a gold-standard data set for subjectivity classifications. In *Proceedings of the association for computational linguistics* (ACL-1999).
27. Hu, M., & Liu, B. (2004). Mining and summarizing customer reviews. In *Proceedings of ACM SIGKDD international conference on knowledge discovery and data mining* (KDD-2004).
28. Jedrzejewski, K., & Morzy, M. (2011). Opinion mining and social networks: A promising match. In *Advances in social networks analysis and mining (ASONAM), 2011 international conference on 2011 July 25* (pp. 599–604). New York: IEEE.
29. Kasthuri, S., Jayasimman, L., & Jebaseeli, A. N. An opinion mining and sentiment analysis techniques: A survey. *International Research Journal of Engineering and Technology (IRJET), 3*(2). e-ISSN: 2395–0056.
30. Agrawal, R., Rajagopalan, S., Srikant, R., & Xu, Y. Mining newsgroups using networks arising from social behavior. In *Proceedings of the 12th international conference on world wide web 2003 May 20* (pp. 529–535). ACM.

31. Vijaya, D. M., & Sudha, V. P. (2013). Research directions in social network mining with empirical study on opinion mining. *CSI Communication, 37*(9), 23–26.
32. Stavrianou, A., Velcin, J., & Chauchat, J. H. (2009). A combination of opinion mining and social network techniques for discussion analysis. *Revue des Nouvelles Technologies de l'Information*. 25–44.
33. Buche, A., Chandak, D., & Zadgaonkar, A. (2013). Opinion mining and analysis: A survey. arXiv preprint arXiv:1307.3336
34. Sobkowicz, P., Kaschesky, M., & Bouchard, G. (2012). Opinion mining in social media: Modeling, simulating, and forecasting political opinions in the web. *Government Information Quarterly, 29*(4), 470–479.
35. Gao, Wei, & Sebastiani, Fabrizio. (2016). From classification to quantification in tweet sentiment analysis. *Social Network Analysis and Mining, 6*(19), 1–22.
36. Montoyo, A., MartíNez-Barco, P., & Balahur, A. (2012). Subjectivity and sentiment analysis: An overview of the current state of the area and envisaged developments. *Decision Support Systems, 53*(4), 675–679.
37. Cambria, E., Schuller, B., Xia, Y., & Havasi, C. (2013). New avenues in opinion mining and sentiment analysis. *IEEE Intelligent Systems, 28*(2), 15–21.
38. Osimo, D., & Mureddu, F. (2012). *Research challenge on opinion mining and sentiment analysis*. Universite de Paris-Sud, Laboratoire LIMSI-CNRS, Bâtiment, p 508.
39. Xia, R., Zong, C., & Li, S. (2011). Ensemble of feature sets and classification algorithms for sentiment classification. *Information Sciences, 181*(6), 1138–1152.
40. Melville, P., Gryc, W., & Lawrence, R. D. (2009). Sentiment analysis of blogs by combining lexical knowledge with text classification. In *Proceedings of the 15th ACM SIGKDD international conference on Knowledge discovery and data mining 2009 June 28* (pp. 1275–1284). ACM.
41. Zhang, Z., Ye, Q., Zhang, Z., & Li, Y. (2011). Sentiment classification of Internet restaurant reviews written in Cantonese. *Expert Systems with Applications, 38*(6), 7674–7682.
42. Tan, S., & Zhang, J. (2008). An empirical study of sentiment analysis for chinese documents. *Expert Systems with Applications, 34*(4), 2622–2629.
43. Ye, Q., Zhang, Z., & Law, R. (2009). Sentiment classification of online reviews to travel destinations by supervised machine learning approaches. *Expert Systems with Applications, 36*(3), 6527–6535.
44. Tretyakov, K. (2004). Machine learning techniques in spam filtering. In *Data mining problem-oriented seminar, MTAT 2004 May 3* (Vol. 3, No. 177, pp. 60–79).
45. Bhatia, S., Sharma, M., & Bhatia, K. K. (2015). Sentiment knowledge discovery using machine learning algorithms. *Journal of Network Communications and Emerging Technologies (JNCET), 5*(2), 8–12.
46. Hildreth, C. R. (2001). Accounting for users' inflated assessments of on-line catalogue search performance and usefulness: An experimental study. *Information Research, 6*(2), 6–2.
47. Jeyapriya, A., & Selvi, C. K. (2015). Extracting aspects and mining opinions in product reviews using supervised learning algorithm. In *Electronics and communication systems (ICECS), 2015 2nd international conference on 2015 February 26* (pp. 548–552). New York: IEEE.

A Modified Hybrid Structure for Next Generation Super High Speed Communication Using TDLTE and Wi-Max

Pranay Yadav, Shachi Sharma, Prayag Tiwari, Nilanjan Dey, Amira S. Ashour and Gia Nhu Nguyen

Abstract In the era of high speed communication, high speed data downloading and fast internet uses increase day to day. This leads to several applications, such as IP telephony, mobile internet access, gaming services, HD mobile television, 3D TV, video Conferencing and cloud computing. In the 4G technology, communication system provides ultra-broadband mobile Internet access, cellphones and for other movable devices, USB wireless modems for laptop. A modified 4G systems based on TD-LTE with Wi-max technology is carried out to improve the 4G network status that uses LTE with WiMax. The current work proposed a comparative survey of LTE technology with Wi-Max and TD-LTE with Wi-Max in 4G. Two 4th generation candidate systems are technically implemented, namely the

---

P. Yadav · S. Sharma
Research and Development Department, Ultra-Light Technology (ULT) Bhopal, Bhopal, India
e-mail: pranaymedc@gmail.com

S. Sharma
e-mail: sharma.shacivds@gmail.com

P. Tiwari
Department of Computer Science and Engineering, National University of Science and Technology MISiS, Moscow, Russia
e-mail: prayagforms@gmail.com

N. Dey (✉)
Department of Information Technology, Techno India College of Technology, Kolkata, India
e-mail: neelanjan.dey@gmail.com

A.S. Ashour
Department of Electronics and Electrical Communications Engineering, Faculty of Engineering, Tanta University, Tanta, Egypt
e-mail: amirasashour@yahoo.com

G.N. Nguyen
Duy Tan University, Da Nang, Vietnam
e-mail: nguyengianhu@duytan.edu.vn

N. Dey et al. (eds.), *Internet of Things and Big Data Analytics Toward Next-Generation Intelligence*, Studies in Big Data 30,
DOI 10.1007/978-3-319-60435-0_21

standard Wi-MAX and the LTE (Long Term Evolution) standard of first broadcast with TD-LTE technology. In order to simulate the proposed structure, Network Simulator (NS-2) can be used.

Keywords TDLTE · Wi-Max · LTE · Wireless USB and 4G

1 Introduction

A wireless sensor network (WSN) is a network which entails many small and self-contained electromechanical devices that are used in several applications, such as monitoring environmental conditions. These sensors can extend beyond a wide geographical area. These nodes can moreover communicate among each other via hops in an ad hoc manner. The sensors have restricted power as they are operated based on batteries. Energy is considered the primary concern among WSN. The WSN are considered the foundation of the entire internet of things (IoT) energy smart solutions. It consists of independent dispersed sensors that can monitor environmental/physical conditions, including pressure, temperature, humidity, sound and motion. This sensor is cooperatively pass and interconnected the collected data through the networks to the principal location and base station. It enables the integration of communications solutions and several technologies.

Several design aspects are critical for the WSN implementation including proper placement of devices, mobility, infrastructure, network topology, size and density of a network, connectivity, service life, node addressability and data collection. Typically, the WSNs can be flat or hierarchical architecture at which the nodes' roles vary. In flat architecture, all nodes are even. The hierarchical architecture involves cluster heads that control the sensor nodes in their clusters. In both architectural, the data collected through each sensor is transmitted to a base station (BS), which is the only processing center by the network. Afterward, the base identifies the surrounding characteristics.

Wireless communications and digital electronics are improved due to evolution of economical, compact, short-distance, small and multifunctional sensor nodes in micro-electro-mechanical systems (MEMS). These small-scaled sensor nodes that used in the detection, processing of data and communication among the components, take benefit of the sensor networks which is depend upon on the cumulative effort of a huge number of nodes. Sensor networks have a significant enhancement over traditional sensors, which are performed in two different ways:

- Sensors can be placed away from the location where actual phenomenon is under process, thus the data is fetched by sensory perception. In this approach, the implementation of large sensors is also required to distinguish targets from ambient noise.
- Many sensors which have the ability of detection can be positioned. The deployment of sensors and communications topologies for them is cautiously

designed. Time series of the detected phenomenon are transmitted by sensors to the central nodes, where the calculations are made and then the data is updated.
- A sensor array is defined to be made up of an extensive number of sensor nodes.

Using the 802.11 standards, the wireless local area network (WLAN) allows wireless communication between any two electronic devices via high speed Internet connection within a concise range. The speed of the network may go up to 54 Mbps within a range of about 30 to 300 meters. Primarily, there are five core WLAN provisions, namely IEEE (802.11a, 802.11b, 802.11e, 802.11g and 802.11n). Previously, IEEE802.11a/b/g has been a popular used specification of WLAN, however, now 802.11 standard which works at the 2.4 and 5 GHz bands is highly utilized. WiMAX is designed by Alvarion and Intel in 2002 for long-range communication and rectified by the IEEE Engineers under the name of IEEE-802.16. Figure 1 illustrated the mesh network using Wi-Fi or/and WiMAX.

Figure 1 illustrated the mesh topology of a network with Wi-Fi and/or WiMAX. WiMAX can run on all layers of a municipal mesh network (hot zone/metropolitan area). The current Wi-Fi mesh network architecture is essentially consists of three layers. There is access to a mesh transport routing, mesh node and nodes of traffic to mesh gateways and to a link network linking a meshed gateway to another content gateways or to an Internet point.

Typically, WiFi is a short-range wireless local area network (LAN) implementation in area, such as in a college campus, university, office whereas WiMAX spread over a larger area of a size that of a metropolitan by connecting houses, buildings and even cities. WiMAX is distinct from WiFi in many ways, namely the covered area and data throughput. Although, both WiFi and WiMAX offer a wireless linking to every problem using different technical work methodology. Unlike WiMAX, WiFi cannot operate at a greater distance as it is operated in unauthorized frequency. Due to the less power, the same effects arrive at the

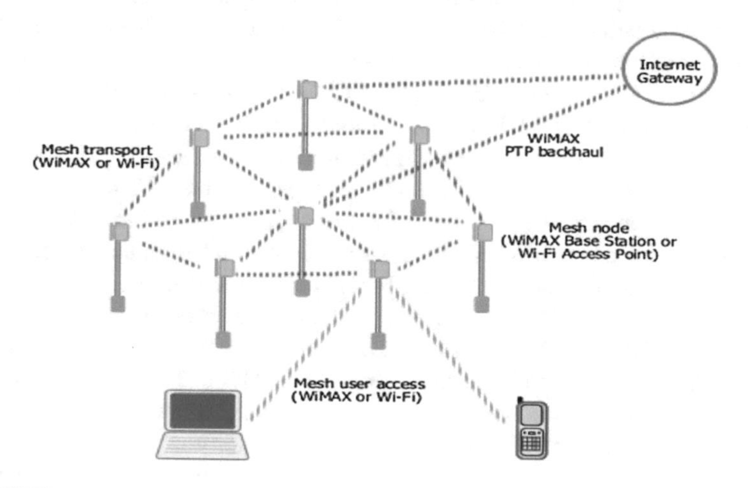

Fig. 1 Wi-Fi and/or WiMAX mesh network

distance. In addition, both the WiFi and WiMAX used different adopted algorithm, the WiFi MAC layer utilizes contention access algorithm, while WiMAX utilizes a planning procedure. In contention mode procedure, users should participate against the data rate at the access point. However, in the planning mode procedure, it tolerates users to compete only once at the AP. As an output, WiMAX has a higher throughput, spectral efficiency and latency as compared to WiFi.

2 Literature Survey

Lin [1] measured and compared the throughput and latency of Time Division-Long Term Evolution (TD-LTE), WiMAX and the third generation (3G) systems by performing some technical tests. Quantitative measurements and comparisons provided guidance to the operators for deployment of their future networks. Analysis of the results indicated that TD-LTE is the enhanced among all of them, followed by WiMAX and WCDMA and TD-SCDMA. Lin [2] considered voice as the major telecommunication service, where the VoIP performance played a major role in implementing the global interoperability and providing all IP network service for microwave access (WiMAX) technology. The results depicted that the field measurements performed outstandingly when the two communication devices were in stationary position and showed satisfactory QoS when both the communication devices were in motion at a speed of 50 km/h.

The thorough study of the analyzed results suggested that the VoIP service by the M-Taiwan program WiMAX-based infrastructure performed well with the standard G.107 requests in the worst and in the most severe scenario, where the CPE VoIP are connected wirelessly to the same WiMAX BS with the two mobile CPEs at speeds maximum upto 50 km/h while both pass out handovers at the same time [3]. Sung [4] provided a software for VoIP operation measurement named NCTU VoIP Testing Tool (NCTU-VT). This research defined the NCTU-VT software development, which is a free tool that offers a solution for evaluating performance of VoIP. Correspondingly, NCTU-VT is associated with two commercial tools, Smart VoIPQoS and IxChariot on the basis of a case study on VoIP enactment in a Wi-Fi environment. Cano-Garcia [5] presented a test configuration to identify the real UMTS networks' end-to-end performance. Using this test bench, a set of measurements was executed for the analysis of end-to-end behavior of Internet access based on UMTS. This research concentrated on the characterization and analysis of packet delay. Periodically, the packets are sent and their end-to-end delays are monitored. Thus, through this various features of the operation and configuration of the UMTS network are deduced. Using these comparatively simple measures, a parsimonious model of end-to-end UMTS behavior is developed that could be used for network simulation or emulation experiments.

Prokkola [6] designed a tool to calculate the quality of service (QoS) unidirectional performance statistics for passive observing of the anticipated applications and also provides information on the QoS delivered by the network. The tool designed is essentially compatible with any type of network construction providing supported IP. In this research work, the use of QoSMeT has been demonstrated to identify the outlined and particular behavior of 3G delay. The QoSMeT is developed for end-to-end evaluation, but for administrators, operators and researchers it is required to extract the subnets effects and even network components. Consequently, the development of the tool capable of processing several measurement points is being worked. Further in future, QoSMeT might be also utilized to offer direct input to QoS applications.

3 Technical Concepts

3.1 Worldwide Interoperability for Microwave Access (WiMAX)

WiMAX technology is described as global microwave access interoperability by the WiMAX environment. It is known for yielding wireless broadband connection to mobile/fixed terminals in an extended topographical area. It is operated on authorized and unauthorized frequencies using LOS and non-LOS technologies. It also ensures to deliver last mile wireless high-speed access to Internet able to giving high-intensity applications.

3.1.1 Architecture of WiMAX Network

The MS, BS, HA Server, AAA Server, ASN Gateway devices are operated on Wi-MAX technology are IP-based nodes. These nodes are directly connected to the main network of computer. The BS is in charge for carrying out the air interfacing and managing the radio resources. The received data from the mobile stations is sent via the interface of air. The ASN gateway has managed the mobility and QOS policy between various base stations. It is also used to link several base stations to the backend basic services network. The IT is open source software that achieves the process of authentication. In addition, High Availability (HA) is also open source software that performs the mobile IP. Its job is to execute the roaming between the ASN gateways.

3.1.2 Standards for WiMAX

IEEE 802.16 is the standard for WiMAX. It is standardized by the IEEE, which is a series of broadband wireless technologies. The IEEE 802.16 group was generated in 1998 for the evolution of wireless broadband services.

3.1.3 WiMAX Protocols

The primary topologies, which are utilized in WiMAX are point to point for the backhaul and point to the point multi point station for the station of subscriber. For the implementation of both the topologies, several inputs and output antennas are used. The defined structure of the protocol of the IEEE 802.16 wireless broadband standard is illustrated in Fig. 2.

Figure 2 illustrated four layers of the protocol, namely the MAC, convergence, transmission and physics, which correspond to the lowest layers of the OSI model, i.e. the data binding layers and physical layer. WiMax implement many applications and user interfaces, such as Ethernet, ATM, TDM, IP and VLAN. The IEEE 802.16 standard is much adaptable protocol that allows time division multiplex (TDM) or frequency division duplex (FDD) deployments that allow both

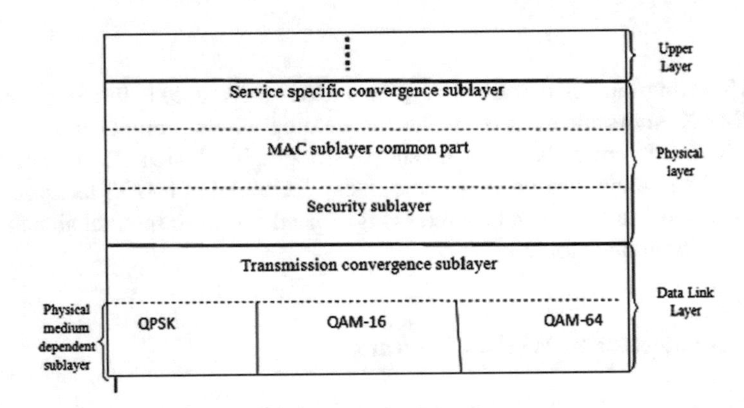

Fig. 2 IEEE 802.16 protocol structure of broadband wireless MAN

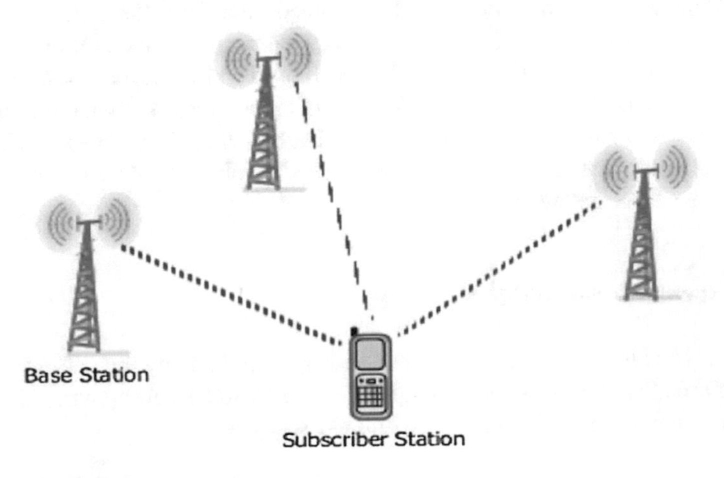

Fig. 3 Soft handoff

half-duplex and full terminals. The IEEE 802.16 supports primarily 3 physical layers, the essential physical mode is FFT OFDM (Orthogonal Frequency Division Multiplexing) of 256 points, the other different modes are 2048 OFDMA (Orthogonal Frequency Division Multiplexing Access) and Single Carrier (SC). The relative European standard - the ETSI Hiperman standard explain a single PHY mode that is approximately equal as the 256 OFDM modes in the 802.16d standard. The MAC has been devised for a point-to-multipoint wireless access to the environment. It accommodates protocols, such as the Ethernet, ATM, and IP (Internet Protocol). The frame structure of MAC comprises the downlink and uplink profiles of the terminals according to the link conditions, which is a compromise between capacity and lustiness in real time.

3.2 *Handover in Mobile WiMAX*

For balancing the load, a process of view transfer is required to distribute the network load with the transfer process. An important attribute of mobile WiMAX technology in the field of mobility is the transfer. The transfer is the process of altering the serving BS to obtain an elevated throughput in the network by the MS. In addition, transfers are considered the necessary component for the efficient utilization of system resources and increased quality of service. In 802.16e, the transfer process is triggered for two reasons.

3.2.1 Handover Types in Mobile WiMAX

This work has lately categorized the transfer types from different points of view in wireless networks and WiMAX. Types of transfer are categorizes on the basis of technology, structure, initiation and delivery mechanisms. Technologically, it will be specified whether the transfer algorithm is executed between two same or different technologies. Structurally, it will be specified whether the transfer is triggered between different BS or among different channels in a BS. The initiation aspect tells about the responsible driving force for directing the transfer i.e. MS or BS. Furthermore, the execution aspect reveals about the diverse type of transfer according to the defined standard transfer types.

- **Technological aspect**: Handovers can take place between different technologies. This implies the transfer may happen between mobile WiMAX and other wireless technologies, or vice versa. It is classified into different classes, vertical handovers and horizontal handovers. Horizontal handovers refers to the transfer in a single technology whereas the vertical handovers means the transfer process involves different technologies, such as mobile WiMAX and wireless LAN.
- **Structural aspect**: Transmission can persist between different BS, but it can also occur in the different channels of a particular BS, which is named as the

intra-cell transfer, whereas in handover, transfer of one BS to another BS takes place by the MS.

- **Initiation aspect**: One is the case of MS Handover in which MS transfers the MS connection to an appropriate target BS (TBS) because of deterioration of the radio signal. Additional is the BS initiated Handover in which BS initiates the handover due to presence of imbalance situation in the system. This handover is transmission triggered due to non-balanced traffic distribution.
- **Performance Mechanisms Aspect**: WiMAX mobile implemented three types of transfer processes for execution. Hard transfer (HHO) is a mandatory process where as the fast switching of the BS and the transfer of macro diversity is optional.

1. Hard Handover (HHO): "Brake-Before-Make" mode is utilized in this procedure, in which the MS disconnects from the service BS before establishing connection with the next BS (TBS). It didn't have any connection during two phases; Synchronization and reintegration of TBS and network.
2. Fast Base Station Switching Transfer (FBSS): In the MDHO and FBSS, the MS is associated to one or more than one base stations during transfer execution. This process is cumbersome as during the transfer execution, phases 3 and 4 are repeated for several BSs. In this mechanism, a list of active BSs is maintained by the MS that have established a connection to it. The MS receives and transmits each frame of any of the BS to all the active base stations. Thus, during this process no interruption takes place. A point to be noted in FBSS is that all the Base Stations in the group does receive the frames addressed to the MS but only on among the lot transmits the data back via air interface. On the contrary, rest of the BSs leave behind the packet.
3. Macro Diversity Transfer (MDHO): In this type of transfer, a list of competent BSs is registered for the MDHO over its coverage area by MS. This group of competent BSs is identified as diversity. A BS still occurs in the diversity game which is expressed as a BS anchor. The MDHO procedure is started by the MS when it wants to transmit or receive from several BS at the same time. By MDHO transfer, the MS interconnects with all BSs allowing the diversity combination to be used to achieve optimal signal quality in both downlink and uplink.

3.3 Wireless Networks Bandwidth Optimization

The slower rate of 2nd generation wireless networks (9.6 Kbps on GSM, 14.4 Kbps with PCS and 19.2 Kbps with CDPD) is considered the main obstacles for the companies and users to adopt wireless Internet applications. The defined standards of 3G networks are expected to provide national coverage with its increased speed and capacity.

3.4 Handoff Mechanism in Mobile WiMAX

For building an efficient mobile network, a smooth transfer mechanism must be defined to maintain an uninterrupted user communication session, while the user is in motion. The transfer mechanism manages the switching of the subscriber station (SS) from one BS to another when the transmission process via one BS becomes weaker than the nearby present BS. All the different techniques developed can be classified into 2 types of handoff, namely the soft/hard handoff.

In handoff, a subscriber station preserves multiple connections with very short delay. Soft handoff is utilized in cellular voice networks, including the Global System for Mobile Communications (GSM) or code division multiple access (CDMA). In this process, first a strong connection is established with a new BS and then the previously established connection is broken.

This procedure is appropriate for processing the voice and various latency-sensitive services like Internet multiplayer and videoconferencing. When used to provide data traffic (net browsing and email), soft forwarding will cause lower spectral efficiency as this form of traffic is congested and does not necessitate continuous switching of one BS to another. A Subscriber Station preserves a linking to a single BS at a given instance. Mobile WiMAX technology was perceived from the beginning as a broadband technology able to offering triple play services which are voice, data and video. Nevertheless, a WiMAX cell network is anticipated to be leaded by delay-tolerant data traffic. Voice is packetized (which is called VoIP) in Mobile WiMAX and then processed as other IP packets categories except that it has priority. Hard handoff (HHO) is thus used in WiMAX Mobile. The hard handover is illustrated in Fig. 4.

In the case of a hard transfer, a linking with a BS is completed beforehand a Subscriber Station passes to another BS. This is referred as a breakout approach before doing. Smooth transfer is much efficient in terms of bandwidth than smooth

Fig. 4 Hard handoff

transfer, but it has a longer delay. An optimized grid-optimized transfer mechanism has been developed for mobile WiMAX to maintain a transfer delay of less than 50 ms.

3.5 Long Term Evolution (LTE)

LTE is a broadband technology for mobile described by the 3G Partnership Project (3GPP). It is the GSM/UMTS standard of 3GPP development. The main aim of LTE is to provide and deliver high-speed data rates, continuous and smooth mobility between heterogeneous networks for cellphone users. The requirements as well as the number of mobile users for high-speed data rates are rapidly increasing. LTE-Advanced, meets all IMT-Advanced requirements and sometimes exceeds requirements. In October 2010, ITU-R decided that the submitted LTE-Advanced proposal meets all the requirements and needs of the first version of IMT-Advanced, which referred to it as the fourth generation (4G) system. The LTE-Advanced is already in testing phase by different countries and even started commercially operating it in 2014. In addition to the salient features including lower latencies, increased data rates, and better spectral efficiency are critical aspects to be considered in the LTE, which is the central based new type of architecture of the entire IP network, known as the Evolved Packet Core (EPC). LTE should extensively use the user-installed femtocells to attain its high-speed and spectral targets for a larger number of users. The users' confidentiality and sensitivity and data in transit in these digital cellular networks are dominant to both business and private users.

LTE uses OFDMA radio access technology. It offers orthogonality between many users for both the downlink and uplink channels so that there is no interference in the same cell but Interference between cells. Power control and coordination between cellular interference are also an essential property of LTE. Figure 5 depicted the evolution of the GSM radio access network to LTE network.

3.5.1 LTE Network Architecture

LTE has an architecture built on IP with slight difference from WiMAX architecture in the respect of security technique. LTE is incapable of meeting the security requirements of the company and it authenticates only on the basis of the identity (IMSI) and entry of the SIM card. An improved security technique was proposed including authentication of identity and key along with corporate certificates. With the use of orthogonal frequency division multiple access (OFDMA) technology, for the highest category terminals, LTE can provide 150 Mbit/s download speeds for multi-input multi-output (MIMO). The incompatibility of 3G standards with present available specifications and the constant request for higher data rates have shifted the attention of industry to the 4G WNS and basically the speed supported is in

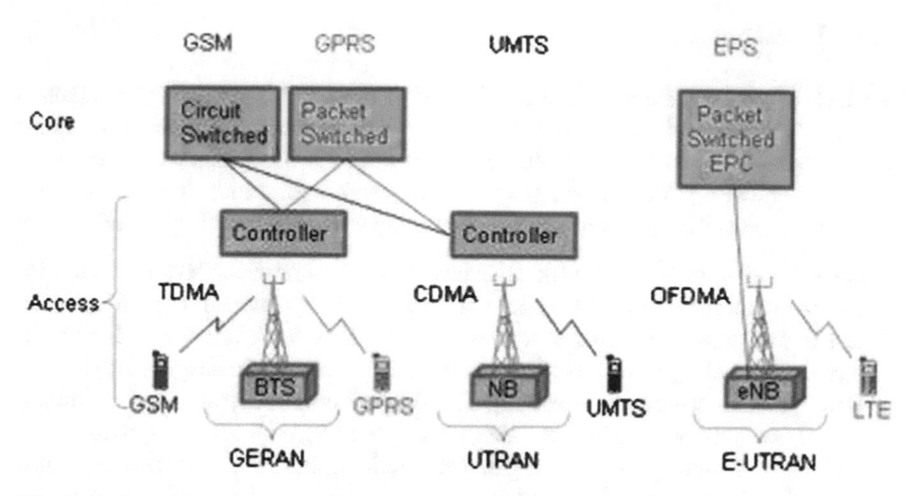

Fig. 5 Radio access network solution from GSM to LTE

surplus of 100 Mbps. It is responsible for the integration of the entire wireless network. The use of high bandwidth offers a perfect mode for data transport. Orthogonal frequency division multiple access (OFDMA) and Orthogonal frequency division multiplexing (OFDM) adequately allots network resources to wide number of users and offer them with superior audio and video. Also, 4G have superior low latency data transmission and security. It is a packet-switched network in which digital network elements are implemented. It supports service portability and global mobility.

3.5.2 Standards for LTE

LTE Advanced is a standard mobile communication that was officially recognized as a 4G system candidate to the ITU-T. LTE Advanced is accepted as a standard by the 3rd Generation Partnership Project (3GPP) as a primary enhancement to the LTE standard.

3.5.3 Features of LTE

LTE supports various features which deed the instant radio situations in a constructive way. A power control mechanism is used in the uplink, to control interference and to provide high average signal quality.

3.5.4 LTE Protocols

The LTE protocols, including radio link control (RLC), Packet Data Convergence Protocol (PDCP) protocols, physical layer (PHY) protocols, and medium access control (MAC) are included. Figure 6 demonstrated a brief idea of the protocol stacks graphically. The control plane stack consists of radio resource control (RRC) protocols. The primary works in each layer are computed up below.

1. **NAS (Non-Access Stratum)**: It establishes a connection between the user interface and the main network and governs sessions between them. Validation and registration of processes is also performed. Also, it has the capability of enabling/disabling Bearer context and position recording management.
2. **RRC (Radio Resource Control)**: All the information related to the broadcasting system of the NAS and Access Layer (AS) is provided by RRC. It is capable of establishing, releasing and maintaining the RRC connection. It also look after security functions, which primarily include the mobility functions, key management, QoS management functions, and Direct transfer of NAS messages between NAS and UE.
3. **PDCP (Packet Data Convergence Protocol)**: The essential job of PDC is compression of header; sequential retransmission and transmission of PDCP Session Data Units (SDUs) for acknowledgment mode radio carriers at the time of transfer. It also caters for duplicate detection, encrypting data and integrity protection.

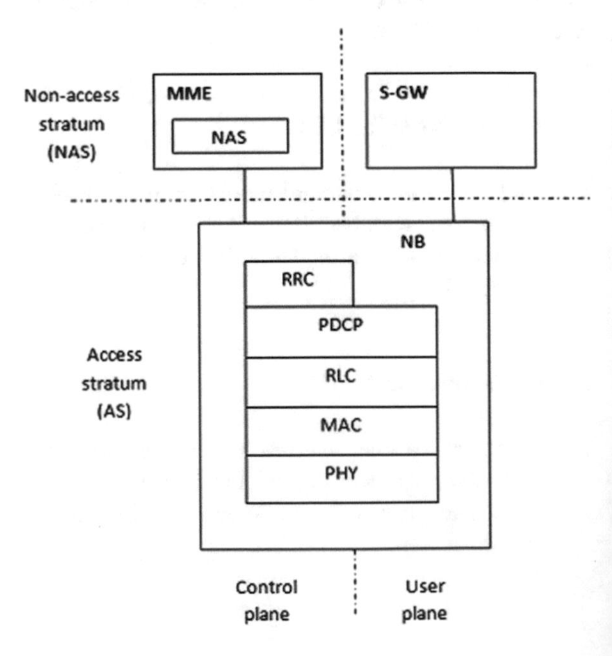

Fig. 6 LTE protocols stack

4. **RLC (Radio Link Control)**: Error amendment by automatic repetition demand (ARQ) and segmentation in consistent with the proportions of the re-selection and transport block if retransmission is required is provided by RLC. The amalgamation of SDU for the alike Radio bearer, retrieval, protocol error identification and delivery in sequence is also executed by this.
5. **MAC (Media Access Control)**: It executes de-multiplexing and multiplexing of RLC packets (PDUs). It also does information planning, developing report of HARQ hybrid error correction, hierarchy local chains and padding.

4 Proposed Methodology

4.1 3G (Third Generation)

The 3G refers to wireless mobile telecommunications technology generation. It is based on the fixed set of standards defined for mobile services and mobile telecommunication usage services, which are in accordance with the International Telecommunication Union's (IMT-2000) International Telecommunications 2000 specifications. The 3G has various applications in day to day life such as wireless voice and video calling, Internet access to mobile as well as fixed wireless equipment and mobile Television. It offers a data transfer rate of 200 kBit/s or above. The enhanced versions of 3G, generally indicated to as 3.5G and 3.75G, also offer multi-Mbit/s mobile broadband access to mobile modems in laptops and smartphones. Thus, the 3G technology is applicable to wireless voice and video calls, mobile as well fixed wireless access to the internet, and mobile TV technologies.

4.2 4G (Fourth Generation)

The fourth generation of wireless mobile telecommunications technology, mainly indicated to as 4G is the enhanced replacement of 3G technology. A particular set of standards have been defined by ITU in IMT Advanced for 4G technology. Some extra potential and current applications includes modified access to high-definition mobile TV, gaming services, mobile web, IP telephony, videoconferencing and 3D television. A 4G candidate system called the Long Term Evolution (LTE) has been commercially implemented in Oslo, Norway, and Stockholm, Sweden since 2009. A long debate took place to decide whether LTE should be considered as 4G or not.

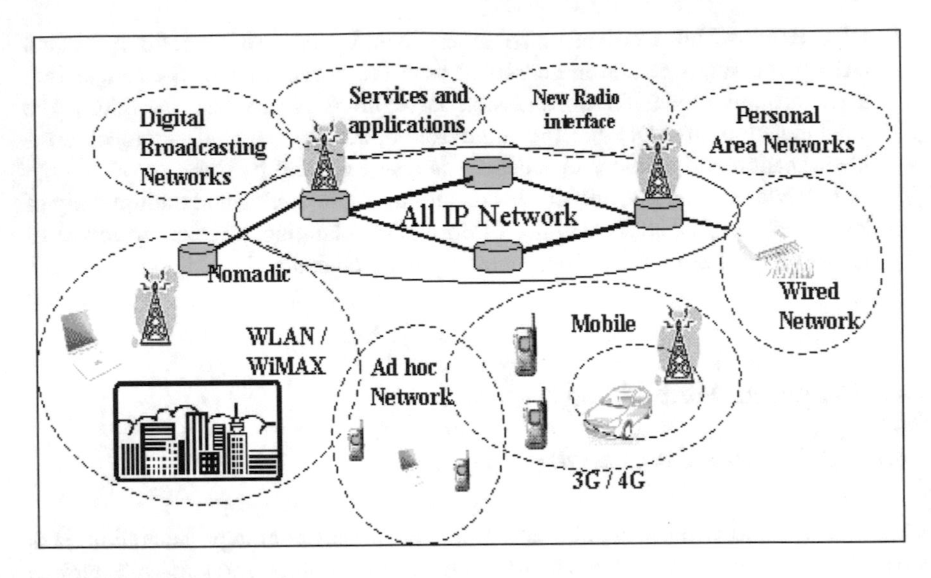

Fig. 7 Shows the proposed structure

4.3 TD-LTE

In FDD, a carrier frequency is used both for the uplink of data and the downlink of data. On the other hand, a single carrier frequency in utilized in TD-LTE for both the actions: downlink and uplink. In TD-LTE, the radio frame is divided into sub-frames, which are then allocated to the uplink or downlink depending on the immediate requirements of the users. This approach of random allocation of sub-frames is best suited to the editable data profile of today's users, who for the most part download more than they download, but rarely transmit data like photos and videos.

The spectrum defined for mobile communications is valuable and the concept of usage of single frequency in TD-LTE rather than a spectrum pair is advantageous for any of the operator where the available spectrum is limited or where an operator has access to only one unpaired frequency as shown in Fig. 7.

The proposed method or system is the combination of Wi-Max and LTE system. For the improvement of bandwidth, data rate and frequency use this architecture. The complete system of 4G communication system is dividing into the three parts. The transmitter end algorithm is reported as follows.

Algorithm: Transmitter end

Input

Number of mobile nodes: Mobile sensor node = N_n
Sender node = S_n representing a sub part of N_n
Receiver Node = D_n representing a sub part of N_n
Time of Simulation = t0/t0=100seconds
Routing protocol for wireless communication = AODV;
MAC Standard= 802.16
Mobile nodes radio range = RR, where the radio range for WiMAX =1000 meters same as consider for LTE

RREQ_Pks_Broadcast (S_n, D_n, RR)
{
If ((RR<=1000m) && (Next_Neighbour >0))
{
Compute route
{
routing_table-> insert (routing_table -> routing_table _nexthop); // send RREQ to Next_hop
routing_table 1->insert (routing_table1 -> routing_table); // send RREQ to destination
If (D_n = = Available)
{
Send ACK with routing_table1;
Data_packet_send (s_no, next_neighbour, type)
}
Else
{
Destination (D_n) is not Exist;
}
}
Else
{
D_n un-reachable;
}
}
End

This algorithm shows the direct wired network. It is implemented for the data rate improvisation with the help of OFDM. In wired network the direct link is available communication of wired and wireless on the basis of 3G and 4G. The total bandwidth feasible for frequency in wired network in Frequency division multiplexing (FDM) methodology maximum of 48 kHz. The OFDM multipath interface utilized by WiMAX and LTE, where a signals from a Sender travels to a detector through two or more than two paths and, under the correct condition frequency of 3 GHz is used as compare to wired network.

If (Frequency < = 48 Khz && data rate <= 10mbps)

{

Wired and Wireless network communication is possible on the basis of 3G Technology.

}

The condition of proposed LTE, WiMAX with 4G communication is work on more frequency and data rate.

If (Frequency < = 3 GHz && data rate <= 50mbps) // the value is decided on the basis of 4G capacity.

{

Wireless and wired network communication is feasible based on 3G Technology.

}

}

Now channel estimation calculation and channel equalization on the basis of frequency, subsequently calculate the required frequency for communication. The communication of frequency range and data rate limit not less than lower limit (consider theory of different research), but consider more as according to different theory mention on internet or research papers.

5 Simulation and Result

In this chapter the result of proposed method is discussed. The proposed method is simulated on network simulator software which is known as NS-2. NS-2 is well known tool in the sphere of network simulators, which is described as follows.

5.1 Simulation Environment

Simulation is the process in which salient features of processes and different kinds of systems are replicated to learn about the characteristics and performance of the system. Some of the network simulators which can be used for simulating the routing protocols are OPNET, Glomo-Sim and NS-2 etc. For this work, network simulator software version 2.34 (NS2.34) is used. It is an open source platform which is easy to use and is also available for free.

5.1.1 Network Simulator

Network Simulator (NS) shown in Fig. 8 is a simulation tool, which is an object oriented and discrete event simulator, advanced primarily for networking research. It is liable for handing substantial bolster over both wireless and wired networks for simulation of routing, TCP, and multicast protocols. The coding of NS-2 simulator is executed in C++ and Object Tool Command Language (OTCL). The scripting language OTCL is utilized by the users for signifying the network and determining other feature like traffic or routing protocols and agents. Figure 8 shows a view of network simulator.

5.1.2 Tool Command Language (TCL)

The TCL is a scripting language that utilized to frame the simulation scenario, node and system traffic configuration. Typically, any TCL script contains traffic scenario, network topology, and event scheduling. TCL scripts are compatible with many platforms and easy to use.

5.1.3 Trace File Analysis

The NS-2 simulation produces trace files as a results form, which contains the network complete data flow information. The traces in NS-2 empowers recording of information connected with packet events, such as packet drop, packet send or arrival happening in a link or queue. Example of wireless trace format for 'AODV' Ad-hoc network routing protocols: r 1.501360188 _2_ RTR — 0 AODV 48 [0 ffffffff 0 800] —— [0:255 −1:255 30 0] [0 × 2 1 1 [1 0] [0 4]].

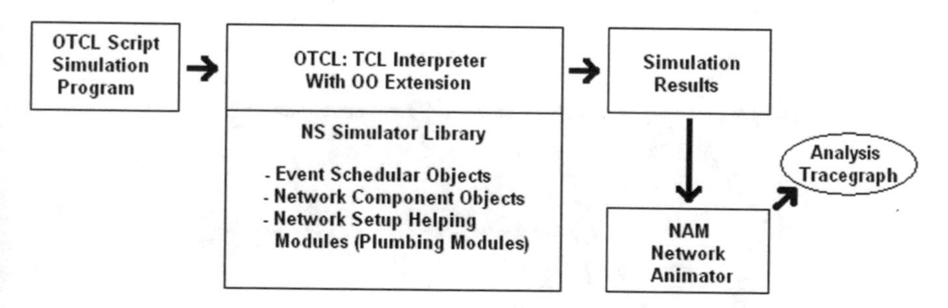

Fig. 8 Simplified users view of network simulator

5.1.4 Network Animator

It is used for the network topology visualization, where the NAM is an animator tool of NS-2 to draw a network topology picture showing the simulation time increase. The visual packets movement view around the network can be examined. Figure 9 shows the view of NAM, where the display of the running window is illustrated in Fig. 10.

The NAM file of 3G included routers, clients, end servers and nodes. Data is transferred from client to end servers through routers. It shows the delivery of packets from one to other. There is congestion of data between them. Figure 11 illustrated the packet delivery ratio (PDR) graph of the 3G.

Figure 11 shows the PDR of the 3G system. It reports the successfully received packets at the endpoint in relation to the total sent number of packets. The PDR is calculated per AWK file. AWK contains all information about all nodes and their performance. The packet delivery report (PDR) calculation is done on the basis of the received and transmitted packets as recorded to the trace file. Normally, PDR is the collected packets ratio at the destination point and the packets transmitted by the input source. Packet Delivery Ratio is determined via AWK script, which produces the final result by processing the trace file. Figure 12 illustrates the packet delivery ratio proposed method.

Figure 12 shows the packet delivery report of the proposed system. The quantity of packets that delivered to a destination point with respect to the total transmitted number of packets is mentioned. The packet delivery report (PDR) calculation is on the basis of the generated and received packets amount mentioned in the trace file. PDR is normally explained as the ratio of the packets total number collected at the destination point and the total number of packets that are generated by the input source. Packet Delivery Ratio is determined using the AWK script that do the processing of the trace file as well as produces the result. The time consumed to

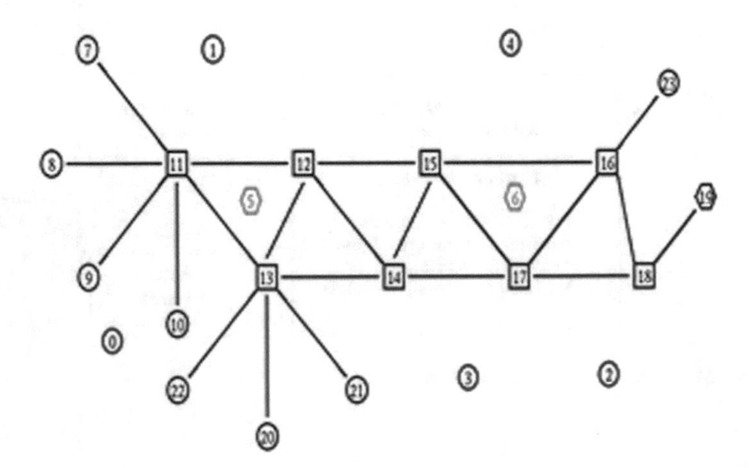

Fig. 9 NAM window of proposed work

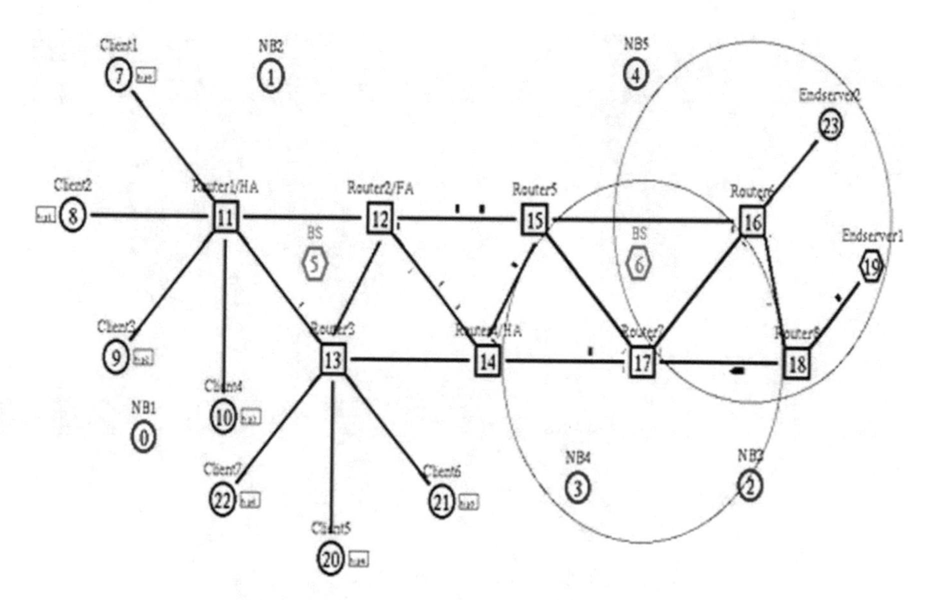

Fig. 10 Running window of NAM

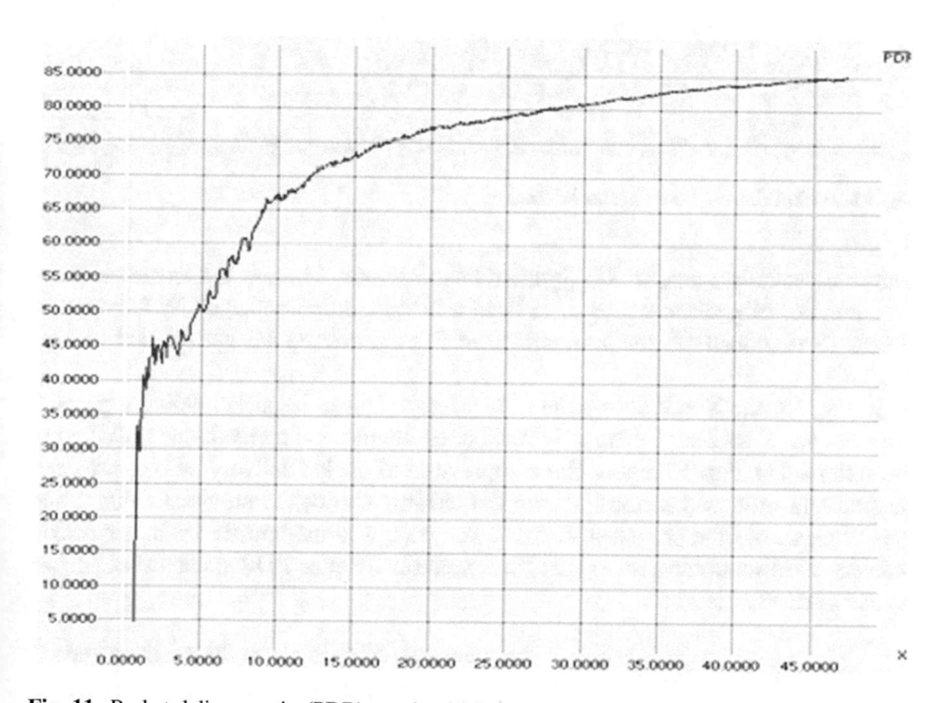

Fig. 11 Packet delivery ratio (PDR) graph of 3G

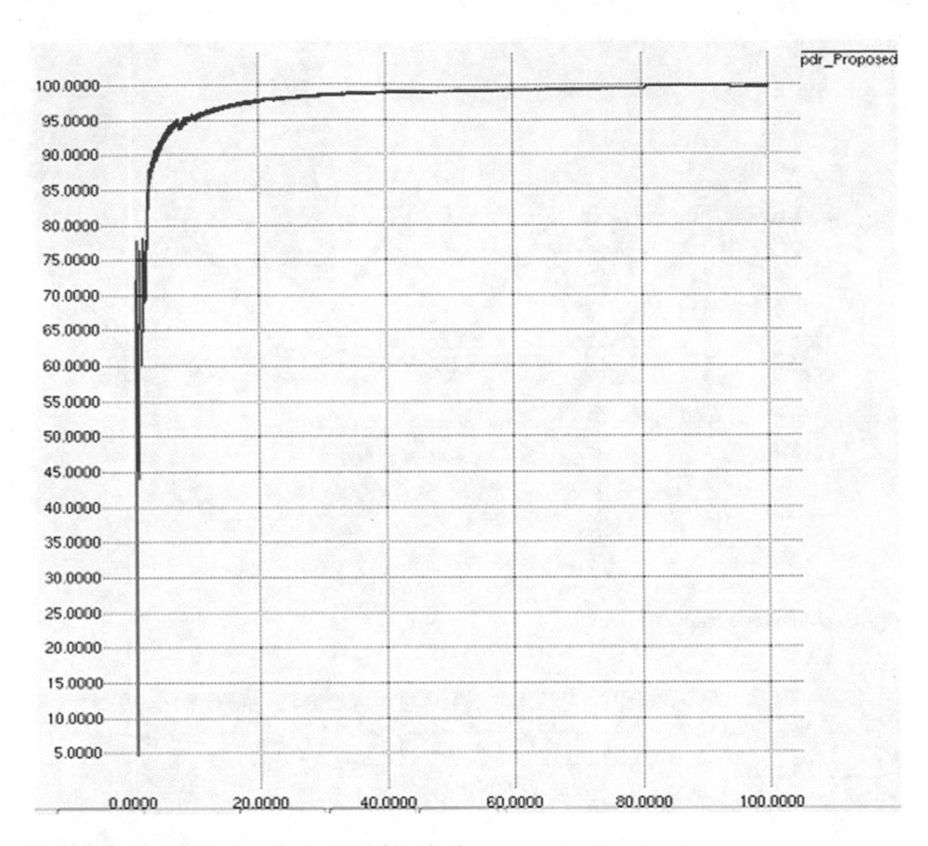

Fig. 12 Packet delivery ratio proposed method

route the packet per unit time is signified by the X axis, whereas Y axis indicates the total number of packets which have been transmitted and collected by the system. Figure 13 demonstrates the comparison of PDR proposed 4G system and old 3G system.

In Fig. 13, the X axis shows the time taken to complete packet delivery per unit time and the Y axis shows the total number of the packets transmit and collected in the system. The Fig. 13 reports the compression of packet delivery ratio of previous method and proposed method, where the red line shows the previous method and green line shows the proposed method. The proposed method shows better packet delivery ratio as compared to the other methods. Better PDR means that performance of system is good.

$$\text{(Packet delivery ratio) PDR} = \frac{\text{Total number of delivery at the receiver end}}{\text{Total number of delivery at the transmitter end}} \tag{1}$$

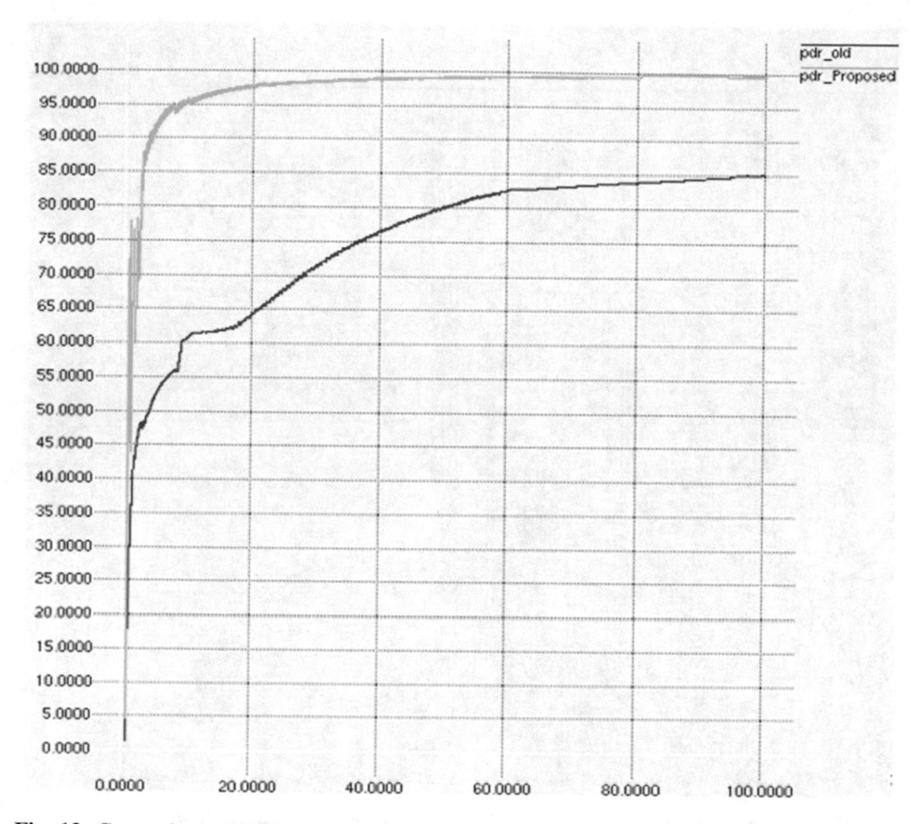

Fig. 13 Comparison of PDR proposed 4G system and old 3G system

Figure 14 demonstrates the comparison of delay in packet delivered at the receiver end in both methods, namely the proposed fourth generation system and pervious system. The time and delay in per packet delivery is signified by the X-axis and Y-axis; respectively. These results clearly show that proposed method achieves low delay compared to previous system in packet transmitted at the receiver end. Figure 15 included the comparison of throughput proposed 4G system and old 3G system.

The throughput illustrated in Fig. 15 is demarcated as the packets total number that is received successfully in a unit time. Throughput is represented in bps. It is computed using AWK script in which the trace file is processed and thus the result is produced.

$$\textbf{Throughput} = \frac{\textbf{received\_data}}{\textbf{Data Transmission Period}} \quad (2)$$

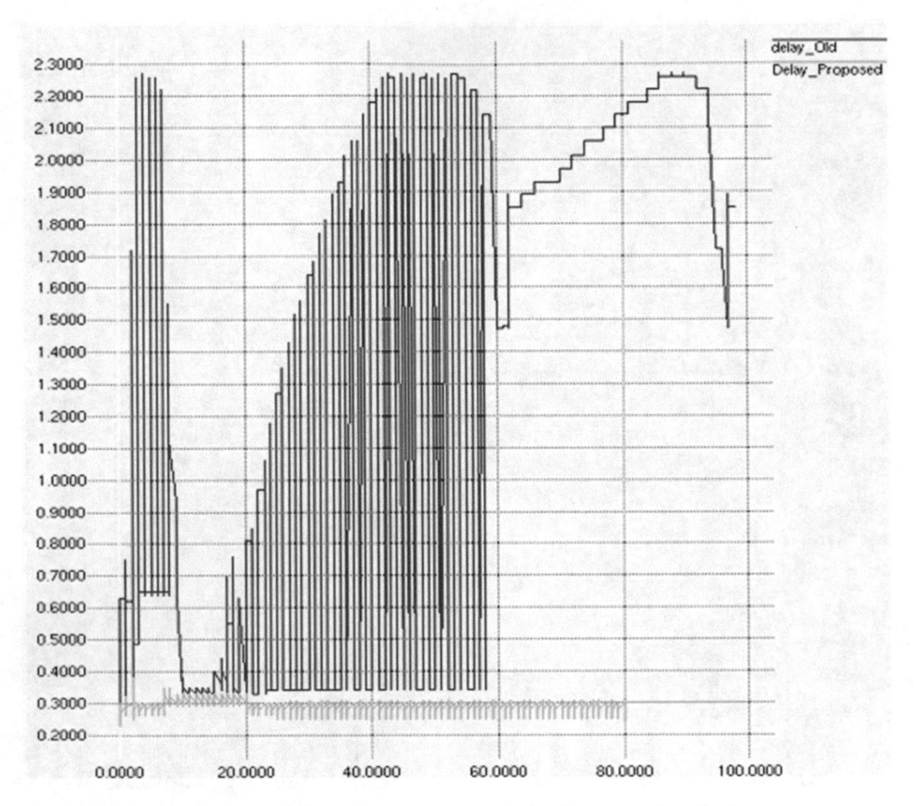

Fig. 14 Delay comparison of delay in previous method and proposed method

where, the Transmission Time = File Size/Bandwidth (s), and the Throughput = File Size/Transmission Time (bps). The result outcome of the proposed TD-LTE is calculated in NS2 as follows in Table 1.

Table 1 reports the results of the proposed 4G TD-LTE system. The basic result parameters calculated for the proposed method, they are the PDR of 99.67%, routing overhead is 2745, normalized routing load is 0.39, throughput of the network (KBps) is 3.50 and the average end to end delay is 0.29 ms. There are basic parameters calculated in the proposed method and further compared the parameters with previous system. The outcome is calculated with the network tool. Table 2 reports the output of previous method.

Table 2 illustrates the previous method result showing the value of parameters, namely the total sent packets is 810, total received packets is 688, total dropped packets is 123, average hop count is 3 hops, packet delivery ratio is 84.94%, routing overhead is 1433, normalized routing load is 2.08, throughput of the network (KBps) is 0.34 and average end to end delay is 2.76 ms. Combining the results in Tables 1 and 2 in Table 3 to report the assessment of the proposed TD-LTE with the previous work results establishes the efficiency of the proposed method.

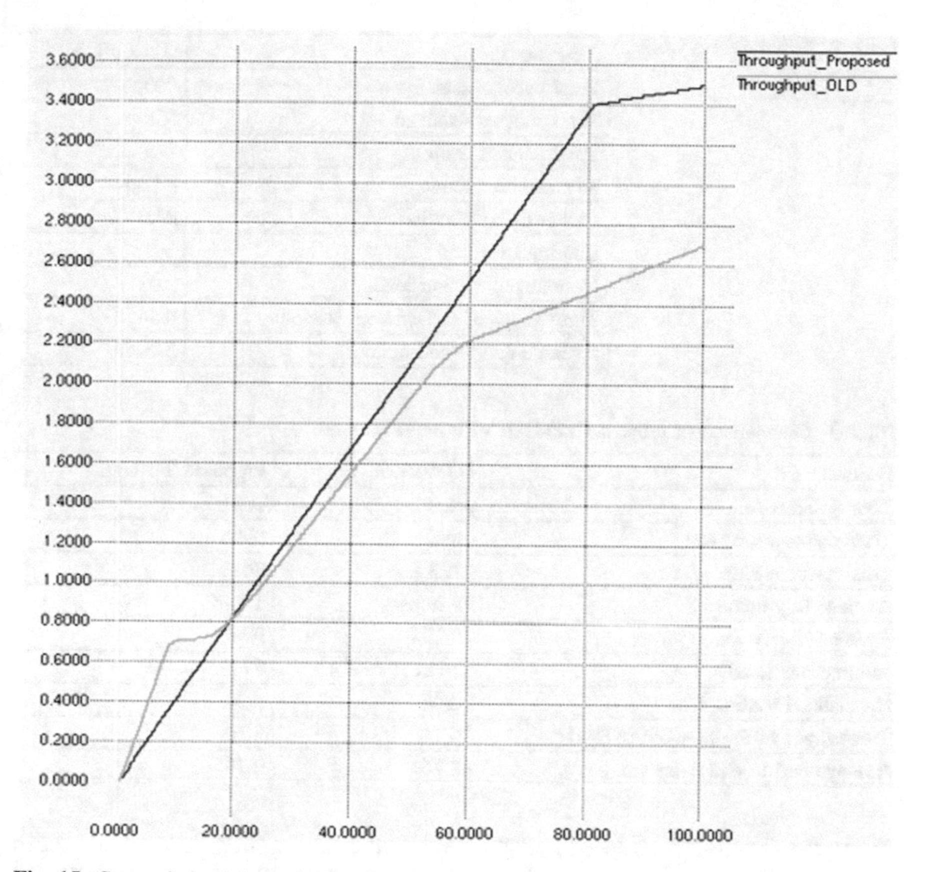

Fig. 15 Comparison of throughput proposed 4G system and old 3G system

Table 1 The proposed TD-LTE results

Parameters	Proposed TD-LTE
Total packets sent	7021
Total packets received	6998
Total packets drop	24
Average hop count	1 hops
Packet delivery ratio	99.67%
Routing overhead	2745
Normalized routing load	0.39
Throughput of the network (KBps)	3.50
Average end to end delay	0.29 ms

The preceding results depict the effectiveness of the proposed wireless communication system. Several studies have been carried out on the ireless sensor networks and Wimax [7–13]. For the improvement of the data communication and

Table 2 The previous work output results

Parameters	Old work
Total packets sent	810
Total packets received	688
Total packets drop	123
Average hop count	3 hops
Packet delivery ratio	84.94%
Routing overhead	1433
Normalized routing load	2.08
Throughput of the network (KBps)	0.34
Average end to end delay	2.76 ms

Table 3 Comparison of proposed TD-LTE with previous work

Parameters	Old work	Proposed TD-LTE
Total packets sent	810	7021
Total packets received	688	6998
Total packets drop	123	24
Average hop count	3 hops	1 hop
Packet delivery ratio (%)	84.94	99.67
Routing overhead	1433	2745
Normalized routing load	2.08	0.39
Throughput of the network (KBps)	0.34	3.50
Average end to end delay (ms)	2.76	0.29

due to the massive data speed and the increased data requirements, a high speed communication connection is required which is possible with the merging of the Wi Max and TD LTE properties.

6 Conclusion

The proposed work presented a TD-LTE system design for next generation wireless communication system. The main motive of this work is the analysis of Wi-Max system with TD-LTE concept for long range high speed data communication. Both the 3G and 4G LTE technologies were compared on the basic of different parameters, namely the total number of packets sent, routing overhead, received and dropped, normalized routing load, average hop count, packet delivery ratio, throughput of the network, and average end to end delay. In addition, different technical aspects for the designing of next generation communication system were discussed.

References

1. Lin, Y.-B., Lin, P.-J., Sung, Y. C., Chen, Y.-K., Chen, W.-E., Lin, B.-S. P., et al. (2013). Performance measurements of TD-LTE, WiMAX and 3G systems. *IEEE Wireless Communications*, *20*, 153.
2. 3GPP TS 25.308. (2011). 3rd generation partnership project; technical specification group radio access network; high speed downlink packet access (HSDPA); overall description; stage 2 (Release 10), June 2011.
3. Lin, Y.-B., et al. (2010). Mobile-Taiwan experience in voice over IP-worldwide interoperability for microwave access trial. *IET Proceedings Communication, 4*(9), 1130–1141.
4. Sung, Y.-C., et al. (2010). NCTU-VT: A freeware for wireless VoIP performance measurement. *Wireless Communications and Mobile Computing*, *12,* 318.
5. Cano-Garcia, J. M., Gonzalez-Parada, E., & Casilari, E. (2006). Experimental analysis and characterization of packet delay in UMTS networks. *Next Generation Tele traffic and Wired/Wireless Advanced Networking (NEW2AN).*
6. Prokkola, J., et al. (2007). Measuring WCDMA and HSDPA delay characteristics with QoSMeT. In *IEEE International Conference Communication (ICC).*
7. Fuentes, Lidia, Gamez, Nadia, & Sanchez, Pablo. (2009). Managing variability of ambient intelligence middleware. *International Journal of Ambient Computing and Intelligence (IJACI), 1*(1), 64–74.
8. Johansson, D., & Wiberg, M. (2012). Conceptually advancing "application mobility" towards design: applying a concept-driven approach to the design of mobile IT for home care service groups. *International Journal of Ambient Computing and Intelligence (IJACI)*, *4*(3), 20–32.
9. Roy, S., Karjee, J., Rawat, U. S., & Dey, N. (2016). Symmetric Key encryption technique: A cellular automata based approach in wireless sensor networks. *Procedia Computer Science*, *78*, 408–414.
10. Kumar, S., & Nagarajan, N. (2013). Integrated network topological control and key management for securing wireless sensor networks. *International Journal of Ambient Computing and Intelligence (IJACI)*, *5*(4), 12–24.
11. Borawake-Satao, Rachana, & Prasad, Rajesh Shardanand. (2017). Mobile sink with mobile agents: Effective mobility scheme for wireless sensor network. *International Journal of Rough Sets and Data Analysis (IJRSDA), 4*(2), 24–35.
12. Binh, H. T. T., Hanh, N. T., & Dey, N. (2017). Improved Cuckoo search and chaotic flower pollination optimization algorithm for maximizing area coverage in wireless sensor networks. *Neural Computing and Applications*, pp. 1–13.
13. Chakraborty, Chinmay, Gupta, Bharat, & Ghosh, Soumya K. (2015). Identification of chronic wound status under tele-wound network through smartphone. *International Journal of Rough Sets and Data Analysis (IJRSDA), 2*(2), 58–77.

Printed in the United States
By Bookmasters